AF753437

OUT OF THIS WORLD

THE NEW FIELD OF SPACE ARCHITECTURE

Edited by

A. SCOTT HOWE

AND

BRENT SHERWOOD

LIBRARY OF FLIGHT

Ned Allen, Editor-in-Chief

LOCKHEED MARTIN CORPORATION

BETHESDA, MARYLAND

Published by American Institute of Aeronautics and Astronautics, Inc., 1801 Alexander Bell Drive, Reston, VA 20191–4344

American Institute of Aeronautics and Astronautics, Inc., Reston, Virginia

1 2 3 4 5

Library of Congress Cataloging-in-Publication Data

Out of this world : the new field of space architecture / edited by A. Scott Howe and Brent Sherwood.

p. cm. -- (Library of flight)

Includes bibliographical references and index.

ISBN 978-1-56347-982-3 (alk. paper)

1. Space stations--Design and construction. 2. Extraterrestrial bases--Design and construction. 3. Large space structures (Astronautics)--Design and construction. 4. Domestic space. 5. Architecture, Industrial. I. Howe, A. Scott. II. Sherwood, Brent.

TL797.O98 2009

629.47--dc22

2009014067

Cover design: Jim Killian

Cover image: *Moon 2000 A.D.*, by Syd Mead, 1976. Used with permission.

Copyright © 2009 by the American Institute of Aeronautics and Astronautics, Inc. All rights reserved. Printed in the United States of America. No part of this publication may be reproduced, distributed, or transmitted, in any form or by any means, or stored in a database or retrieval system, without the prior written permission of the publisher.

Data and information in this book are for informational purposes only. AIAA is not responsible for any injury or damage resulting from use or reliance, nor does AIAA or the authors warrant that use or reliance will be free from privately owned rights.

To future generations of space architects, who will build on our foundation, learn from our mistakes, and accomplish the astounding things we all dream of.

Foreword

Whereas the field of architecture for human communities in non-terrestrial and space environments has roots going back as far as Darwin and, some argue, into Hellenistic science and philosophy, it blossomed only with the rise to prominence of ecology, the study of relationships between humankind and our Earth environments, evidenced by the transformation of the early 20th century conservation movement into the post-World War II environmental movement and the founding of the Environmental Protection Agency in 1970. Environmental systems theory and human space flight engineering were synergistic: "I believe" spoke President Kennedy in 1961, "this nation should commit itself to . . . landing a man on the Moon and returning him safely to Earth. No single space project in this period will be more impressive to mankind or more important in the long-range exploration of space; and none will be so difficult or expensive to accomplish." As part of that goal, NASA, organized just two years before, boosted its efforts to identify and characterize the environmental systems required to support human life in space. And, in large part because of NASA's work, the academic and engineering communities began related studies in manned space science, technology, and engineering. The interdisciplinary field of human space flight was born.

Early space capsules supplied human needs from depletable stocks, but once spacefaring became an affair of more than a few orbits over a single day, systems were needed that recycled resources too heavy to boost into orbit in sufficient quantities, oxygen and water especially. Contemplation of even longer missions, to the planets and perhaps beyond, required more thorough-going recycling concepts that encompassed everything for humans to flourish, except energy. The complete recycling of energy is barred by the second law, and so a second field was born: that of harvesting energy from the environment, initially via solar cells, but ultimately via mining natural resources along the space journey.

An inevitable, if unheralded benefit of space architecture work is highlighted in the last chapter of this book: presenting insights on bringing these identified space system needs to our understanding of, and our action-oriented policies for, managing our homeland spaceship—Earth. There is a growing realization across many nations that we must begin treating "spaceship Earth" as a finite system. Adlai Stevenson first said in 1965 that "we travel together, passengers on a little space ship, dependent on its vulnerable reserves of air and soil." In the seminal *Only One Earth*, economist Barbara Ward and biologist René Dubois wrote "careful husbandry of the Earth is *sine qua non* for the survival of the human species and for the creation of decent ways of life for all the people. . . ." Nearly a half century later, today we are upon the threshold of a world renaissance, abandoning the old cowboy economy that treated our precious resources as inexhaustible stocks, and working out the measures and plans we will implement to manage our finite but renewable resource flows. The study of space architecture is then what the physicists call a "toy model" of the larger science, economics, and technology of spaceship Earth. It gives us the precious opportunity to work a "practice problem" or two before we take on the existential and irrevocable challenge of managing our spaceship Earth where we might not have many chances to get it right. Scott Howe and Brent Sherwood's *Out of This World: The New Field of Space Architecture* is a "must read" reference for today's energy and climate change policymakers and students.

The Library of Flight series is part of the growing portfolio of information services from the American Institute of Aeronautics and Astronautics. It extends the Institute's publications with the best from a growing variety of topics in aerospace from aviation policy, including history and law, to case studies on aerospace systems, to management of aerospace enterprises, and beyond, complementing the two other AIAA book series: Progress in Astronautics and Aeronautics and the AIAA Education Series. The Library of Flight documents the crucial role of aerospace in enabling, facilitating, and accelerating global commerce, communication, defense, and the technical management of our spaceship Earth, and so, this new volume occupies a significant position on the Library shelf.

Ned Allen
Editor-in-Chief
Library of Flight

table of contents

preface

Depending on when we start counting, this book's genesis was in 2002, or the 1970s, or the late 19th century, or 10,000 years ago. Humans have been using permanent materials to make architecture—the willful shaping of the physical human environment—for about 10 millennia. But the earnest notion of inhabiting outer space awaited the rapid advancement of industrial technologies around the dawn of the 20th century. Jules Verne, Konstantin Tsiolkosvky, and other visionaries started looking up, seriously considering how humankind might travel into the novel environment of space. The dreams became real within only decades, as humans left Earth orbit for the first time in 1968. Shortly thereafter, architect Maynard Dalton and industrial designer Raymond Lowy designed the interior of NASA's first space station, *Skylab*, and incidentally established the field of space architecture. Long-duration missions forced space agencies in the United States and the Soviet Union to solve real problems posed by humans living and working in space.

In the last two decades of the 20th century, diverse professionals—some networked and some isolated but all motivated by the vision of a significant human future in space—began to develop theory and principles of space architecture. They compared ideas, published concepts, conducted design studies under contract to NASA, established educational tracks at universities, and organized technical committees within professional aerospace societies. Finally, in 2002, this community came together in Houston to produce two seminal events. At the World Space Congress, they hosted a peer-reviewed Symposium on Space Architecture. And at a workshop following the congress, 46 professionals—architects, engineers, industrial designers, aerospace managers, technologists, and researchers—drafted a manifesto they called the *Millennium Charter*, defining the substance and role of this new field of space architecture.

The diversity and depth of presentations at the symposium made it clear that this new field should present itself as a professional resource for humanity's future in space. The integration of technical responsibility and humanistic sensibility that is unique to architecture is essential for shaping a safe, productive, and ennobling physical human environment in outer space.

We have solicited chapters from many of the prime movers of this field, in an effort to proffer a comprehensive introduction to the topic. To the chapter authors is due the credit for vision, ingenuity, focus, and persistence in defining what space architecture is all about. Many of them have labored for decades, often alone, convinced that they have glimpsed the spark of something profound for humanity's long-term future. In the oldest tradition of architecture, they stand ready to answer this deep calling by serving. Thanks are owed to many, but special thanks are due the organizers of the 2002 Houston events: space architects Marc Cohen and Constance Adams.

A. Scott Howe
Brent Sherwood
August 2009

PART 1
Introduction to Space Architecture

Welcome to the world of space architecture: the theory and practice of designing and building inhabited environments in outer space.

This introduction to the field contains a wealth of ideas and images that explain how humans live in space now, and it explores how they may do so in the near and far future. It provides a concise reference of our knowledge about the remote, hostile space environment, outlines issues central to designing space architecture, describes the most advanced space architecture of today, and proposes far-ranging designs for an inspiring future.

Part 1 opens with "What Is Space Architecture?" by Brent Sherwood (*Chapter 1*), followed by "Vernacular of Space Architecture" (*Chapter 2*), where Kriss Kennedy introduces the fundamental language of the field. The rest of the book is divided into parts that reflect the three principal domains of the field: orbital architecture, planet surface architecture, and Earth-based space architecture.

1

what is space architecture?

BRENT SHERWOOD

BROADENING ARCHITECTURE

One hundred years ago, no one would have questioned what architecture is or what architects do. At that time—and even throughout the dramatic modern transition from Burnham and Sullivan to Wright, Mies and Corbusier, to Venturi, Meier, and Gehry, and beyond—it was clear: architects design buildings.

The noble legacy of the master builder goes back virtually to the beginning of civilization. Imhotep and Sinan, famed designers of the Saqqara stepped pyramid and the Suleyman Mosque, respectively, are still known to us down through the ages, along with the names of legendary rulers and warriors.

Today, however, *architecture* has become a popular word in major industries emergent in the 20th and 21st centuries that have nothing to do with building: war making, aerospace, computer, and software. Whole conferences, textbooks, and standards manuals are devoted to architecture, with little thought of the long tradition of building places for human activity.

This has happened because no other term adequately captures the act and product of grasping and manipulating a complex design problem characterized by thousands of parts, mutually conflicting requirements, diverse specialties, and the willful creation of order out of chaos. This is the essence that architecture shares across all fields, whether the elements are bricks, subsystems, contractors, logic gates, or coded instructions.

RECLAIMING ARCHITECTURE

Yet it is rare for an entire profession to find its very name co-opted—perhaps architecture is unique in being both one of the oldest professions and one of the most broadly adapted. Ironically now, even as some attention turns to designing places for human activity in outer space, architects are not the first professionals called to help.

The international community of space architects aims to change that. We believe that the training, skills, knowledge, experience, and outlook unique to architects in the building tradition are of vital value as humans expand into space. Consider three nested scales of space exploration. The very largest scale is accessible only through telescopes that expand our observational reach almost 13 billion light-years, approaching the beginning of the universe. Deep inside that universe is a second scale, only light-hours across: a tiny bubble within physical reach of our machines. Robotic probes extend human sensing and action throughout our solar system and now just barely outside it. Finally, deep inside that bubble is the almost inconceivably tiny exploration domain very close to Earth, where humans can explore space directly.

This innermost scale of direct human experience is where space architects contribute. The lure of space is strong for designers of the human environment. For the first time in human history, we are reaching, and therefore must learn to live and thrive in, environments so alien that our very weight is reduced or canceled. Different gravity levels, lack of air, strange lighting, extreme temperatures, killing radiation, hardscrabble material resources, vast distances, and psychological remoteness characterize

Copyright © 2008 by Brent Sherwood. Published by the American Institute of Aeronautics and Astronautics, Inc., with permission.

places in space. For space architects, this is more inspiring than discovering a new continent. The multitude of challenges entailed, which at first seem overwhelming, motivates us as designers. How can humans live in such places? We intend to find out and to conquer the challenges. We aspire to be a new breed of architects versed both in the traditions of building for human activity and the new subjects unique to space systems.

Consider Peggy Deamer's (2006) description of what architects do: design something in detail, imagining the process of it coming together; enhance and share these design decisions with others; structure an office that processes, synthesizes, manages, and markets this information; guide contractual relationships with other organizations—contractors, construction managers, subcontractors, fabricators, lawyers—to turn this information into a building; and wonder what it all means. Space architects do these same things in space.

In 2002, following the first International Space Architecture Symposium at the World Space Congress in Houston, 46 practitioners and students interested in space architecture convened a workshop (SAW 2002), where we adopted this concise definition of space architecture:

> "The theory and practice of designing and building inhabited environments in outer space."

SPACE-ARCHITECTURE DISCIPLINES

To architects, the "practice of designing and building" is understood to mean grappling with myriad topics and integrating dozens of distinct specialties to define, understand, and control the creation of physical environments that respond simultaneously to complex, conflicting, and dynamic needs.

What are these specialties? Table 1 lists constituent topics of architecture and urban planning for both Earth and space. Note first that every terrestrial subject also applies to space sooner or later. Larger-scale, urban considerations that apply principally to large populations are inchoate even for small communities. Note further that several essential space subjects have no direct terrestrial analog (denoted by the gray box), although "green" and "smart" trends in current architectural practice are beginning to introduce advanced and even space-derived technologies into power, thermal, and data "building systems." Finally, note that for many subjects applicable both in space and on Earth the space environment adds complexities. Indeed, space architecture and urban planning include about half again as many new subjects, and many of these are highly technical.

On Earth, architects are the professionals who orchestrate the disciplines necessary to create coherent, built solutions that meet the needs of human use. Thus, to establish practical and noble habitable environments, space architects must master many subjects, and this will challenge historical traditions within both the architecture and space industries.

WHAT ABOUT SYSTEMS ENGINEERING?

The most integrative discipline in traditional aerospace practice is *systems engineering*. Systems engineering is responsible for ensuring that all of the requirements, performance models, subsystems, and technical plans work together to achieve what is needed. So what would be different about space architecture and why should we need it? Two distinctions arise.

First, the central focus of space architecture is designing to support human activity. Most aerospace systems are not habitable, and the ones that are involve fairly primitive habitation. However, this will change significantly as mission duration, distance, crew size, and mission purpose become more challenging. Systems engineers are not equipped to address such issues, whereas architects are.

Second and more fundamental is that most—but not all—systems engineering activity is analytical, responsive to the decompositional flowdown and verification of requirements and seeking to uncover technical integration flaws that could result in mission failure. By contrast, architecture is what we may call *design-directive*. It envisions an integrated design solution, from parti through detail, and then creatively guides supporting discipline analyses to achieve that vision. The closest analog in the world of systems engineering is a kind of systems engineering called "systems architecting" used in the initial stages of an aerospace project to shape the overall solution. Aerospace systems architects typically are most active in the conceptual stages, whereas traditional architects typically lead the complete unfolding and realization of their designs. Contemporary aerospace projects do not use the architect's method, even though systems architects and systems engineers might be kindred talents.

TABLE 1 Space architecture and urban planning comprise even more subjects than their terrestrial counterparts.

TRADITIONAL TERRESTRIAL DISCIPLINES	ADDITIONAL SUBJECTS FOR SPACE
ARCHITECTURE	
Human activity analysis and programming	Mission operations planning
Psychology and sociology	Psychology of remote isolation and sensory deprivation
Comparative historical analysis	
Abstract and representational modeling	
Structural engineering	Aerospace structures and mechanisms, including pressure containment and vacuum tribology
Materials development and testing	Aerospace materials and space environments
Environmental control engineering	Life support systems
Design for sustainability	
Site engineering	Planetology including alien engineering geology, weather, atmospheres, chemical environments, diurnal cycles, and gravity
Landscaping	
Construction engineering, safety, and quality inspection	
Interior design	
Color and lighting design	
Fire safety	
Power generation, management, and distribution	
Acoustic engineering	Vibration and noise control
Environmental impact and wilderness management	
Furniture design	
Industrial design	
Art	
Economics and finance	
Negotiation and contracting	
Construction management	Aerospace project management
	Astrodynamics; attitude control; and guidance, navigation, and control Propulsion, launch, and landing Vacuum thermal management Command and data handling Autonomy Reliability, safety, and mission assurance
URBAN PLANNING	
Potable water supply and distribution	Advanced, closed, and biologically based life-support systems
Waste management	Material recycling
Agriculture and processing	Biomass production
Power production and distribution	
Industrial production	Space mining and *in situ* resource utilization
Mass transit and industrial transportation	
Material supply and distribution	
Commerce	
Crime and law enforcement	
Environmental protection	
Communication networks and media	
Public recreation and spectator events	
Parks management	
Public health management	
Death accommodation	
Defense	
Urban growth	

PAST AND FUTURE HISTORY OF SPACE ARCHITECTURE

Over the decades, certain engineering managers within NASA and its major contractors have appreciated the value introduced into the team-based aerospace project development environment by trained architects.

> Architects are trained in design, which engineers for the most part are not. Engineers are basically trained in analysis . . . Architects understand what design is, and how to do it . . . Many engineers have a natural knack for design and eventually learn how to do it well, but architects come with that training when they graduate (Woodcock, 2008).

Thus a few architects have worked inside the aerospace establishment dating back to the *Apollo* program. Some still do in various capacities, sometimes involved with design decisions but almost always involving integration across technical, business, and strategic issues. As practicing space architects, these pioneers are the vanguard of the field.

We can speculate that decades hence, perhaps in the second half of the 21st century, when large numbers of people are traveling, working, and living in space, architects might evolve to a more central role. By then, the technically challenging issues that dominate design and operation of habitable space systems today will have been largely solved and reduced to practice. As a result, and as spacefaring populations grow, more atavistic human needs will come to the fore. The disciplines of aerospace engineering could then become subsumed into the panoply of subjects architects have historically coordinated.

In the short term, however, and into the foreseeable future of our working lifetimes, space architects will continue to support key design decisions of space system development projects. We hope our continued engagement can lead the way for subsequent generations of space architecture professionals. |

References

Deamer, P. (2006), "Introduction to the Symposium 'Building (in) the Future: Recasting Labor in Architecture,'" Yale School of Architecture, New Haven, CT, Oct.

SAW (2002), *Millennium Charter, Fundamental Principles of Space Architecture*, Space Architecture Workshop, Houston, TX, Oct. 12, available at http://www.spacearchitect.org/.

Woodcock, G.R. (2008), Private Communication, Aug. 10

2

vernacular of space architecture

KRISS J. KENNEDY

INTRODUCTION

SPACE ARCHITECTURE is a field not yet fully developed. Many portrayals of future space architecture in comic books, cartoons, and movies have conjured up fanciful, massive stations and 100-km Moon or Mars domes. However, realistic space architecture began when the 1960s space race put humans into orbit and then progressed to the *Apollo* Moon missions and several generations of space stations. The current acme of actualized space architecture is the International Space Station (ISS) orbiting 400 km above Earth. As NASA once again looks beyond low Earth orbit to interplanetary human spacecraft, space facilities at libration points, and Moon and Mars bases, a true and real space architecture vernacular—a specialized language—begins to take shape. This chapter introduces the vernacular of space architecture in the early 21st century.

SPACE ARCHITECTURE ELEMENTS

The basic vernacular of space architecture draws its vocabulary from a modular kit of elements. Orbital stations, interplanetary spacecraft, and planet-surface outposts all have common elements. At this time, space architecture for the near-Earth space environment comprises *launch vehicles* to put materiel into space, *pressure vessels* (modules) to contain breathable air and provide habitable conditions, and *support infrastructure*. These three broad categories include many subtypes of elements and systems.

The transportation system that carries the modular elements into space and to their destination, places significant constraints on their dimensions, mass, and configuration, much as ground transportation carriers and vehicles constrain terrestrial building components.

Pressure vessels for human activities, stowage, and logistics functions include several subtypes: living, laboratory, and working modules; spacecraft docking or berthing adapters; airlocks; and transition nodes interconnecting other modules.

Support infrastructure includes all of the ancillary technical functions and utilities required to keep the habitable volumes viable: support structure; power generation and distribution; thermal control; computing and telecommunications; and spacecraft basics like propulsion, attitude control, guidance, navigation, and control. The support infrastructure for orbital facilities or interplanetary spacecraft begins with a structural frame or truss to carry structure loads from the habitable vessels into the rest of the vehicle system. The support structure also carries the utility system distributed components, for example, large solar arrays for power generation, large radiator panels for rejecting waste heat, antennas or optical components for telecommunications, and spacecraft propulsion subsystems and guidance components. The support infrastructure for a planet-surface base includes structure foundations and utilities as well. Spacecraft propulsion and guidance are not needed, but surface transportation and applications equipment are.

The space environment heavily influences the design of all space architecture elements. To date, all pressure vessels have been launched from Earth largely intact and preintegrated. Eventually, as space architecture progresses, we will develop the capability to "live off the land" on the surface of the Moon and other bodies, manufacturing habitats, outfitting, and consumables from what we find on site. This advanced technology is called in situ resource utilization (ISRU).

This material is declared a work of the U.S. Government and is not subject to copyright protection in the United States.

SPACE ENVIRONMENTAL FACTORS

Space habitats are designed to sustain human life in the inhospitable environment of space. The dominant, common feature of these habitats is that they are pressure vessels in the vacuum conditions of orbital space and the lunar surface. (Mars's atmosphere is so thin that it is essentially a vacuum as well). Other driving characteristics of space environments include orbital debris, radiation, microgravity (on orbital space stations and transfer missions), partial gravity (on the Moon and Mars), and planetary dust. These pose major design challenges for space habitation. Highlights regarding gravity level and planetary dust are described next; see Chapters 3 and 14 for more in-depth discussion of the attributes of these two key space environments.

The weightlessness of microgravity presents both challenges and opportunities for designers of orbital and transfer habitats. The microgravity environment eliminates the physical relevance of normal "up-and-down" orientation and renders moot the traditional area method of allocating activity space. It forces space architects to use volume rather than area as the governing measure of accommodation. The architecture of ISS shows how the use of volume can be maximized in space (see Chapter 4 in this volume).

Inducing artificial gravity on space transportation systems through rotation incurs a significant mass and systems penalty over microgravity systems, depending on the propulsion type and overall configuration selected (Capps et al. 1991). It is not yet known whether artificial gravity would be required for human missions to Mars, but present technologies for limiting microgravity deconditioning appear unlikely to enable productive surface operations by astronauts after six to nine months of interplanetary transit. If artificial gravity is required, the mass penalty alone translates to at least 5–15% additional cost. See Chapter 12 for an in-depth discussion of designing human environments in rotating systems.

Environments with less gravity than Earth's pose interesting habitat design challenges. Human motion and performance in reduced-gravity environments differ from those in Earth gravity, but our only experience base—the *Apollo* surface crews—was confounded by the stiff, bulky spacesuits they wore. We expect that because microgravity places less restriction on human locomotion, both vertical and reach envelopes call for increased volume allowance.

Planetary dust is a potential problem for planet-surface elements subject to repetitive activity or long-term exposure. The Moon's regolith (planet "soil") has a high fraction of extremely fine dust, which is very abrasive and electrostatically "sticky" and can cause severe problems for mechanical equipment or human lungs. The *Apollo* missions discovered the challenges associated with lunar dust contaminating vehicle and surface equipment. Mars dust is less understood than lunar dust, but is beginning to be characterized chemically and physically by missions such as the Mars exploration rovers and *Phoenix*.

Space architects must design for dust control to limit dust penetration into mechanisms and habitable environments to limit mechanism failures; increased maintenance, repairs, and resupply; and habitat contamination. The latter is of particular concern because of damaging effects on crew health and crew accommodation systems.

SPACE HABITATS

Space habitats are categorized into three classes:

- Class I habitats are pre-integrated—entirely assembled, integrated, functionally verified, and ready to commission when delivered to space.
- Class II habitats would be prefabricated, but require expansion, outfitting, or assembly when deployed in space or on a planet surface.
- Class III habitats would be produced *in situ*, with their structure manufactured using local resources developed on the Moon or Mars.

Figure 1 shows the relationship of habitat technology, habitat classes, and time.

Class I: Pre-integrated

- Earth-manufactured
- Earth-constructed
- Fully outfitted and tested prior to launch
- Space-delivered with immediate capability
- Volume and mass limited to launch payload size capability and mass capability

Class II: Prefabricated—space/surface-assembled

- Less restricted to launch-vehicle size or mass capability
- Allows larger volumes
- Earth-manufactured
- Requires assembly prior to operability
- Space assembly or deployment
- Partial subsystem integration
- Critical subsystems Earth-based and tested prior to launch
- Some or all internal outfitting done in space
- Robotic and/or crew time needed for completion and commissioning

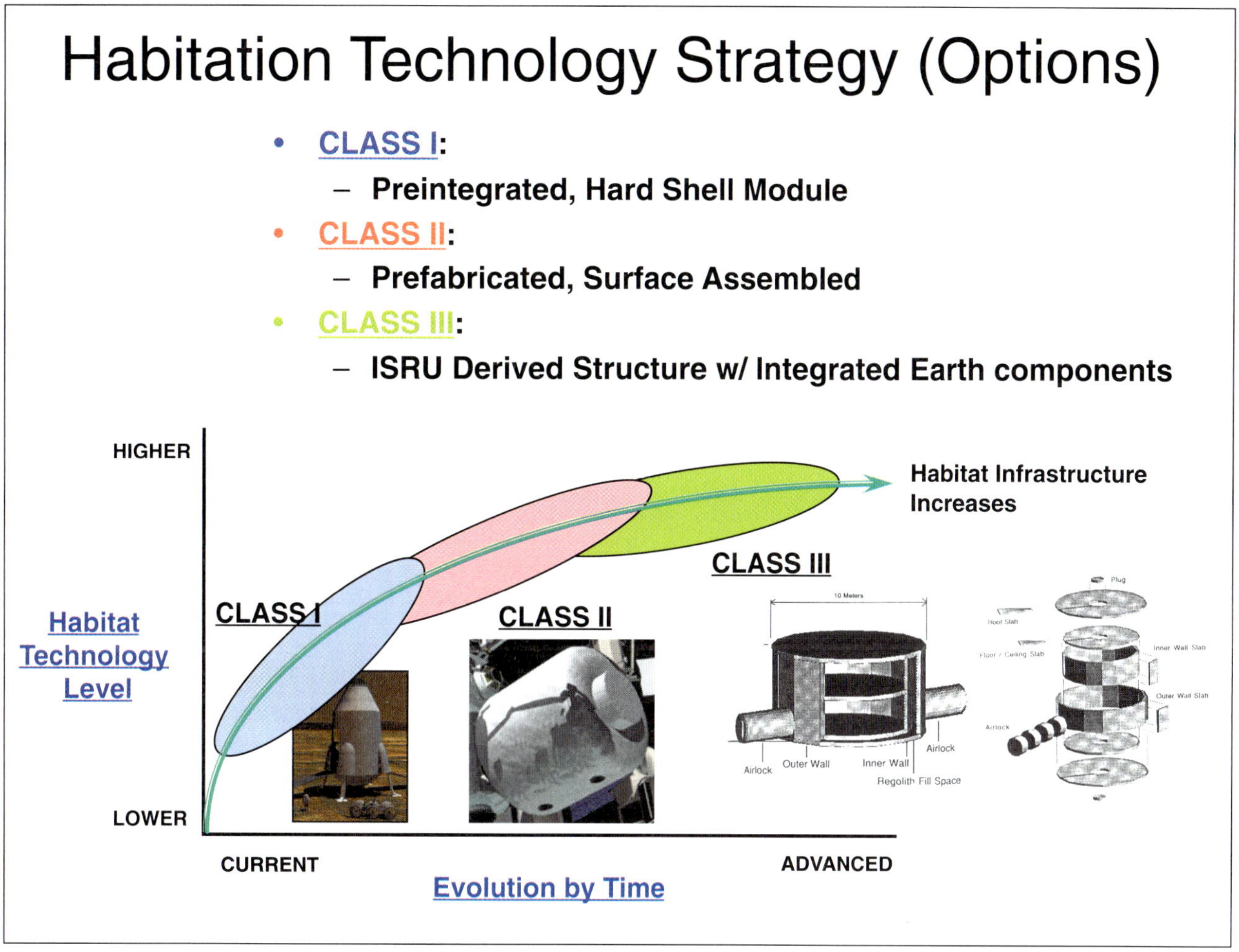

FIGURE 1 Space habitat technology evolves to larger volumes and use of local materials.

Class III: *In situ* derived and constructed

- Not restricted to launch-vehicle size or mass capability
- Allows largest volumes
- Manufactured *in situ* with space resources
- Space-constructed and space-tested
- Requires space manufacturing capability and infrastructure
- Requires significant robotic and/or crew time for construction and completion
- Requires total integration of subsystems
- All internal outfitting done in space
- Critical subsystems are Earth-based and tested prior to launch

Space habitats represent the beginning of a new architecture for space-faring civilization. They naturally attract great public interest for a human-exploration program because they are the "house in space" and great professional interest because they are sophisticated pressurized structures containing and protecting the ultimate payload: people. They are complex, heavy, expensive elements at the heart of all human spaceflight systems, so their concept development requires careful consideration.

Space habitats can be grouped based on duration: short (days to weeks), medium (weeks to months), and long (months to years). Volume requirements vary based on crew size, mission duration, and functional activities; the notion of standards for determining appropriate volume undergoes continuous debate. Historical habitat total pressurized volumes are illustrated in Figure 2.

Long-duration space habitats impose especially stringent requirements on space stations, transfer-vehicle systems, and planet surface systems. Multiyear stay times represent an order-of-magnitude increase in mission duration over typical ISS expedition durations. For transfer and surface systems, direct escape is impractical, and given the high cost of space transportation, resupply and crew rotation schedules will be sparse. The duration and distance from Earth

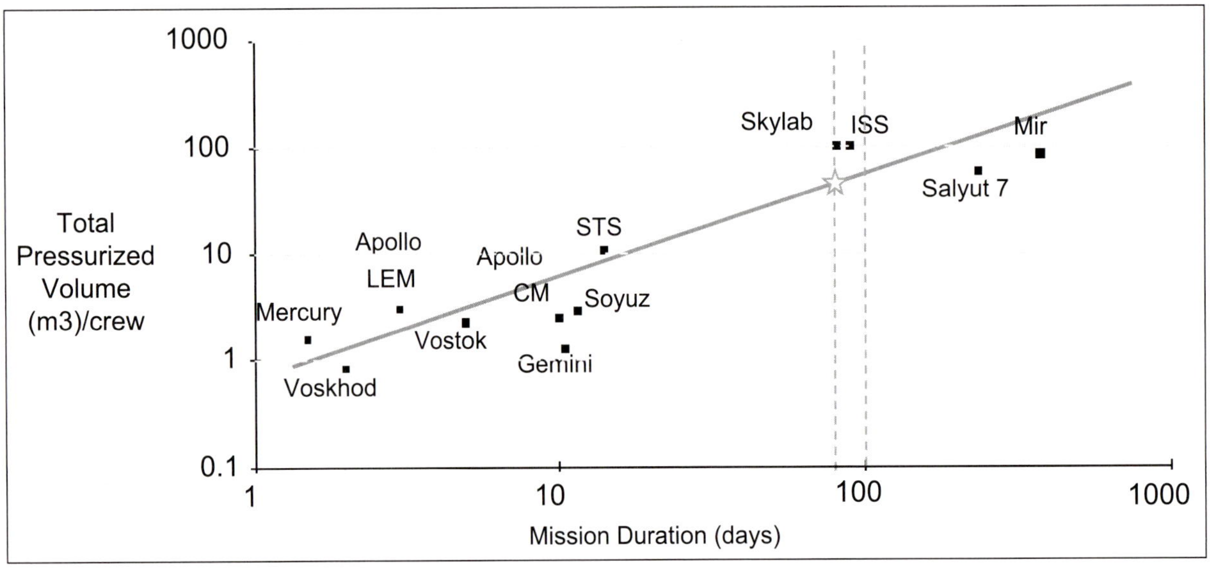

FIGURE 2 Historical space habitat pressurized volume.

compound the problem of crew isolation and confinement, exacerbating concerns about human psychological needs. Commonality—defined as using the same element design in multiple settings throughout the program architecture—can become a key driver. The extremely high cost of developing space hardware is a strong motive for multi-use elements. See Chapter 11 for a detailed discussion.

Each habitat type requires a different design approach, but all have to meet the requirements of providing a pressurized environment for the humans to live and work within. Common requirements regardless of destination include the following: crew safety, acceptable physiological and psychological support for humans, successful accommodation of mission objectives, reliable structural integrity with adequate safety margins, forgiving failure modes (e.g., leak before rupture), ability to be tested to a high level of confidence before being put into service, ability to be integrated with available launch systems, straightforward outfitting and servicing, easy maintenance, long design life, and commonality at the system or subsystem level.

Space habitat configurations vary according to user requirements, destination, and mission. However, core human needs must be provided for: air, water, food, temperature control and ventilation, personal hygiene, and waste management. Factors in the space environment not inherently conducive to human habitation must be compensated to create as Earth-like an environment as possible. Table 1 shows five principal environmental considerations and how they vary with circumstance. Space architects must understand these environments.

A typical ISS module section demonstrates emphasis on modularity and absence of a normal up-and-down relationship (Figure 3). However, most astronauts since *Skylab* have favored establishment of a local vertical within each module to minimize disorientation, and this is now the common approach. Equipment configuration, lighting, airflow, and interior colors are used to provide orientation cues. The ISS equipment configuration is very efficient because it maximizes use of module volume and provides the largest single open space.

The module diameter was designed to take full advantage of the Space Shuttle Orbiter payload bay diameter (4.5 m), and its length was determined by the largest single integrated mass that could be launched to the ISS orbit with the shuttle (15,851 kg). The equivalent area of an ISS module is approximately 16 m^2 or the size of an average bedroom on Earth. On the ISS, up to six people live for about 90 days at a time in five modules. This demonstrates the efficient volume usage of microgravity-optimized designs as well as the cost premium of habitable space volume. See Chapter 4 for a detailed discussion of lessons learned from the ISS design.

PSYCHOLOGICAL AND PHYSIOLOGICAL CONSIDERATIONS

The psychological aspects of habitat design are affected by mission duration and crew size, among other considerations: the longer the mission, the greater the need for crew privacy and recreation. Increasing crew size increases the need for provisions for solitude, increases the complexity of human interactions, and adds social structure.

TABLE 1 Space habitat design environment considerations vary with location.

CONSIDERATION	EARTH ORBITAL	INTERPLANETARY TRANSFER	LUNAR/MARS SURFACE
Vacuum	Pressure vessel	Pressure vessel	Pressure vessel
Debris	Micrometeoroid and orbital debris (OD) OD is a growing problem requiring heavy shielding	Micrometeoroid only; no OD	Micrometeoroid and blast ejecta More micrometeoroid protection required for Moon than Mars due to absence of atmosphere
Gravity	Microgravity	Microgravity; rotating artificial induced gravity	Partial gravity (< 1 g); different proportions for interior spaces
Radiation	Solar proton event (SPE); galactic cosmic radiation (GCR) Low Earth orbits shielded by geomagnetic field, but South Atlantic Anomaly and polar orbits yield some exposure SPE protection advisable for long durations	SPE and GCR shielding required for long duration. Lunar missions with short loiter time may accept the risk of avoiding the mass penalty of shielding[a]	SPE protection required; GCR protection required for long duration Lunar shielding: regolith and/or water Mars shielding: partial protection from atmosphere, best at low elevations but degree unknown
Dust	None	None	Lunar: severe design challenge for human health and equipment Mars: chemical and biological impact unknown

[a]The issue of SPE shielding for lunar missions is a persistent design challenge because SPEs have low likelihood but high consequence. Most missions lasting weeks will never see such flares, yet they could be fatal for unshielded crews. The unshielded *Apollo 16* and *Apollo 17* missions closely bracketed the August 1972 SPE. The shielding mass required can easily render a transportation architecture nonviable. Given the probability, lethality, mass penalty, and inaccurate forecasting of SPEs today, no clear consensus exists yet regarding how firm a "requirement" for shielding should be.

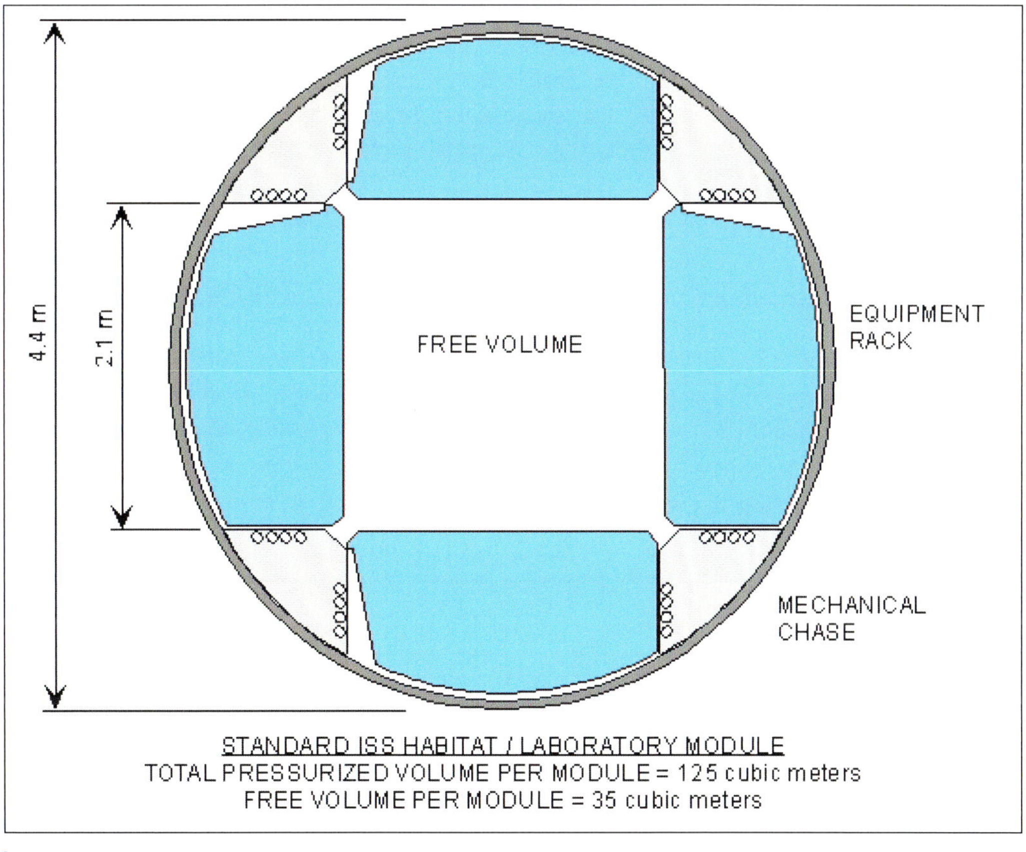

FIGURE 3 ISS rack-based, four-standoff module section configuration.

Short-Duration Missions

For missions that last up to a few weeks, crews can share personal quarters by rotating shifts, as was done when the space shuttle carried *Spacelab*. Each crew member does not need much volume for recreation, exercise, dining, etc., because of the "camping-like" conditions and the ability to rotate shifts, which reduces the need for redundant spaces.

Medium-Duration Missions

For missions lasting up to six months, crews require private personal quarters for sleeping and private recreation (e.g., reading and communicating with relatives on Earth) and more volume for grooming and personal hygiene. Crews on such missions tend to work standard shifts, resulting in more volume needed for dining, recreation, exercise, and meeting areas. For microgravity conditions, significant daily exercise becomes critical at this mission duration to limit degradation of both bone mass and muscle conditioning.

Long-Duration Missions

For years-long missions, crews should have more "comforts of home." Each member of the crew needs a private cabin for sleeping, personal storage, dressing, and "sitting." More generous recreational and exercise facilities are required, as well as a comprehensive health maintenance facility. Long-duration crews on planet surfaces presumably can explore their surroundings or engage in other outside activities through extravehicular activity (EVA). It is unknown on year-long interplanetary missions, where there is no Earth to look at and no planet to walk on, what activities can mitigate the negative effects of boredom, confinement, and limited social interaction.

Gross volume required for space habitats is estimated based on quantitative considerations of equipment and supplies and on historical data about human space exploration and remote environments on Earth. A first-order parametric volume estimation based on crew size and mission duration gives the designer a starting point for the space habitation system. Although it is clear that short-duration missions can be provisioned roughly analogous to shuttle missions, and Earth-orbiting stations can be provisioned based on lessons learned from ISS, few data are available to make a solid determination for long-duration planetary missions.

Physiological deconditioning of the human body in microgravity affects the cardiovascular system, musculoskeletal system, immune system, and endocrine system; it also causes body fluids to shift upward. These effects occur because the human body evolved in a gravity environment and adapts to microgravity conditions. The heart and other muscles weaken, bones lose density, the sense of balance is upset, lung and kidney functions change, taste and smell are compromised, and appetite is diminished. The effects might be reduced in partial-gravity environments like the Moon or Mars, but some deconditioning is still expected. Whether deconditioning stabilizes in partial gravity is still unknown. Countermeasures such as exercise and pharmaceuticals are currently assumed. Pharmaceutical mitigation of space adaptation syndrome (space motion sickness caused by the transition to microgravity) has been quite effective, but pharmaceutical control of bone mass loss has been tried only in bed-rest studies. Exercise countermeasures require significant daily time, as well as equipment, facilities, and volume.

HABITAT SYSTEMS

Table 2 lists typical habitat subsystems, and Figure 4 shows a top-level interface diagram for how they relate to the interior and exterior of the habitat system.

The habitat always interfaces with external systems in all architectures. Just as on Earth, energy and other utilities, views, and exchange of gases, fluids, and physical objects represent key external interfaces. For a house, these interfaces might simply be to a yard or farm or to urban infrastructure. In space, all of those functions are included in the integrated system, and so space architects must understand them to be able to design with them. Other types of external systems can include experiments, sensors for monitoring the system infrastructure, and EVA systems that support the crew outside the pressurized shirt-sleeve environment of the habitat. EVA systems include habitat elements like airlocks, spacesuits the crew members wear while working and exploring outside the habitat, and life-support backpacks [personal life-support system (PLSS)] that make their suits habitable. All of these systems are designed to meet limiting requirements for crew safety, performance, mass, power, volume, reliability, and robustness.

HABITAT DESIGN APPLICATIONS

Space habitat configuration combines all of the subsystems required to provide and maintain a living and working environment in space as well as to support the purpose of the habitat's mission. The configuration can vary from the simple open volume(s) of a short-duration spacecraft, to divided volumes for a medium-duration facility, to complex and

TABLE 2 General description of typical habitat subsystems.

SUBSYSTEM	DESCRIPTION
Structure/enclosure	Pressure vessel
Environmental Control and Life Support System (ECLSS)	Provides air pressure, oxygen partial pressure, CO_2 removal, potable water (degree of recycling "closure" depends on mission length and remoteness), and waste management (storage, stabilization, and recycling)
Thermal Control System (TCS)	Heat collection, removal, and rejection to maintain habitable temperature
Power and power distribution	External power source (solar photovoltaic, solar thermal, beamed, or nuclear), power storage (batteries, fuel cells), and power regulation and distribution
Data Management System (DMS) and Communications	Computing, system monitoring and control, data storage, remote telecommunications
Internal Audio/Video	Internal communications
Crew Accommodations	Crew quarters, galley, wardroom equipment, recreation equipment/facilities
Science Accommodations	Mission-specific science and experimental equipment
Stowage	Storage volume, containers, and inventory-control for personal and mission-related items
Radiation Shelter	"Storm shelter" with hydrogen-rich shielding for episodic solar proton events

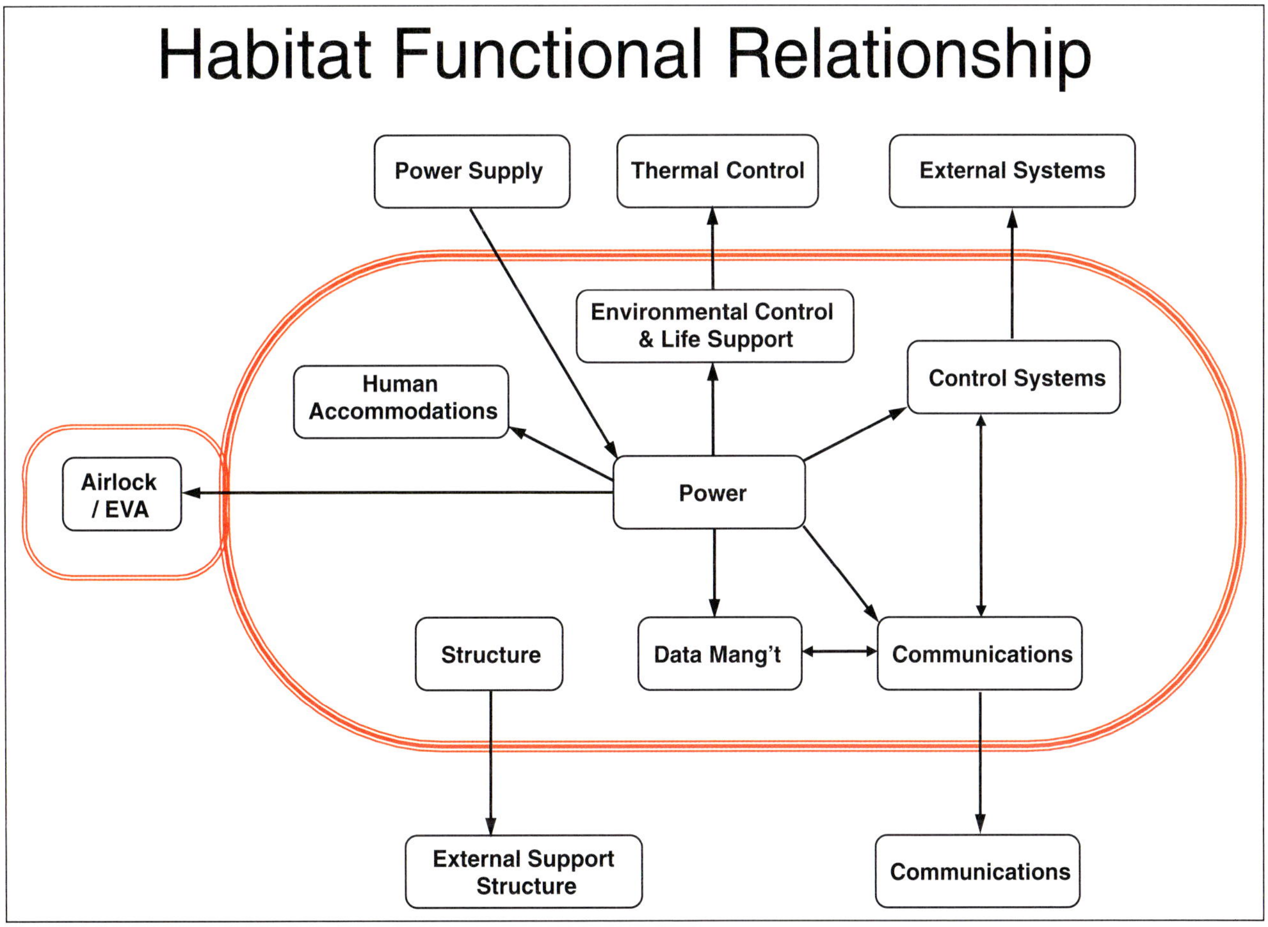

FIGURE 4 Habitat element top-level system interfaces.

TABLE 3 Top-level space habitation design trades and guidelines.

CONFIGURATION DRIVER	TRADE	GUIDELINE
Habitat function	Habitat layout and functional allocation	Separate habitat and laboratory functions for long-duration missions.
Number of crew	Volume required	Volume per crew member increases significantly for typical mission crew sizes.
Structure	Aluminum, composites, inflatables	Aluminum for preintegrated, short-duration habs. Inflatables for larger-volume situations.
Life support	Loop closure, technology heritage	Open-loop for short missions. Increasing closure for longer durations.
Data handling and management	Redundancy, automation, dependence on ground crew	Fault-tolerant architecture with multisystem redundancy. Habitat health monitoring automation increases for longer duration.
Thermal control	Body-mounted vs deployable radiators	Body-mounted on transportation vehicles. Deployable for large-capacity heat rejection. May be able to use regolith thermal sink and/or convection cooling on planets.
Power	Source: solar vs nuclear Storage: battery vs fuel cell	Long-duration, ISRU-intensive, or deep-space applications may require nuclear.
Crew accommodation	Social interaction, privacy, exercise, recreation	Increased privacy and social-interaction provisions required for deep-space and/or long duration.
Protection from environment	Radiation, orbital debris, micrometeoroid, dust	Earth orbital hab requires some radiation protection and shielding for orbital debris, micrometeoroids. Transfer hab requires radiation shelter and micrometeoroid shields. Planetary hab requires some radiation and micrometeoroid protection and dust control.
Risk	Level of redundancy	Fail-op, fail-op, fail-safe on critical systems (i.e., two-fault tolerant with full functionality, three-fault tolerant to safe condition).

multiple, interconnected volumes of a long-duration habitat. Table 3 summarizes key design trades and considerations.

Orbital Habitats

Since the early 1970s, humans have been living and working in space. Medium-duration orbiting facilities have notably included *Skylab* (Figure 5), *Spacelab*, *Salyut* 7, *Mir*, and the ISS. These systems demonstrate incremental evolution in habitation design and technology. Except for *Skylab*, early space habitats provided only the necessities to survive in low Earth orbit. Little was understood about effects of space on humans or how to accommodate them. Overall, the Russian Space Agency has the most experience with long-duration space habitation systems.

ISS modules (Figure 6) embody lessons learned from previous experiences and research on isolation and space effects. These designs provide increased free volume for each member of the crew, private spaces such as crew quarters, dedicated laboratory

FIGURE 5 *Skylab* was the first U.S. space station.

FIGURE 6 ISS, U.S. laboratory module *Destiny*.

facilities, areas for group functions, recreation and exercise, and high-quality hygiene facilities. Previous space habitats each had some combination of early versions of these features, but ISS has had more emphasis on addressing human physiological and psychological well-being to maximize productivity. TransHab (Figure 7) is a new technology for orbital habitat modules, designed to improve living in space. (See Chapter 8.)

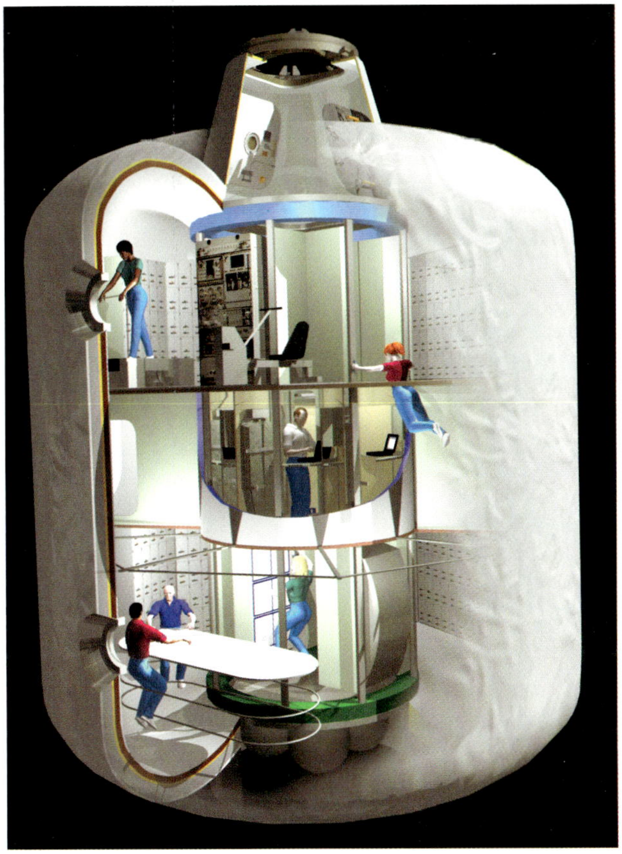

FIGURE 7 TransHab module configuration concept.

Transfer Habitats

There are numerous examples in the literature of space habitats—both historical and conceptual—designed for transportation systems. Transfer habitats for Earth-to-orbit and return-to-Earth applications tend to be small, such as the conical Apollo command module (CM) (Figure 8) or the Space Shuttle Orbiter cabin (Figure 9). The only interplanetary transfer habitat flown so far is the *Apollo* CM. At the time, mission architecture feasibility dictated using the same capsule for launch, transfer, lunar orbit, and Earth entry. The transit time was short (about three days one-way), and the only crew member to occupy it continuously was the CM pilot, who had the volume to

FIGURE 8 *Apollo* command module was a conical capsule supported by an unpressurized service module.

FIGURE 9 Space Shuttle Orbiter cabin squeezes seven astronauts into a space far smaller than its unpressurized cargo bay.

himself during the surface mission (except in the contingency case of *Apollo 13*). However, interplanetary transfer habitats Mars or asteroid missions would require more volume (Figure 10). See Chapter 11.

Planetary Habitats

The only planet-surface habitat flown so far is the cabin of the Apollo lunar module (LM) (Figure 11). It provided arguably the minimum possible volume needed to survive on another planet. Even more than the CM, the LM was governed almost completely by the Saturn V launch system capacity and lacked enough space even for the surface crew to sleep lying down. Anticipating longer surface missions, designers have proposed diverse ideas for future space habitats for the Moon and Mars. These concepts range from Class I, preintegrated, ISS-derived modules (Figures 12–14), to Class II inflatable structures (Figure 15) and hybrid systems (Figure 16), to Class III systems derived and constructed from *in situ* materials (Figure 17). Each habitat concept tends to focus on specific features; none of these diverse ideas represent committed programmatic choices, but it is reasonable to assume that funding constraints will emphasize ISS heritage for early lunar surface systems.

FIGURE 10 Conceptual design for a near-Earth asteroid mission module accommodates multimonth deep-space transit.

FIGURE 11 *Apollo* lunar module cabin minimally supported two people for two days.

FIGURE 12 ISS modules provide the foundation for many contemporary lunar surface system habitation concepts to take advantage of engineering development heritage: node (left); ISS airlock system (right).

| **FIGURE 13** Pressurized rover based on ISS module heritage.

| **FIGURE 14** Lunar outpost based on pressurized rovers, working in conjunction with fixed, shielded core module.

FIGURE 15 Inflatable habitat system would be deployed on the surface, shielded, and then outfitted.

FIGURE 16 Airlock concept shows integration of Class I and Class II structures.

FIGURE 17 ISRU-derived lunar concrete allows Class III structures after mining, beneficiation, and processing infrastructure is sufficiently capable (artist: Pat Rawlings).

NEXT-GENERATION TECHNOLOGIES

In the future, habitable architecture is less likely to be limited to just assemblages of cylindrical, hard-shell modules. Inventions like TransHab mean that many wonderful and architecturally sophisticated configurations will emerge in the 21st century. Advances in materials development and manufacturing techniques are enabling for large-scale planetary operations, the eventual settlement of Mars and other locations, and migration of humans into space. These advances include durable, lightweight structures; techniques to emplace, erect, deploy, or manufacture habitats in space; "smart" structures and materials that "self-heal"; and integration of automated components to enable self-deployment.

Requirements for space pressure vessels and unpressurized shelters will challenge us to innovate structure systems that combine high-strength and lightweight materials to achieve reliability, durability, repairability, radiation protection, packaging efficiency, and life-cycle cost effectiveness. Inflatable structures lead in the search for promising new technologies for habitable space structures. Inflatable envelopes open new possibilities for habitat shapes and sizes and therefore change how we will design habitats, laboratories, hotels, and space resorts. And a number of techniques are being proposed, based on advances in robotics and autonomy, to allow delivery and setup of such habitats to space and planet surfaces, or even in situ manufacturing and construction, before humans arrive.

Future research will yield fully integrated skins that use embedded sensors, processors, and mechanical and chemical effectors to comprise smart structures that autonomously and continuously monitor their own condition, and detect, analyze, and repair incipient structural vulnerabilities. Eventually, smart structures can become essentially alive, not just self-healing, but self-regulating systems that integrate life-support technologies within them. Breakthroughs in biotechnology have opened up many exciting possibilities already; combining biotechnology with fabric and matrix structures might someday produce a kind of protective, self-healing, self-regulating system analogous to human skin but optimized for space conditions and providing life support for the people and plants it shelters.

We can expect NASA's long-term research and development into such habitation technologies to continue yielding "spin-offs" that benefit humankind on Earth, including lightweight robust structures; self-diagnostic and self-healing systems; robotic technologies for fabrication, construction, and maintenance; extreme recycling; and life-support closure.

FUTURE OF SPACE ARCHITECTURE

Architects and engineers shape the human environment by building, and so they will shape our future on Earth and in space. By working "within the system," space architects are helping ensure sound architectural design principles and practices are included as national space agencies and the aerospace industry develop mission scenarios, habitats, space vehicles, planetary bases, supporting systems, and commercial businesses based on them. Groundbreaking work is laying technological foundations for decades to come. Although much progress has been made since Yuri Gagarin's seminal flight, there remains a great deal of work on the ground and in space to prepare for living and working on the Moon and on Mars (Figure 18).

Space architecture is a fascinating field, replete with high technology and limitless boundaries. Its basic vernacular of space vehicles, pressure vessel habitats, and support infrastructure has become clear. Today's space architects are at the forefront of the evolutionary step where architecture breaks free of terrestrial limits to explore how humankind will settle space and other planets (Figure 19). |

| **FIGURE 18** Complex human exploration scenarios will benefit from the training and skill of space architects.

FIGURE 19 The vernacular of space architecture extends into a limitless future.

References

Capps, S.D., Fowler, R., and Appleby, M. (1991), "Induced Gravity Mars Transportation Systems," Space Manufacturing 8, *Proceedings of the Tenth Princeton/AIAA/SSI Conference*, AIAA, Washington, D.C., pp. 126–131.

Kennedy, K. (1992), "A Horizontal Inflatable Habitat for SEI," *SPACE 92: Engineering, Construction, and Operations in Space Conference*, American Society of Civil Engineers.

Kennedy, K. (1992), "Dust Control Research for SEI," *SPACE 92: Engineering, Construction, and Operations in Space Conference*, American Society of Civil Engineers.

Kennedy, K. (1994), "Alternative Habitat Concepts for the First Lunar Outpost," *SPACE 94: Engineering, Construction, and Operations in Space Conference*, American Society of Civil Engineers.

NASA (1999), "Lunar Outpost," Internal Document, NASA Johnson Space Center, Houston, TX.

PART 2
Orbital Architecture

Spaceships have ventured into orbit for a half-century, and Russian and U.S. orbiting stations have enabled sustained operations there for almost as long. The orbital architect's challenge is like that of the submarine designer: ensure survivability, efficiency, and habitability in a lethal environment, fairly close to home, within vehicular forms.

In "Design Constraints for Orbital Architecture" (*Chapter 3*), Brent Sherwood sets the stage for Part 2 by laying out inescapable conditions faced by all orbital architecture. These "facts of life" constrain the technology options available and apply at all scales from spacecraft subsystems to eventual orbital cities. Three sections follow, focusing in turn on 1) the architecture of the International Space Station; 2) next-generation orbital architecture using expandable-volume TransHab technology; and 3) orbital architecture for future deep-space missions, artificial gravity, and large-scale development.

INTERNATIONAL SPACE STATION ARCHITECTURE

As the premier contemporary example of orbital architecture, the International Space Station (ISS) offers many lessons. It is vital for space architects to understand in depth what has been accomplished, how the project came about, and how it might be adapted or improved upon for future applications. Its module interior architecture was defined in the mid-1980s. A quarter-century of lessons from the development, construction, and use of the built solution can shed new light on architects' thinking about "roads not taken."

In *Chapter 4*, "Performance of the International Space Station Interior," Rod Jones describes the ISS architecture, how it is manufactured, and how it performs in use. James Broyan et al. detail in "International Space Station Crew Quarters" (*Chapter 5*), the most personal level of ISS architecture: private crew cabins developed after the rest of the station architecture. In "Retrofitting the International Space Station" (*Chapter 6*), Susan Fairburn proposes a kit-of-parts solution, compatible with ISS rack structures, for an alternative type of crew quarters outfitting. Finally, in

Chapter 7, "Alternative Space-Station Module Architectures," David Nixon and Jun Okushi take us back to a seminal study in the mid-1980s that investigated alternatives to the architecture chosen for the ISS.

TRANSHAB ARCHITECTURE

For decades, NASA has considered inflatable modules to be the next step in space habitation system technology. Inflatables would enable us to decouple the usable volume attainable on orbit from what can be fit inside the "mold line" of a launch-vehicle payload bay to get it there. TransHab was a major step forward in the development of space-rated inflatables. Developed by a NASA tiger team in the early 1990s for Mars mission applications, TransHab almost made it onto the ISS instead, and then found a third life as the foundation of commercial endeavors currently underway.

Kriss Kennedy explains the conception, architecture, technology, and test program of the seminal TransHab prototype in *Chapter 8*, "TransHab Project." In "Design of a TransHab–Based System" (*Chapter 9*), Matt Herman describes a privately funded, orbital-station concept that uses TransHab technology as a point of departure. Bigelow Aerospace launched a prototype system in 2006. *Chapter 10*, "Space Hotel Based on the TransHab" by Paola Favata, shows how TransHab's larger size opens possibilities for new human environments. Favata proposes a small, luxury resort hotel to accommodate initial, high-end space tourism. Full use of microgravity three-dimensionality allows a startling combination of compactness and spaciousness.

FUTURE ORBITAL ARCHITECTURE

One of the major benefits from the ISS is detailed knowledge of how to assemble a large, complex system in Earth orbit and modify its configuration over time, even as it is continuously used. This experience paves the way not only for deep-space human exploration but also for more elaborate Earth-orbital applications, including eventual large-scale commercial development.

In "Habitats for Long-Duration Missions" (*Chapter 11*), Brent Sherwood and Stephen Capps describe a vintage trade study comparing module architectures for multiyear Mars missions. By comprehensively measuring and scoring multiple criteria over thousands of options, this chapter provides an example of how complex space architecture decisions are approached. Ted Hall provides a concise tutorial about the features of rotating systems and what it would be like perceptually to occupy them in "Artificial Gravity" (*Chapter 12*). This discussion challenges common assumptions and yields principles designers must know as they envision inertial-gravity applications. Finally, in *Chapter 13*, "Orbital Cities," Brent Sherwood extends what we have learned about engineering systems for the Earth-orbital environment to the macroengineering scale, which results in a pattern language for eventual space cities overflying the home planet.

3

design constraints for orbital architecture

BRENT SHERWOOD

INTRODUCTION

FRANK LLOYD WRIGHT spoke about "organic architecture," by which he meant architecture that appears to grow naturally out of its constraints, rather than fighting them. Indeed Wright claimed that, "Constraints are an architect's best friend." Constraints emerge from various sources, and Wright infamously paid variable attention to them in fact.

Constraints on the design of the human environment arguably comprise the following five types. Architecture tends to focus more on the first three, whereas urban planning tends to focus on the last three. Good architecture and good urban planning attend to all five:

- Conditions levied by the environment, site, and context
- Capabilities limited by technology and budget
- Requirements imposed by the owner, users, neighbors, and community
- How people actually live, work and use architecture
- Accommodation of growth and evolution of both the design program and its context.

This chapter proffers and discusses important constraints from these five categories as applied to low Earth orbit (LEO). These issues are largely familiar to aerospace engineers and essential for space architects to understand.

ENVIRONMENTAL CONSTRAINTS

Environmental and site conditions in LEO are as axiomatic as are gravity and weather on Earth. Environmental constraints in LEO include Keplerian and non-Keplerian orbital dynamics, microgravity, hard vacuum, temperature extremes, raw sunlight, Earth's tenuous atmosphere, orbital debris, rotational dynamics, radiation, and variable views. Although the causes of these phenomena cannot be detailed here, they can be found in texts on spacecraft systems design. This section simply summarizes their principal architectural implications.

Orbit

In LEO, there is no such thing as a fixed location. Separate objects are in separate orbits moving at about 7 km/s; separate orbits diverge into different planes with time. It is extremely costly to bridge orbits not in the same plane. Many subtle forces act together to determine an object's actual orbit at any given time. Orbits that differ in average altitude become noncoplanar over a span of days; if allowed to drift naturally, objects orbiting together at first might eventually approach each other head on at 14 km/s. The only zero-energy way to keep objects together is to link them mechanically. Externally applied forces change an object's orbit, not just its position; application of forces to achieve rendezvous is not intuitive. Launching from a given site on Earth into a given orbit can only occur during two brief launch windows per day.

Microgravity

Objects in LEO are in continuous free-fall; behavior of fluids is dominated by surface tension; behavior of solids is dominated by friction, electrostatic, electromagnetic and elastic forces. Fluids, whether Tang, fuel, or flame, tend to spherical shapes. Dust and objects drift on air currents if inside or orbit individually if outside. The human body takes on a neutral body posture, with all joints slightly bent; the spine stretches; fluid collects in the upper body; taste and smell are subdued because of nasal congestion; physical and chemical deconditioning occur over time. The $1/r^2$ gravity field requires continuous application of torque to maintain desired attitude of an object whose inertial axes are neither parallel nor normal to the nadir vector.

Copyright © 2008 by Brent Sherwood. Published by the American Institute of Aeronautics and Astronautics, Inc., with permission. This chapter is based on material presented in AIAA Paper 90-3737, reprinted with permission of Boeing Company.

Hard Vacuum

In LEO, the typical vacuum exceeds by several orders of magnitude the quality of vacuum attainable in laboratories on Earth. Rejecting waste heat can only be done radiatively, or by sacrificing fluid in "flash evaporators." Sounds are conducted and reverberate throughout a system until dissipated within the structure and its contents, because sound waves cannot dissipate directly into vacuum. All spacecraft gradually leak atmosphere. The vacuum near spacecraft is "dirtier" than in the wake of a ram shield or in free space. But for the cost of makeup air, interior vacuum cleaning is easy. The external vacuum is lethal, but not instantly. The joints of inflated spacesuits are hard to bend and therefore cause fatigue.

Extreme Temperatures

Influenced by many factors, object temperatures in LEO can vary hundreds of degrees depending on whether they face the sun, the Earth, each other, or deep space. Terminator passage (up to 32 times per day) is the dominant periodic constraint. Space-system designs must include a combination of clever configurations, tolerant materials, passive shields, heat pipes, or active cooling loops to redistribute heat, heaters/radiators to add/reject heat, and "barbecue" rotation to even out the heat load.

Unfiltered Sunlight

Objects in sunlight are exposed to 1389 W/m^2 of unfiltered solar spectrum, including ultraviolet (UV) wavelengths that can embrittle or degrade materials, blind sensors including retinas, burn tissues and cells, and induce thymine-dimer DNA damage. Systems require UV-tolerant materials and coatings and UV filters for sensors, visors, and windows.

Tenuous Atmosphere

LEO includes the uppermost, rarefied region of Earth's atmosphere, which is rich in reactive monoatomic oxygen. Coatings must be used to control erosive effects on materials of the atomic oxygen, particularly on surfaces exposed to the ram flux. Electrical contactors exposed to the conductive plasma can control the charge potential of spacecraft or use the geomagnetic field for propulsive benefit. Drag losses must be compensated to prevent orbit decay.

Orbital Debris

LEOs contain enormous populations of artificial orbiting objects, from dust and paint flecks, to shrapnel and loose parts, to dead spacecraft. All are projectiles and are hazardous in proportion to their kinetic energy $(1/2 \cdot mv^2)$, and their orbits propagate uniquely. The flux probability peaks at incoming angles roughly 45 deg to starboard and port off the bow. Risk is proportional to the area exposed to the flux. Shielding is practical only for particles of ~1 cm size or smaller. Shielding space suits, windows, sensors, radiators, and solar arrays during use is not practical at all.

Rotational Dynamics

Weightlessness can theoretically be compensated by a rotating flight system; centripetal acceleration induces pseudoweight. No human-scale system has been developed or flown. Rotating systems with internal energy-dissipation (friction) settle naturally into the lowest-energy state, which is rotation about the axis of maximum moment of inertia—pancakes are stable; spindles are not. The rotation axis remains fixed in inertial space unless acted upon by an external torque. Out-of-plane motions within the rotating system generate Coriolis accelerations, which cause vestibular disturbances in animals if the ratio of rotation rate to radius is high. Trajectories of objects thrown inside rotating systems appear counterintuitively curved when viewed within the rotating frame of reference.

Radiation

Earth's atmosphere attenuates incoming cosmic rays (high-energy atomic nuclei) through absorption; objects above the atmosphere, from LEO up, are exposed to this flux. However, charged particles from solar emissions (high-energy protons) are generally diverted in LEO by the geomagnetic field. Some exposure occurs where highly inclined orbits pass through the field's polar regions and where lower-inclination orbits pass over weak areas in the field (e.g., the South Atlantic Anomaly). Without dedicated (hydrogen-rich) shielding, LEO residence totaling a few dozen months yields lifetime exposures of the same order as permitted for radiation workers.

Variable Views

In LEO, the view vectors to targets of interest for different purposes—sun, Earth, dark sky, astronomical objects, nearby hardware, clean vacuum, beamed-power sources, oxygen ram flux, debris flux maxima—are generally mutually incompatible and change with time. Without intervening air, the view of space objects is clear, limited only by diffraction and glare. The Earth view varies constantly because of Earth's rotation and weather and is beautiful and poignant; spacefarers report never tiring of it.

Taken together, these constraints govern the unavoidable environment for LEO architecture.

TECHNICAL CAPABILITIES

Human capability to build in the LEO environment just described is constrained by available technology and financial resources. The current state of both determines initial conditions for possibilities in the coming decades.

State of Practice—Qualified Technology

The International Space Station (ISS) embodies the most advanced technologies yet qualified by the major spacefaring nations of Earth: Russia, the United States, Europe, Japan, and Canada. Chinese human spaceflight capability is just now being developed, and other countries like India are not far behind.

Habitable pressure vessels for in-space use are prefabricated, preoutfitted, and modular (Class I structures). Cylindrical, welded metal-skinned modules, they are adapted from, or fit on or inside, rocket-powered launch vehicles, so that they are limited to launch-vehicle diameters (generally a maximum of 4.4 m). Complex adapter mechanisms form a seal when mated in space, allowing hatches to be opened between adjoining modules. Joining occurs by hard docking (flying the modules together slowly, e.g., Russian systems) or by being soft-berthed using a manipulator arm (e.g., U.S. systems). Subsystems and distributed systems (wiring, ducts, and tubing) are mostly factory installed on Earth, with only minor outfitting being done in space. Windows do not exceed 0.5 m in diameter, contain many layers of glass, and are few.

The environment is noisy. Sleep schedules are maintained on a 24-hour cycle, despite the absence of normal external diurnal cues. Sleeping occurs in microgravity restraints, either pitched in quieter corners or contained within sleep compartments.

Food is individually packaged; preparing hot meals means microwaving vacuum-packets. Dishes and utensils are disposed of after use. Hand washing occurs inside a glove-box-type container. Shaving occurs with a vacuum razor. Body cleansing occurs in a vacuum-dried shower stall or with wet wipes. Clothing is discarded after several uses. Elimination is done with devices that use air currents to guide and capture the waste. Food and solid wastes are chemically stabilized for destructive reentry or cargo return to Earth along with trash.

Oxygen is introduced from cryogenic bottles, exhaled carbon dioxide is absorbed by chemicals, potable water is recovered from condensation and urine, and wash water is recovered from "gray" water already used for washing. Leakage makeup air is stored as liquid nitrogen.

Spacesuits have pressurized fabric limbs attached to a hard torso. Size is adjusted using spacer rings in the limbs. A bubble helmet allows head movement for viewing. Environmental equipment is mounted in a backpack. Gloves are cumbersome and tiring. Inside, long johns circulate fluid to control temperature. Waste elimination is via diapers.

External truss structures extend the "real estate" on which to integrate flight systems for attitude control, power, thermal control, communications, payload support, vehicle parking and housekeeping. Power is provided by attached solar arrays that track the sun; for an Earth-oriented platform, sun tracking requires two rotational degrees of freedom. Heat rejection is accomplished by exchanging heat from internal water loops to external ammonia loops that pass through radiators. Radiators are oriented continuously normal to the sun vector. Modules are wrapped in "Whipple bumper" debris shields, multiple layers that disintegrate and absorb the impact energy of impinging particles. All external surfaces are covered with thermal-control finishes or coatings: polished aluminum, white paint, or gold-coated plastic multilayer insulation.

Concerning the current state of resources, note that the ISS "assembly complete" configuration comprises 14 modules of various size and function, a seven-segment truss, and five solar-array assemblies. The U.S. portion alone cost roughly $25 billion to develop, build, test, certify, and deploy over 25 years; altogether, ISS fully stretched the global economic capacity devoted to human spaceflight. Although ISS represents a Herculean nonrecurring investment, the cost to further adapt these infrastructure elements to meet future, specialized applications and technology improvements will not be small, either. Developing other types of elements will cost even more.

State of Design—Foreseeable Technology

Habitable vessels with outer hulls that are inflated after launch (and with pre-integrated utilities in the core) permit diameters larger than launch vehicles. Private enterprise has begun launching prototypes of such vessels, although none have yet been outfitted or occupied. The next step is vessels assembled in orbit from prefabricated panels and then welded and outfitted in situ. The final step is pressure vessels manufactured in space, e.g., metal vapor vacuum deposited inside temporary, inflated, or spin-stabilized forms and then finished and outfitted. As O'Neill postulated decades ago, enormous enclosed volumes are achievable via the assembled or manufactured methods. Until then, however, large volumes could only be approximated by stringing together modules of limited diameter and with diameter-constricting connectors between them. In all cases, the architecture achieves a large habitable volume by clustering simple volumes. Modules can be isolated by closing hatches between them in the event of accidental depressurization; the smaller the modules, the greater the need for rapid isolation.

Until fully mature space manufacturing is attained, that is, precision machining of large, heavy assemblies, mechanisms connecting habitable modules are launched from Earth and therefore are limited to launch-vehicle shroud dimensions. Such mechanisms are designed to withstand hundreds of berthing cycles.

Utilization and utilities are functionally separated: larger "utilization" modules are supported by smaller, attached "utilities" modules. This enables flexible design of unencumbered volumes for functional uses. It averts the current state of the art, so that utility systems are housed outside the main modules, albeit still within the pressurized envelope. Only the distribution systems take up space "inside."

Mature life-support systems evolve away from physicochemical technology except for specialized, small, or temporary backup applications, and toward "ecological" technologies that use soil-bed reactors and estuarine-flow reactors, populated by microorganisms and plants, to reclaim atmosphere and water, and for the concentration and removal of toxic chemicals.

Although actual pressure hulls are thin, the apparent wall thickness is substantial (of order 0.5 m) because of deployable debris shielding. This makes hull openings reminiscent of medieval castle wall fenestration. Windows, hatches, and other vulnerable mechanisms are covered by debris shields except when in use.

Sunlight is admitted through filters that remove UV frequencies and regulate brightness. Sunlight can be simulated using solar-spectrum lamps during dark-side passage but is energetically expensive and therefore typically used only where functionally necessary (e.g., in work areas). The design of the environment therefore submits to a natural rhythm in which light and shade alternate every 45 minutes. Earth light is also admitted for ambient lighting.

Power is generated by attached solar plants (photovoltaic or dynamic heat engine) or shielded nuclear plants or beamed to receiving arrays on the facility from remote powerplants that use these sources. In all cases, heat rejection requires radiators directly connected to circulating-fluid heat transport systems.

Large, long-lived LEO debris (defunct spacecraft and spent stages) is de-orbited or captured for salvage. This significantly reduces the risk to orbiting infrastructure of catastrophic accidents, by eliminating the major source of random collisions that exponentially increase debris populations. (*Mir* was de-orbited because it was not cost effective at the time to boost it to a safe storage orbit for salvage.) As the space population grows, cannibalizing defunct systems becomes more practical and economically attractive than launching everything anew. (This is likely to occur before Earth-to-orbit traffic grows so economical that launch once again becomes the cheapest option.) In-space salvage therefore becomes at least a contingency, probably a hobby, and perhaps a business. A mature in-space support operations market guarantees a steady stream of systems to be repaired or recycled. The end state is full-fledged materials recycling: organic wastes, gases and solvents, polymers, metals, semiconductors, and glasses. Prospects for this industry are enhanced by the inexhaustible energy and vacuum of space.

Tether structure systems are used for facilities that require ready access despite being physically separated (e.g., for local vacuum cleanliness, for view factor, for vibration isolation, for various *g* levels, or for security). They are strung along the gravity gradient (parallel to the nadir vector) like pearls on a string, linked by a common elevator.

Economically, capabilities at any given time are bounded by the level of investment corresponding to a nominal or modest "stretch" for the participating nations and industries. At the turn of the 21st century, this is ~$100 billion for a multiyear government research project like ISS. *Apollo*-type, "spike" investments by governments (of order ~$1 trillion when inflated to turn-of-the-century currency) are probably an upper bound. However, once LEO development is driven by commercial investment, growth is likely to become more rapid than typically imagined.

Future Requirements That Can Be Anticipated

A mature human spaceflight market segregates into four sectors: *utilization* (using the properties of space or performing deep-space missions), *security* (a subset of utilization specialized for defense of space-based or Earth-based assets), *support operations* (building, running, maintaining, and expanding space systems themselves), and *passenger travel* (for business or leisure). Each introduces unique driving requirements, but grouping the first three sectors

TABLE 1 Requirements for two classes of mature LEO architecture can be anticipated.

FACILITY OPERATIONS AND INDUSTRIAL UTILIZATION	PASSENGERS AND CREW
• R-bar (parallel to the orbital radius vector) and/or V-bar (parallel to the orbital velocity vector) access kept clear for incoming and departing vehicles • High concentrations of power input and thermal rejection capability • "Exposure facilities" for direct access to natural space environments (e.g., ram flux, solar wind, particle radiation, wake, Earth view) • "Vacuum hangars" with sunshields and controlled lighting, provisions for remote manipulation and spacewalking (EVA), utilities umbilicals, direct and video viewing from a shirt-sleeve-environment (IVA) operations gallery • Large-volume airlocks adjacent to vacuum hangars for shirt-sleeve/cleanroom operations on equipment • Industrial space cranes • Depot for propellants and other gas, fluid, or solid consumables • Materiel depot for sorting and environmentally controlled stowage of working stock • Access to vacuum and clean vacuum • Selected views for specific applications • Ability to reconfigure industrial infrastructure: volumes, fixtures, utilities • Room to experiment	• Large amounts of power input and thermal rejection capability • Comfort, quiet • Volumes that can accommodate public assembly • Means of witnessing "interesting" operations • Selectable awareness of terminator passage • Big windows with mitigation of solar glare • Unobstructed, uninterrupted nadir (Earth) and zenith (star) views; dark space views without washout by locally reflected sunlight. • Earth viewing of temperate and polar latitudes • Views of the facility exterior from the interior • Spacewalks and space rides outside and away from the facility • Plants and pets • Fresh food • Interior "outdoor" areas • "Ground" and "sky" cues • Recreational and team sports • Ability to personalize living quarters and redecorate common spaces • Provisions for hobby activities • Room to grow
COMMON TO BOTH	
• Antiproximity among work, living, and social areas • Antiproximity between populated areas and operations or industrial hazards • Access to weightless and weighted environments and easy passage between them • Safety, privacy, and security • Standard system operations and interfaces (including international and intercompany) • Broadband connectivity • Regularly scheduled transportation	

separately from the fourth reveals the two likely sets of architecture drivers summarized in Table 1.

HUMAN DESIGN DRIVERS

Humans are extraordinarily adaptable, as evinced by habitation designs worldwide. Yet, as spacefaring populations grow, and especially as those populations comprise an increasing fraction of business and leisure passengers, tolerance for highly abnormal environmental conditions will decrease. Therefore mature LEO interior accommodations must include many features considered normal on Earth: sound muffling (aided by the functional separation of utility equipment from serviced spaces, just described), varied environmental lighting and interior finishes, provisions for genuine privacy, individual broadband connectivity, onboard laundry, actual cooking and washing in kitchens and bathrooms, fresh food (grown onboard or delivered routinely), and modern medical care including surgery.

Throughout social, recreational, and residential architecture, hydroponically grown plants will provide the psychological relief of proximity to living nature. Soil, although heavy, will also be used as a rooting medium to host substrate-bed reactors to rejuvenate atmosphere and purify water in ecological life-support implementations.

Water will likely be a particularly interesting feature of LEO architecture. Because of its many uses (radiation shielding, thermal moderation, potable reservoir, life-support buffer, recreation, art), it will likely become a key architectural element despite its heaviness. Volumes in which people spend a large fraction of their time (e.g., sleeping areas) will likely be surrounded by water-filled shields. Fully filled chambers will allow conventional underwater sport swimming with breathing apparatus. Partially filled chambers will afford sporting and recreational opportunities that take unique advantage of the large-scale behavior of liquids in microgravity. Fountains in rotating artificial gravity zones will display eerily compelling trajectories.

The most fundamental architectural feature will be the addition of a third layer of enclosure. On Earth, most architecture is characterized by "indoor" and "outdoor" spaces. Prescient exceptions include Roman urban architecture and modern shopping malls, in which an intermediate layer is introduced: the "controlled outdoors," a kind of faux exterior in which space, light, sound, plants, water, and illusion are used to induce the positive psychological benefits of the outdoors while being controlled and even climate protected. In LEO, such indoor exteriors, including "pocket parks," will be essential for inhabitants constrained to never directly sense the lethal exterior.

People in LEO will expect to be able to watch Earth—a lot. They will also expect to be able to observe celestial targets. They will expect to witness interesting, dynamic operations. They will seek solitude and quiet, as well as congregation. They will expect both weightless and weighted conditions and to be able to pass easily from one to the other. They will expect to "get outside" [i.e., spacewalk, or extravehicular activity (EVA), as close to raw space as physically possible] expeditiously and safely. They will want the facilities to function without much attention; they will expect the architecture to not get in their way. They will long for large volumes, environmental vistas, open ground, and skies above. They will insist on plants and animals to live around, play with, and probably also to eat. They will decorate their quarters. When together in a space, they will want a common orientation.

Taken together, many of these human environmental and behavioral drivers are so incontrovertible as to appear atavistic. In specialized or temporary circumstances, they can be suppressed or approximated in varying degrees according to urgency. But long-term activity, and especially urbanism, cannot ignore them. A mature human society is as far removed from a military submarine as New York City is from an encampment. The architecture—not the people—must accommodate.

LONG-TERM GROWTH

Over the coming centuries, LEO infrastructure will grow, following the same pattern as terrestrial urbanism, from encampment to outpost to settlement to village to city to megalopolis. Even at the microscale of the ISS, inchoate urban principles can be discerned: coexistence of construction with operations, activity zoning, "preferred real estate" locations, diurnal and seasonal rhythms, and even multilingualism.

By projecting end-states of the large-scale future, we can begin to identify patterns useful now as design constraints to facilitate organic growth. Anticipating such growth, including the evolution of design programs and environments, is key to designing a scheme that stays relevant and therefore persists. Over the long term, access and physics govern everything in LEO: proximity relationships (for efficient technical operations and a commodious human environment), view factors (for operations, industrial users, and occupants), physical access (for transportation systems, freight, and refuse), and substrate properties (in the case of LEO, the inescapable rules of astrodynamics and orbital flight described earlier). All must be accommodated.

These constraints are as basic as gravity, climate, and tradition on Earth. They arise from characteristics of the LEO environment, how systems operate within it, how people live, and the need for growth. Taken together, they give space architects a framework within which to address, compare, and integrate alternative concepts. |

INTRODUCTION

IN TERRESTRIAL BUILDING CONSTRUCTION, the product of "architecture" is drawings and specifications, which identify hardware requirements and depict the integrated design. The International Space Station (ISS) program architecture was established through the specification of key hardware features and constraints. These features imparted inherent capabilities used to help manufacture, assemble, test, and activate the hardware.

4

performance of the International Space Station interior

ROD JONES

ISS ARCHITECTURE

The original ISS architecture requirements were established early in the development phase (Figures 1–3). Jones (2000a) describes the requirements selection process used to define the quadrant or "four-post" architecture of the ISS pressurized elements. The key features were pressure vessel envelope, standoffs, racks, and hatch shape and size.

Module

The ISS module size was dictated by the space shuttle cargo bay capacity: the maximum module envelope diameter while vibrating during a shuttle launch is 180 in. (4572 mm). Once fabricated, the primary module structures were installed in element rotation stands (Figure 4) held by the same longerons and trunnion pins used later to secure them in the cargo bay for launch. The stand allowed the module to be rotated 360 deg, facilitating access to its external and internal surfaces during assembly and test.

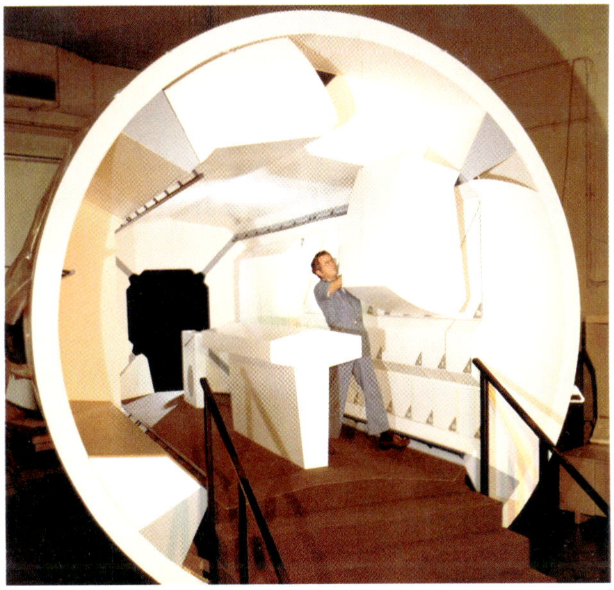

FIGURE 1 Don Margargee demonstrating the 1985 quadrant design proposed by McDonnell Douglas. (Courtesy of NASA Digital Imagery Management System [DIMS] and photo library.)

This material is declared a work of the U.S. Government and is not subject to copyright protection in the United States.

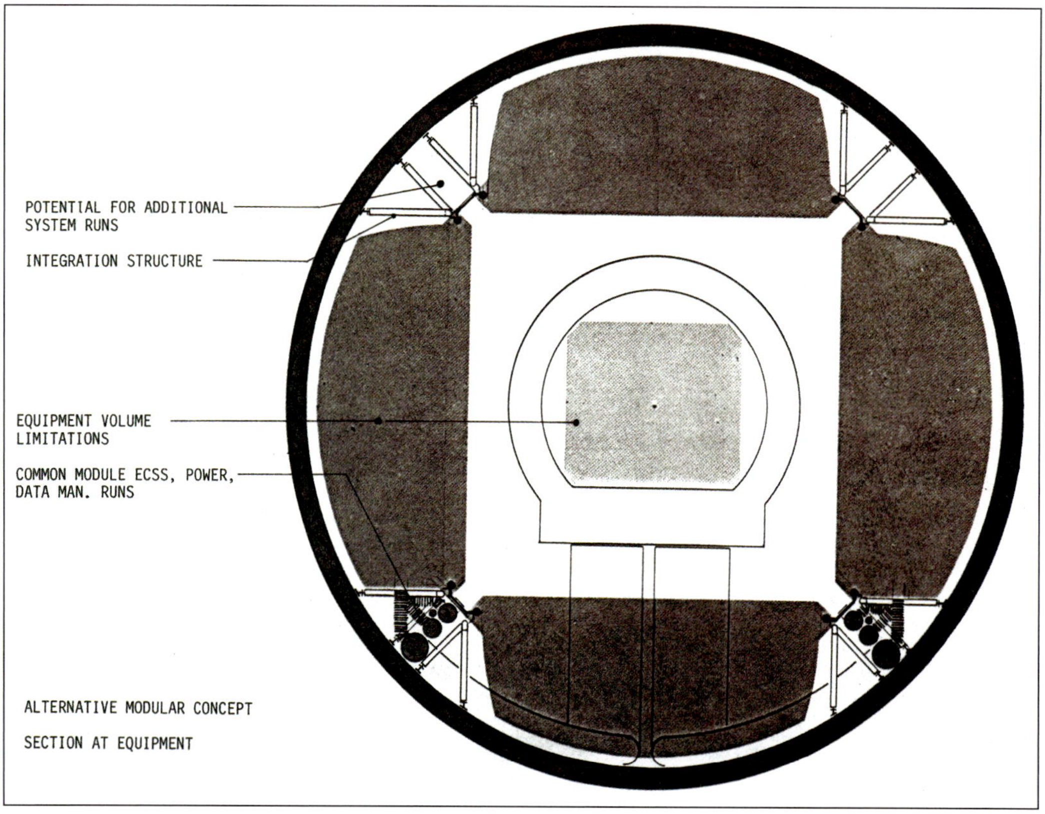

FIGURE 2 Sketch depicting key architectural features of the four-standoff concept for NASA 1985 Crew Station Review No. 2. (Courtesy of NASA DIMS and photo library.)

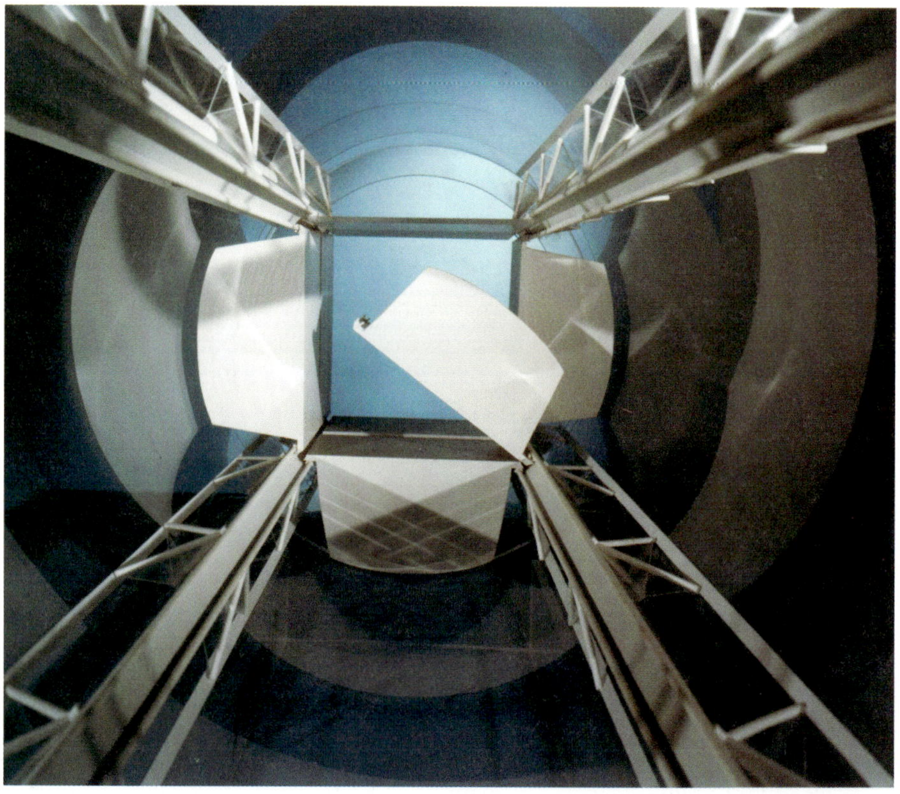

FIGURE 3 Early model depicting standoff and rack architecture and rack rotation. (Courtesy of NASA DIMS and photo library.)

FIGURE 4 External view of the *Destiny* module installed in the element rotation stand prior to installation of multilayer insulation and meteoroid debris shields. (Courtesy of NASA DIMS and photo library.)

Standoffs

The standoffs are the areas for structure to support distribution of utilities, attach core hardware such as lights and vents, and provide on-orbit attachment points for racks. To allow various teams of technicians to work simultaneously on the four standoffs, they were preassembled outside of the module (Figure 5). Each standoff is a self-contained element. This allowed power and fiber-optic harnesses, hoses, and vacuum lines to be tested and verified while there was still adequate accessibility. Once assembly and testing were complete, the standoffs were inserted into the module, suspended from a rail passing through both ends of the module, lowered into place (Figure 6), and installed. Interestingly, this preintegration scheme was not the original plan, but resulted from a 1993 trade study that documented the impracticality of conducting standoff detailed assembly inside the module.

Endcones

The endcone areas of the module provide the zone where subsystem distribution lines and cables transition from the standoffs to utility feed-throughs surrounding the hatch, so that utilities can be interconnected among modules. Although much subsystem hardware is packaged into racks, endcones afford valuable outfitting space as well (Figures 7A and 7B). Several major subsystem components are located here, including data-management mass memory units, power system controllers, and emergency response equipment including portable fire extinguishers and portable breathing masks. All hardware layouts had to conform to ISS accessibility requirements for maintenance and changeout. Unlike the standoffs and racks, the endcones were outfitted inside the module; therefore, assembly and test of this area was in the critical path to completing the module.

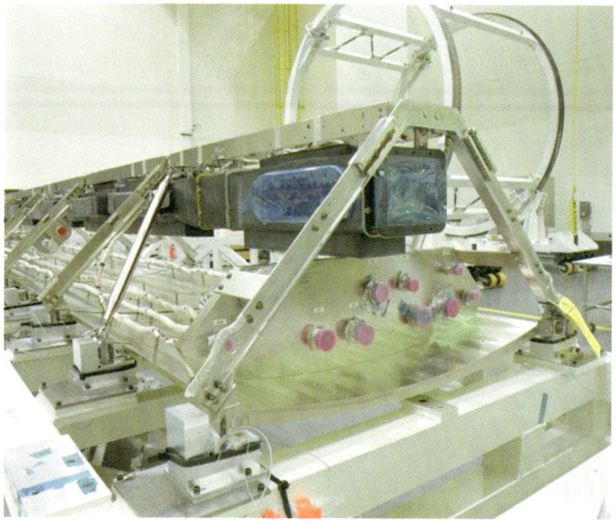

FIGURE 5A *Destiny* module standoff during preassembly.

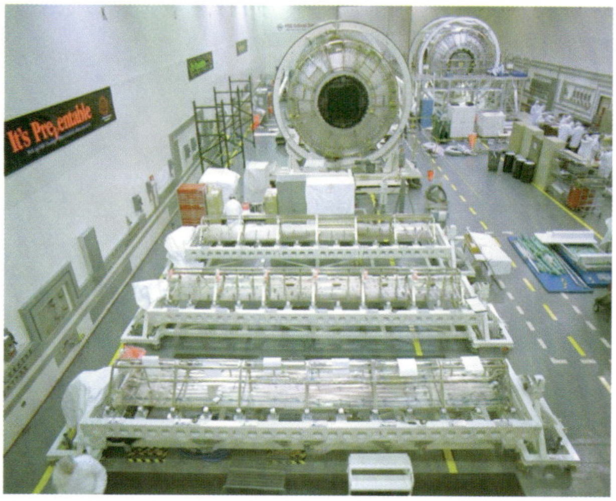

FIGURE 5B *Destiny* standoffs (foreground) and module (background) early in the assembly process. (Courtesy of NASA DIMS and photo library.)

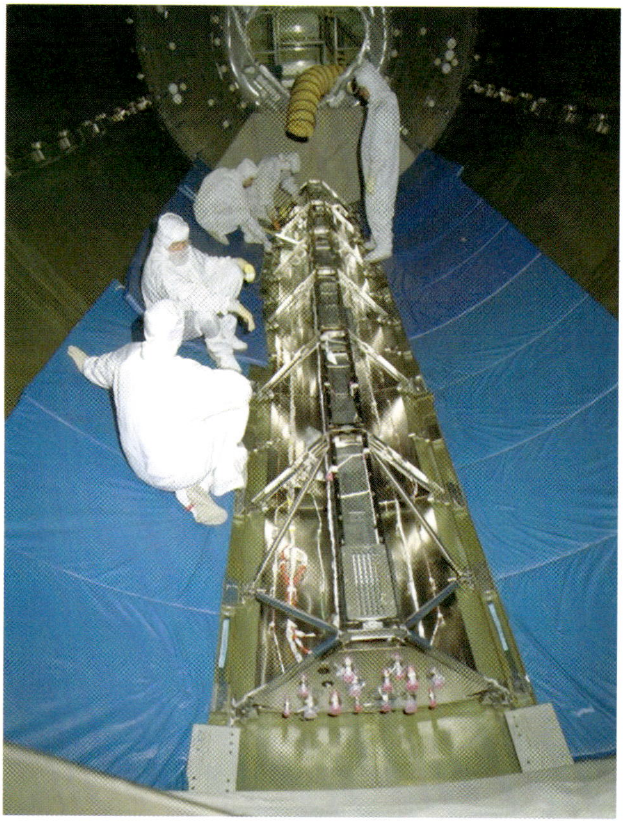

FIGURE 6 Installation of the first standoff into the module. (Courtesy of NASA DIMS and photo library.)

Racks

Racks were designed to be the primary method of packaging hardware system and payload for on-orbit component changeout. Although done primarily so systems could be upgraded during operations on orbit over the life of the ISS, the rack architecture allows independent assembly and checkout of each rack prior to installation in the module (Figures 8A and 8B), enabling a significant portion of the subsystems to be integrated in parallel just as with the standoffs. Racks were installed and removed numerous times during the test phase of the program to resolve anomalies and replace failed components.

Rack-level packaging also allowed the program to respond to significant requirements changes without having to redesign the entire module. The original Space Station *Freedom* program baseline was a low-inclination orbit. When Russia joined and the program became the ISS, orbital inclination was raised to 51.6 deg to accommodate the latitude of the Russian launch site in Kazakhstan. However, this higher inclination significantly reduced the shuttle cargo upmass. The impact was minimized by off-loading nonessential racks from the module's launch configuration. These racks were instead launched on subsequent missions in the multipurpose logistics module (MPLM) and installed on orbit.

Hatches

The hatch was sized to accommodate on-orbit transfer of racks, cargo, and crew between modules. Special ground handling equipment is required to pass racks through the hatch and install them. Because of the tight tolerances between hatch and rack, a dolly and track system is used (Figure 9). Unlike the primary ISS modules, the MPLM has a large, aft-access hatch operable only on the ground (Figure 10). The large opening provides greater clearance between the rack and the hatch for use of a specialized boom crane called the rack insertion device.

(a) (b)

FIGURE 7 Endcone outfitting: a) inside the module, prior to installation of closeouts and b) in use on orbit, showing closeouts and handrails in place. (Courtesy of NASA DIMS and photo library.)

FIGURE 8A An equipment rack during assembly.

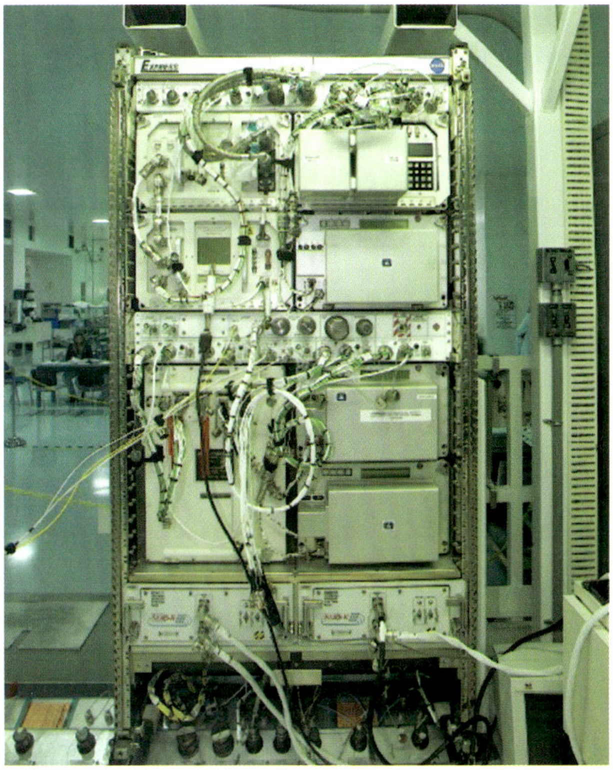

FIGURE 8B An EXPRESS payload rack during standalone checkout testing. (Courtesy of NASA DIMS and photo library.)

FIGURE 9 Rack being installed into the module, showing the hatch opening, fulfilling a key design requirement. (Courtesy of NASA DIMS and photo library.)

FIGURE 10 MPLM aft access closure used to facilitate ground installation of racks for launch. (Courtesy of NASA DIMS and photo library.)

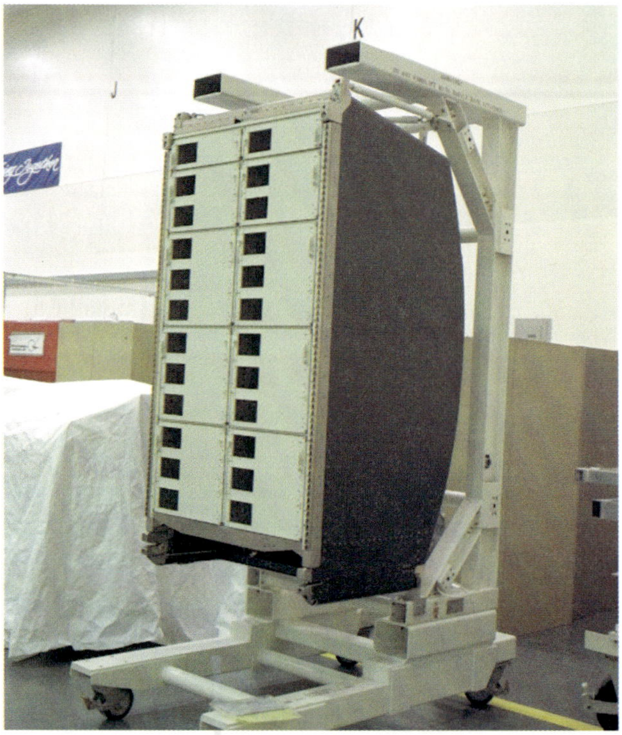

FIGURE 11A Resupply stowage rack attached to ground support equipment.

FIGURE 11B Hard-walled stowage locker inside the RSR. (Courtesy of NASA DIMS and photo library.)

Stowage

The original primary resupply and orbit stowage approach was the resupply stowage rack (RSR) (Figures 11A and 11B), a hard-walled locker system with each locker designed to protect its contents from the rest of the rack cargo. Although cargo friendly, the RSR is quite mass inefficient; its tare mass (mass of the rack as a container) is close to its cargo-carrying capability.

The resupply stowage platform (RSP) was developed for reduced structure mass, thereby allowing more mass to be allocated to the useful cargo. The RSP consists of a center plate with attach points so that bagged cargo can be strapped to both sides (Figures 12A and 12B). Although rack mass is significantly reduced, bags provide less cargo protection than the RSR lockers so that more padding is required, resulting in reduced volumetric efficiency. Each resupply mission is made up of a complement of RSRs and RSPs depending on the cargo requirements (Figure 13). To provide additional, mass-efficient stowage on orbit, the

FIGURE 12A Resupply stowage platform for use with bagged cargo.

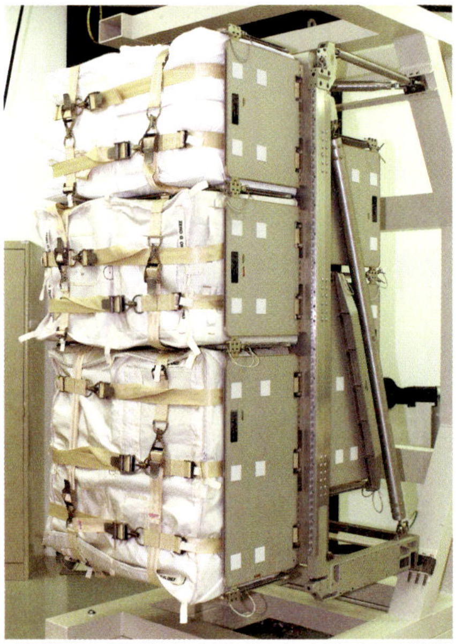

FIGURE 12B RSP packed for a mission. (Courtesy of NASA DIMS and photo library.)

program developed a fabric zero-*g* stowage rack (ZSR) (Figure 14).

LAUNCH AND DEPLOYMENT

Shuttle mission STS-98 carried launch package 5A, the *Destiny* laboratory module, to the ISS in February 2001 (Figure 15). The mission successfully installed and activated the module (Figure 16). The mission went according to the plan outlined in Jones (2000b) (Figure 17).

FIGURE 13 Interior of MPLM on orbit, showing launch location of RSPs. (Courtesy of NASA DIMS and photo library.)

FIGURE 14 Fabric zero-*g* stowage rack installed in the *Unity* node. (Courtesy of NASA DIMS and photo library.)

Destiny was launched with five system racks, eight ZSRs, a minimum number of crew restraints, and 10 rack-front closeout panels, fabric partitions required to maintain proper aisle airflow (Figure 18). After the module was berthed to the ISS, the crew's primary task was setup and activation of the lab systems (Figure 19). One key task was relocating the atmosphere revitalization system rack. It had been installed temporarily in a location that optimized the module's center of gravity for launch, so that the crew had to relocate it on orbit prior to activation. Although mockups depicting various portions of the task had been used to demonstrate the concept, it was not possible to simulate the end-to-end task on Earth. The 5A and subsequent outfitting missions have shown that rack transfer and installation is a relatively easy job for the crew to perform and does not require use of specialized handling aids (Figure 20).

FIGURE 15 STS-98 carrying the *Destiny* module during approach to ISS. (Courtesy of NASA DIMS and photo library.)

FIGURE 17 *Destiny* in context: the ISS assembly configuration at the time of the module's installation. (Courtesy of NASA DIMS and photo library.)

FIGURE 16 *Destiny* being removed from the cargo bay. (Courtesy of NASA DIMS and photo library.)

FIGURE 18 *Destiny* interior after on-orbit activation during ingress by the crew. (Courtesy of NASA DIMS and photo library.)

FIGURE 19 Initial lab outfitting. (Courtesy of NASA DIMS and photo library.)

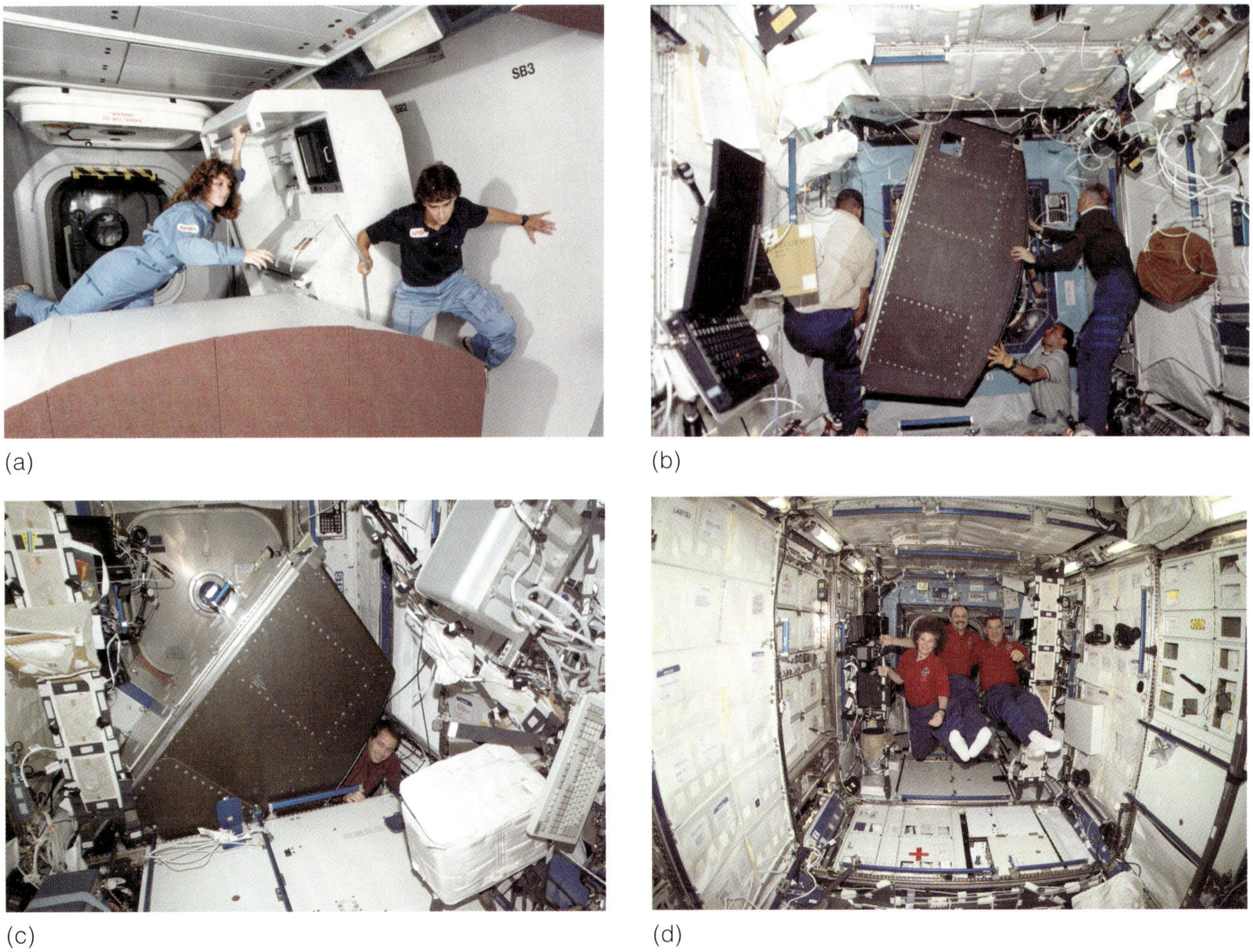

FIGURE 20 Rack installation: a) early mockup of the operation; b) crew members relocating the ARS rack on orbit; c) rack tilted out for standoff maintenance; and d) *Destiny* lab fully outfitted. (Courtesy of NASA DIMS and photo library.)

ADAPTING THE ARCHITECTURE THROUGH USE

As with all architecture and all space habitation systems, ISS crews have made numerous on-orbit additions and modifications to the as-designed environment to suit their particular needs (Figure 21). This adaptation is not limited to the U.S. segment: the Russian service module was launched without its table because of limited ascent performance, and so the crew improvised one from leftover flight-support equipment (Figure 22). Upholding Navy and NASA traditions, the crew added a ship's bell to announce arrival of each new crew to the station, adapted a flat panel to display crew mission patches, and started a ship's log to document crew visits (Figure 23).

Initial requirements specified the Russian partner would provide three crew quarters, yet the service module was designed for a crew of only two. The two program agencies could not come to agreement on the adequate implementation of a third on the Russian segment, and so the temporary sleep station (TeSS) was developed and installed in *Destiny* (Figure 24). TeSS embodied lessons learned from several generations of crew accommodation mockups (Figure 25) and presaged many features of the more permanent crew quarters (CQ) design, including a "bump out" to provide adequate internal volume, wall-mounted sleep restraint, workstation, and surface for posting personal mementos. See Chapter 5 for a more detailed discussion of the CQ. In any case, the program did not address accommodations for visiting crew. Despite

(a)

(b)

FIGURE 21 Adaptation of the ISS based on "real life": a) deployment of a vacuum hose and b) robotics workstation. (Courtesy of NASA DIMS and photo library.)

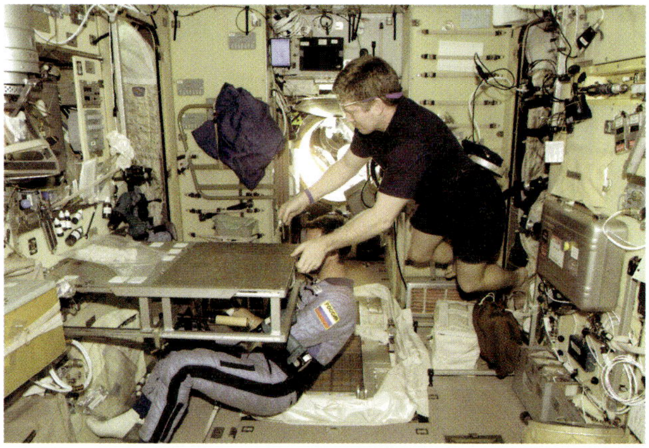

(a)

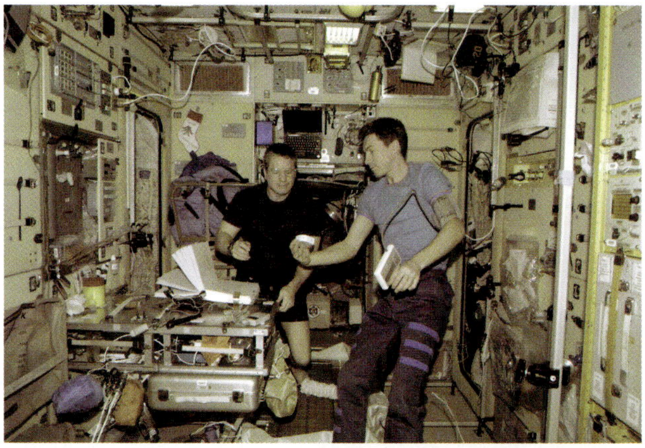

(b)

FIGURE 22 "Shep's table" a) being installed in the ISS service module; b) it provides work surface and some stowage. (Courtesy of NASA DIMS and photo library.)

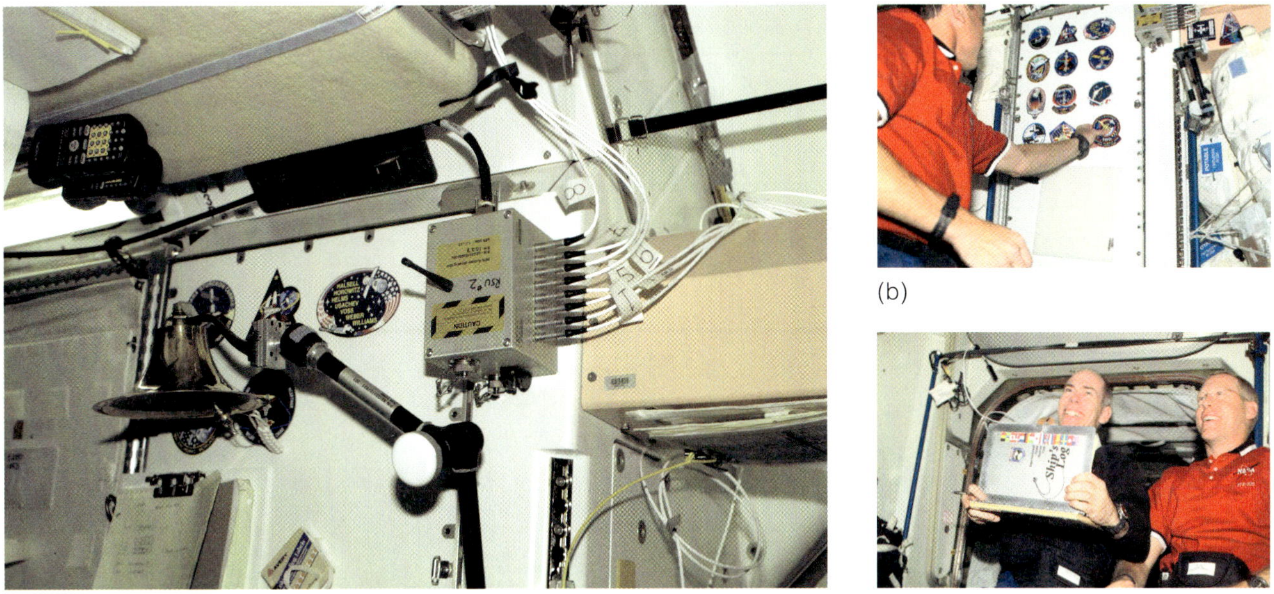

(a) (b) (c)

FIGURE 23 On-orbit additions: a) ship's bell; b) visiting crew patches; c) ship's log. (Courtesy of NASA DIMS and photo library.)

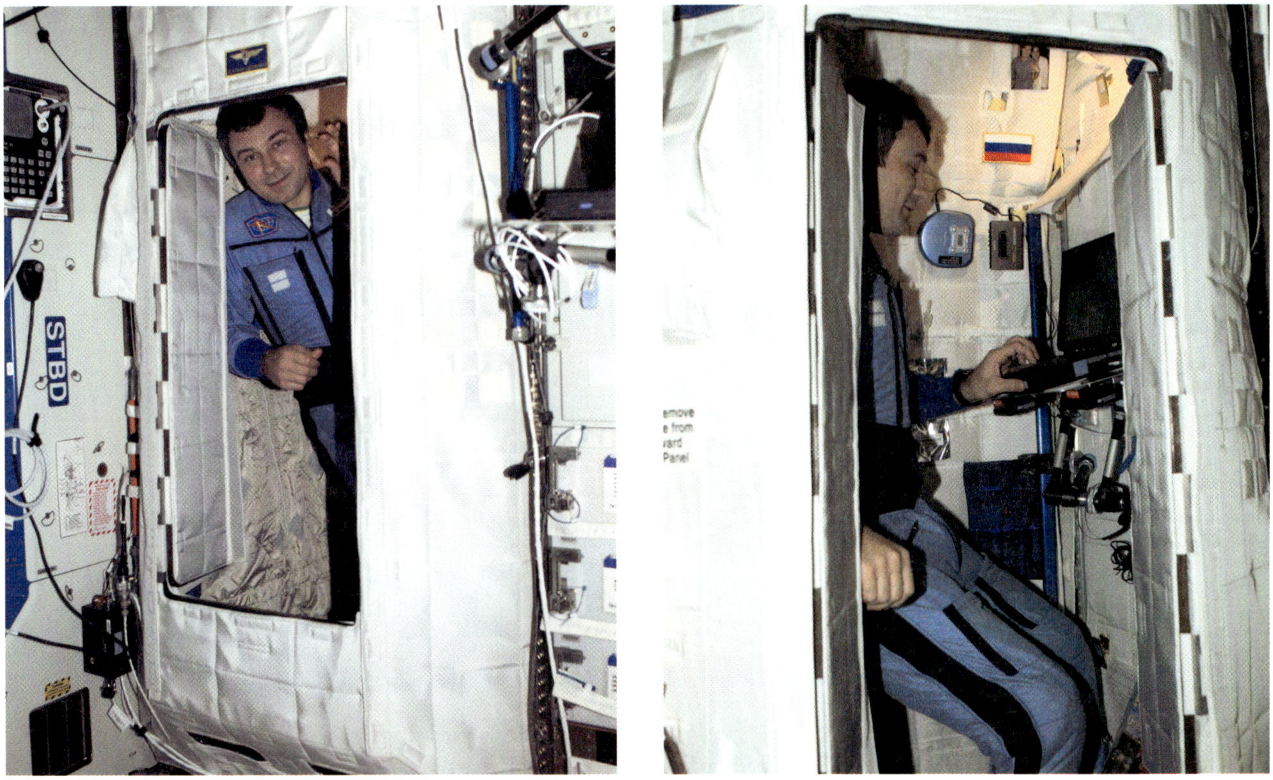

(a) (b)

FIGURE 24 Temporary sleep station: a) deployed in the *Destiny* module and b) TeSS interior. (Courtesy of NASA DIMS and photo library.)

early studies of the impacts of crew sleeping in the aisle, such visiting crews have used the *Destiny* aisle as their makeshift crew quarters (Figure 26).

Designers envisioned the ISS interior to be clean, organized, and functional (Figure 27). In actual usage, the interior generally has been adapted by the crew to be organized and functional for sure, but the clean look is long gone (Figure 28). ISS looks quite a bit more lived in—like the interior of *Mir*—than was planned.

FIGURE 25 Crew quarters mockups designed for the U.S. habitation module. (Courtesy of NASA DIMS and photo library.)

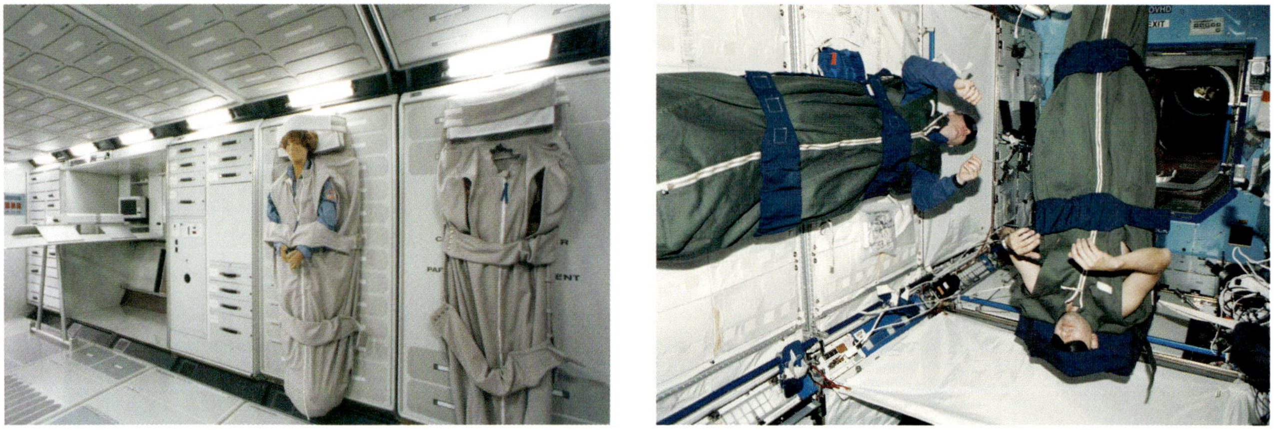

(a) (b)

FIGURE 26 Sleeping in the aisle: a) it was rejected in early mockup studies, b) but occurs anyway when shuttle crews visit. (Courtesy of NASA DIMS and photo library.)

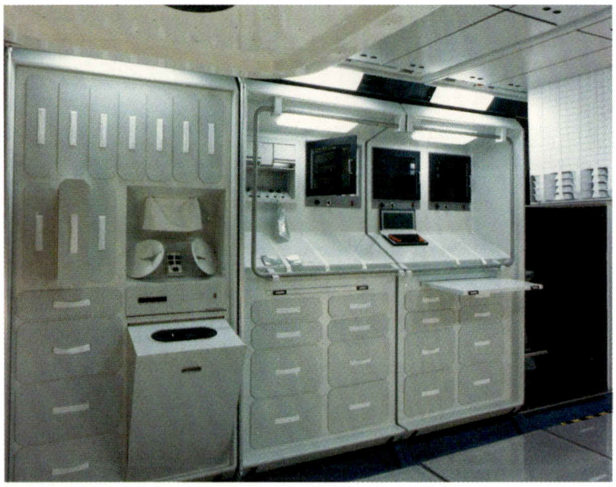

FIGURE 27 Interior module mockup, the way the designers envisioned it. (Courtesy of NASA DIMS and photo library.)

FIGURE 28 The way the crew really lives—visiting crew in *Unity* node; note stowage in radial port. (Courtesy of NASA DIMS and photo library.)

CONCLUSION

The ISS U.S. segment has benefited from adherence to key, fundamental architectural requirements. These requirements adequately addressed fundamental functional architecture issues including modularity, manufacturing, maintainability, and hardware change-out, and should ensure operation and safe occupation of the ISS for at least 20 years.

However, the habitable aisle volume allocated by the original architectural concept has not been protected or managed well as the design has transitioned into operational use. ISS operations demonstrate that the original architectural requirements are not sufficient by themselves to manage the ongoing, changing environment. As designers, we failed to predict this impact and provide adequate internal features to anticipate crew modification of the environment for day-to-day needs (Figure 29). Despite the years spent designing and redesigning ISS, we failed to establish controls to manage addition of new hardware requirements including temporary cabling, data networks, payload-design worksite setup and operations, and impacts from visiting crew. Development of a full complement of planning and operations constraints and tools to track and manage the interior remains a work in progress, as we gain more experience living and working within the architecture we specified at the beginning. Future space programs can define operations requirements and architectural features that provide these controls.

FIGURE 29 STS-98 and Expedition 1 crew in the *Unity* node; note the significant amount of nonstandard stowage. (Courtesy of NASA DIMS and photo library.)

References

Jones, R. (2000a); "Early Decisions for Space Station," Society of Automotive Engineers, Paper 2000-01-2329, July.

Jones, R. (2000b), "Architecture in Mission Integration, Choreographing Constraints," Society of Automotive Engineers, Paper 2000-01-2331, July.

INTRODUCTION

AS SPACEFLIGHT duration increased from several days on Gemini and Apollo, to weeks-to-months on the shuttle and Skylab, and then to many months on *Mir* and the International Space Station (ISS), dedicated spaces for crew-member privacy eventually became essential. Private crew quarters (CQ) provide a relatively quiet retreat from cabin equipment noise and allow private medical consultations, conversations with family, and restful sleep. However, requirements for CQ acoustic, visual, and light isolation also pose design, habitability, and safety challenges because a private volume introduces potential confined-space hazards.

During ISS assembly, the U.S. operational segment provides a temporary sleep station (TeSS) as crew quarters for one crew member in the *Destiny* laboratory module. The Russian segment provides permanent CQ (called *kayutas*) for two crew members in the service module. The TeSS provides limited electrical, communication, and ventilation functionality.

This chapter describes development of a permanent, rack-sized CQ to upgrade the ISS assembly-complete configuration. Four of these CQs get installed in the *Harmony* Node 2 element; when added to the two existing kayutas, these support an ISS crew size of six. The new CQs provide private crew-member space with enhanced acoustic noise mitigation, integrated radiation reduction material, controllable airflow, communication equipment, redundant electrical systems, and redundant caution and warning systems. The rack-sized CQ system has multiple crew-member restraints, adjustable lighting, controllable ventilation, and interfaces that allow each crew member to personalize his or her CQ. Providing an acoustically quiet and visually isolated environment, while ensuring crew-member safety, is critical for obtaining crew-member rest and comfort to enable long-duration crew performance. The chapter covers numerous human factors, engineering, and programmatic issues addressed during the concept, design, and prototyping phases of the project.

5

International Space Station crew quarters

JAMES LEE BROYAN, JR.
MELISSA ANN BORREGO
AND
JUERGEN F. BAHR

HISTORY OF CREW QUARTERS IN HUMAN SPACEFLIGHT

The history of CQ has been documented previously (Dietz and Doerre 1990; Adams 1998), but a few key milestones provide insight. Design and development efforts of long-duration crew-member accommodations began with the U.S. *Skylab*. *Skylab* featured visually private space for each crew member but lacked acoustic and light isolation and had very limited ventilation control. The *Skylab* crew indicated these deficiencies, and the lack of headroom for taller crew members reduced effectiveness of the private volumes (Adams 1998).

The Russian kayutas were introduced on *Salyut 6*, and the basic configuration was used in the *Mir* base block and ISS service module. Kayutas provide a visually private volume with a 20-cm-diam window, but the window increases crew-member space radiation exposure. Kayutas draw air from the cabin but are generally too warm and do not provide sufficient acoustic

This material is declared a work of the U.S. Government and is not subject to copyright protection in the United States.

attenuation of cabin noise (Adams 1998). Valuable lessons learned from participating in the *Mir* program were incorporated into the ISS CQ.

In the early-to-mid-1990s, Space Station *Freedom* (SSF) CQs were planned to provide approximately 1.5 times the volume of an SSF equipment rack, totaling 3.2 m^3. This would have provided more storage, thermal control, and work space than the 2.1 m^3 ISS CQ design described next. Four such SSF CQs were to be located in one end of the habitation module. The SSF CQs were not developed beyond the early preliminary design stage. However, ISS CQ requirements maintained most of the functionality and basic anthropometric envelope as the SSF CQ, albeit with reduced stowage, office utilities, and thermal adjustability.

In the late 1990s, concept development for deployable crew quarters (DCQ) resulted in crew proposed for Mars missions (Kennedy 1999). In 2001, the reduced-functionality TeSS unit was protoflighted (i.e., developed as a prototype that was then used operationally) in only nine months. TeSS was launched on Flight 7A.1 and located in the *Destiny* module to increase crew size from two to three.

TeSS was intended only as a short-term solution, providing limited functionality until fully functional CQs could be delivered with the habitation module. Given the short development time, many habitability and maintainability requirements were waived or downgraded to goals (Keener 2002). A brief description of the TeSS design serves to compare and contrast the advanced CQ design described later. For photos of TeSS, see the preceding chapter.

TeSS provided limited functionality within a private volume consisting of a rack volume with a 31-cm bump-out protruding into the module aisle, acoustic panels, and attachments for crew-member items. The TeSS had pass-through openings for electrical cables and for ensuring that cabin alarms could be heard. These openings thus limited the effectiveness of acoustic and light isolation. TeSS did not incorporate independent ventilation, but instead redirected one pair of cabin ventilation ducts into and out of the rack. The ventilation flow rate and direction was limited, and some crew members reported the inability to control airflow direction as a source of discomfort.

The TeSS structure was folded flat on a resupply stowage platform (RSP) to minimize launch loads and had to be assembled on orbit. Radiation-shielding blocks were launched separately and assembled inside the TeSS to reduce crew-member exposure (Zapp 2001). Both initial installation and subsequent behind-rack operations have resulted in substantial crew-member time to assemble, disassemble, and reassemble the unit. Still, TeSS was a substantial improvement in crew-member on-orbit living compared to *Skylab* quarters and Russian kayutas and provided valuable lessons in acoustics, fabric liners, and crew-member preferences over a period of eight years. TeSS operational life was extended far beyond the originally intended two years, after the habitation module was canceled from the ISS program in 2002. Instead, the *Harmony* element was outfitted with provisions to accommodate four CQs.

ENVELOPE

The CQ is designed for space shuttle launch in the multipurpose logistics module (MPLM). Each CQ is custom designed because the on-orbit configuration depends on the interaction of all four; however, each CQ occupies a standard U.S. rack volume and uses the standard structural attachment points. Compared to TeSS, significantly less assembly work is required to achieve the final on-orbit configuration and outfitting with radiation-reduction panels and electronics.

ISS crew quarters are located in the *Harmony* Node 2, equipment Bay 5 locations. *Harmony* connects *Destiny* with the Japanese experiment module *Kibo* and the European laboratory module *Columbus*. The four CQs are located in a ring configuration around equipment Bay 5 (Figure 1). The four CQs are very similar but unique, even though the structural attachment points are identical. To maintain

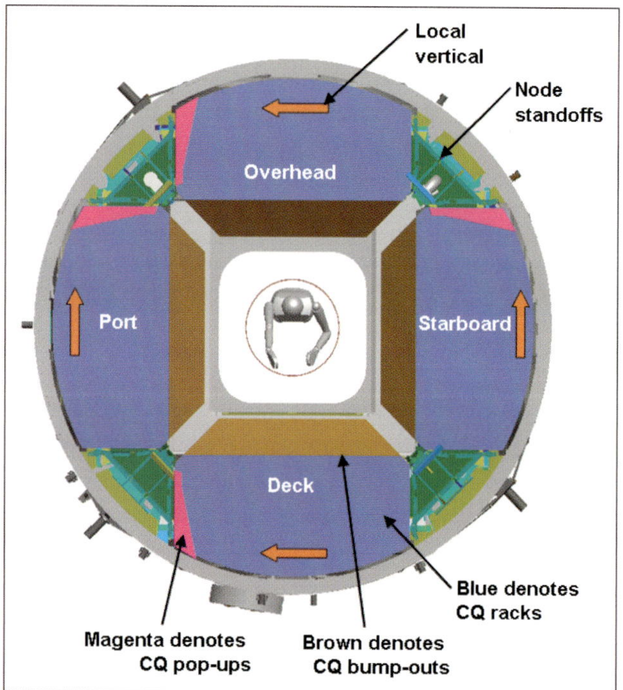

FIGURE 1 *Harmony* Node 2 module accommodates four crew quarters units. Bump-out and pop-up volumes increase usable rack space.

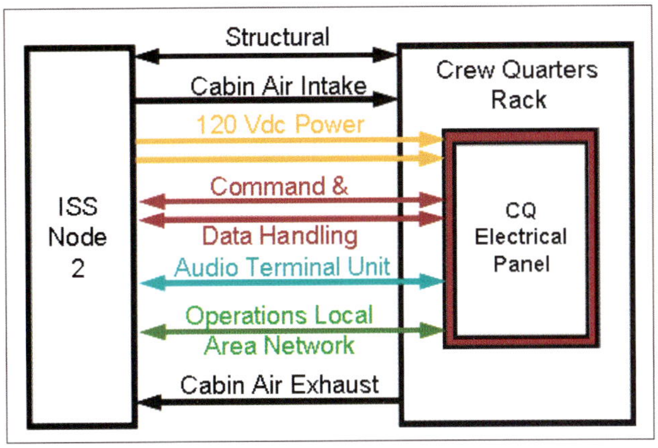

FIGURE 2 Functional block diagram shows interfaces between CQ and host module Node 2 (*Harmony*).

a local-vertical heads-up orientation in port, starboard, deck, and overhead locations, the CQs facing each other are partially mirrored in the bump-out and location of crew attachment surfaces.

The CQ envelope is basically limited to a standard rack volume by the ISS vehicle architecture. The pop-up takes advantage of the piece of volume above the top attachment points. This small volume adjacent to the module standoff allows the rack's rear upper corner to clear the standoff as the rack is rotated into position during installation, but is then typically unusable by racks after their installation. In the case of an occupied CQ, however, this valuable volume can be recovered after installation. The extent of the bump-out protrusion is limited by the ISS requirement to maintain a minimum clear aisleway of 152 cm to allow passage of rack-size hardware. Because the bump-outs come to that boundary, the CQ doors must rotate inward to allow simultaneous crew egress in the event of a Class I alarm.

CQ electrical utilities are routed from the standoffs beneath the racks through the floor panels to minimize utility draglines. Each rack has six power and data cables (Figure 2).

DESIGN OVERVIEW

The primary purpose of the CQ is to give crew members a place in which to retreat from the noise and busy activity levels in shared areas of the ISS. The CQ design is highly efficient in limiting noise to NC-40 levels, similar to what is experienced in a public library. External light is shut out almost completely. The result is a private environment, but one that is also a potentially hazardous confined space. Sufficient airflow must be maintained to prevent asphyxiation and to remove body heat, and artificial light must be provided.

Each CQ starts with an international standard payload rack (ISPR)-sized aluminum frame with composite panels. The majority of the back wall and ceiling are made of ~6.5 cm of ultra-high-molecular-weight polyethylene, which shields the crew-member's head and blood-forming organs from ever-present cosmic radiation and occasional solar flares.

The CQ has the same mounting interfaces as an ISPR. For launch, the bottom-rear corners are pinned or bolted to the spacecraft, and z-shaped knee-braces support the top front corners. On orbit, the rack is attached by its four accessible front corners. Pivot pins at the lower attach points allow the rack to be rotated in and out of its bay. The bump-out is reversed and the pop-up retracted for launch. Each CQ provides 2.1 m^3 of interior volume.

Inside each CQ, one side wall is designated as the sleeping and resting surface (Figure 3). The wall across from the sleep wall provides mounting for electrical assemblies, computer workstation, lighting, and other utilities. The back wall opposite the entrance is curved to follow the module hull. This back wall provides stowage space with elastic bungees and hook-and-loop-fastener patches. The ventilation subsystem is located in the bump-out and is packaged efficiently around the inside of the CQ door.

The CQ interior actively exchanges air with the conditioned ambient cabin air using two independent, parallel fans with a single speed control. Cabin air is drawn into the CQ through a fan with acoustic

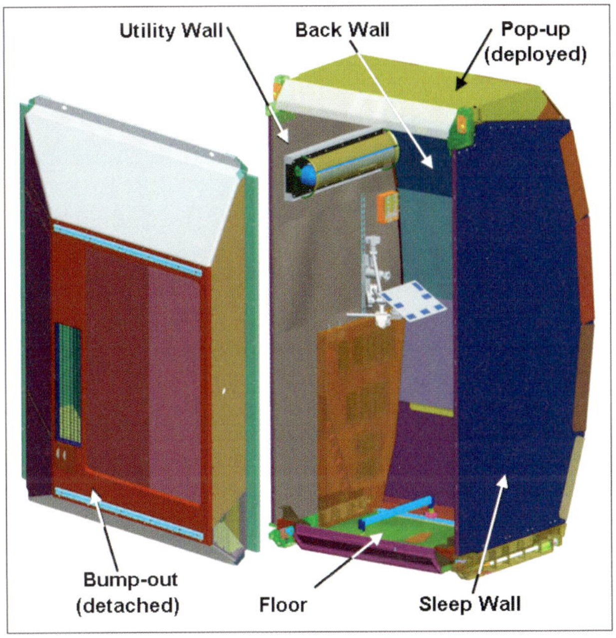

FIGURE 3 CQ major layout maximizes efficient use of minimal space.

muffler and enters the CQ interior through a large diffuser at the crew-member's head position. Airflow is drawn from the CQ interior near the crew-member's feet via a separate fan and returned to the cabin. The returned air carries heat away and allows smoke detection by existing node instrumentation. Three fan speeds provide adjustability for comfort and day/night operation, but the fans cannot be shut off completely without removing power from the rack. Ventilation status is continually reported to the ISS vehicle monitoring subsystem. Even if one fan should fail, airflow is sufficient to prevent CO_2 buildup that might endanger the crew member.

Illumination is provided by the standard ISS fluorescent light called the general luminaire assembly. It is installed at the top of the utility wall and can be dimmed and shaded to provide customized lighting. It can be turned off to achieve near total darkness: only two power LEDs on the electrical panel remain on, and in case of an alarm, four LEDs illuminate the door.

The center of the utility wall has rails for mounting an adjustable shelf to support a laptop computer. Two 120-volts power outlets and an ethernet port are provided. The back wall has two speaker boxes that annunciate Class I alarms controlled by the ISS vehicle. A third box on the utility wall annunciates audio and alarms relayed from one of the *Harmony* audio terminal units.

DEVELOPMENT AND KEY TRADES

The CQ project followed the traditional aerospace systems engineering approach with distinct phases and control gates: system requirements, preliminary design, and critical design reviews (SRR, PDR, and CDR). The SRR phase was thorough in identifying and incorporating all stakeholder requirements, in part using an early volumetric mock-up. During the PDR and CDR phases, an integrated team approach was taken: standing working groups in six functional areas (radiation, interior, structural-mechanical, avionics, thermal-ventilation, and operations) produced a broad team discussion and consensus on all major design decisions. Design concepts were reviewed at integrated product team meetings, which included representatives from NASA engineering, crew office, mission operations, human factors, and safety organizations, and functional specialists as required.

Several early CQ trade studies determined the basic CQ architecture. DCQ and TeSS, foldable structures, were considered for the ISS CQ because a collapsible structure requires only straightforward launch loads consideration. TeSS was launched on an RSP along with several cargo bags for cabling, acoustic blankets, and radiation reduction bricks. However, CQ goals for integrated electronics, ventilation, relatively thick radiation reduction panels, and stringent acoustic/light isolation make a foldable scheme impractical. TeSS could be folded flat because all of these functions were added via draglines or even launched separately, requiring significant crew-member assembly time as well as postassembly rack rotation. Additionally, *Harmony* does not have sufficient cabin air diffusers and returns to support a TeSS-like ventilation system. Hence, the CQ needed acoustically isolated fan/duct systems.

The use of a common ISPR was also considered for the basic CQ structure. The ISPR has multiple internal structural components that would require modification to accommodate 95% male vertical stature. In particular, the lower utility interface panel and upper torque tubes would need to be extensively modified. Additionally, almost all panel surfaces would require modification to accommodate radiation reduction materials, pop-up attachments, bump-out attachments, electrical assembly attachments, and provisions for acoustics. Such extensive modifications largely neutralized the benefit an ISPR-based design.

The option remaining was a custom rack tailored to the functional requirements but still using common rack attachment interfaces and overall envelope requirements. A custom rack would allow maximum habitable volume, more headroom, radiation protection panel integration, an integrated pop-up mechanism, efficient redistribution of structural loads, and least potential crew-member installation time. Additionally, mounting provisions for electrical components and crew items could be located based on anthropometrics rather than driven by location of sufficient ISPR hard points. The trade study concluded a custom rack was the best rack approach to provide the required functionality and structural integrity for launch loads including natural frequency.

Although the volume of the aisle bump-out was defined from the outside in by cabin operations constraints, functionality was traded to determine the best allocation of crew space and equipment space within the bump-out envelope. The door, ventilation intake, and ventilation exhaust locations were previously set as requirements based on translation path and overall Node 2 ventilation patterns. After radiation reduction material, the ventilation subsystem occupies the most volume as a result of acoustic treatments to meet interior noise levels. The ventilation ductwork was located entirely in the bump-out envelope to minimize on-orbit assembly and maximize perceived head room during CQ occupancy. The ventilation system electrical and safety monitoring circuits require connection during installation.

The electrical connectors, duct work, and door structure are maintained within the bump-out envelope, so that the bump-out assembly can be simply reversed and mounted to the front of the rack structure for launch. This enables launch integration to be done with the standard rack envelope and acts as a structural stiffening plane for launch. The standard rack attachment interfaces allow use of standard rack handling, shipping, and ground-support hardware for launch preparation operations.

Ease of CQ installation into *Harmony* was a high-priority goal. On orbit, the CQ standard ISPR attachment mechanisms are unbolted from the MPLM. Rack handling fixtures can be attached to transport the CQ into *Harmony*. The Bay 5 rack locations are prepared by removing the zero-*g* stowage racks launched with *Harmony* and detaching the data and electrical cables stowed in the standoffs. The CQ rack is mounted on the standard rack rotation points. Launch-phase shims and bolts for pop-up, door, and floor panel are removed and stowed. The CQ is rotated and locked into place using standard ISPR procedures. The bump-out launch bolts are removed and stowed, and the bump-out is removed from the CQ rack. The launch bag containing acoustic blankets, interior cables, speakers, and general luminaire assembly (GLA) is unstowed. The CQ floor panel is removed and node cables routed through the floor. The bump-out is then reattached (with the protrusion facing out) using quick-turn fasteners. The CQ doors are then reinstalled, pop-up deployed, acoustic blankets unfolded and attached, and electrical cabling and GLA mounted. The CQ then requests application of power from the ISS via a station laptop. The CQ is then ready for final outfitting with crew items (sleeping bag, laptop, and personal items). Acoustic isolation, ventilation, and crew-member comfort are immediately noticeable; no instrumentation, check-out, or initialization period are required.

DOOR

Significant effort went into designing the CQ doors, beginning with the term "door" itself. Some household words automatically connote expectations that must be redefined in the NASA lexicon. CQ doors have no need for latching, and they are different also from standard hatches and area closures. However, the CQ doors need a good light/air seal at all hinge points and seams to achieve the challenging darkness and NC-40 noise-reduction requirements. Several hard- and soft-construction design options were developed, all driven by one-handed operation.

The close proximity of adjacent CQ units made soft-fabric panel/flap solutions with zippers or hook-and-loop fasteners undesirable because their sharp impulse noise when being opened or closed would disturb neighboring crew-members' sleep. (Campers are quite familiar with this disturbance noise.) Also, attached acoustic blankets must fold up and stay completely out of the way when the doors are operated. Given the topology of the CQ entrance, this resulted in two independent door halves. One door rotates inward 90 deg, and the other door rotates inward 120 deg. This provides a 51 cm × 102 cm unobstructed door opening based on favorable feedback from the TeSS design. A bifold door was not selected because it would have required double hinging to accommodate the bulky acoustic blankets when folded.

The door halves overlap so that only one side needs to be secured. That side is latched with a simple detent capture latch at top and bottom. This type of latch provides enough resistance to prevent the door from accidentally being pushed open by a passerby, yet low resistance (about 30 newtons) to allow easy opening by a fifth-percentile female crew member.

The door hinges are designed to allow easy and rapid removal if the door fails in the closed position. The door can simply be lifted off the hinge pins when in the closed position. Additionally, for several types of CQ maintenance (when power, and thus light and ventilation, are shut down) it is preferable to remove the doors completely.

VENTILATION/ACOUSTIC

Temperature and Flow Control

The CQ uses node cabin air for ventilation, as opposed to ducted air or ISS cooling-loop fluid connections. Consequently, reducing the temperature delta between cabin and CQ interior can only be done by increasing the airflow rate (Figure 4). Increasing airflow up to ~2.4 m^3/min (85 centifermi) is effective for decreasing delta temperature. Beyond that rate very little additional cooling occurs, but acoustic noise continues to increase.

The common cabin air assembly (CCAA) conditions the node cabin air. The CCAAs are programmed to reduce the temperature to 18°C about

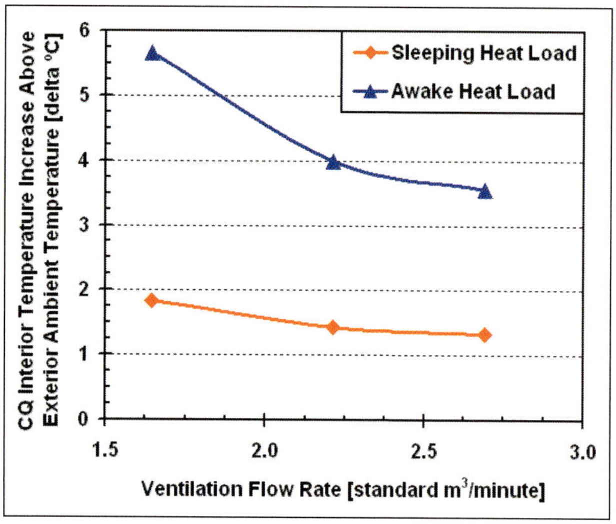

FIGURE 4 Airflow manages interior CQ temperature for awake- and sleep-condition metabolic loads.

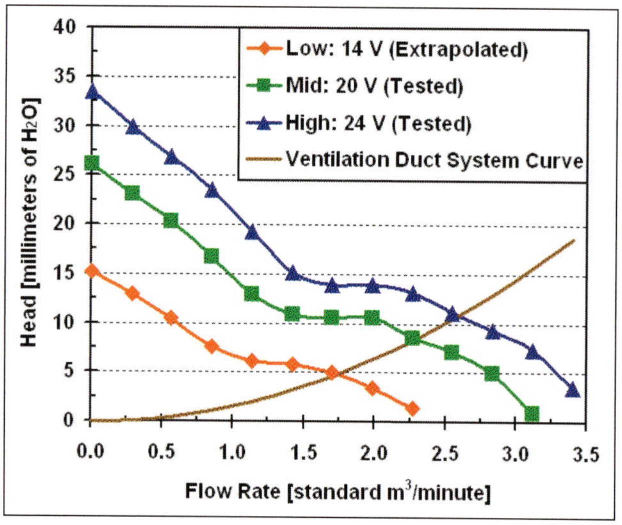

FIGURE 5 Flow rate compensates CQ ventilation duct pressure losses.

one hour before sleep and maintain 22°C for wakeful operations (Balistreri 2007). This would enable the CQ interior temperature to get as low as 20°C. Because of the specific location of CCAA diffusers in the Bay 5 general area and resultant local aisleway temperature variations, the port and starboard CQs are approximately 0.6°C cooler than the deck and overhead CQs.

Air circulation serves two main purposes: flushing carbon-dioxide concentrations to prevent crew-member asphyxiation and heat exchange for crew-member comfort. The following discussion summarizes development-phase status of the highly integrated and complex CQ ventilation design. The external configuration is straightforward because of the node airflow topology. The CQ air intake is located on the front, upstream of the door, where it receives the largest amount of fresh air from the cabin. After passing through the CQ interior, air is exhausted at the foot, directed toward the CCAA return. This prevents air circulation back into the CQ or into neighboring CQs. It also maintains safe operation should the doors be left open.

Internally, the ventilation design was driven by two factors: circulation of fresh air from head to feet and reduction of noise. Virtually all noise generated inside the CQ is caused by the ventilation system. Noise transmitted from the cabin or by the CQ fans travels with the air toward the crew-member's head and is therefore especially noticeable. This challenge is met by implementing a push–pull fan system: one fan pulls air into the CQ, while another pushes air out the CQ exhaust. The exhaust fan noise is emitted to the cabin rather than to the crew-member's hearing. Use of two fans provides more flow rate at reduced noise and size and allows greater pressure drop within the ducts (acoustic treatments and flow bends in the ducts cause pressure drop), which in turn enables more acoustic treatments (Figure 5).

Flow Detection and Safety

The two-fan-design safety requirement was developed for DCQ and kept for the ISS CQ. Preventing asphyxiation requires that virtually all systems associated with ventilation and alarms be redundant and independent. The CQ does not have redundant airflow paths but does provide redundant fans, power supplies, monitors, and alarms. Absolute CO_2 levels in the CQ are determined by the cabin environment plus the CO_2 added by the occupant. The CQ interior has a CO_2 level higher than cabin-ambient but is limited to a ≤0.8-mm Hg delta. At 0.85-m³/min airflow and 4.5-mm Hg ambient CO_2, the CQ interior CO_2 delta is 0.72-mm Hg. Carbon-dioxide levels decrease with increasing flow, similar to the delta temperature relationship just described.

Computational flow analysis shows that with doors open, passing aisleway ventilation provides sufficient turbulence to prevent hazardous CO_2 buildup; however, the CQ would become uncomfortably warm. No additional air circulation would be needed, were it not for the discomfort caused by this heat buildup. With doors closed (typical sleep scenario), the fans must move at least 0.85 m³/min (30 centifermi) to prevent an asphyxiation hazard. The CQ meets this requirement even with one fan failed. The fans are always on and have a single three-way control switch located near the occupant's head, which can be set to low, medium, or high speed (not off). This accommodates various

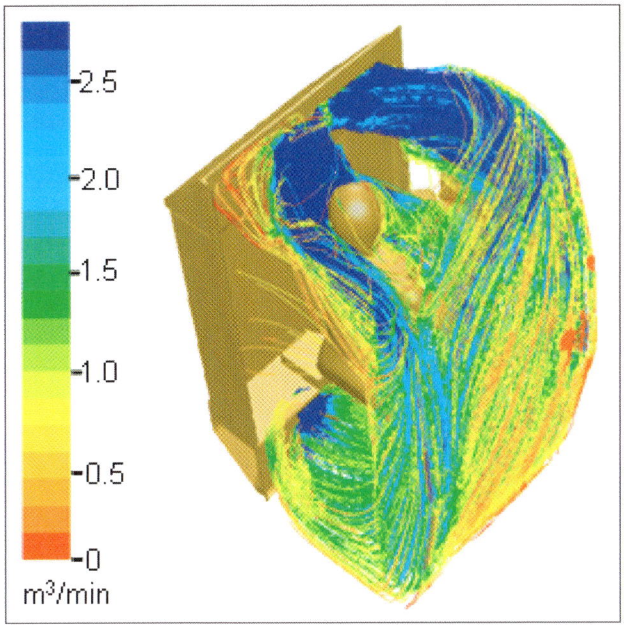

FIGURE 6 Ventilation design ensures complete circulation (high-speed fan setting illustrated).

heat load scenarios and completely circulates air within the CQ volume (Figure 6).

Heat load is an important factor in keeping the closed CQ environment habitable. There are five major heat sources in CQ: crew-member metabolic load, fans, GLA, laptop computer, and CQ power supply.

The heat-generating elements are positioned as far downstream in the airflow as possible. This reduces the amount of heat the ventilation air acquires before contacting the crew member. Incoming air acquires only the intake fan heat load. Approximately half of the incoming air stream is directed toward the crew-member's head, where cooling is most efficient, especially if a sleeping bag is used. After the head, the air flows down along the body toward the feet, removing heat and increasing in temperature. Similarly, the other half of the inlet air is directed away from the crew member to remove heat generated by light and electronics on the utility wall. Air then exits the interior crew volume and flows across the electrical power supply in the exhaust duct. The power supply and exhaust fan heat are dissipated inside the exhaust duct—not the rack volume, which keeps the interior approximately 0.6°C cooler than if the power supply were inside the main interior CQ electrical panel. As seen from Figure 4, a 0.6°C reduction is equivalent to approximately 0.4 to 0.7m^3/min of airflow, depending on total CQ heat load. Avoiding this additional airflow significantly decreases fan noise that would otherwise need to be attenuated.

Industry employs push–pull configurations to reduce noise of fan assemblies. A parallel system was not chosen for the CQ for two reasons: doubling heat and noise in the intake duct and the complexities of implementing redundancy. A dual fan could not be located in the exhaust system because airflow would bypass the occupant's head if the CQ door were open. A parallel fan configuration also creates additional safety concerns. If one fan failed, air from the surviving fan might circulate back through the failed one instead of ventilating the CQ. With the push–pull system, should one fan fail, the surviving fan provides sufficient circulation through the entire system, so it is not necessary to awaken a sleeping crew member for this contingency. Also, locating one fan at the foot yields more perceived headroom.

Air Intake Description

The CQ accepts air through the intake alongside the door (Figure 7). From there it flows up across the door and through an adjustable diffuser. Noise from the cabin environment (52 decibels) and the intake fan is mitigated as the air follows a serpentine path past foam/fabric abatements. The intake air takes one 180-deg and two 90-deg turns. Optimal placement of the fan was determined experimentally in a wood mockup of the bump-out and ductwork. Abatement materials were chosen through testing and analysis.

Air Exhaust

The exhaust system is located beneath the door to minimize noise near the occupant's head. To minimize accidental alarms caused by airflow blockage,

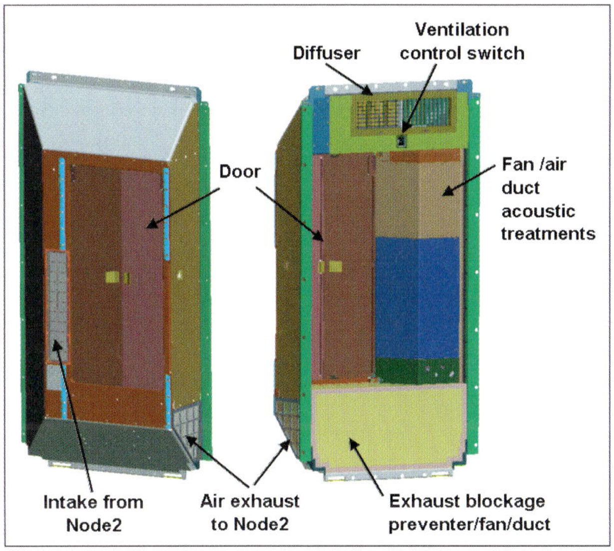

FIGURE 7 CQ bump-out (exterior view on left; interior view on right) accommodates general airflow and ventilation components.

the vent is covered with a semirigid, perforated screen. This increases the effective exhaust inflow area by approximately 2.5, and prevents loose objects (e.g., towel or shirt) from blocking it to create a low flow (CO_2 rise) condition that would activate a nuisance alarm. The exhaust duct employs a 90-deg and 180-deg turn for noise reduction. It contains a football-shaped diverter that causes the airflow in the abatements to branch and then reunite, precluding line-of-sight noise paths into the cabin.

ELECTRICAL

The electrical subsystem comprises electrical panel, power supply, egress light, test panel, audio terminal unit (ATU) loudspeaker, audio annunciation device (AAD), inlet and exhaust fans, and inlet and exhaust flow sensors. Unlike TeSS, which only had pass-through openings for electrical cables, the CQ electrical and data interfaces are routed through the standoffs beneath the racks. Each CQ receives six electrical interfaces via cables between *Harmony* and the electrical panel that conditions and distributes them within the CQ system (Figure 8):

- Two separate 120 volts dc, 3.5-ampere power lines for primary and redundant power
- Two separate command and data handling (C&DH) lines for *Harmony* MDM1 and MDM2 signals
- One operations local area network (LAN) line to connect to the *Harmony* ethernet
- One Harmony ATU line

The CQ power supply converts 120 volt dc input power to ±15 volt dc to power the fans and regulates voltage to +5 volt to power fan status circuits, audio circuits, and egress circuits. One-hundred-twenty volt dc power is routed directly to the GLA, laptop, and an extra connection for crew-preference items.

If one 120-volt dc power bus fails, redundancy in the electrical panel and power supply allows fans, egress lights, MDM logic, and caution-and-warning alarm speakers to continue automatically. However, the noncritical laptop and GLA need to be manually switched to the surviving 120 volt dc line via a select switch on the electrical panel. The egress light illuminates the interior door area whenever there is a loss of both primary and secondary power or annunciation of a Class 1 alarm. Backup 9-volt batteries located in the electrical panel power the egress light and redundant audio in the event of loss of both primary and secondary power. The test panel allows the occupant to regularly test the health of the backup battery system, the egress light LEDs, and the annunciation of Class 1 alarms.

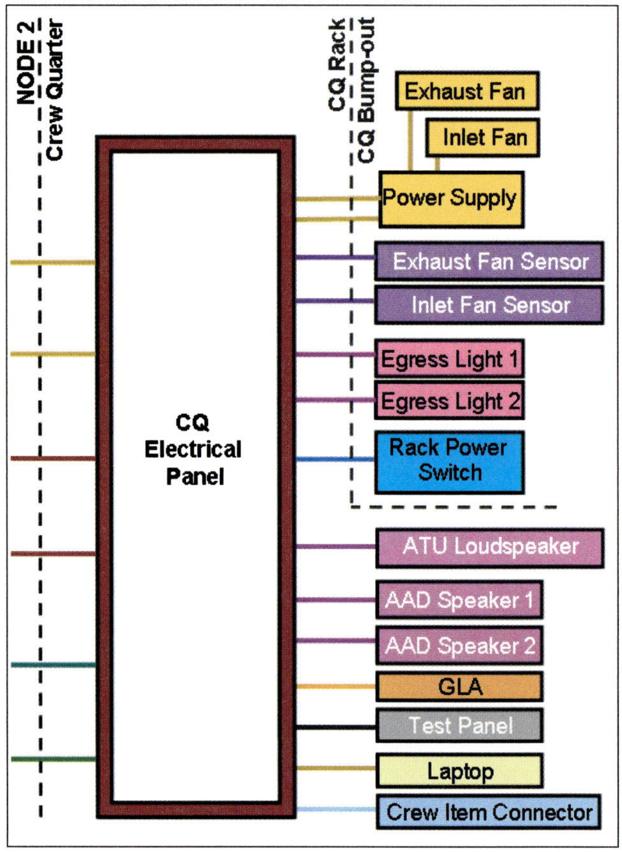

FIGURE 8 CQ electrical block diagram shows power and signal functional distribution within the CQ system. Six electrical interfaces on left side are identified in Figure 2.

CAUTION AND WARNING

Ironically, the CQ internal acoustic environment is too quiet for the occupant to reliably hear cabin alarms, especially during sleep. Consequently, the CQ design provides a caution and warning (C&W) system that includes two dedicated speakers (ADD) to annunciate Class I alarms and uses the ATU loudspeaker as a third leg of redundancy. The ATU loudspeaker annunciates Class II and III alarms to only two of the four CQs at a time.

Should an ISS Class I alarm occur, the two Harmony MDMs send independent signals to each CQ via redundant C&DH data lines. The redundant speaker system produces a tone to alert CQ occupants to the emergency. The ATU loudspeaker is the only means by which the CQ occupant receives Class II and III alarms; thus, the loudspeaker has neither volume control nor on/off switch to avoid the occupant missing alarms by inadvertently turning the speaker down or off. The volume of all three CQ speakers is permanently set to meet the C&W requirement of 20 decibels above ambient noise.

A CQ occupant can also interface to the ATU via a headset; however, when the headset is plugged into the loudspeaker, the crew member receives alarms through both headset and speaker. Again, no "speaker off" function is implemented in the speaker design to prevent the occupant missing alarms because of inadvertently leaving the headset plugged in, nor does the headset have a volume control. Audio volume is controlled at the *Harmony* ATUs.

All four CQs have ATU lines routed from the *Harmony* standoffs to a connector patch panel nearby. However, the node only has two ATUs connected to this patch panel. Any combination of two CQs can receive the ATU signals via the patch panel; not all alarms can be patched into all CQs at the same time. Operationally, this is resolved by a "buddy system" in which two on-duty crew members can receive Class II and III alarms and determine if off-duty crew members should be awakened. A candidate improvement for future CQ designs would be to provide independent data lines for each type of alarm, so that every CQ received each type of alarm signal, and the ATU line volume could be adjustable and muted with headset usage.

Crew-member asphyxiation is a hazard for any confined space, and physiological considerations are well documented. CQ airflow monitoring devices provide health status of the ventilation subsystem. Because it is a Criticality 1 subsystem, monitoring fan health to manage the potential for hazardous CO_2 buildup caused by low flow is itself a critical requirement. There are a total of four airflow monitoring devices in each CQ. Each fan has a tachometer and independent flow sensor. The tachometer detects mechanical failure of the fan while the flow sensor detects airflow failure. Either failure triggers the MDM with the appropriate failure indication, resulting in an alarm broadcast to the occupant via the ATU loudspeaker.

Early CQ design included a smoke detector in the exhaust duct. However, based on computational-fluid-dynamics analysis and as concurred by the NASA Safety Office, this detector was deleted from the final design. The Node 2 smoke detectors are located in the CCAA inlet, less than two bays away from the CQs. The analysis concluded the worst-case time of flight for smoke from the CQ fan inlet to the CCAA inlet, assuming a minimum (one fan failed) flow rate of 0.85 m^3/min would be approximately 85 s. The Safety Board agreed this was adequate detection time. Deletion of the smoke detector from the CQ design left more space in the exhaust duct for acoustic abatement material, reduced overall CQ maintenance time on orbit (cleaning), and potentially reduced the number of false smoke detector alarms caused by airborne particles (e.g., hair, skin, and clothing lint). Fire suppression within a CQ is accommodated via a standard ISS fire extinguisher hole in the CQ bump-out. CO_2 discharge through this port is sufficient to reduce oxygen levels and extinguish combustion of a single standard fire (Zang 2002).

LIGHTING

The CQ has three lighting requirements. The first is for cabin light isolation for privacy and sleeping. The light isolation requirement for sleeping, 54 lux at head level when the node is illuminated, requires fairly light-tight sealing around the door frame, pop-up hinges, and through the ventilation ducts. This requirement also limits the number of status LEDs on interior CQ equipment. The second requirement is that the low lighting level during sleep drives the need for egress indication lighting (0.5 lux) to automatically illuminate in case of a Class I alarm or loss of electrical power. Although the CQ is a relatively small volume, disorientation could occur during an alarm situation without visual cues, especially given the microgravity environment. Redundant LED assemblies are located on the top inner surface of the doorway so that the crew member can operate the door quickly.

The third lighting requirement is for general illumination during wakeful operations. The CQ must provide up to 108 Lux in general areas and 323 Lux on reading surfaces. These lighting levels must be adjustable per crew-member preference. The standard ISS GLA was selected for commonality and because it is brightness adjustable. For added adjustability, the GLA is mounted to the CQ utility wall on seat tracks so the occupant can raise or lower it (Figure 9). An add-on fabric shade over the GLA allows adjustment for direct lighting, indirect lighting, or glare control. The CQ interior acoustic liner surface is a relatively bright white color to aid distribution of the limited GLA output throughout the CQ. Finally, mounting provisions for portable reading/task lighting are located over the right shoulder on the sleep wall. Lighting assessments were performed on the low-fidelity and midfidelity mockups and used to calibrate CQ lighting models to validate the requirements and verify the design.

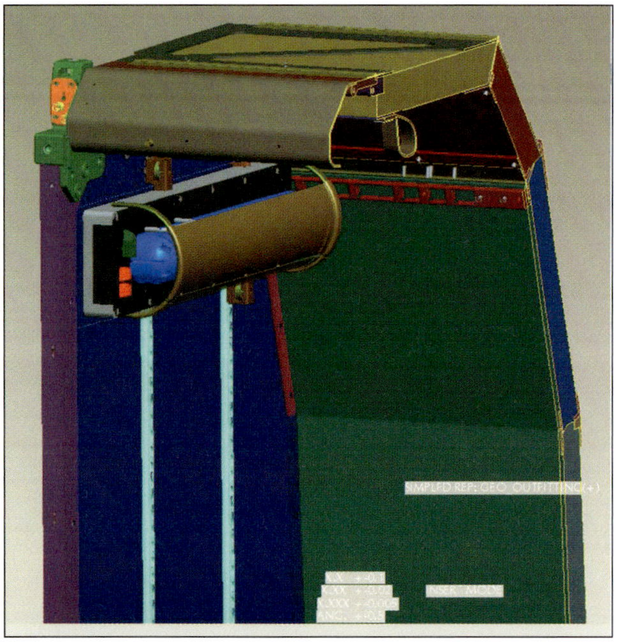

FIGURE 9 GLA location is adjustable on seat tracks (light blue). Adjustable fabric shade (gold) helps control quality of the light. Bump-out and sleep wall omitted for clarity.

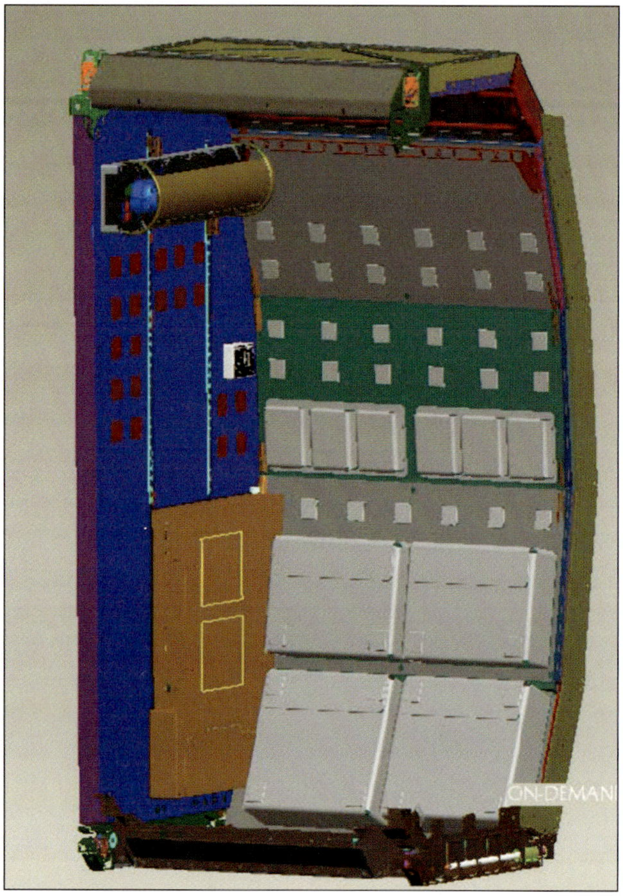

FIGURE 10 General CQ interior layout shows utility wall (blue) with electrical panel (gold) and back wall with hook-and-loop fastener patches and CTBs (gray) for storage.

CREW ITEMS

The layout of crew items, attach points, and electrical systems inside the CQ is largely based on human factors analyses to meet accessibility requirements and recommendations resulting from crew evaluations. One such recommendation was to keep crew items and equipment assemblies separate from the sleep area, similar to the TeSS layout. To accommodate this recommendation, one of the side walls is designated as the sleep wall. The sleep wall provides 12 D-rings for adjustable sleeping-bag attachments. A 15-cm length of seat track is also provided on the sleep wall for attachment of a portable light.

Opposite the sleep wall is the utility wall. The utility wall provides a variety of attach points (hook-and-loop fasteners and attachment rings) for display of personal items. There are a total of 6 D-rings on the utility wall and 14 D-rings on the back wall for bungee attachment. Although highly desired by the crew, spacing constraints limited the number of hook-and-loop fastener patches to approximately 125 5-cm × 5-cm patches dedicated for occupant usage. Two lengths of seat track on the utility wall accommodate both a Bogen arm for the laptop desk and the GLA. The test panel, electrical panel, and ATU loudspeaker are also located on the utility wall.

The crew member can stow approximately 0.09 m^3 of personal items inside the CQ using cargo transfer bags (CTBs) or similar devices. These items are typically stowed at the bottom of the CQ volume to maximize arm and headroom (Figure 10).

The CQ interior walls are covered with acoustic blankets to provide additional sound isolation. The modular blanket configuration saves operations time because it enables individual sections to be removed for cleaning or replacement. Stain testing was done on several materials to determine which material should be exposed to the crew. This testing addressed the materials' resistance to soils and staining as well as their ability to be cleaned. Gore-Tex® was found to be sufficiently stain resistant and is used as the outer layer of interior and exterior (bump-out) blankets. The blankets have an inner layer of Thinsulate® for sound absorption and a Nomex® back layer. Exterior blankets have additional sound adsorption layers.

STRUCTURE AND RADIATION MITIGATION

The CQ structure provides the conventional functions: distributing launch and user loads, attachment to transporter and cabin, component mounting, and dissipation of thermal loads. The CQ structure also reduces crew radiation exposure.

NASA astronaut radiation exposure standards are federally mandated, and in addition to fixed values include the as-low-as-reasonably-achievable (ALARA) standard. A critical element of ALARA compliance is hardware design optimization, particularly for vehicle areas where crew members spend significant portions of their time. In round numbers, a crew member spends at least one-third of the time in the CQ's relatively small dedicated volume. Consequently, it is quite beneficial for CQ materials selection and placement to mitigate radiation exposure. The CQ project established an ALARA process consisting of regular discussion, analysis, and evaluation between the design engineers and the NASA Johnson Space Center Space Radiation Analysis Group (SRAG). ALARA considered the impacts of radiation protection materials on usable volume via crew-member evaluation of mock-ups. Typically space radiation is most effectively blocked by materials containing high percentages of hydrogen (Zapp 2001). Ultra-high-molecular-weight polyethylene (UHMWPE) was selected. Based on SRAG recommendations and the results of crew-member evaluations, 6-cm thick panels were located on the ceiling and back wall surfaces. Overall, 125 kg of UHMWPE is integrated into the CQ structure, reducing crew-member exposure to galactic cosmic radiation by approximately 9% and solar flares by approximately 74%.

Having a noncollapsible rack volume to minimize crew installation time requires the wall surfaces to be stiff enough to avoid natural frequencies below 25 Hz to avoid launch vibration mode coupling. Machined aluminum isogrid and composites were both considered. However to limit the amount of metallics close to the occupant and to minimize structural weight, flat carbon honeycomb panels comprise the side walls and flooring. The weight thus saved was then allocated to UHMWPE panels. Primary loads are carried by an aluminum frame that captures the flat composite panels (Figure 11). The bump-out does not require radiation panels because the CQ across the aisle provides protection along this surface (see Figure 1). The bump-out contains numerous chamfers, penetrations, and component mounting provisions that make flat-panel composite construction less practical. Consequently, the bump-out is fabricated

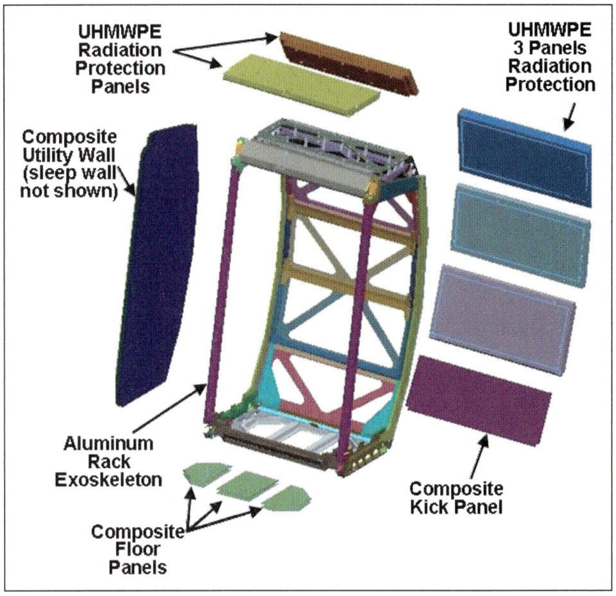

FIGURE 11 CQ structural design consists of aluminum exoskeleton, composite panels, and UHMWPE radiation-shield panels.

of aluminum, an approach that simplifies manufacturing by avoiding curved composite surfaces and joints.

HABITABILITY EVALUATIONS

In each CQ design phase crew members evaluated a full scale mock-up. In coordination with the NASA Johnson Space Center (JSC) Habitability and Human Factors Group, the CQ design went through three separate, formal crew-member evaluations.

Low-Fidelity Foam Mock-up

The first crew-member evaluation was conducted during the SRR on a low-fidelity mock-up. The purpose of the evaluation was to evaluate interior volume, not detailed placement or layout of crew items. Each test subject evaluated two CQ design concepts focused on the amount of radiation protection that could be provided to meet the ALARA requirement.

The mock-up was built of plywood and foam. Interior hardware was simulated using foam blocks to represent components such as GLA, speakers, ventilation subsystem, laptop, and stowage (Figure 12). Each test subject evaluated a CQ volume with 7.6-cm-thick walls, ceiling, and floor, which represented the maximum amount of radiation protection proposed (leaving minimum usable crew volume). The mock-up was then reconstructed with 1.3-cm-thick walls, ceiling, and floor to represent the minimum amount of

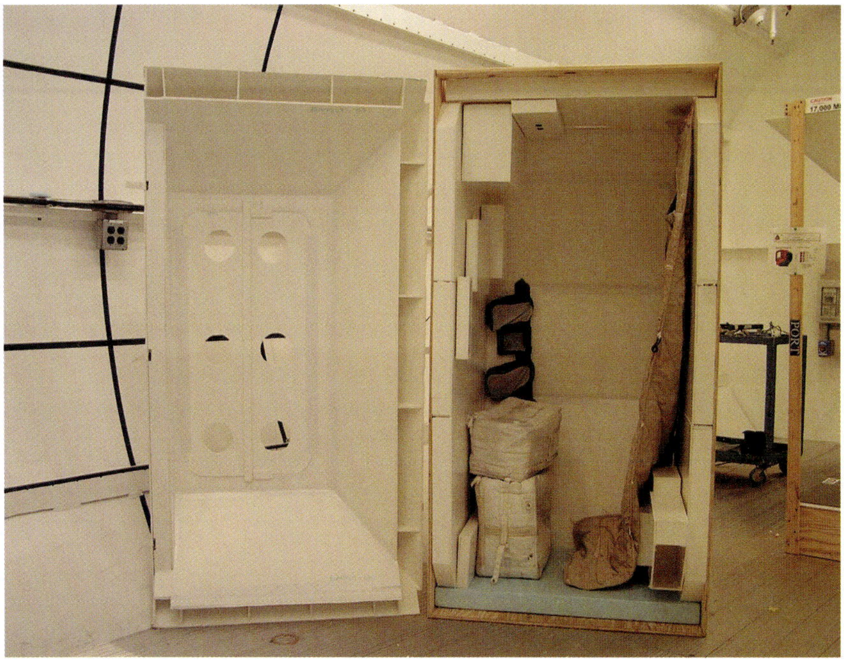

FIGURE 12 SRR-vintage volumetric foam mock-up had 7.6-cm-thick walls for radiation protection.

radiation protection proposed (leaving maximum usable crew volume). The evaluation questionnaire included 25 questions, and over a three-week period 23 crew members participated.

The crew-member evaluations concluded that the crew member could live with either configuration, using the CQ for sleeping and other activities. However, they recommended radiation reduction materials be placed only on necessary walls to maximize usable volume. The 7.6-cm-thick walls excessively compromised head room. Based on evaluation results, SRAG completed several modeling and analysis iterations and determined the best placement for structurally integrated radiation reduction materials (6.4-cm back wall panels and pop-up), which became the basis for the next design phase.

Other suggestions included 1) separating crew-member interaction items from items requiring no interaction; 2) addition of attach points on exterior and interior for handrails to help with ingress/egress; and 3) addition of several attach points on all walls to allow occupants to personalize their CQ.

Low-Fidelity Design Review Mock-up

At PDR, a second crew-member evaluation was conducted using a plywood mockup (Figure 13). The primary assessment areas were 1) design implementation of the radiation protection solution developed by SRAG; 2) general equipment layout; and 3) volume constraints given preliminary volume layout and internal outfitting, prior to detailed structural design. Internal outfitting included volumetric representations for the C&W panel, fan assembly control panel, outlet panel, GLA, seat track, laptop desk, approximately 0.09 m^3 of personal stowage CTBs, and smoke detector. A Russian sleeping bag and foot restraints (handrails) were also provided for the evaluation.

A total of 23 crewmembers evaluated the mock-up. Each participant was asked to evaluate volume, accessibility, and usability of items as well as any potential impacts of the proposed habitable volume and design. Ninety-one percent of the crew members felt that volumetrically the CQ felt very spacious and that the volume was adequate to accommodate all personal items and most activities.

FIGURE 13 PDR-vintage, low-fidelity, wood mock-up had preliminary component layout. Upper white and lower gray blocks external to mockup represent adjacent CQ bump-outs.

Concerns were expressed with the concept for a removable/disposable liner, specifically the impact on crew-member time to replace it on orbit. Other recommendations included providing additional seat track length and quantity to allow more adjustability of the laptop desk and using non-hook-and-loop fastener mechanisms for holding the door open and closed. There was overwhelming negative feedback on the GLA position and recommendations to move the GLA higher and incorporate a directable shade to customize the light level. Comments on the fan control assembly (nonfunctional for the mock-up) location were that it was too close to the crew-member's head. The recommendation was to move the fan assembly further away from the head position as well as to conduct additional human-factors analyses on placement of items inside the CQ (primarily placement of the control panel and its design).

Mid-fidelity Mock-up

At CDR, a final crew-member evaluation was conducted on a mid-fidelity CQ mock-up. The primary focus of this assessment was to evaluate potential volumetric impacts and solicit specific design suggestions regarding volume constraints, accessibility and location of items, and basic usability of components within the CQ volume. The mock-up was constructed of aluminum and composite structural materials, and flight-like acoustic blankets lined the interior, as well as the bump-out exterior (Figure 14). The internal outfitting (Figure 15) included the same items as the low-fidelity mock-up but in their proposed final locations. A GLA shade, functional flow control switch, and simulated C&W alarms were mocked up.

Twenty-one subjects evaluated the mock-up. Each participant was asked to evaluate the mock-up regarding habitable volume, habitability, accessibility, and internal outfitting. They were also asked to comment on the overall design and usability of items inside. One hundred percent of the test subjects indicated the volume was adequate for sleeping and other activities. Because the ventilation subsystem was not flightlike, several negative comments were made regarding the high noise level and inadequate airflow. The participants commented that the test panel and egress light designs were adequate but believed these subsystems were not necessary in the CQ design. (However, these features are necessary per safety requirements and are incorporated in the final design.) Positive feedback was received on the GLA shade; however, several recommendations were received to improve the shade's ease of use. Recommendations were made for additional bungee attachments and hook-and-loop fastener patches; however, the crew members were pleased overall with the layout and design of the equipment restraints.

FIGURE 14 Exterior of CDR-vintage, mid-fidelity mock-up shows acoustic blankets and double-door design. CQ is mounted on rack handling assembly.

FIGURE 15 Interior of mid-fidelity mock-up (utility wall on left, sleep wall on right) was outfitted with flight-like acoustic blankets, lighting, and crew-member items.

SUMMARY

The design heritage, trade studies, performance analyses, and mock-up evaluations described in this chapter provided the basis for the CQ's final configuration. Typical of the design process, apparently simple functional requirements resulted in rather complex and mutually conflicting derived requirements to maintain crew-member safety, structural integrity, minimal installation/maintenance time, and reliability. Resolving the conflicts greatly increased the complexity of development and necessitated close coordination across disciplines. Balances had to be struck, for example, between radiation protection and crew-member volume, and between acoustic abatement and ventilation backpressure caused by it. Achieving satisfactory balance resulted from prioritizing system functionality early during requirements development, employing good systems engineering, coordinating with habitability stakeholders, and using multiple crew-member evaluations.

The ISS CQ is a pivotal system that enables a full crew complement of six. It is the most advanced space crew-quarters system developed to date. It provides an acoustically quiet, visually isolated area for a crew member to sleep, relax, and enjoy a private retreat away from the busy ISS onboard environment. The design fully integrates radiation protection and crew-member provisions into its structure. Though its range does depend on *Harmony's* ambient set point, the CQ's ventilation subsystem provides independent control. Acoustic control is extensive on all interior surfaces, and these materials provide longevity in terms of stain resistance, cleanability, and replacement in sections. Acoustic abatement in the ventilation subsystem provides substantial noise attenuation from both the fans and exterior cabin environment. The acoustic blankets and abatements are fully accessible, removable, and cleanable of dust and dander by vacuum cleaner. The fully integrated, redundant electrical and C&W systems ensure crew-member safety without the need for interaction. LAN hook-up, adjustable lighting, and flexibility for locating various types of crew items all help make the CQ a place ISS crew members can truly customize to maximize their comfort and productivity. |

Acknowledgments

This chapter summarizes hard work by numerous NASA Johnson Space Center (JSC) and Engineering Support Contract project managers, design engineers, and analysts, as well as contributions by JSC functional specialists, engineers, and crew members whose frequent input was invaluable in ensuring as much anthropometric and habitability functionality as possible. The CQ project is funded by the NASA JSC ISS Vehicle Office.

6

retrofitting the International Space Station

SUSAN M. FAIRBURN

INTRODUCTION

FEEDBACK FROM expedition crews who have lived aboard the International Space Station (ISS) include requests for an improved living environment. Designers can improve the living environment in part by learning from the experiences of these crews. By developing solutions that can be retrofitted into the existing architecture, designers can offer environment improvements that enrich crew-members' experience.

Crew-quarters (CQ) design was key to *Skylab* habitability, with a specific goal of providing a private space for each crew member, who might spend six to eight hours a day there. Despite the privacy afforded by a designated place for each crew member, *Skylab* crews reported poor sleep because of noise, light leaks, or disturbances by fellow crew members. Adams (1998) noted that *Skylab* lacked attachment points for relocating sleep restraints, thereby effectively precluding crew members from sleeping elsewhere. Generally speaking, *Skylab's* interior outfitting was not designed for modularity or reconfigurability.

In contrast, one of the principal ISS design features is the basic structure of the modules and the rack volumes they accommodate, the international standard payload rack (ISPR). The ISPR is intended to allow interchangeability and reconfiguration. Chapter 5 described in detail the various ISS CQ designs. This chapter presents a conceptual alternative CQ design intended to be highly flexible. Crew feedback has cited flexibility of use as a desirable feature for long-duration missions. Flexibility offers numerous advantages for space applications where living volume is limited, and delivery and maintenance costs are major concerns. It allows objects or environments to be used in different ways, benefiting crew members who desire visual stimulation and variety in their environment without requiring extensive amenities. Avoiding extra amenities in turn averts unnecessary transportation demands and costs.

A design solution based on a common mounting structure and "kit of parts" could offer increased flexibility. Moreover, such an approach is durable; any part could be detached and updated, improved, or replaced. The design process described here began with a series of self-directed, empirical exercises that provided insight and spurred concept generation, followed by a review of relevant ISS specifications. The resulting design is compatible with the basic elements of existing CQ equipment, but offers adaptability over time using a proposed kit of parts, and thus allows crew members to tailor the layout and use of their private environment at any time.

Copyright © 2008 by Susan M. Fairburn. Published by the American Institute of Aeronautics and Astronautics, Inc., with permission.

BACKGROUND

ISS CQ Designs

The primary objective of the ISS is to provide a manned outpost for scientific research in a reduced-gravity environment. As terrestrial creatures, we are continually exposed to the gravitational field of the Earth. Gravity shapes our perceptions and expectations of how we interact with our environment. Because we develop with it, we can intuitively anticipate the effects of weight from gravity and adjust accordingly. Orbital microgravity is an environment that poses both opportunities and restrictions. Weightlessness must therefore lead to readjustment and reorientation.

A complete history of the evolution of ISS CQ designs can be found in Chapter 5. Here a brief summary is included for context. NASA defines an ISS CQ rack as follows:

> The crew quarter[s] rack shall provide each individual crewmember with private space isolated from external light and sound to sleep, don and doff clothing, read, write, perform recreational activities, and for personal/private and medical consultations. The crew quarter[s] shall also provide limited storage for personal items, clothing, and computer accessories. Interface locations shall be provided for handholds, mobility aids, and restraints. (NASA 1999)

Significant cost growth and increasing budgetary constraints reduced the many original, ambitious space station concepts to the ISS core for initial operations (Figure 1). This interim station would provide formal CQ accommodations for a crew size of two rather than the eventual six. In the original plan, the final module to complete the ISS configuration would have been the U.S. habitation module, containing four private CQ and supporting crew habitability needs like hygiene, meal preparation, medical conferencing, personal stowage, privacy, and socialization. These four CQ, plus two kayutas in the Russian segment, would allow private accommodations for a permanent crew of six.

FIGURE 1 NASA digital model of the ISS during assembly, in its April 2002 configuration.

The temporary sleep station (TeSS), located in the *Destiny* U.S. laboratory module, became the interim solution to accommodate a third crew member during the assembly phase (NASAexplores 2001), approximating the baseline ISS goals at reduced cost. TeSS offers a personal living volume, but not the same degree of privacy or crew amenities as the permanent CQ originally to have been developed for the habitation module.

When the U.S. habitation module was canceled in 2002, the final ISS CQ design was derived from the original CQ and informed by lessons learned from using TeSS and the kayutas. Four of these ISS CQ are installed in the *Harmony* node. They are comfortable, sheltered havens, albeit designed with limited outfitting options.

Habitability

Living on a space station means being in a confined, limited-volume place, in close proximity to fellow crew members, with no chance to "get away." Individual responsibilities are significant, and stress levels can be high. In the closed environment of a space habitat, privacy is necessary for psychosocial health; individuals need a private place to rest and relax. The closer one is forced to be with others, and the more time one spends in close proximity, the greater the need for a means to retreat.

Clearwater (1985) cites research and experiences derived from analogous environments, including prior space stations, as supporting the need for a holistic approach that takes into consideration various issues, for example, psychological, social, and physical, in the design of an effective space environment. Providing crew members with places of their own to personalize and use as they choose would recognize the psychosocial significance of having personal space in the first place. Indeed, some crew members might prefer a personal living volume, whereas others might prefer a temporary sleep enclosure that can be secured in various locales around the station.

One of the key challenges designing the interior configuration of living spaces with such limited volume is to provide a sense of spaciousness while offering enough flexibility to adapt to different uses, and users, over time. A good design would therefore consider crew interaction, privacy, proximity, lighting, sight lines, acoustics, temperature, circulation, crew activities and schedules, and many other issues. All are relevant to achieving a truly habitable interior, that is, one that promotes productivity, well-being, and desirable occupant behaviors (Bedini and Perino 1999).

Addressing CQ Habitability Through Design

FLEXIBILITY Extended space missions beg for a flexible environment with outfitting that can be reconfigured for cleaning and maintenance and for accommodating changing crew composition, preferences, and activities. This approach is consistent with the ISS ISPR standard for exchangeable racks. Flexibility is a common attribute in terrestrial furnishings, with examples including objects and environments that offer multiple options for arrangement and flexibility of function. Flexibility can be achieved via a group of objects that can be configured to achieve different functions by a system that supports a family of objects, each possessing a unique function, or by arriving at a reduced collection of elements whose functions are fused.

Italian designer Gianantonio Mari conceived of a modular environment made up of a dynamically organized space to house modular and combinable furnishing units (Ambasz 1972). The system is based on a grille framework that permits horizontal or vertical extensions via a support structure and collection of modular interior equipment. Design of the interior elements addresses various functional requirements (e.g., containment, support, separation) while allowing personal expression. Mari defined use through a study of rituals and ceremonies in five major categories: privacy, sleeping, dining, leisure, and sensory.

More recently, Rashid's Surfacescape (2001) furniture system reduces clutter by fusing functions into four "seating" units: chaise/seat, multilevel seat, carpet/seat, and booth/couch. A floor plan as structure permits multiple arrangements, in combination with the four seating units, resulting in a system that is modular in function.

KIT OF PARTS Flexibility can be achieved through a kit of parts. Howe (personal communication, 20 May 2002) defines a kit-of-parts design as one based on "assemblies of standard, easy-to-manufacture components, sized for convenient handling or according to shipping constraints . . . ." One of the strongest advantages of the kit-of-parts design and construction approach is that it not only ". . . achieves flexibility in assembly and efficiency in manufacture, but also by definition requires a capacity for demountability, disassembly, and reuse" (Howe et al. 1999). Adopting a kit-of-parts philosophy adds flexibility to a building or artifact by using elements that allow arrangement and rearrangement. The ISS is itself based on a kit of parts: modules, nodes, docking adapters, truss segments, solar arrays, thermal radiators, and payload attachments (Messerschmid and Bertrand 1999). Although some ISS elements are not intended for disassembly and reuse, the basis of the ISS truss and standardized pressure-module system is consistent with kit-of-parts design.

RETROFITTING Retrofitting provides an existing machine or structure already in use—such as a jet, computer, space station, or rack volume—with parts, devices, or equipment that did not exist, or were not available, at the time of the original design or manufacture. Retrofitting is likely at various stages of a long ISS life cycle, as its mission (and perhaps even its owner) changes and as technology advances. It is certainly a suitable approach to CQ design and outfitting. A plausible way to design for habitability within the ISS pressure module architecture is to retrofit the standard rack space for crew quarters. Indeed the baseline ISS CQ essentially takes this approach. The alternative design described next, called retrofitted crew quarters (RCQ), aims to improve habitability through flexible retrofitting of a standard rack volume using a kit of parts. A rack shell that allows adaptation and evolution would allow outfitting to meet multiple crew needs over time, including crew quarters.

HUMANIZING SPACE

The author prepared the following guiding principle at the outset of the design project:

> In your private crew quarters, at any time, you may need: light, control of body position, a work surface, a place to put things, ventilation, privacy and darkness—so that you can sleep, display photos/treasures, store clean/dirty clothes, stow rubbish—you need a place to rest, to relax, to read or to listen to music, to meditate, to do personal work, to communicate, or to just be.

Providing crew members with environments that can accommodate their personal needs and preferences is challenging given the limited volume, payload restrictions, safety specifications, and objective of maximizing acceptability of the environment for a heterogeneous group of users.

Design Process

Insight into the issues and challenges associated with designing a personal CQ is obtained in numerous ways including literature review and a creative process of idea generation. In this case, a series of three, self-directed empirical exercises explored issues of confinement, simulated weightlessness, and personal space as well as the personalization of a "living volume." Whereas these exercises provided valuable insight into some issues discussed in the human space exploration literature, ISS specifications provided tangible realities to guide the final design. Key assumptions were made about how the RCQ would interface with existing ISS module structure and equipment:

1) The RCQ would be attached to the habitation module and secured in place using existing rack hardware and seat-track attach fittings
2) Access to the module pressure shell and standoffs for maintenance would be achieved through removal of the whole RCQ.

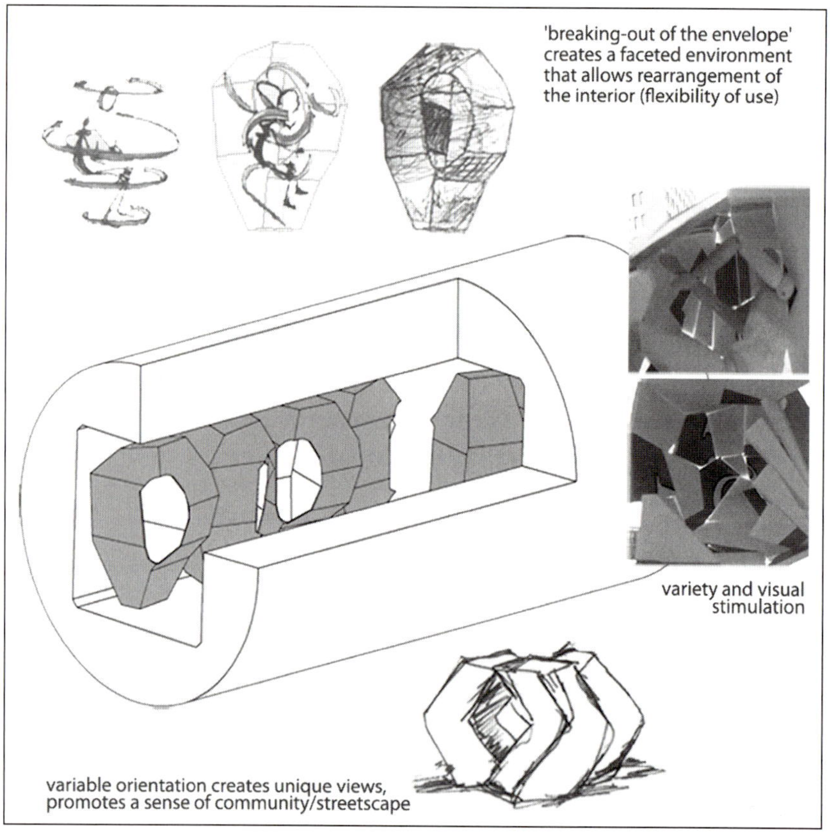

FIGURE 2 Montage of a concept based on a faceted, multi-orientation RCQ volume.

Figure 2 illustrates a concept based on breaking the RCQ envelope into discrete surfaces: a faceted form as an expression of the multifunctional nature of the constrained personal living volume. The concept evolved from simulated weightlessness underwater, where achieving a position and orienting oneself was experienced as three-dimensional interaction with a multiplaned sphere.

The first empirical exercise included construction of a full-scale ISS rack volume, a basic shell that could be entered and experienced to promote better understanding of the limited space. It also allowed the addition and arrangement of interior elements to aid exploration and development of the RCQ interior concept.

Proposed Design—Shell

The shell design evolved from the premise that the interior layout should be dynamic, that is, in a continual state of transformation determined by a crew-member's preferences and patterns of use. At this point, the form such a proposal would take within the context of the ISS rack (Figure 3) was a shell with an integrated array of attachment points to support the arrangement and rearrangement of a family of functional elements.

As shown in Figure 4, the design is asymmetrical, offering crew members visual variety, stimulation, and relief from the symmetrical, rectilinear space-station environment. The exterior is made up of three walls, a floor and ceiling, and a front bump-out and access section. When assembled, the tapered form provides a variety of surfaces for tailoring the arrangement of interior features.

Supporting a crew-member's adaptation is important to environment habitability. Providing an interior that facilitates a crew-member's microgravity movement and positional control is an exciting challenge for designers. The second empirical exercise comprised underwater sessions to provide insight and design inspiration. The essence of the exercise was to experience the natural state of human body and loose objects in simulated weightlessness. The predominant conclusion was that in weightlessness this natural state is motion without rest, whereas in a gravity-weight environment, the natural state is rest.

Designing for Movement in Space: Control vs Freedom

Positioning oneself in space can be active or passive. For the purpose of this RCQ design, active restraint is defined as control of body position requiring the expense of energy to engage surfaces and/or edges. An example of active restraint is reaching for something by hooking a toe under a surface and grasping a hold (two-point restraint). Bracing between surfaces is

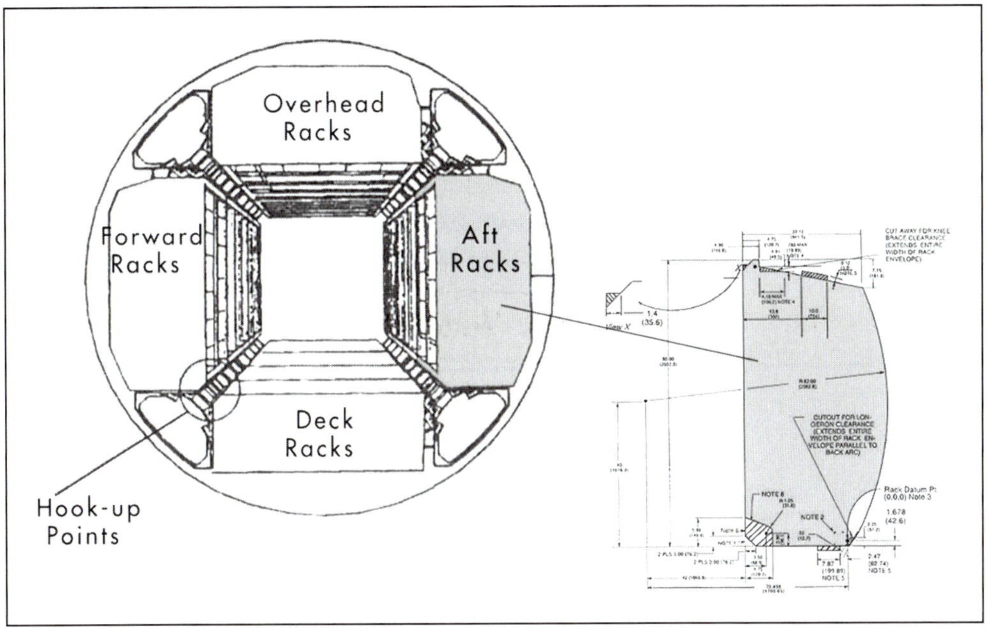

FIGURE 3 ISS rack structure (ISPR).

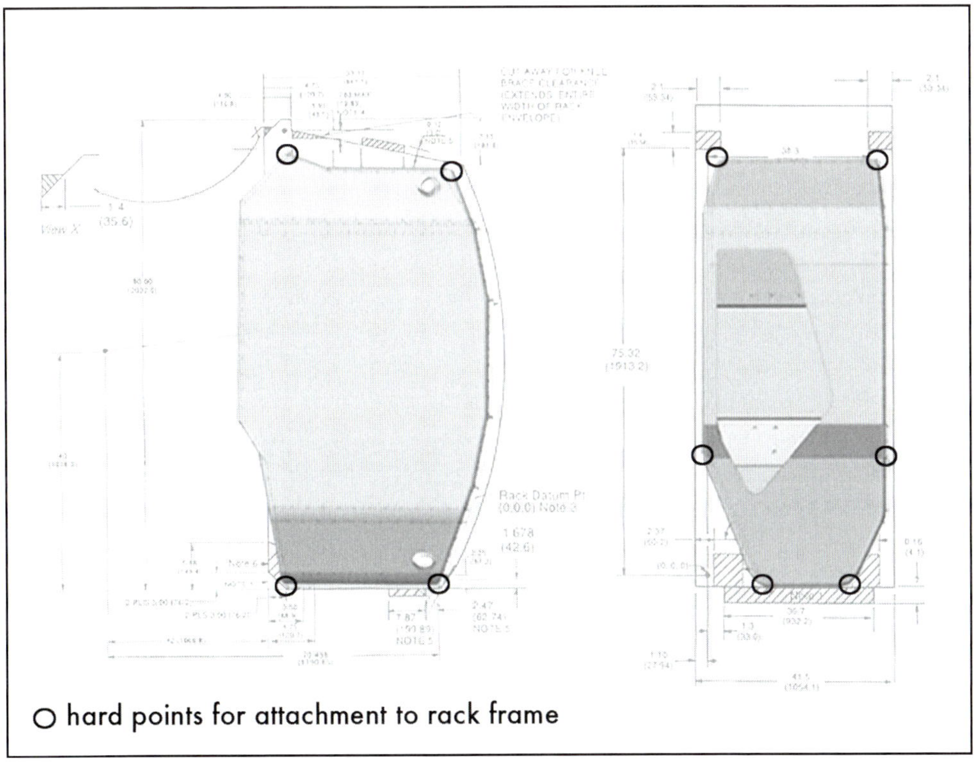

FIGURE 4 RCQ shell with attachment points for securing it to the rack frame.

also active restraint. Passive restraint allows maintaining a position without expending energy. Examples of passive restraint include being in a sleep restraint, or being "wedged" between objects or surfaces. While living in space, crew members adapt to weightlessness, learning how to move about and position themselves to perform tasks. A crew member can retrieve an object inside the RCQ by bracing between surfaces, an approach that requires a combination of planes the individual can span comfortably. Bracing is commonly used within the ISS.

The faceted exterior shell of the RCQ design facilitates positioning by offering a crew member a choice of surfaces with varying dimensions, angles, and shapes, for controlling body position through bracing. The RCQ shell also provides the frame and housing for a family of interior panels and holds. The kit-of-parts design philosophy supports use of a standardized mode of attachment to maximize the possibilities for arrangement of parts and flexible utilization. However, the number and distribution of attachment sites is an important design consideration for ensuring the system meets its functional requirements while enriching the personal living environment. The array of attachment sites in the RCQ shell was defined by the system interface (Figure 5). It was apparent that this array had to balance flexibility, economy, usability, and aesthetics.

Although a reduced-gravity environment offers opportunity for three-dimensional usability, it is not necessary to provide complete coverage of the interior surface with attachment points. Some areas of the RCQ, such as the lower corners, are awkward to access and therefore serve designated operational purposes (e.g., ventilation, power, and data intakes/outlets).

Whereas a distribution of attachment sites offered broad functionality and flexibility for a given user, it is important to consider the possible range of user dimensions so that the array would be practical for all crew members. Attachment points are based on the ergonomics of positioning and mobility for various activities that take place within a personal CQ. The analysis takes into account reach envelopes and the user range specified by NASA: 5th-percentile Japanese female to 95th-percentile American male. This range is large; for example, stature varies from 148.9 cm to 190.1 cm, a 41.2-cm spread (NASA 1995). The key dimensional indicators are stature, biacromial (shoulder), and trochanteric (hip) height, and hip/chest breadth (Figure 6).

While the issues of reach envelope and user dimension were considered, development of the array and kit of parts also evolved from the third empirical exercise, a study of usage patterns. Figure 7 illustrates the patterns of use for two hypothetical users referred to as "the organizer" and "the free-formist." This evaluation was done during design development, and so the images are based on an early version of the interior elements.

Proposed Kit-of-Parts Design

An integral part of the design process evaluated the appropriateness of different combinations of interior elements. Considering the activity types and the position

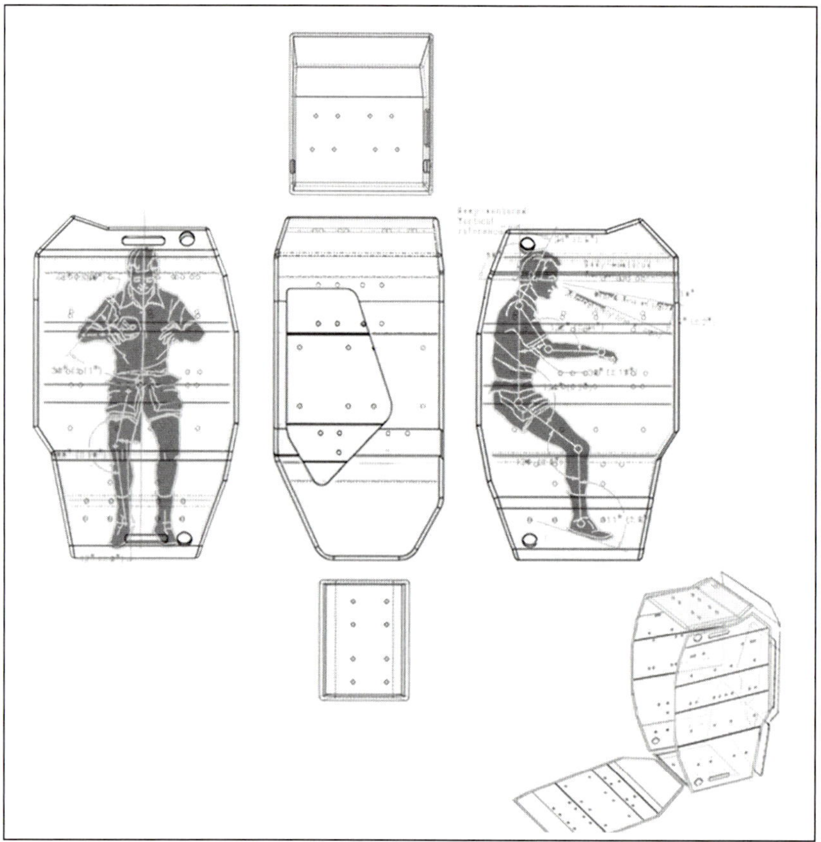

FIGURE 5 RCQ shell surfaces showing interior array of attachment points. Images of crew member in neutral body position convey scale and proportion.

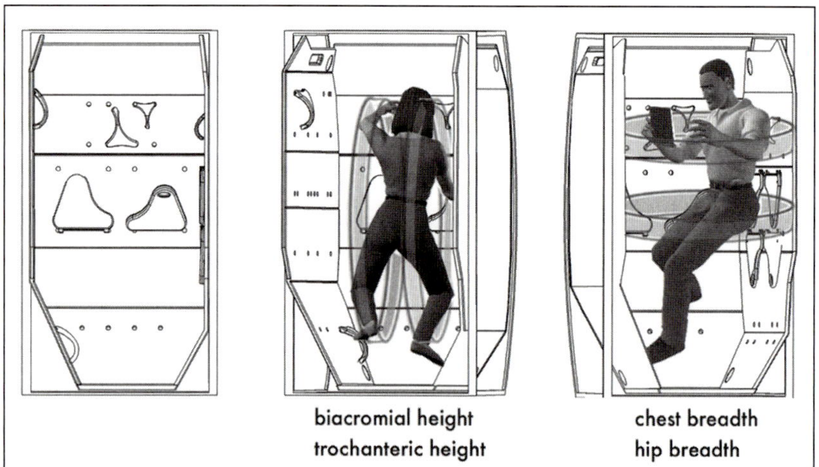

FIGURE 6 An anthropometric array showing attachment points and previewing the RCQ architecture of panels and holds.

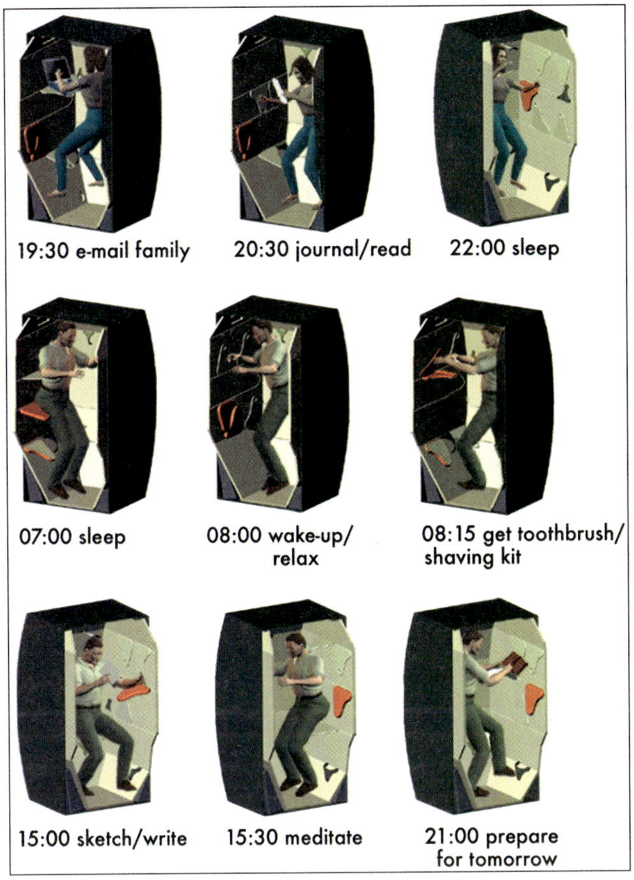

FIGURE 7 Patterns of use during a 24-hour period, showing layout diversity for different crew members: the organizer (top) and the free-formist (center and bottom).

of the user when performing them allowed the exploration of opportunities for fusing functions. For example, while reading, an individual might prefer some way of controlling body position. To perform a personal work activity, such as writing a letter, an individual requires a hard surface. Combining desirable features presented an opportunity to design one element that could control body position while also providing a surface for writing or displaying images like photographs. Fusing these functions into a single *panel* element with edges for positioning and a surface for working yielded a flexible and efficient design solution. Figure 8 illustrates various uses for the panel form, shown for different tasks and users of differing dimensions.

Detailed design development led to the variety of panel types shown in Figure 9. The design family is informed by the need to address a range of functional requirements that are then answered by various edge contours, thicknesses, and surfaces. Two basic types of panels, both hinged, are for storage and for working. The storage panels offer volume as a primary feature, while the working panels provide surface and attachments. Although some functions could be fused,

FIGURE 8 RCQ layouts showing interior functions for a variety of crew-member living activities: 1) sleep, 2) personal work, 3) store/retrieve, and 4) rest/read.

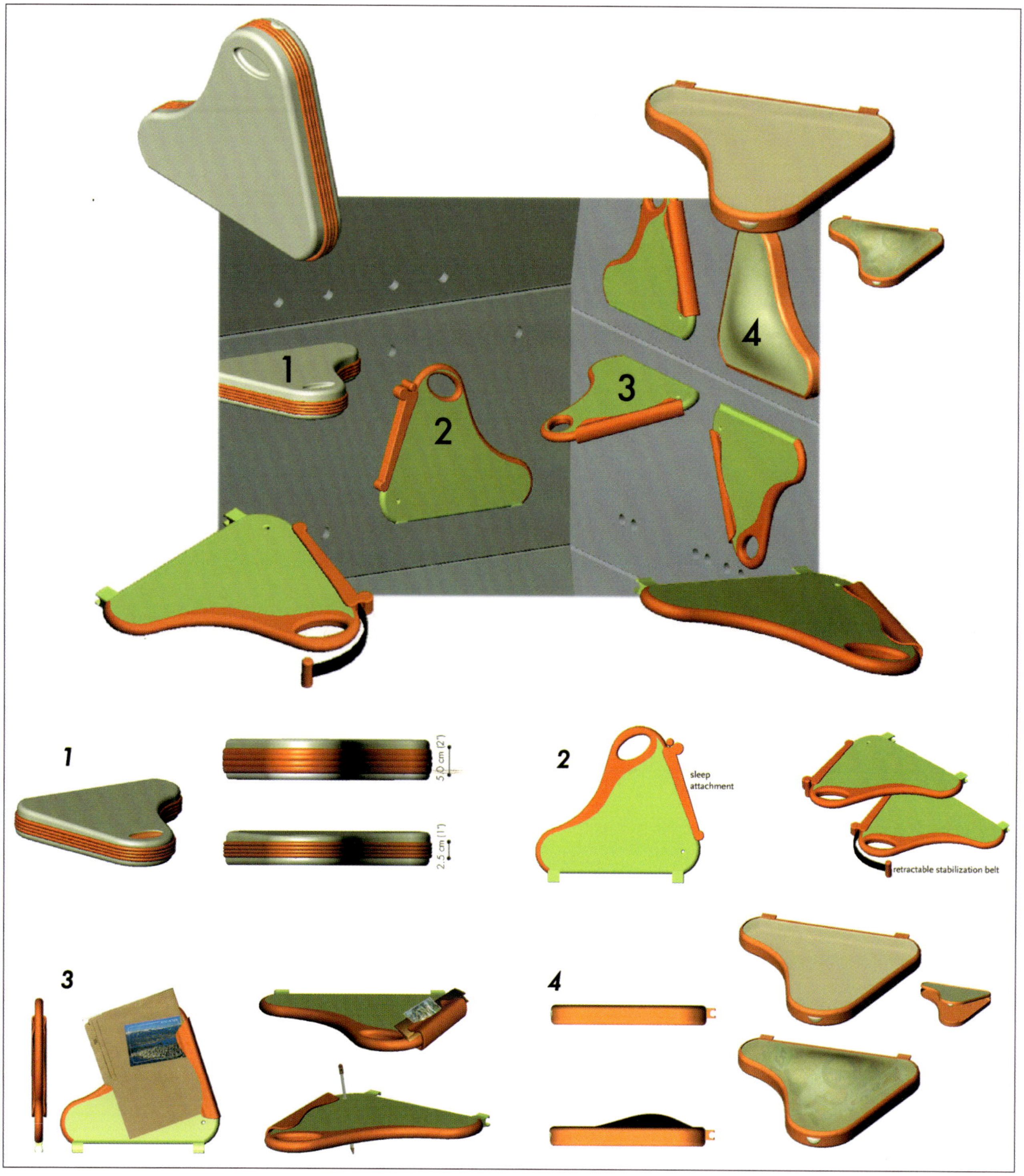

FIGURE 9 Panel components: 1) storage panel with expandable bellows; 2) work panel with retractable sleep attachment; 3) work panel with clip/artifact attachment; and 4) transparent storage panel.

other functions required unique design features such as clips for securing papers/artifacts, reveals for grasping and lodging items, and openings for containing objects. The end result is four panel components.

Panels are moved around by the crew member to accommodate personal storage, work surface, and body position needs. All of the panels use a common mode of attachment; a discontinuous hinge provides both dynamic attachment and a grasping slot (Figure 10). The hinge-and-panel combination allowed creation of intersecting planes in the RCQ environment.

The kit of parts was completed by a *hold* element that complements the panels. The hold was developed after the evaluation process revealed the panel element alone could not fully satisfy the functional requirements. Figure 11 illustrates two sizes of

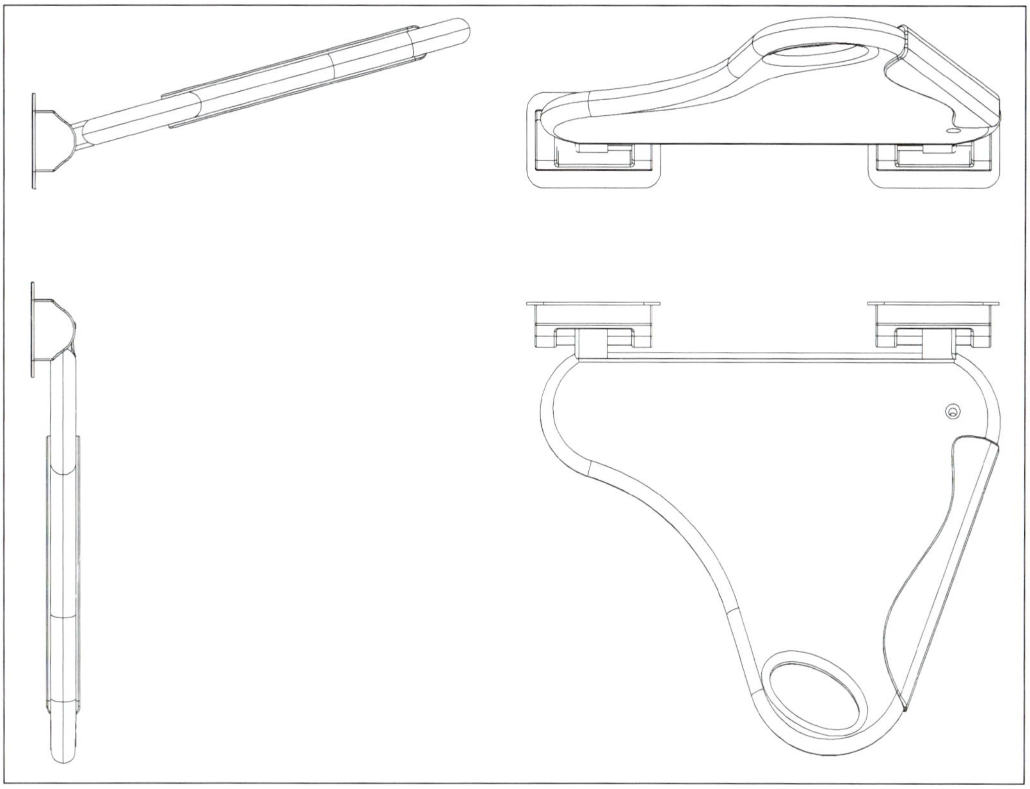

FIGURE 10 Work panel shown with hinge attachment (left to right: side and front view, with panel in two positions).

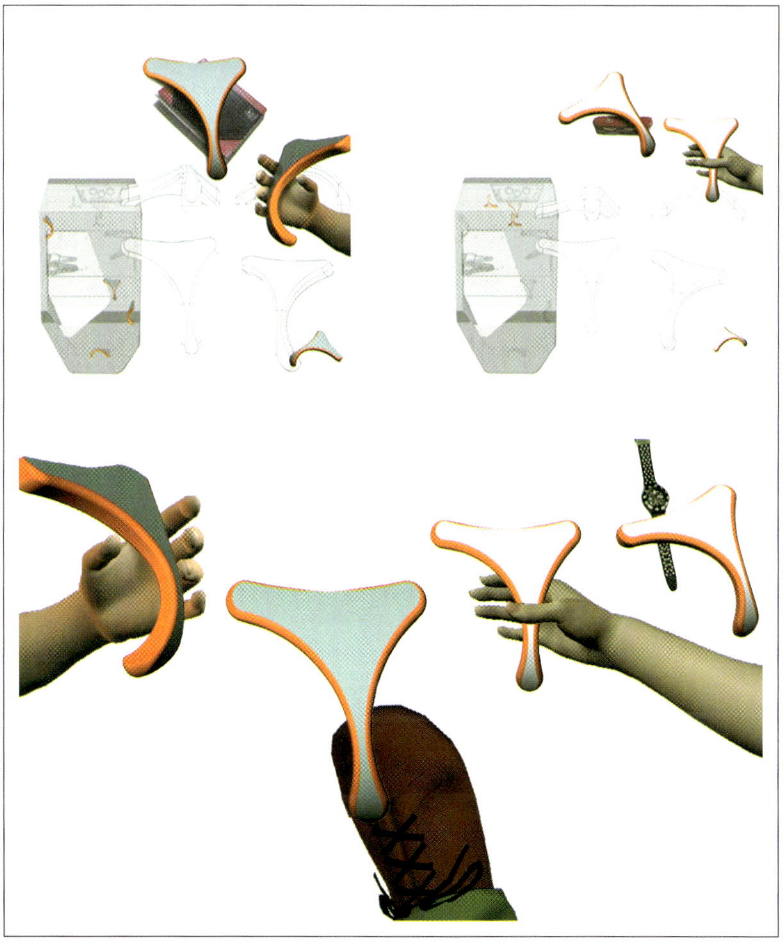

FIGURE 11 Close-up view of the two sizes of holds showing grasp, hook, and quick storage/display options.

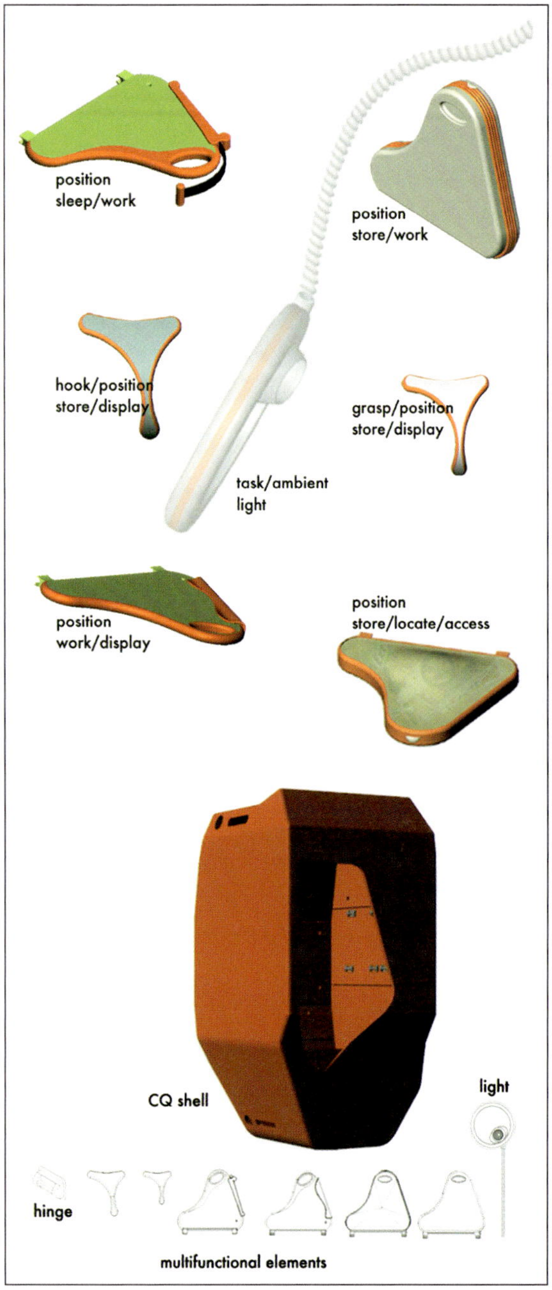

FIGURE 12 Complete RCQ kit of parts: CQ shell, hinge attachment, four panel types, two hold sizes, and a compatible task/ambient light.

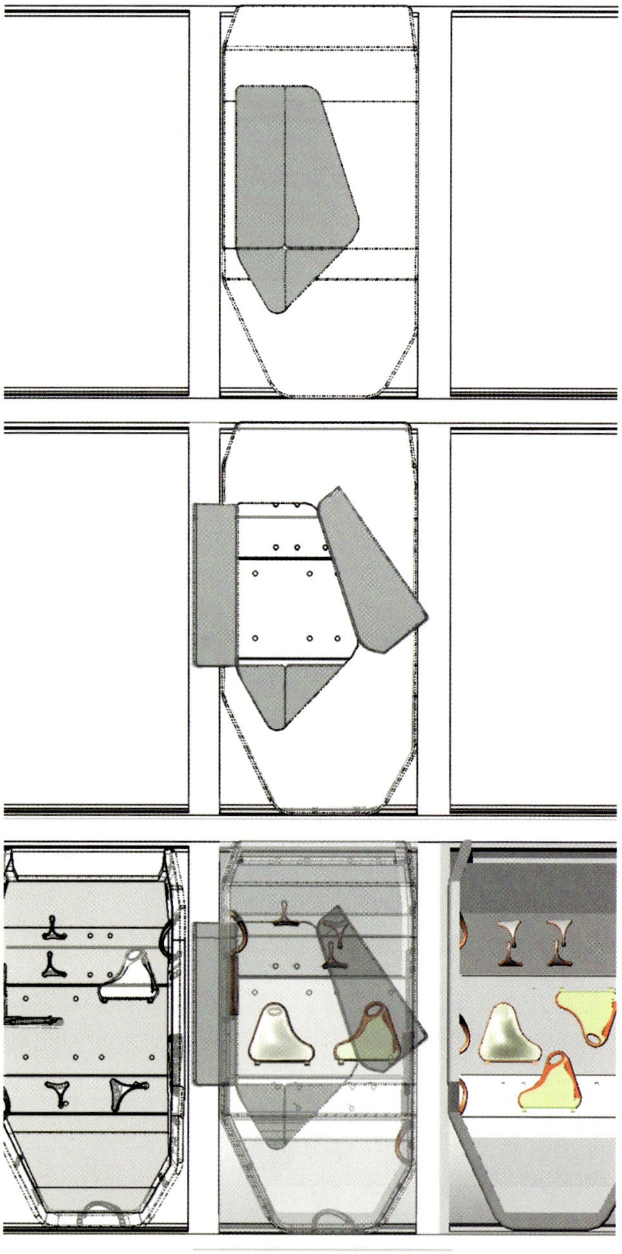

FIGURE 13 RCQ system shown in stages of outfitting: top—the shell in the rack volume; middle—RCQ with doors open but without parts; and bottom—various kit-of-parts combinations.

holds that offer static hook/grasp points to facilitate body position and motion control. The holds offer temporary storage, like hooks on a bathroom door; the variety of arcs accommodates a range of object shapes and sizes by using one hold or wedging objects between two or more holds.

System in Use

The integrated RCQ kit-of-parts system includes multiple panel and hold components (Figure 12). A light that uses its own connection modality is included in the outfitting for completeness. This simple kit can be used in various combinations to meet functional requirements while still allowing a crew member to personalize the living environment.

Figure 13 illustrates the system in stages. Assembly begins with the deployable shell, whose wall sections are sized for convenient breakdown, packaging, and transport. On orbit the shell walls slot together for assembly with the completed enclosure attached to the rack hardware and secured in place. The shell is preoutfitted with the standard, embedded hinge

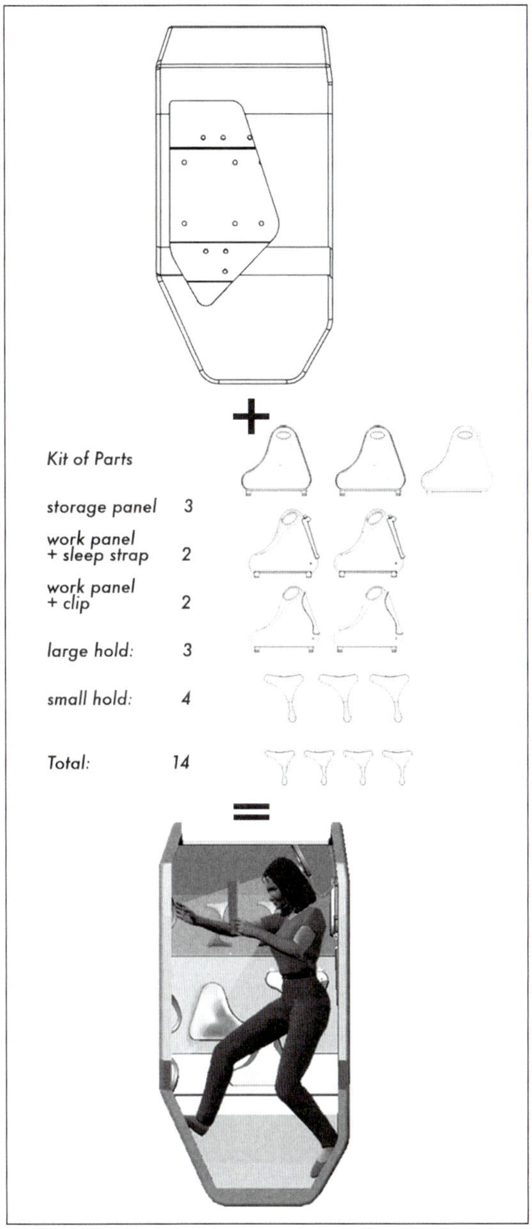

FIGURE 14 RCQ outfitting example 1, with parts list. Crew member in active reading position, braced between the lower surfaces of the shell.

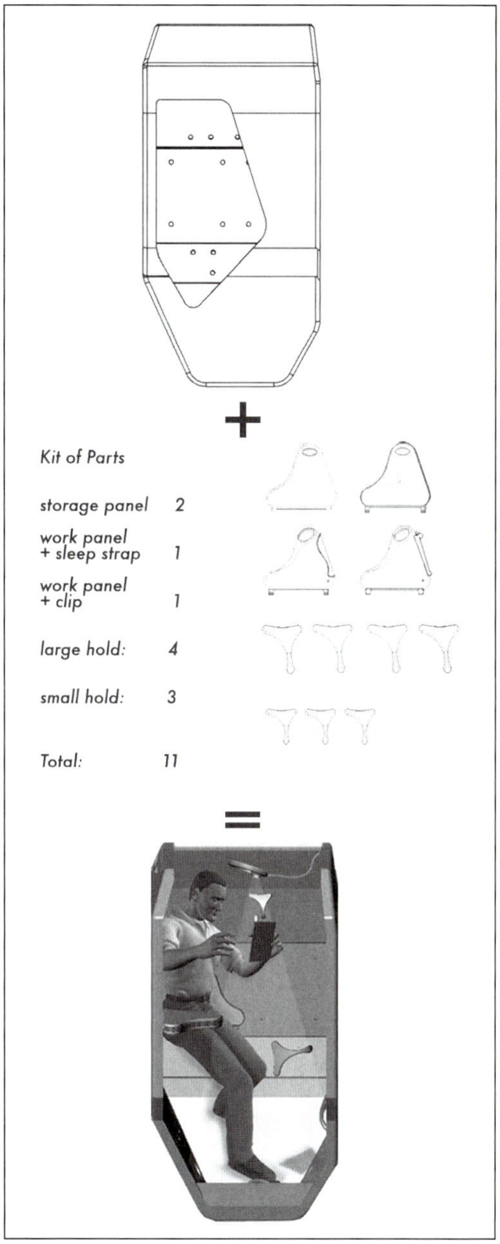

FIGURE 15 RCQ outfitting example 2, with parts list. Crew-member reading in a passive position, wedged in the corner.

connections. In keeping with kit-of-parts theory, once a standard connection is used, the possibilities are limitless for variations in form, scale, appearance, and the numbers of parts. Figures 14 and 15 show examples of personal outfitting arrangements and the parts lists required to configure them.

OTHER APPLICATIONS

The proposed system—a shell and "kit of parts"—was developed for a single set of use requirements: crew quarters in an ISS module. Yet it is plausible the same approach could be utilized for other ISS habitability purposes, for example, sick-bay medical or psychological consultation, library, workshop/studio, music booth, on-line education room, or movie screening booth. And the kit of parts itself could be widely applicable throughout the station.

ISS might have a long and interesting life, ultimately hosting a variety of government and private uses beyond the conduct of microgravity science that is its prime mission now. Is it too late to improve ISS habitability through economical design as the future unfolds? The simple, modular crew-systems design approach proposed here might offer a viable response. The system of holds and panels is a flexible, easy-to-use, fun, intuitive, visually minimalist, and nonrestrictive way to control position. The attachment array is informed by microgravity anthropometry, reach envelope, and a balance between flexibility of use, economy of parts, and aesthetics. Various arrangements can offer both passive and active restraint, depending on individual preference or activity. Arrangements intended for one use can easily also accommodate others. Because the design is compatible with the ISS interior architecture, it allows adaptation over time and with changing requirements. |

References

Adams, C. M. (1998), "Defin(Design)ing the Human Domain: The Process of Architectural Integration in Long-Duration Space Facilities," *Proceedings of the 28th International Conference on Environmental Systems*, Society of Automotive Engineers, Warrendale, PA pp. 992–999; also SAE paper 98–1789.

Ambasz, E. (ed.) (1972), *Italy: The New Domestic Landscape: Achievements and Problems of Italian Design*, Museum of Modern Art, New York, in collaboration with Centro Di, Florence, pp. 268–275.

Bedini, D., and Perino, M. (1999), "Space Architecture for Human Systems," *Keys to Space: An Interdisciplinary Approach to Space Studies,* International Space Univ., McGraw–Hill, Boston, Sec. 6.3, p. 6.17.

Clearwater, Y. (1985), "A Human Place in Outer Space," *Psychology Today*, July 1985, pp. 34–43.

Howe, A. S., Ishii, I., and Yoshida, T. (1999), "Kit-of-Parts: A Review of Object-Oriented Construction Techniques," *Proceedings of the International Symposium on Automation and Robotics in Construction,* International Assoc. for Automation and Robotics in Construction, London.

Messerschmid, E., and Bertrand, R. (1999), *Space Stations: Systems and Utilization*, translated by T. Freyer, Springer, Berlin.

NASA (1999), "End Item Specification (EIS) for Crew Quarters," Rev. NC, 17 Feb. 1999, International Space Station Program SSP 50356, Sec. 3.1, p. 8.

NASA (1995), "Man-Systems Integration Standards," NASA-STD-3000, Vol. 1, Rev. B, Secs. 3.0, 3.2–3.3.

NASAexplores (2001), "Make Room for One More Astronaut," 13 Dec. 2001 www.nasaexplores.com/lessons/01-086/9-12_index.html [retrieved Feb. 2001].

Rashid, K. (2001) www.canadianinteriors.com/kmj01ns.html. [retrieved 11 Nov. 2001].

7

alternative space-station module interiors

DAVID NIXON
AND
JUN OKUSHI

INTRODUCTION

THE NASA SPACE STATION REFERENCE DESIGN of 1984 provided the original guidelines for interior design of the pressurized modules for a permanent U.S. space station—what would become the ISS a decade later. In 1984, NASA proposed two habitability modules, one for day-shift and the other for night-shift activities. The guidelines showed an interior layout for habitability and laboratory modules based on a central access corridor running the length of each module, with modular racks and compartments lining the corridor on each side similar to the arrangement inside the European Space Agency (ESA) *Spacelab* modules in use at the time on space-shuttle missions. As space-station development moved into Phase B, NASA contractors studied several alternative interior designs in which they arranged corridors, aisles, racks, and compartments in different configurations. By this time, NASA had combined the two habitability modules into one for cost reasons, resulting in the need for an interior configuration with maximum rack and compartment capacity. A design emerged from the contractor studies, called the "four standoff" configuration, that offered maximum capacity and simplicity. It comprised a central corridor with a square cross section. Racks and compartments of repetitive shape and size lined all four corridor sides down the module with little variation. It still bore similarities to ESA's utilitarian *Spacelab* design.

The Southern California Institute of Architecture (SCI-Arc) began a study project for NASA Ames Research Center (ARC) in 1986 at the time the four standoff configuration had just been selected. Although the four standoff design resolved functional requirements very effectively, it was less desirable in terms of long-duration crew habitability (not a problem for *Spacelab*) and less successful in terms of human-factors potential. In pursuit of maximum capacity, efficiency and economy, the design had overlooked architectural quality and variety. The Aerospace Human Factors Division at ARC decided to commission studies to investigate ways of improving living standards inside the modules, which led to the SCI-Arc/ARC project. The study was conducted between 1985 and 1988, performed by a student/faculty team at SCI-Arc under a cooperative agreement with ARC. Much of the reference material used in the project was drawn from NASA and contractor progress documents of the time. The project is fully described in two NASA reports (Nixon 1986; Nixon et al. 1989).

PROJECT SEQUENCE

The project consisted of three separate and consecutive phases. Figure 1 shows the project sequence. Phase 1 occurred from 1985 to 1986 and comprised the development of nine different concepts for the habitability module interior. Each student team member produced a design with plans, sections, and a scale model. Review of the nine concepts took place with input from NASA at the end of Phase 1. Phase 2 took place from 1986 to 1987 and comprised the development of three different concepts for the habitability module interior. Students worked in

Copyright © 2008 by the authors. Published by the American Institute of Aeronautics and Astronautics, Inc., with permission.

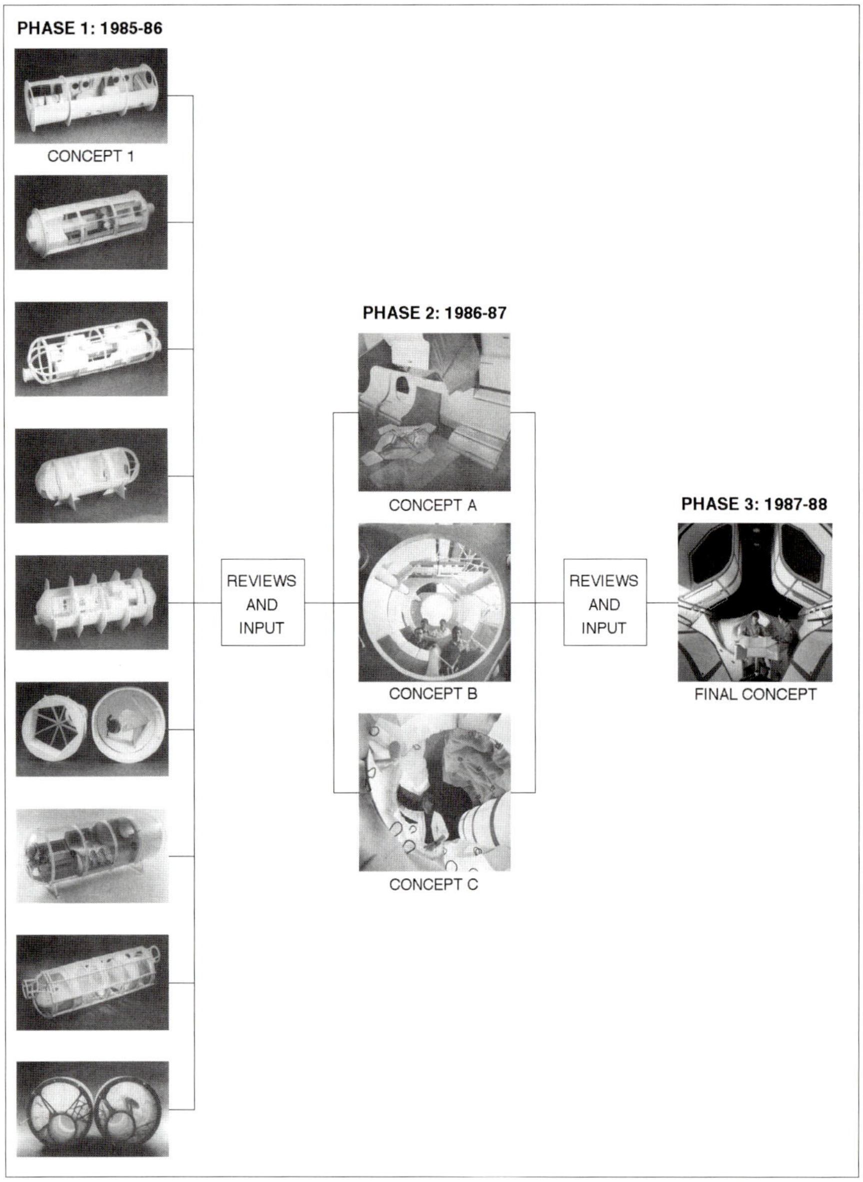

FIGURE 1 Project sequence downselected in three phases.

teams with each team producing plans, sections, and a full-scale mock-up of part of the module's interior. Review of the three concepts took place, again with input from NASA, at the end of Phase 2. Phase 3 occurred from 1987 to 1988 and comprised the development of a single concept for the wardroom portion of the habitability module. The project concluded with a NASA and industry review at the end of Phase 3.

Phase 1 Design

Phase 1 began with research on habitability module accommodation requirements derived from anticipated crew activities: meetings and teleconferences, planning and training, relaxation and entertainment, eating and drinking, food preparation and cooking, exercises and games, housekeeping and hygiene, space-station operations, library and study, and shift and crew handovers.

The accommodation requirements identified all major equipment and outfitting items to be incorporated in the habitability module. Work continued on the development of design guidelines, covering crew timelines and activity sequences, activity proximities and compatibilities, and individual and group ergonomics.

The Phase 1 research provided the necessary information for the Phase 1 design, which developed nine individual concepts for the interior. The aim was to propose and test alternative design approaches based on individual interpretations of the requirements and guidelines related to volumetric constraints. The concepts ranged substantially in character, from conventional and fixed configurations with dedicated activity volumes, to experimental and multipurpose configurations with adaptable activity volumes. Each concept comprised longitudinal and transverse sections through the module and a scale model with exterior skin cut away to show interior arrangement.

Phase 1 Review

Review of the Phase 1 concepts evaluated each concept using a standard design analysis sheet developed for the purpose. The analysis process utilized 10 design factors. Each factor addressed a key issue essential for consideration at a conceptual level. Together, the design factors provided a comprehensive means of comparing and scoring the design concepts at this early stage of design development. The design factors were

1) Communal organization
2) Spatial perception
3) Internal circulation

4) Compartment adaptation
5) On-orbit completion
6) Life-cycle modification
7) Ergonomic utilization
8) Exterior observation
9) Equipment rationalization
10) Structural inspection

Figure 2 shows a typical design analysis sheet for one of the concepts. The wide central column contains comments on how the design resolved each of the 10 factors. The intermediate column on the right indicates whether the resolutions resulted in a significant advantage, disadvantage, or neither. The far-right column scored the resolution on a five-point rating scale, where 1 was optimum and 5 was minimal. In the concept example shown in Figure 2, the design optimally resolved spatial perception, internal circulation, and exterior observation but minimally resolved compartment adaptation. Table 1 tabulates key aspects of the review results for all nine concepts.

Phase 1 Results

The Phase 1 research, design, and review stages yielded some major conclusions and recommendations:

CONCLUSIONS

1) Highly adaptable configurations perform effectively in responding to day-to-day activities and routines.
2) Crew perception and physical movement benefit from horizontal ("chocolate éclair") rather than vertical ("boudin slice") configurations.
3) Orbital assembly, life-cycle modification, and hull inspection are related design issues.

RECOMMENDATIONS FOR ELEMENT AND EQUIPMENT

1) Compactness and miniaturization are needed to minimize volumetric allocation and maximize available habitable volume.
2) Multifunctionality and versatility are needed to minimize performance inflexibility and maximize mass- and cost-efficiency of hardware.

DESIGNER:	CONFIGURATION:	CONCEPT:		
DESIGNER:	CONFIGURATION:	CONCEPT:		
DESIGNER:	CONFIGURATION:	CONCEPT:		
DESIGNER:	CONFIGURATION:	CONCEPT:		
DESIGNER:	CONFIGURATION:	CONCEPT:		
DESIGNER: JUN OKUSHI	CONFIGURATION: HORIZONTAL	CONCEPT: 9	BENEFIT	RESOLUTION 1 2 3 4 5
COMMUNAL ORGANIZATION	Activity areas clustered along circulation spine. Activities separated into communal wardroom (active) functions and semi-private library/ workstation (passive) functions.		•	○●○○○
SPATIAL PERCEPTION	Direct utilization of anthropometric geometries and movement patterns in developing activity area configurations achieves interesting and exciting spatial environment.		A	●○○○○
INTERNAL CIRCULATION	Direct perimeter circulation path from module end to end.		A	●○○○○
COMPARTMENT ADAPTATION	Compartmental adaptability not clearly defined and requires extensive development. Extensive adaptability unlikely to be realized due to nature of concept.		D	○○○○●
ON-ORBIT COMPLETION	Internal skeletal and enclosure elements capable of on-orbit completion. Divisibility and itemization of internal configuration requires examination.		•	○○●○○
LIFE-CYCLE MODIFICATION	Internal enclosure elements and equipment may not be capable of life-cycle modification. Nature of life-cycle changes requires substantial clarification.		•	○○●○○
ERGONOMIC UTILIZATION	Considerable potential for effective ergonomic utilization of interior envelope elements and equipment. Requries further development.		•	○○●○○
EXTERIOR OBSERVATION	Windows potentially free of obstructions. Choice of window location fairly extensive. 360° anthropometric rotation requires windows clear of internal structure.		A	●○○○○
EQUIPMENT RATIONALIZATION	The unique nature of interior configuration combined with specialized approach to design of structure and envelope linings substantially limits possibility of rationalization.		•	○○●○○
STRUCTURAL INSPECTION	Elements and equipment could be designed to be detachable from module shell. Free-form communal area would aid accessibility - structural members may reduce it.		•	○○●○○

A = ADVANTAGE B = DISADVANTAGE 1 = OPTIMUM 2 = ACCEPTABLE 3 = AVERAGE 4 = DEFICIENT 5 = MINIMAL

FIGURE 2 Example of analysis and evaluation sheet used in Phase 1 review.

TABLE 1 Summary of Phase 1 concept features across the nine concepts

CONCEPT	ORIENTATION	SPATIAL PERCEPTION	CIRCULATION	MODULATION	FEATURED ELEMENTS
1	Constant horizontal	Multicolor scheme with flexibility of window arrangements	End-to-end overhead translation path	Compartment outfitting and storage system	Wardroom table
2	Constant horizontal	Transparent and enhanced greenhouse wall	End-to-end underfoot translation path	Utilizes platonic solids	Greenhouse and workstation
3	Multi-axial interfaces	Solid and void enhanced interplay	Central spine path	Anthropometrics/ ergonomics driven	Series of dedicated singular function zones
4	Constant horizontal	Generous configuration of internal volumes	Segmental translation	None (overall architectural configuration)	Multifunctional table/ workstation
5	Multiple vertical	Variable volume via retractable elements	Off-centric spine path	Variable (sliding bulkheads)	sliding bulkheads
6	Multi-axial interfaces	Transformable volume achieved by retractable racks	Random	Variable	Standoff configuration
7	Constant horizontal	Large enhanced curvilinear elements	Segmental translation	Interlocking curvilinear elements	Large curvilinear elements
8	Multi-axial interfaces	Enhanced freedom	Segmental translation	None (large bulkheads configuration)	Free-form membrane soft-surface volume envelope
9	Constant horizontal	Three-way enhanced-curve cellular volume	End-to-end underfoot translation path	None (large bulkheads configuration)	Free-form membrane soft-surface volume envelope

3) Ergonomic efficiency and user friendliness are needed to minimize operational inconvenience and maximize user comfort.
4) Autonomy and self-containment are needed to minimize systems interdependence and maximize individual functional durability.

Phase 2 Design

Phase 2 began with definition of objectives:

1) Simulate and evaluate the physical form and environmental characteristics of the wardroom and its constituent elements and equipment.
2) Generate and experiment with innovative architectural/industrial design alternatives for potential incorporation in full-scale mock-ups.
3) Obtain experience in design and construction of full-scale mock-ups.
4) Apply anthropometric and group ergonomic design criteria to architectural interior configurations.

For cost and size reasons, Phase 2 focused on a portion of the habitability module that contained the wardroom and three full-scale configuration concepts. The first task fabricated a shell to simulate a portion of the habitability module sufficiently long to incorporate the wardroom. Shell dimensions were 2134 mm (84 inches) radius by 4877 mm (192 inches) long. The shell was open at each end. The three configuration concepts were the following:

- **Concept A**

  Approach: Four perimeter standoff spines that provided attachment and support for deployable and interchangeable modular racks, compartments, and ergonomically adaptable workstations.

  Features: Two crew workstations, one wardroom table, two personal hygiene units, two library/study compartments, two life-support equipment (ECLSS) units, eight fold-out sleeping compartments.

- **Concept B**

  Approach: Triangulated core with central access corridor and three structural/utility spines providing support and attachment for specific and interchangeable modular elements and equipment.

Features: Two galley food preparation stations, two galley hygiene stations, two "greenhouse" units (gloveboxes), one soft "storewall" unit, one wardroom meeting table, radial storage compartments.

- **Concept C**

  Approach: An accessible, off-center utility route and modular, curved-geometry racks and compartments providing anthropometrically responsive, soft interior fascias for crew station functions.

  Features: Radial contoured racks/elements continuous, modular utility spine.

Phase 2 Review

Review of the Phase 2 concepts involved the evaluation of each concept using a standard design analysis sheet developed for the purpose, similar to the one used in Phase 1. The analysis process utilized 57 design factors organized into nine groups. Each factor addressed a key issue essential for design consideration. Together, the design factors provided a comprehensive means of comparing the three full-scale wardroom concepts. The nine design factor groups were as follows:

1) Architectural concept
2) Utility systems
3) Architectural subsystems
4) Perceptual quality
5) Ergonomics
6) Wardroom activities
7) Associated features
8) Orientation/translation
9) Crew group uses

Figure 3 shows a typical design analysis sheet for one of the nine design factor groups for one of the concepts. The nine design factor groups are listed in the top horizontal tabs. This example is for the utility systems factors group. The left column contains the utility systems design factors. The four identical columns on the right show the reviewers' evaluations using a five-point rating, where 1 is optimum and 5 is minimal. In the concept example shown, the design optimally resolved primary utility cores and utility systems distribution, but minimally resolved utility systems attachments and pressure wall access.

Phase 2 Results

Phase 2 concluded with a summary of the most successful design features of the three full-scale concepts

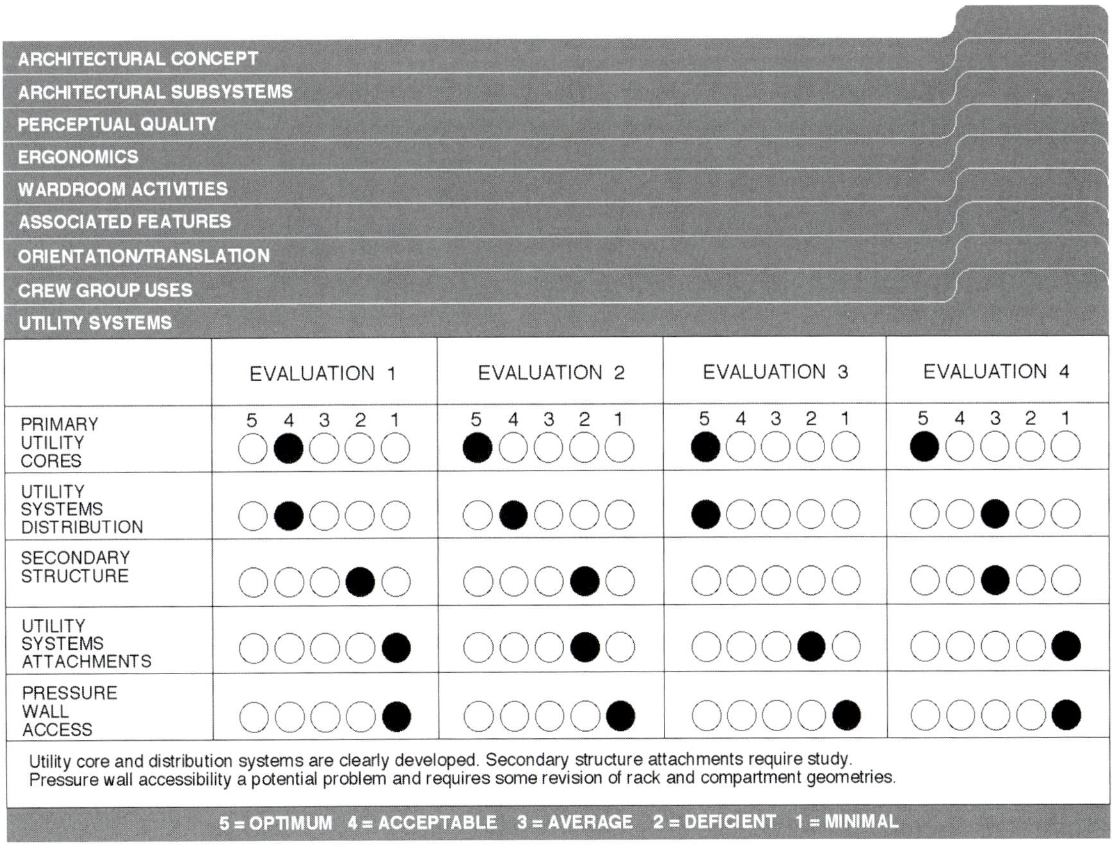

FIGURE 3 Project sequence and data sources.

under their respective design factors. Chief among these were the following:

- Architectural concept: Two levels of crew accommodation and activity, and functionally dynamic racks and compartments, can make the most of a limited internal volume.
- Utility systems: Different ways of incorporating intramodule utility routes can include perimeter utility ducts, central utility spines, and crew-accessible utility tunnels.
- Architectural subsystems: A single module of fixed shape and size can accommodate different rack and compartment geometries, increments, and attachment methods.
- Perceptual quality: Fold-away compartments and variation of forms, surfaces, lighting, and textures can improve the sense of interior spaciousness.
- Ergonomics: Interfaces between crew members and enclosures, consoles, and surfaces must be responsive to crew anthropometric variables.
- Wardroom activities: The wardroom must accommodate crew groups of different size, engaged in different types of activities, at different times of day.
- Associated features: Compartments and equipment items with deployable/retractable operational capabilities can fold and stow away when not in use to recover valuable volume.
- Orientation/translation: Well-defined orientation and translation routes inside the module can become integral and positive features of the architectural concept.
- Crew group uses: Internal configurations must be able to accommodate a range of simultaneous activities with appropriate community/privacy gradients.

Phase 3 Design

Phase 3 began with definition of objectives:

1) Consolidate and continue research into space-station wardroom habitability based on relevant criteria drawn from previous or parallel programs or studies.
2) Define and develop a feasible and innovative architectural/industrial design proposal for the configuration of the crew wardroom in the habitability module.
3) Contribute to the space-station design effort by providing a life-size wardroom mock-up for use by NASA as a research tool for continuing habitability studies.

Phase 3 concentrated on design and development of a single concept for the wardroom constructed as a full-size, medium-fidelity mock-up and based on a synthesis of criteria drawn from five principal sources: 1) research program requirements determined by the ARC Aerospace Human Factors Research Division; 2) appropriate recommendations derived from concepts developed in Phase 1 of the project; 3) selected architectural and industrial design features drawn from concepts developed in Phase 2 of the project; 4) selected architectural and industrial design features drawn from concepts developed at the beginning of Phase 3; and 5) appropriate data drawn from NASA space-station contractor team studies during Phase B (definition and preliminary design).

The synthesis process and its position in the project sequence are shown in Figure 4. The synthesized data resulted in formulation of 20 major design guidelines for the development of the final concept, which illustrates a typical range of design considerations for module-based space architecture:

1) Habitability module 4.2 m (166 in.) internal diameter and 11.8 m (464 in.) effective length
2) Eight-person, dual-shift crew organization
3) Double-height/dual-level module accommodation configuration
4) Compliance with Phase B rack and compartment outfitting inventory
5) Definitive configuration organization and activity adjacencies
6) Feasible life-cycle modification and reconfiguration
7) Flexible/modular rack and compartment longitudinal fit
8) Adequate free wardroom volume for large crew group uses
9) Clear module translation route and horizontal cueing
10) Distinctive perceptual quality of interior environment
11) Variable decor/finishes within interior environment
12) Rationalized ECLSS and utilities systems distribution
13) Reduced number of full-depth structural standoffs
14) Improved functional and operational structural standoff design
15) Exercise compartments and galley food preparation facilities
16) Planning/station operations and window/observation workstations
17) Deployable/retractable dedicated crew activity compartments
18) Advanced microgravity anthropometrics and ergonomics features

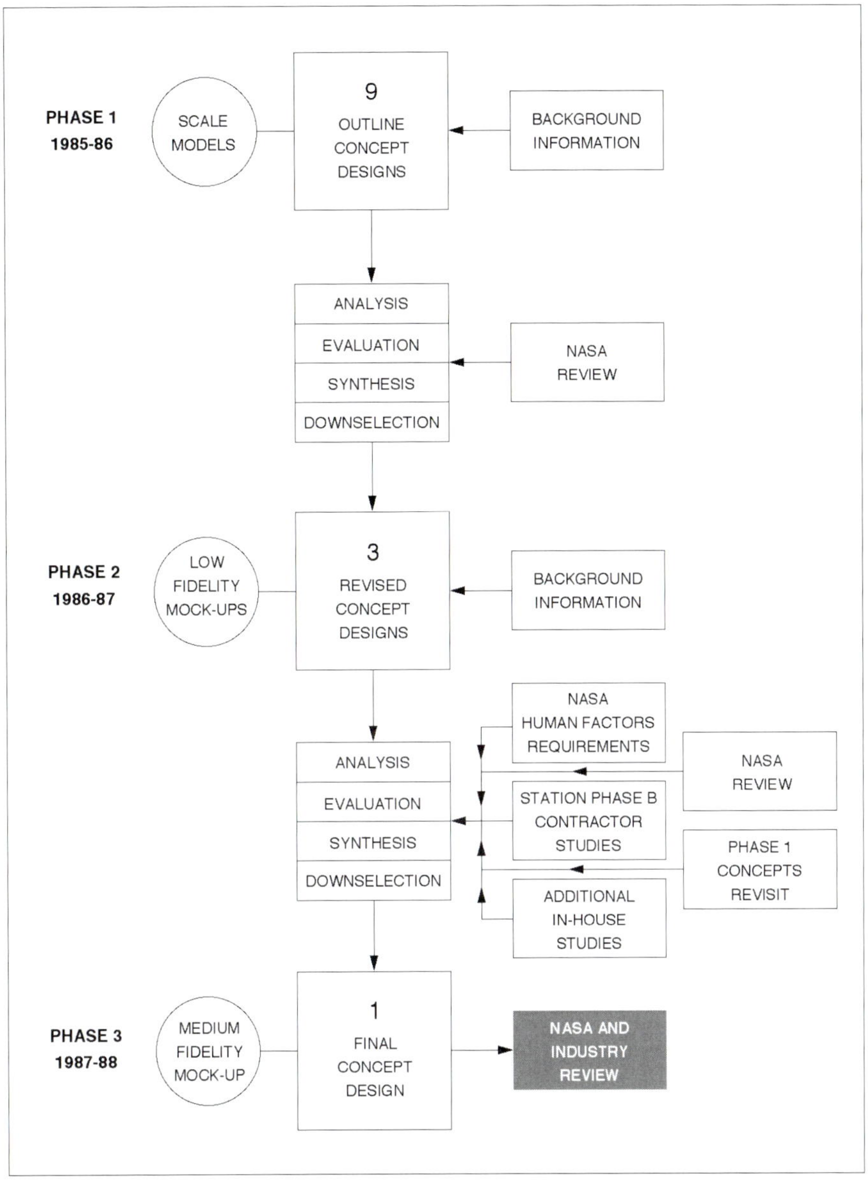

FIGURE 4 Phase 2 analysis and evaluation of three design options.

19) Adaptable/extendable wardroom table and soft stowage system
20) Folding/enclosing workstation operations techniques

Final Concept Mock-up

A full description of the final concept design and mock-up is given in the second of two NASA reports on the project (Nixon et al. 1989). What follows is a brief description. The mock-up was approximately 50% of the habitability module length and focused on wardroom, galley, and exercise facility (Figure 5). The main features incorporated in the mock-up were two exercise compartments, one command-and-control workstation, two window workstations, one soft-stowage bag system, one wardroom table, four passive body restraints, four galley racks, six equipment racks, and a lighting system.

Phase 3 Results

NASA and aerospace industry representatives carried out a review of the mock-up at the end of Phase 3, following a presentation by the Phase 3 team. The reviewers did not use analysis sheets for the review; instead, they made comments directly to the team at the presentation.

Phase 3 ended with a series of conclusions and recommendations, including the following:

- Life-cycle modification: Life-cycle reconfiguration and upgrading options are constrained by initial accommodation, standoff, and utilities design.
- Organization and zoning: A dedicated buffer zone separating day and night accommodation increases noise attenuation and improves personal privacy.
- Architectural configuration: Dual-level configurations improve operational and translational efficiency and generate enhanced perceptual interest.
- Standoff structural systems: Demountable standoff structure contributes to reduced physical obstruction and simpler on-orbit modification.
- Utilities distribution systems: Variable-depth standoff structure contributes to rationalized utilities distribution and improved systems accessibility.
- Rack and compartment sizes: Variable-width racks and compartments contribute to improved organizational versatility and operational performance.
- Rack and compartment functions: Deployable/retractable compartments provide valuable additional free volume and improved occupant performance.
- Crew equipment features: Adaptable and conformable crew equipment features improve workstation ergonomics and facilitate routine tasks.

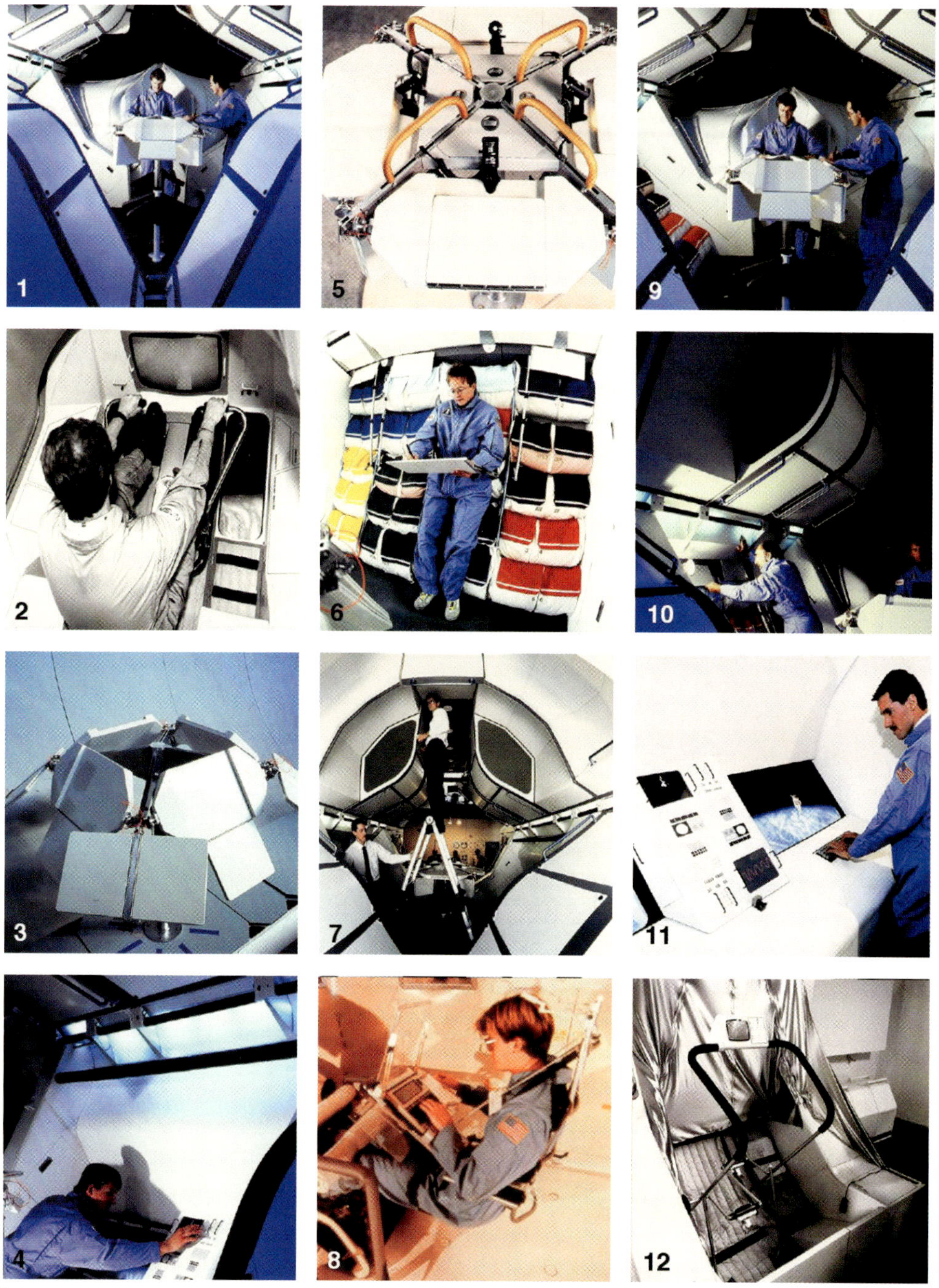

FIGURE 5 Final concept mock-up: 1) view through two equipment racks into wardroom area; 2) rowing machine in exercise compartment; 3) wardroom table with work surfaces fully angled; 4) adjustable lighting mounted onto spines; 5) wardroom table prototype with work surfaces stowed; 6) soft-stowage bag system; 7) view along module with galley in top foreground; 8) wearable workstation prototype on NASA weightlessness flight test; 9) central wardroom area with table; 10) curvilinear equipment racks in wardroom; 11) window workstation with flat screen; and 12) bicycle ergometer in exercise compartment.

CONCLUSIONS AND LESSONS LEARNED

The design outcome of the SCI-Arc/ARC study is of historical interest as a record of an approach to optimizing the interior architecture of a long-term, livable environment inside a highly constrained volume. The ISS design itself is a *fait accompli* based on the four standoff architecture and constructed in orbit over more than a decade beginning in 1998 and described in detail in Chapter 4. United States human space endeavors have moved on to other goals, like replacing the shuttle, returning to the Moon, and making progress toward eventual human missions to Mars. Hence, the cycle of design development begins again for major habitable module systems. What main lessons learned from the SCI-Arc/ARC study could benefit new cycles of design effort?

The first lesson is not about design decisions taken and their architectural outcome in response to a particular design problem. Rather, it is about the value of systematic design inquiry using a series of measured steps of increasing fidelity, accompanied by reviews, to distill and refine a single end product from an initial group of ideas. Applying the same systematic, step-by-step approach to the development of the interior architectural configuration of the space-station habitability module during Phase B in the 1980s could have resulted in design improvements to the livability of the four standoff approach without loss of functional efficiency. Figure 6 shows the five major module interior concepts (four standoff, dual-level corridor, central core wall, double side wall, offset core wall) at the time of space-station Phase B, as well as the contoured form design produced by the SCI-Arc/ARC study.

It is evident from the concepts that a considerable variety of internal architecture solutions is possible within a common module geometry and size. It is also clear that from the standpoints of crew anthropometry, ergonomics, perception, and accommodation quality, several of the concepts were superior to the four standoff design. For example, in the contoured form the standoffs are rotated to the four cardinal points in cross section, with the majority of utilities grouped through two "wall" standoffs, enabling the "floor" and "ceiling" standoffs to be reduced in depth. This in turn permits introduction of an upper "loft" level of accommodation and an increase in the free volume width across the module diameter. The result is greater internal spaciousness, first in the critical upper body zone where physical distance and longer sightlines between crew members occur from side to side across the module, and second in an upper crew translation and movement route that by-passes group activities below and avoids conflict with them, as shown in Figure 7. The other design concepts can claim design ideas of equal merit, but these were not put to the test by means of a systematic design inquiry; upon early decision of the four standoff configuration, they were dropped.

The second lesson is about the ability of a module interior to adapt to new requirements throughout its lifetime in response to new operational conditions or circumstances. This was of concern in the SCI-Arc/ARC study. The contoured form demonstrates high potential for adaptability because from the outset it was conceived as an irregular and asymmetrical configuration unconfined by the rigid, modular geometry of the four standoff approach. The essence of the benefit is reduced dominance of the functional standoffs, permitting introduction of a variety of racks, compartments, and linings that can be rearranged or changed out during the module's life cycle to create new architectural interiors as desirable or necessary. This remains an important consideration for the future of the ISS. Early in design development, module interiors were outfitted to the fullest

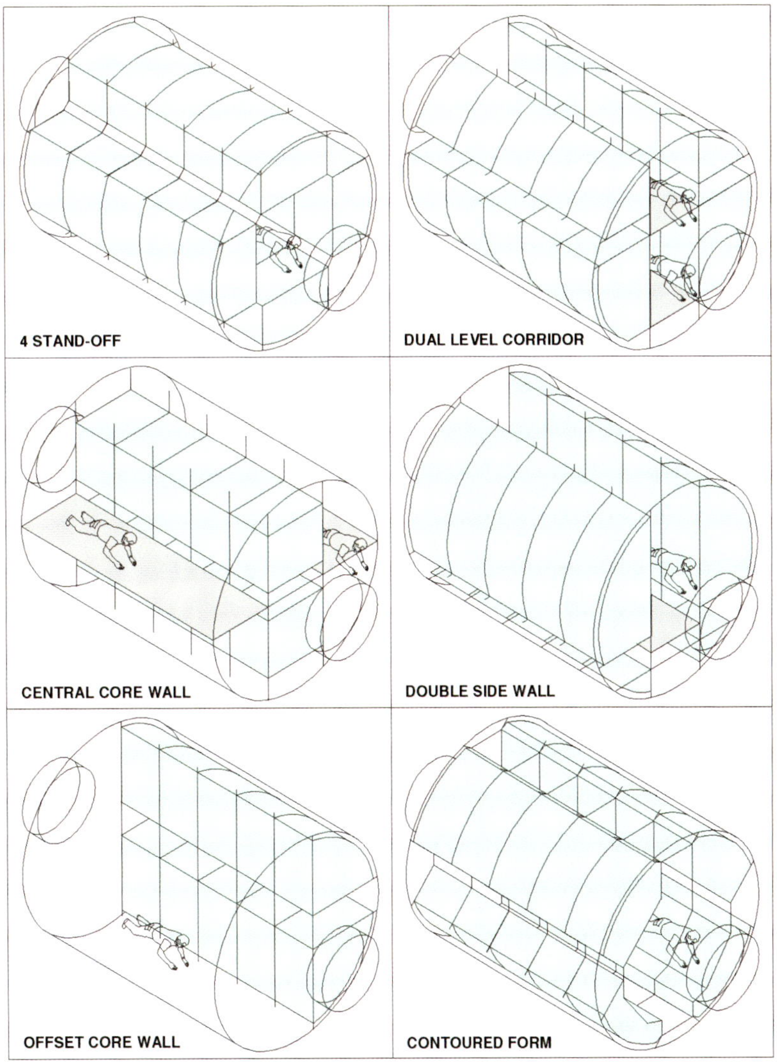

FIGURE 6 Six archetypal interior configurations of a space-station module.

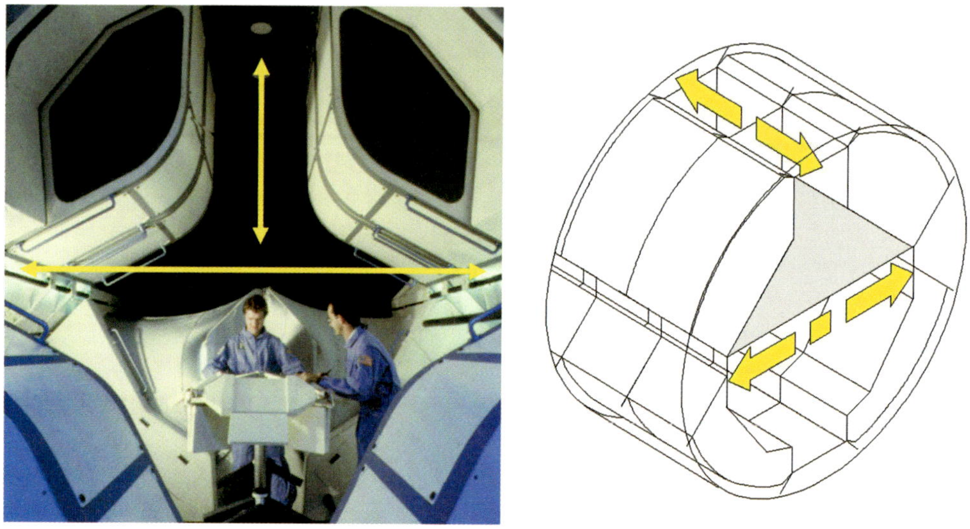

FIGURE 7 Nonaxisymmetric layout contributes to spaciousness and separates circulation from work activities.

extent possible with racks as part of the four standoff approach. This was necessary as the initial reduction of two habitability modules to one, and then the elimination of them entirely, meant the remaining modules had to be outfitted to full rack capacity.

Today, future ISS utilization and the precise nature of activities that will take place onboard are an open question. NASA's post-ISS focus beyond Earth orbit, combined with Russian receptiveness to nonscience uses and the emergence of space tourism as a vibrant market, suggest the ISS future might be quite different from what was originally intended by government planners. The ISS partners might wholly or partly privatize or commercialize the station, to offset life-cycle operations costs to enable new projects, or to generate revenue from market-orientated applications in response to growing market interest. The result could be a need to remodel module interiors for other applications through elimination of redundant racks and their replacement by quite different equipment and outfitting.

The third lesson is about the value of building flexible, full-scale mock-ups of design concepts to analyze and evaluate their advantages and disadvantages at "hands-on" and "walk-through" scale during the design decision process. Full-scale mock-ups are often used by the aerospace industry, and several were built by NASA and its contractors during space-station phases B and C/D to assess module interiors. These ranged from low-fidelity versions fabricated from foam board to medium-to-high fidelity versions fabricated from aluminum. In many cases, they were built to demonstrate design and engineering solutions already defined, rather than as a tool to help analyze and evaluate different design concepts before the solutions were defined. The SCI-Arc/ARC mock-up was different. It was built from a kit of parts. The cylindrical module shell comprised a series of identical modular elements to enable the mock-up to be reconfigured, lengthened, shortened, dismantled, or moved to a new location. The elements were designed and sized to enable manual construction of the mock-up using a simple elevated working platform, without need of a crane or lifting tackle. The mock-up shell could be assembled by four people. Elements forming the lower portion of the mock-up, required to be robust to take live floor loads, were mobile and easily moved across a flat floor by two people. All module elements were sized to fit on flatbed trucks or inside shipping containers. This modular and mobile approach allowed the mock-up to be moved three times during the life of the project—first from SCI-Arc to a new warehouse location in the Los Angeles area for final assembly and review, second from Los Angeles to NASA Johnson Space Center in Houston for display, and finally from Johnson to ARC at Moffett Field, California, for long-term storage. |

References

Nixon, D. (1986), "Space Station Group Activities Habitability Module Study," NASA CR-4010, www.spacearchitect.org.

Nixon, D., Miller, C., and Fauquet, R. (1989), "Space Station Wardroom Habitability and Equipment Study," NASA CR-4246, Dec., www.spacearchitect.org.

INTRODUCTION

AS OLD AS ARCHITECTURE ITSELF, fabric structures have been interwoven throughout humankind's history of building. Nomadic early humans created portable housing as they followed migrating herds, their main resource. They used animal skins stretched over bones and tree limbs to create shelters. This type of habitat led to sewn-together hides combined with erectable structures for easier deployment and breakdown. Over millennia, yarns and fabrics were developed, further enhancing these tent structures. Robust, tensile fabric structures with dynamic, bold, sweeping shapes remain at the revolutionary forefront of modern architecture. So it might not be too surprising that a team of architects and engineers at NASA designed and tested applications of this ancient architectural approach as a way to create habitats for orbital and planet-surface use. This chapter describes the design of the TransHab space habitat module and tells the story of its development.

8

TransHab project

KRISS J. KENNEDY

HISTORY

NASA had considered tensile fabric structures in its past, having designed and tested several inflatable structures for space applications in the late 1960s. Langley Research Center led efforts to develop a 24-ft (7.3-m)-diam torus space station, prototype lunar stay-time-extension module, and large space-station module nicknamed Moby Dick. All were successfully tested. It took many years of persistence, and a few failures, before the textile industry introduced high-tech fibers like Kevlar®, Vectran®, and Polybenzoxazole (PBO). (Note: Throughout this chapter, trade names are used for identification, not to indicate NASA endorsement.)

Over the years, the idea of using inflatable structures for space habitats began to catch on. Several key studies, including the Exploration Synthesis Group in the early 1990s, identified inflatable structures as a technology that could allow NASA to accomplish lighter-weight structures at lower cost. NASA continued to refine ideas and concepts, preparing for an opportunity to prove the enabling utility of inflatable structures for advanced missions. Finally, a tiger team at NASA Johnson Space Center was challenged to design an interplanetary vehicle habitat for a crew of six to travel to and from Mars, using existing launch vehicles. The logical choice for meeting the volume requirement driven by crew activities, food, spares, and other provisions was an inflatable structure. Mars mission studies had already defined several types of habitats for transit (interplanetary travel) and surface mission phases. So when the team began design work, the author coined the name "TransHab" as a contraction of transit habitat. The nickname caught on and quickly became a widely known NASA technology brand.

TransHab pushed the technological envelope beyond previous design work on inflatables. The team of architects and engineers innovated a revolutionary concept that yielded a competitive alternative to aluminum hard-shell architecture by designing and testing prototypes in anticipation of rigorous technical review. Since 1997, the TransHab concept has undergone numerous design iterations. The culminating NASA design was a module derived from the Mars TransHab concept and proposed for addition to the International Space Station (ISS) (Figure 1).

TransHab is a hybrid space structure that synthesizes a rigid central core with an inflatable exterior shell. It is thus differentiated from all previously developed space habitats, which have traditionally used a hard external shell as both main structure and pressure vessel. Therefore TransHab technology is revolutionary both in overall concept and in the development of each of its primary components; it represents a leap from the exoskeletal archetype into a new generation of endoskeletal modules.

This material is declared a work of the U.S. Government and is not subject to copyright protection in the United States.

FIGURE 1 Module derived from the Mars TransHab concept and proposed for addition to the ISS.

Beyond its technological innovation—and in no small part because of it—TransHab is also breaking new ground in support of the human system. The process by which the structure involved human engineering from the early conceptual stages allowed it to achieve a unique level of efficiency as a human-rated space system. Its dimensioning and layout are optimized for flexibility and long-term use by a diverse crew. Because TransHab can be packaged into a smaller volume for launch and then deployed on orbit to provide a much larger volume, this module architecture offers both design opportunities and technical challenges.

Because of congressional budget action regarding ISS activities, the ISS TransHab module development was ultimately canceled. Systems integration and detailing of interior elements ceased, along with the aggressive testing program that had already proved the technology to consistently meet or exceed known requirements. The project successes up to the point of cancellation—unique technology, high level of habitability, and outstanding test record—are attributable to the work of a strongly integrated project team. Test engineers, structures and subsystem engineers, architects, and human-factors experts collaborated intensively from the project's outset. In the end NASA made the TransHab technology available for further development and use by the commercial space industry (see Chapter 9).

TRANSHAB ARCHITECTURE

The TransHab module proposed for the ISS (Figure 2) is approximately 35 ft (10.5 m) long overall. With 25 ft (7.3 m) internal diameter and 23 ft (7.0 m) of open interior length from bulkhead to bulkhead, it provides 12,077 ft^3 (342 m^3) of pressurized volume. Levels 1 and 3 are 8 ft (2.4 m) high at the central core, and level 2 is 7 ft (2.1 m) high at the core. The 8-ft ceiling

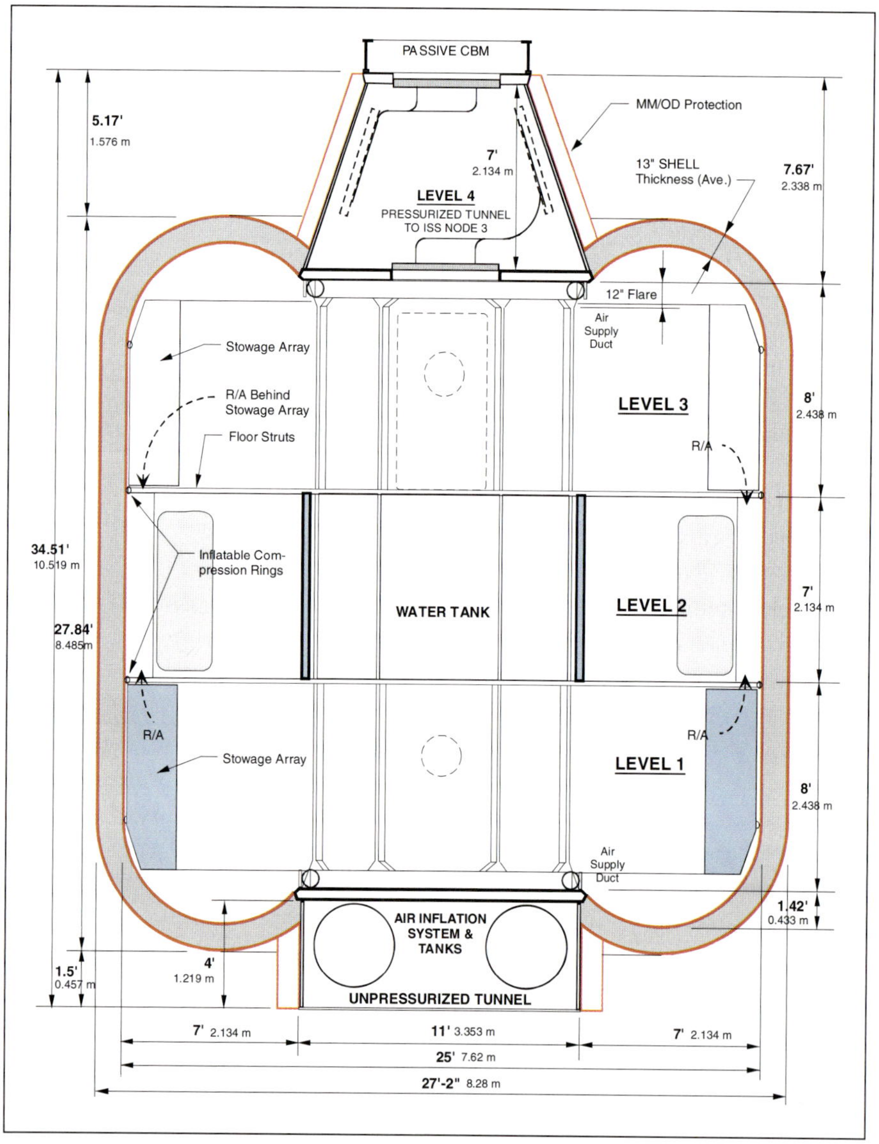

FIGURE 2 ISS TransHab section, showing elements of basic module architecture.

height of level 3 came from human-engineering analysis deriving the ceiling height requirement for crew members using a treadmill. The 7-ft ceiling height of level 2 was derived by combining the minimum head clearance for sleeping crew with room to lock in equipment shelves between floor struts.

The module would be packaged and folded on Earth and launched in the Space Shuttle Orbiter for delivery to ISS. After the orbiter docked with ISS, the module would be removed from the payload bay and berthed into place using the station's remote manipulator system (SSRMS). Once attached, the module shell would be deployed and then inflated to an internal operating pressure of 14.7 psia (101 kPa); during inflation, the air system would be activated to condition the environment prior to crew entry for outfitting. Several days would be required for assembly crew to activate all systems and complete preliminary outfitting. To the greatest extent possible, internal structure, utilities, and other complex systems would be preintegrated within the rigid central core.

The TransHab architecture provides a habitable environment that integrates both private and social spaces. This feature is very important for crew social and interpersonal relationships, particularly during long-duration confinement aboard a space station or interplanetary vehicle. Functional and physical separation of the crew health care area, crew quarters, and galley/wardroom area creates a home-like design for the crew living in space, while allowing each functional area to remain permanently deployed for regular use.

Figure 3 shows an overview of the ISS TransHab architecture. Given its larger volume compared to a traditional hard-shell module, TransHab provides more stowage volume, two unobstructed circulation paths within the module (central passageway in the core and forward-side passage, both large enough to pass an ISS rack), and permanently deployed equipment in the primary activity centers. Other important design objectives were maintaining a local vertical configuration, separating exercise and dining areas, and providing comfortable crew quarters.

The interior is divided into four functional levels: the first, second, and third are for living space, and the fourth is the connecting tunnel to the ISS hatch. Providing a consistent local vertical orientation in keeping with operational requirements established in all programs since *Skylab*, TransHab's architecture separates conflicting functions while enhancing the usability of each area. Level 1 is the galley/wardroom and soft stowage area. Level 2 houses the crew quarters between the core's water tanks and an enclosed mechanical room in a half-toroid of the outer area. Level 3 contains crew health care and more soft stowage.

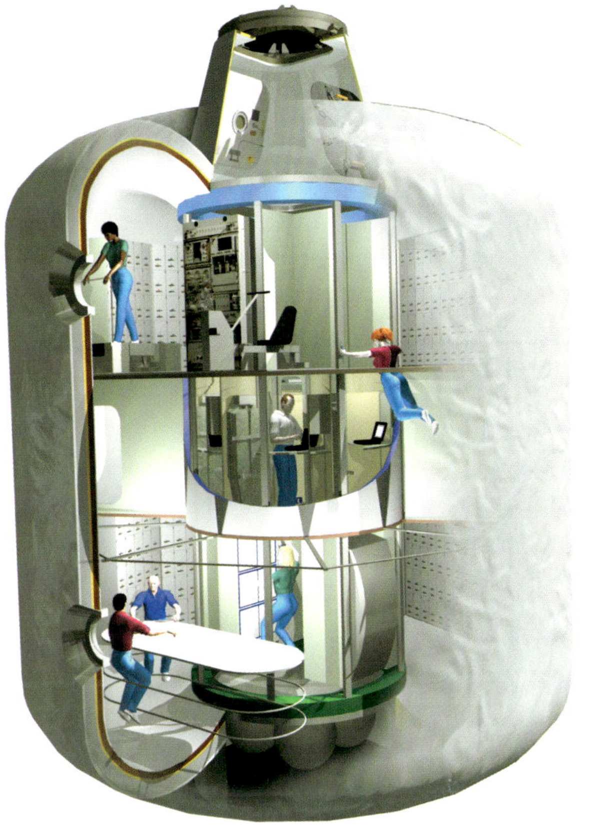

FIGURE 3 ISS TransHab internal view from bottom to top: level 1—galley/wardroom; level 2—crew quarters and mechanical room; level 3—crew health care; and level 4—pressurized tunnel to the ISS.

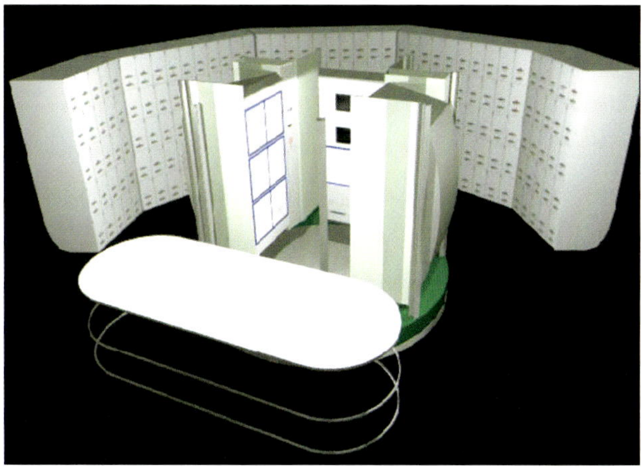

FIGURE 4 Double-height galley/wardroom shares level 1 with stowage.

Level 1 is the main social and professional meeting place (Figure 4). It incorporates the galley, wardroom, and a portion of the stowage array, and an Earth-viewing window. The galley incorporates a rack-based ISS galley and two rack-based ISS refrigerator/freezers installed after the module is activated. Designed to accommodate 12 people during an ISS crew changeover, the wardroom is a double-height room featuring a large table and Earth-viewing window. It is used for meals, meetings, conferences, daily planning, public relations gatherings, and socializing. Having one large common area in which to gather all crew members for important crew briefings and photo opportunities is an important design feature (that has had to be improvised on the ISS), as is the psychological benefit of the large open space provided here. *Skylab* and space shuttle–*Mir* experience confirmed an open, communal area as very important for crew morale and productivity during long-duration isolation and confinement in space. A wardroom or conference area is an important way of meeting the challenges of working and living in space.

Level 2 houses the mechanical room and crew quarters (CQ). The CQ cluster contains six crew quarters and a central passageway located within the central core structure (Figure 5), surrounded by a 2.5-in. (6.4-cm)-thick water jacket for radiation protection during solar flares. Access is only from level 1 (below) or level 3 (above), via the 42-in. (1.1-m) central passageway. The mechanical room outside the core structure uses half the floor area, leaving the other half open to the wardroom area below.

The configuration as shown is after assembly and outfitting following module inflation. Equipment shelves are used as CQ partitions. CQ door panels and doors are installed on orbit. Sized at over 81 ft^3 (2.3 m^3) of volume (CQs five and six are slightly smaller), with a full height of 7 ft (2.1 m), each CQ has personal stowage, personal workstation, sleep restraints, and integrated air, light, data, and power in a volume 27% larger than the original ISS rack-based CQ.

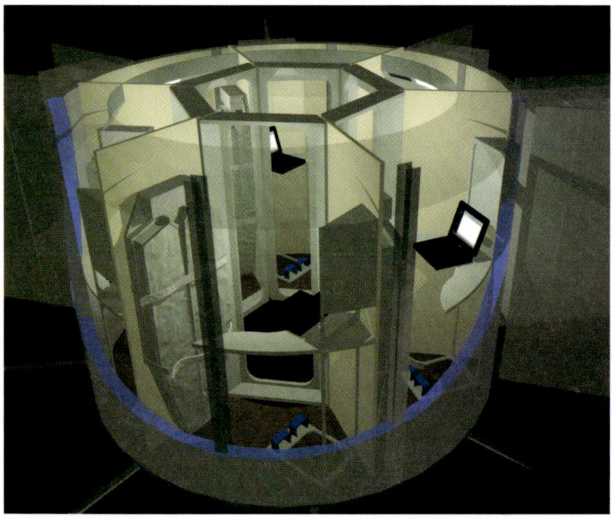

FIGURE 5 Six crew quarters surround central passageway on level 2, enclosed by water-jacket radiation shield. Mechanical equipment fills half the remaining space, adjacent to the upper part of the galley/wardroom.

Level 3 is for crew health care and soft stowage. The health care area incorporates two ISS crew health care system (CHeCS) racks, full-body cleansing compartment (FBCC), changing area, exercise equipment (treadmill and ergometer), partitionable area for private medical exams and conferencing, and an Earth-viewing window (Figure 6). Four movable partitions provide visual privacy for crew members during all activities associated with full body cleansing or with private medical exams. The soft stowage area on this level is identical to that on level 1. Circulation passages are included because this level is the primary entry into the module.

During assembly, exercise equipment items are permanently mounted in their deployed position, avoiding the time and hassle of deploying and restowing exercise equipment with each daily use. Positioned near the window to allow Earth viewing during exercise, the equipment is mounted to two launch shelves attached to the floor struts.

Level 4 is the pressurized tunnel vestibule connecting the ISS (workplace) and the TransHab (home). The tunnel provides structural connection to ISS, houses critical equipment required during inflation, and provides a transition area between Node 3 and TransHab. It has two ISS-standard hatches, avionics, and power equipment. Utilities are fed through the Node 3 bulkhead and routed into the TransHab utility chaseways. It is the only volume pressurized during launch; the packaged central core vents to vacuum during launch. Once the module is berthed and bolted to Node 3, the level 4 vestibule is immediately shirt-sleeve accessible for making the power

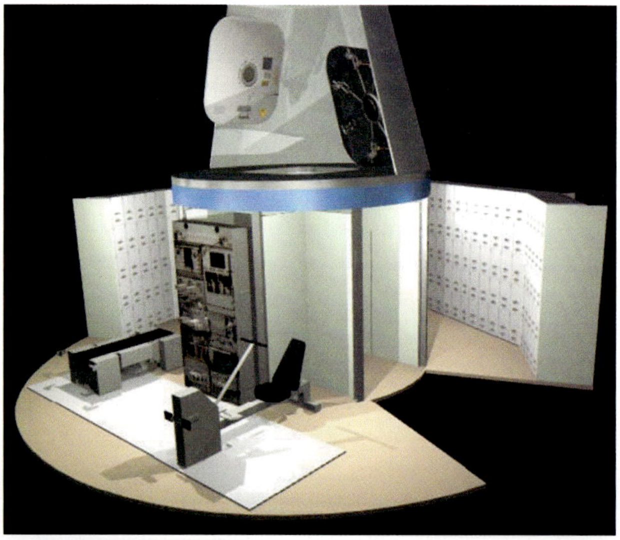

FIGURE 6 Exercise and medical functions share level 3 with stowage.

and data jumper connections to initiate deployment and inflation. A detailed functional operations concept and crew timeline have been completed for launch and activation operations.

The design takes advantage of the greater ceiling height in levels 1 and 3 for easier integration of air ducts and local-area utility distribution. Soffits attached to the core structures both there and in the crew quarters in level 2 combine the air-supply system with an enclosed chaseway for all power, data, and coolant lines so that each area is easily served with minimal exposure to utility connectors within the cabin. This approach also saves valuable on-orbit assembly and preflight checkout time by allowing these structures to remain fixed within the core so that they operate in both predeployed and post-deployed module configurations.

Another example of this integrated systems approach is the design of the stowage array to also serve as a plenum for return airflow. A subsidiary structure attached to the floor struts after module deployment, this stowage array accommodates ISS-standard stowed items in an organized inventory system, while forming a gap between outfitting and the shell walls through which return air is channeled. Thus, this system meets a functional requirement at the same time it helps TransHab to "breathe."

TRANSHAB TECHNOLOGIES

The TransHab project achieved breakthroughs in several technology areas: flexible, high-load composite structures; optimized, independent pressure shells (including inflatable and micrometeoroid shielding); and integration of both technologies into a single, reconfigurable habitat system. This hybrid structure system combines the packaging and mass efficiencies of an inflatable structure with the structural and subsystem-integration advantages of a load-carrying hard core.

A multicomponent spindle element that carries the principal longitudinal and shear loads during launch, the core is stripped down during on-orbit outfitting to its permanent role as a tensile stabilizer. Its internal trusswork pieces are reused as interior framing elements for module outfitting. To make this possible, the trusswork comprises modularized "shelf" units with a universal system for attachment to one another and to other core elements. Thus, the hard structures of the vehicle comprise a modular, reconfigurable system that allows them to respond efficiently to two very different loading conditions.

The only permanent elements of the core are longerons, central toroid shear panels, and the two end tunnels. Whereas the lower tunnel is an unpressurized ring designed to house the module inflation system, the upper tunnel (level 4) remains pressurized during the entire launch-to-activation sequence. This allows it to serve as an internal airlock during the berthing, inflation, and activation. Critical electrical switching units are mounted within this cone so that the crew can access them as soon as the module is berthed. Once assembly is complete, the level 4 tunnel becomes the functional and human interface to ISS.

Longerons provide the primary load path through the core, reacting both pressure loads and launch loads. They are 23 ft (7.0 m) long, with flares at each end for attachment to the bulkheads. The shear panel visible around the level 2 central core also incorporates annular water tanks that provide the dual functions of potable water storage and radiation safe haven.

The relocatable elements of the core system are the core shelves. There are 36 shelves in two sizes: 30 × 84 in. (0.76 × 2.1 m) and 50 × 84 in. (1.3 × 2.1 m). For ground operations and launch, the shelves provide structural support and lightweight mounting for preintegrated life-support and other subsystems equipment. All of the shelves are located in the central core for launch (Figure 7). Once the module is activated on orbit, approximately half the shelves are relocated into the deployed habitat

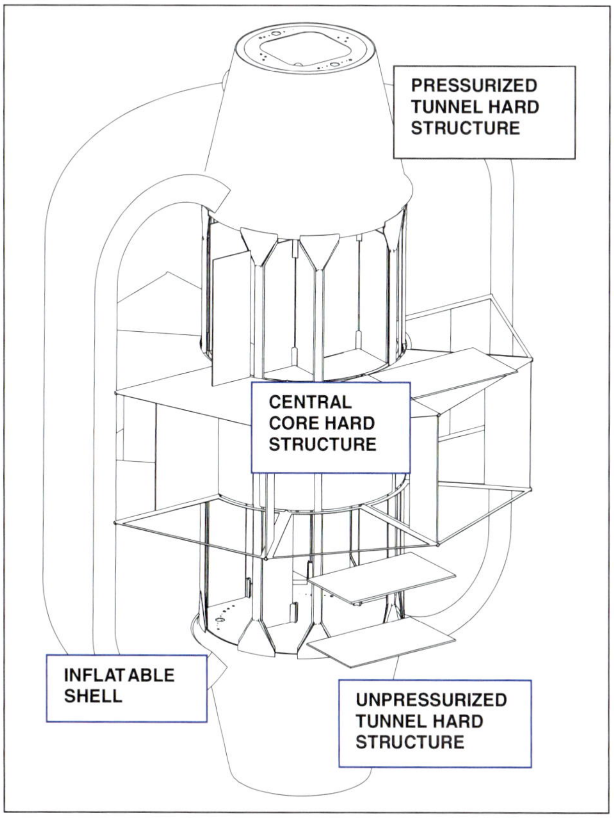

FIGURE 7 Core structure elements reassembled within the inflated shell in the on-orbit configuration.

volume as floor panels and equipment supports, that is, converted from primary into secondary structures. Because the core elements are made of graphite composite in a limited set of shapes, the structure is remarkably low in mass.

About 50% of the module's total structure mass is the pressure shell, a combination of robust inflatable restraints and high-performance debris and thermal shielding. The inflatable shell structure system is separate from the primary and secondary structure and thus is optimized for its function as a pressure shell. Folded and compressed around the core for launch, it is deployed and inflated on orbit. The shell provides orbital debris protection and thermal insulation as well as atmosphere containment. It is composed of four functional layers (Figure 8): internal scuff barrier and pressure bladder, structural restraint layer, micrometeoroid/orbital debris shield, and external thermal protection blanket. The shield uses the classic Whipple-bumper approach; and hypervelocity particles disintegrate upon contact and dissipate their impact energy through successive Nextel® layers spaced apart by open-cell foam. An additional degree of protection is provided by the Kevlar® restraint layer. Woven from 1-in. (2.5-cm)-wide Kevlar® straps, the restraint layer is designed to contain up to 4 atmospheres pressure. Each portion of the layer is optimized for its local load configuration, and strap seams are developed that achieved over 90% seam efficiency. Three Combitherm bladders form the redundant pneumatic barrier. Four layers of felt ensure evacuation between the bladder layers during launch packaging. An inner liner of Nomex provides fire-retardant and abrasion protection against incidental damage inside the habitat.

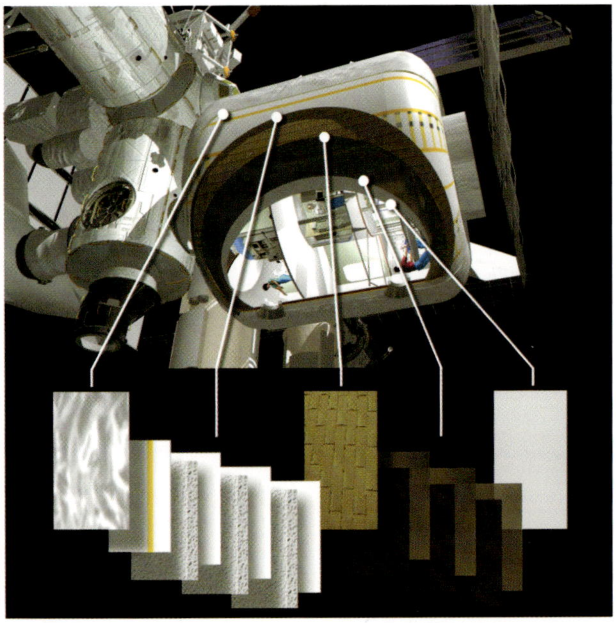

FIGURE 8 TransHab composite shell section: multilayered insulation (left), then four bulletproof layers separated by open-cell foam, then main restraint layer webbing, then redundant air-containment bladder layers, then scuff-barrier interior wall (right).

TEST PROGRAM

TransHab's design concept was based on a relatively unproven space-inflatable structure technology. To prove it would work in space—and be safe—the team set three goals:

1) Determine how to protect an inflatable structure from being ruptured by micrometeoroid and orbital debris impacts.
2) Prove that a large-diameter fabric inflatable structure can hold 1-atmosphere pressure against vacuum.
3) Prove that TransHab can be folded, packaged, and then deployed in vacuum.

To achieve the first goal, the team built a representative shell layup and performed hypervelocity impact testing at NASA Johnson Space Center (JSC) and the White Sands Test Facility. This series of tests proved critical, because if the debris shield could not stop meteoroids, TransHab would have no chance of being qualified. The 1-ft (30-cm)-thick orbital debris shield survived shot after shot, exceeding all expectations (Figure 9). In fact, the test engineers who set up the shot (who like to destroy test targets with their hypervelocity guns) were at first disappointed at not causing target failure, but then got excited when they realized the breakthrough they were now party to. The dramatic tests were shown

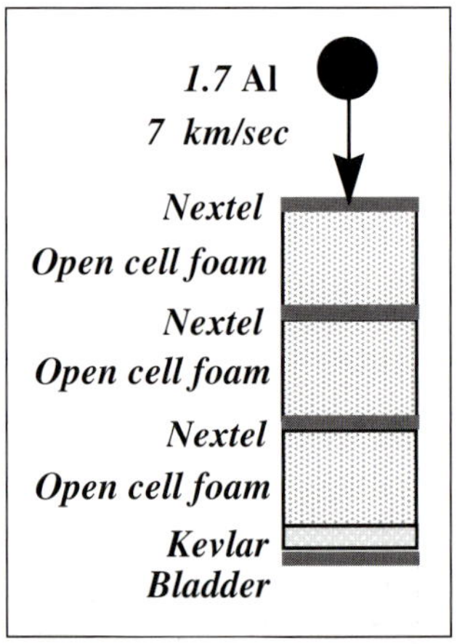

FIGURE 9 TransHab debris-shield test configuration exceeded expectations for surviving orbital debris impact.

by the Scientific American Frontiers television series on Mars mission technology. In continued testing, TransHab's shell survived a 1.7-cm aluminum sphere at 7 km/s (15,600 mph).

Two shell development test units were built and tested at JSC to prove the second and third goals. The first unit was built to prove that the inflatable restraint design would safely contain a sea-level-normal, 14.7-pounds per square inch absolute (101-kPA) atmosphere for the crew operating environment. This unit was 23 ft (7.0 m) in diameter and 10 ft (3.0 m) tall. (Unit height was incidental for this test of hoop stress.) NASA used a safety factor of four (typical for tensile fabric structures in the airship and blimp industries) for this test, known as the "4.0 test." Because of the stored energy in such a large, pressurized object, the only safe test configuration was as a hydrostatic test in the Neutral Buoyancy Lab at NASA JSC. The test was completed in September 1998, another historic milestone for inflatable habitat structures. Figure 10 shows the test article being lowered into the Neutral Buoyancy Lab pool.

For flight, the 1-ft (30-cm)-thick debris shield would have to be vacuum packed and bound to compress its folded thickness so that the module would fit into the orbiter payload bay for launch. Once on orbit, the debris shield would then be released to relax to its operational thickness. So a second test unit needed to prove that the inflatable shell could be folded and then deployed successfully in a vacuum environment. This test unit reused the hydrostatic test article bulkheads with a full-height restraint layer. Also included was the orbital debris shield already proven in the first test goal. Figure 11 shows the test unit in NASA JSC Thermal-Vacuum Chamber A in December 1998. It was successfully folded and then deployed in vacuum.

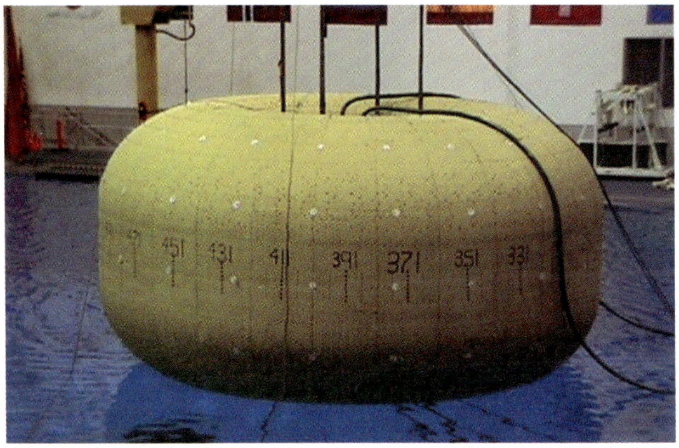

FIGURE 10 Hydrostatic test verified containment of atmospheric pressure with a safety factor of four.

FIGURE 11 Vacuum deployment test in 1998 completed validation of the TransHab shell design.

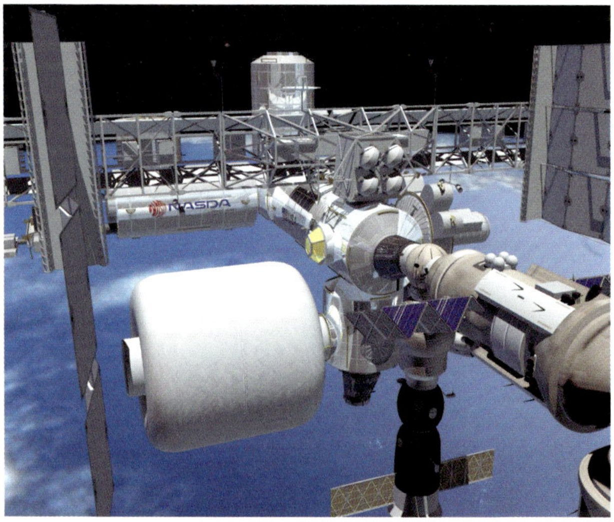

FIGURE 12 Proposed ISS TransHab was to be the final habitation element launched to ISS in 2004.

With the successful completion of hypervelocity impact and inflatable shell development tests, the TransHab project proved inflatable structure technology is ready for use in space. The design meets or exceeds ISS habitation system requirements. Had the TransHab project not been aborted, a flight unit could have been launched as the final habitation element of the ISS by late 2004 (Figure 12).

LESSONS LEARNED

Of many lessons learned during the TransHab project, one of the most important is to keep the project focused on the technology development. TransHab moved too quickly from being a technology development project to becoming a potential substitute for the planned aluminum hard-shell ISS habitation module. This "created antibodies," inviting congressional scrutiny and subsequent cancellation. A similar fate occurred to the X-38 crew-return-vehicle project.

From the technical standpoint, the approach of "build a little and then test, evaluate, and learn"—sometimes referred to as the spiral approach to engineering—was of great value technically and programmatically. It allowed us to mature the design by incorporating what we learned along the way. What follows is a summary of lessons learned as viewed from the author's architect's perspective.

Technical Lessons

- Incorporate what is learned as the design matures. Build a little, and then test, evaluate, and learn (spiral approach to engineering).
- Give technology development a real focus with hard requirements and a schedule.
- Collocate tiger team technical leads, which improves performance and communication.
- Trust the other leads. Let their teams do their jobs, and believe in them.
- Be technically honest. If you can't do something, say so. Don't hide or skew the results.
- Build test articles early and often. Build on small successes.
- Build a mock-up early.
- Design drives the systems required; the systems Don't drive the design because in engineered systems form follows function.
- Delicately balance what is possible and what is convenient.
- Perform independent assessments; get outside peer review.

Management Lessons

- Establish a charter, mission, goals, and objectives early and get personal commitment from the team members.
- Have one person in charge and accountable. Do not allow multiple managers with conflicting objectives.
- Trust your leads both technical and management.
- Build a "four-dimensional team" with a balance of personalities and skills.
- Use a mix of skills and experience levels to enhance team success.
- Collocate matrixed personnel.
- Have a weekly status meeting; however, do not try to solve technical issues there (three-minute rule).
- Give awards, celebrate accomplishments, have get-togethers outside of work. Blow off steam.
- Share the glory in media coverage: it's about the project and team.
- Build a constituency of supporters: team, supervisors, crew, management, and U.S. Congress.
- Know your competition. Don't underestimate political agendas. Don't let the fox into the hen house.
- Keep the team members focused. Shield them from the "politics."

CONCLUSION

The TransHab development was a classic "rapid prototyping" project that contributed many technical achievements to the field of space systems architecture. It broke the launch-vehicle-constrained volumetric barrier of the exoskeleton module archetype by innovating an entirely new, endoskeletal typology. It demonstrated the advantage of combining human engineering with aggressive structural innovation and testing at the conceptual stage. The integrated effort through which this system was conceived and developed proved its virtue in meeting developmental challenges.

The product combined innovative design with new technologies and proved them ready for use in flight projects. The design meets or exceeds state-of-practice space requirements. It puts the "living" into living and working in space, providing improved accommodations for sleeping, eating, cooking, personal hygiene, exercise, entertainment, leisure, storage, and radiation storm shelter on orbital, interplanetary, or planet-surface missions. |

Bibliography

Kennedy, K. J. (1999), "ISS TransHab Architecture Description," *Proceedings of the 29th International Conference on Environmental Systems,* Society of Automotive Engineers, Warrendale, PA.

9

design of a TransHab-based system

MATTHEW HERMAN

INTRODUCTION

THIS CHAPTER DESCRIBES THE DESIGN PROCESS from the point of view of an architect working at a commercial aerospace company on a privately funded project.

Building on lessons learned during the TransHab program at NASA, research and development of a manned inflatable module continued in the private sector. Starting in 1999, a team was assembled to pursue research, design, and development of inflatable structures for commercial manned space operations. This team is composed of engineers, architects, and industrial designers. Assisting the design team with the development of prototypes, ground-support fixtures, and test articles is an in-house machine shop. Additional design and manufacturing capabilities are provided by a series of consultants.

The team was assembled by an investor client whose end goal is to develop a commercially available base model for a manned space vehicle that could be outfitted to perform a variety of functions and operations. Because of the novelty of nongovernment manned commercial space operations, the design team was tasked to develop concepts that were robust, adaptable, safe, economical, and repeatable in order to meet a variety of possible mission requirements. This meant avoiding reliance on single sources for materials, processes, equipment, and launch services where possible. Additional constraints were placed on the design team in the form of International Traffic in Arms Regulations (ITAR), Federal Aviation Administration concerns and requirements, environmental concerns, and economics.

DESCRIPTION AND OBJECTIVE

General Description

The initial goal of the design team was to establish performance, design, development, and verification requirements for an inflatable space vehicle and propose a series of preliminary architectural design solutions. The inflatable vehicle, referred to as the XV, and related structures are intended for use as a manned orbital facility located in low Earth orbit (LEO). Design altitude ranges from 150 to 300 nautical miles. Orbital inclination is approximately 51.6 deg to take advantage of two existing launch sites that provide services for heavy-lift launch vehicles capable of delivering payloads to LEO.

The purpose of the XV is to provide a pressurized vehicle with shirt-sleeve environment for on-orbit crew. The XV receives utility services from internal and external systems and distributes them within the vehicle. These services include electrical power, thermal conditioning, data and audio/video communications, cabin air circulation, carbon-dioxide removal, atmosphere monitoring, condensate water collection, water processing, and fire detection and suppression.

To reduce development cost, the design team assumed use of common or existing hardware designs where appropriate. This required trade studies to compare development cost savings using existing hardware to development costs of improved systems that reduce resupply mass and thus generate life-cycle launch cost savings over 15 years of operation. Systems repair beyond basic maintenance would be done through replacements, necessitating a modular system architecture.

To meet industry standards, all spaceflight system designs, manufacturing, and tests must comply with JSCM 8080.5 as well as other applicable aerospace standards. Crew habitability and life-support systems must comply with SSP 50005, JSC-26882, and JSC-28354 when applicable.

The XV habitable volume is composed of two aluminum endcones connected by longerons (Figure 1). These longerons support launch loads and form the core structure for mounting utility lines, equipment racks,

Copyright © 2008 by Matthew Herman. Published by the American Institute of Aeronautics and Astronautics, Inc., with permission.

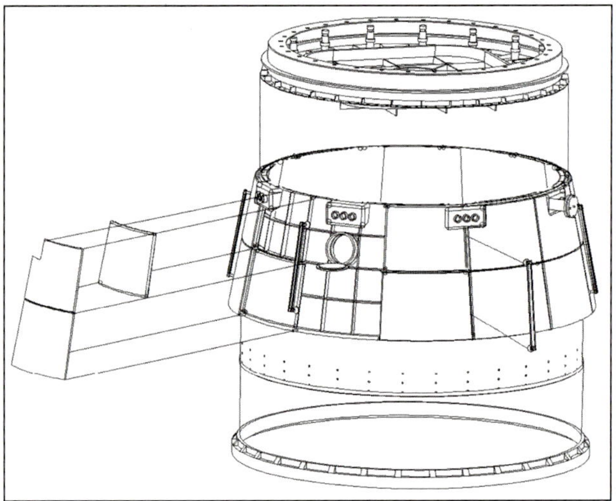

FIGURE 1 Rigid endcone assembly for the inflatable XV structure.

and crew-support elements (Figure 2). Surrounding the core is a flexible air barrier that is protected by multilayer insulation and a flexible debris shield (Figure 3). The air barrier, thermal insulation, and debris shield are compressed and folded around the core prior to launch. On orbit, the thermal insulation and debris shielding expands to a thickness of approximately 40.6 cm (16 in.) as the air barrier is inflated (Figure 4). Functional dimensions driving the architecture are shown in Figure 5.

Key Problems and Characteristics

The design team began the research-and-development phase by collecting information about prior and proposed manned space vehicles. This included relevant drawings, computer models, prime-item development specifications, interface control documents, and lessons-learned documents. Through this research, requirements and specifications for the XV were extracted and expanded to form a design development outline organized into 16 sections:

1.0 Introduction (including references, definitions, materials and processes, and manufacturing guidelines)
2.0 Launch delivery systems and requirements
3.0 Orbital and environment definition and requirements
4.0 Spacecraft operations, logistics, and ground support
5.0 Communication systems
6.0 Guidance, navigation, and control systems
7.0 Propulsion
8.0 Data distribution, storage, and transmission
9.0 Electrical power systems
10.0 Thermal control system
11.0 Environmental control and life-support systems
12.0 Structures and mechanisms
13.0 Payloads
14.0 Robotic systems
15.0 Extravehicular activity systems (EVA)
16.0 Habitability systems

This design development outline structure was based on three sources: Messerschmid and Bertrand (1999), NASA (1998), and Boeing (1998). The outline served as a reference for the design team to ensure their consideration of relevant programmatic elements while developing the vehicle architecture and components. It included spatial requirements for various pieces of equipment, clearance zones, and translation paths. As additional information was collected or existing information refined, the outline headings were expanded. The intent of the outline was to organize drawings, specification documents, and a growing database to avoid confusion and misunderstandings between the investor client, design team, and multidisciplined subcontractors and manufacturers. It was intended to be flexible and generic to accommodate modifications as the XV program evolved. The design development outline was a precursor to formal written specifications and interface control documents. Formal specifications, policies, and principles were drafted according to ISO 9001 standards, AS 9100 standards, and structured according to European Cooperation for Space Standardization (ECSS) requirements and standards for management, engineering, and quality assurance.

During this early phase, information was collected relating to NASA's TransHab program. Although the TransHab design characteristics provided an initial baseline, the XV project has several key differences in function that led to unique architectural design solutions. Key XV requirements included the following capabilities:

1) Be assembled with components each from a minimum of two viable vendors
2) Be transported to the launch site by truck, crane, and plane
3) Fit inside the payload envelope of two commercially available heavy-lift launch vehicles, without excluding the possibility of a space shuttle launch given adaptations such as trunnion pins and weight reduction
4) Withstand worst-case launch loads from the candidate launch vehicles
5) Upon reaching the desired orbit, be an autonomous vehicle with guidance, navigation, and control (GNC); communications and data; electrical power subsystem (EPS); temperature control subsystem (TCS); inflation systems; and environmental control and life-support system (ECLSS) functional on the first launch

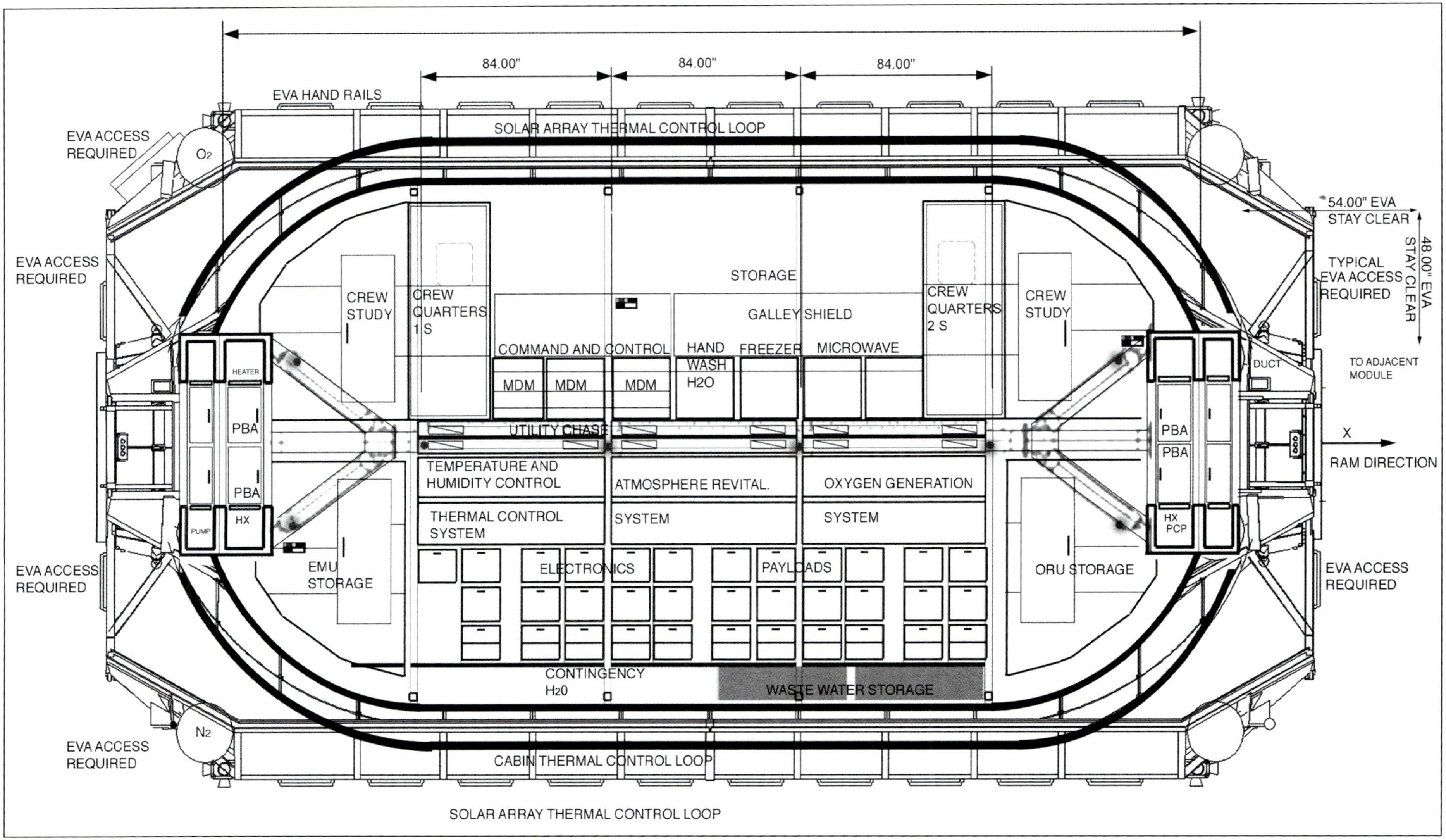

FIGURE 2 Overall XV system configuration, showing functional layout and integration of the inflatable module with other vehicle systems.

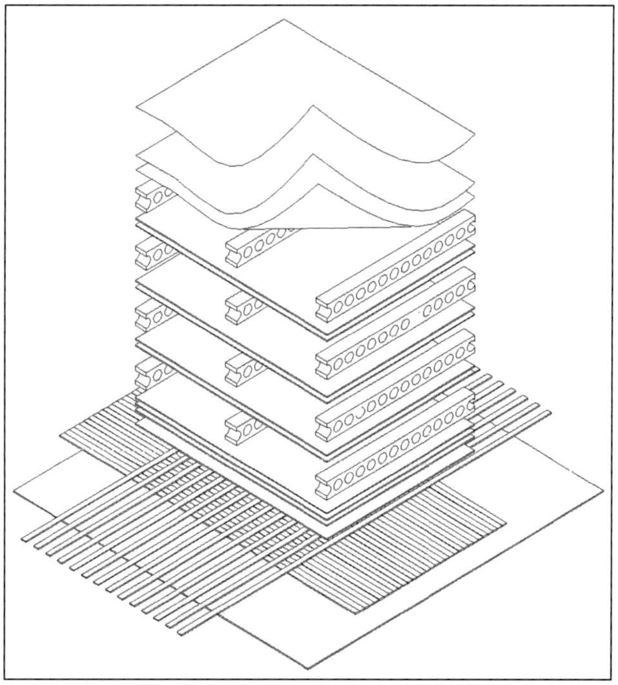

FIGURE 3 Exploded isometric cutaway view of flexible skin assembly, showing air barrier, structural restraint layer, debris shield shock layers, and spacers with multilayer thermal insulation.

6) Provide EVA translation paths and exterior payload mounting points
7) Provide space for an integrated airlock on first launch
8) Provide docking and berthing capabilities for station growth
9) Provide interior structure that accommodates systems outfitting along the core during launch and can be reconfigured by the crew after inflation
10) Provide interior structure to aid the construction and servicing of functional areas for crew dining, conferencing, exercise, hygiene, galley, medical services, and crew quarters for a crew of six (maximum) and three (permanent)
11) Provide interior structure that allows the possibility of large open spaces
12) Provide windows

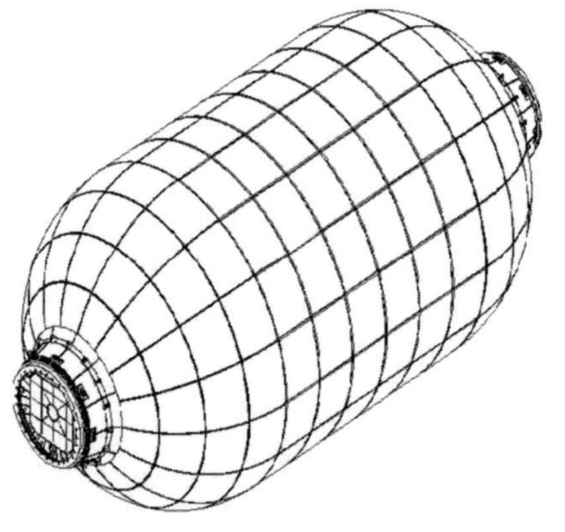

FIGURE 4 Isometric view of pressure vessel in inflated configuration.

PRIOR ART

After information relating to existing and prior concepts for manned space vehicles was collected, the design team performed a series of informal analysis exercises to determine strengths, weaknesses, and relevance of these designs based on the XV requirements written in the design definition outline. The XV development adopted three major architectural precedents from this prior art.

The first major architectural precedent came from a Lawrence Livermore study done during the first Space Exploration Initiative in the early 1990s (Hyde et al. 1990a). The text and illustrations of this study describe an orbital station concept composed of several inflatable modules berthed end to end with body-mounted solar arrays. These were identified as desirable characteristics for the XV. The Livermore study also provided a vision for design discussions relating to future use of inflatable vessels for lunar and Mars missions. Although such missions are not part of the XV program, the design team agreed that a future version of XV might be adapted to serve these roles.

The second major architectural precedent evolved from NASA's TransHab project described in Chapter 8 (Figure 6) (Kennedy 1999). This project provided the design team with a point of departure for inflatable technology. Although many specific TransHab design solutions did not support XV requirements, TransHab did point the team in a productive direction. Documentation and discussions of the guidelines and strategies for constructing and outfitting TransHab helped generate many of the XV architectural concepts (NASA 2000). This project, in conjunction with the documentation, design, and construction of the International Space Station (ISS), allowed the XV project to be envisioned.

The third major architectural precedent evolved from Russian strategies for the *Mir* space station and the Russian segment of the ISS. The design of the *Zvezda* service module (SM) and functional cargo block (FGB) provided architectural concepts for an integrated vehicle that could be delivered to orbit intact and integrated on a heavy-lift launcher and operate all major spacecraft housekeeping systems prior to first crew

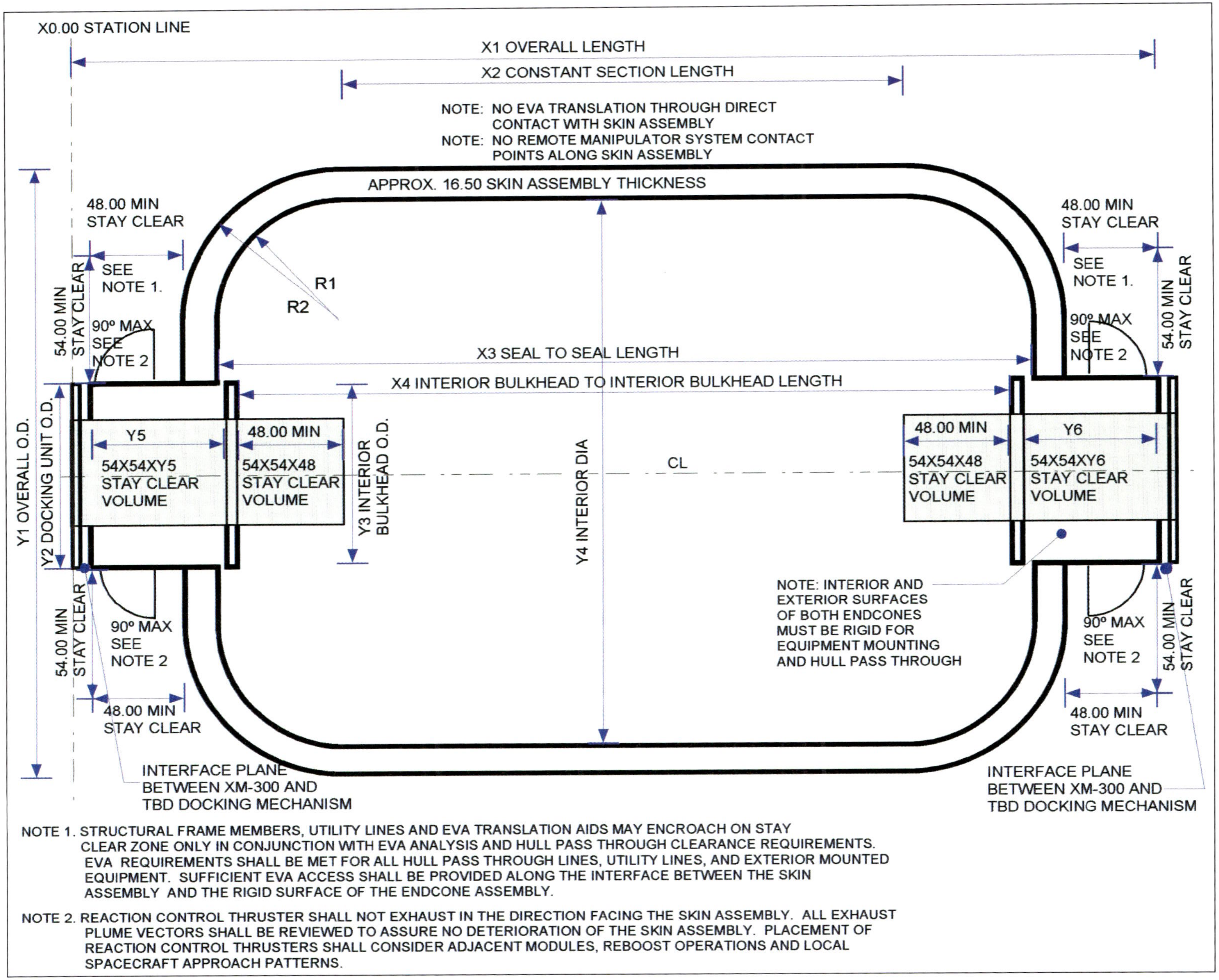

FIGURE 5 Developmental diagram for XV architecture, showing driving functional dimensions.

FIGURE 6 Composite photo of NASA's TransHab vacuum chamber test: before (left) and after (right) inflation (NASA 1998).

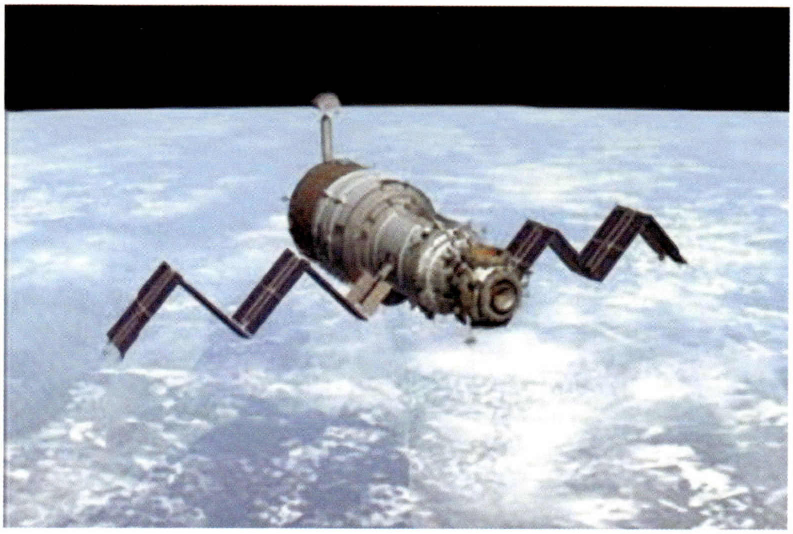

FIGURE 7 Artist rendering of *Zvezda* deploying solar array (data available online at http://spaceflight.nasa.gov).

arrival (Figure 7). Many SM and FGB capabilities were used as top-level requirements for the XV, including use of the exterior vehicle surface to provide attachment points for radiators, gas tanks, thrusters, antennas, external payloads, and EVA handrails.

MAJOR CONFIGURATION DESIGN TRADE

Development of the XV design proceeded via a series of design proposals. The first step of the design process was to establish a generic commercial mission profile. The mission assumes resupply by a *Progress* cargo vehicle with a frequency of 30 days or less (RCS Energia 2002). Reboost is also provided by *Progress*. A *Soyuz* TM is the crew escape and delivery vehicle for the crew of three. This requires that a minimum of two docking ports be available at all times and that crew transfer and cargo flights be carefully coordinated. An airlock is required as part of the initial XV configuration (first launch) for EVA and contingency operations.

The first design proposal was based on the ISS legacy (Figures 8 and 9). Although this proposal met most of the requirements in the design definition outline, it had several faults:

1) It required a minimum of three launches before core systems were complete and the station was ready for crew arrival.
2) Each launch package contained equipment grouped by function rather than by a need-based hierarchy.
3) It failed to address protection of the inflatable module skin during launch.
4) It did not provide adequate stationkeeping services at initial launch.

FIGURE 8 Early study based on ISS architecture. Although these studies were part of a successful process, early efforts were unrealistic because of functional compartmentalization in a collection of assembled modules.

5) It failed to support EVA across the surface of the inflatable module.
6) It limited station growth options by occupying a berthing mechanism with nonhabitable structures.

The weakness of this proposal identified the need to consider the configuration and construction sequence from a commercial client's point of view rather than a mission-specific point of view. The design team was confronted by several questions:

- How to provide maximum capability, maximum infrastructure, and maximum flexibility in one launch?
- How to allow a staged increase of station infrastructure in a manner that provides maximum capability for all possible outcomes?
- How to address undefined future needs of clients while avoiding high redesign costs?
- How to transport and launch an inflatable module with a flexible skin?

Answers to these questions led to the selection of core systems and capabilities that would be required at first launch regardless of mission and function. These systems were given priority in terms of mass budget, location, and integration sequencing. This change in design strategy allowed the architects and engineers to concentrate on systems-engineering the entire XV vehicle rather than designing a collection of modules and systems divided by functions. These influences led to new configuration proposals.

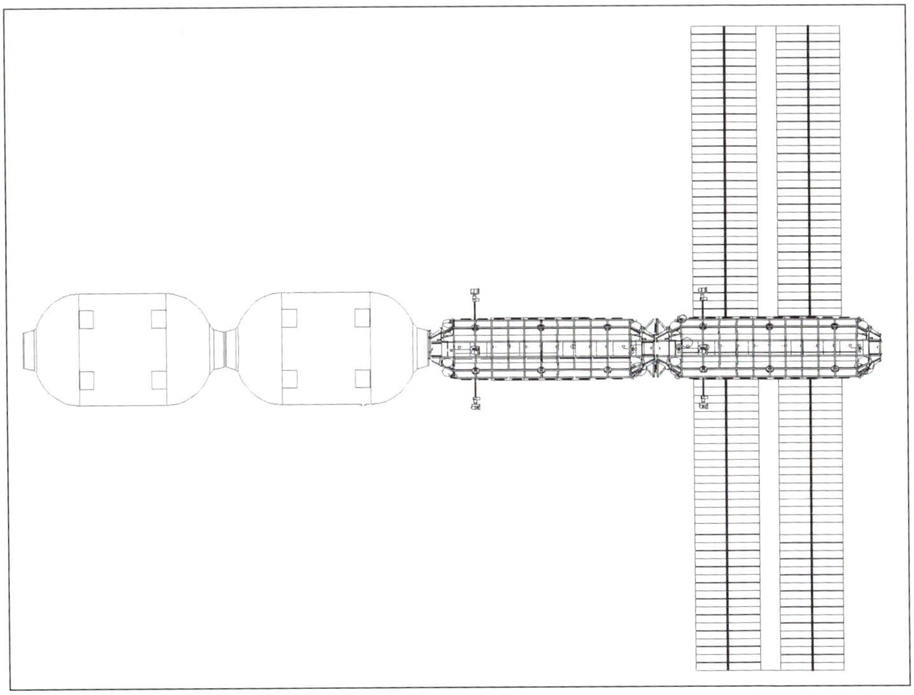

FIGURE 9 First design proposal, a series of segments and modules divided by function.

The catalyst for the design change was the need to protect the inflatable skin during launch. The concern was that residual atmosphere in the interior core volume could begin to pressurize the skin as the ambient pressure rapidly dropped in the payload shroud during launch. This could cause contact between the two vehicles that would damage or destroy the XV's folded flexible skin. The proposed solution was an external truss with in-fill panels to protect the inflatable skin from the payload shroud and vice versa (Figures 10–12). The external truss would then open in three or four segments during inflation on orbit. This concept provides rigid attachment points

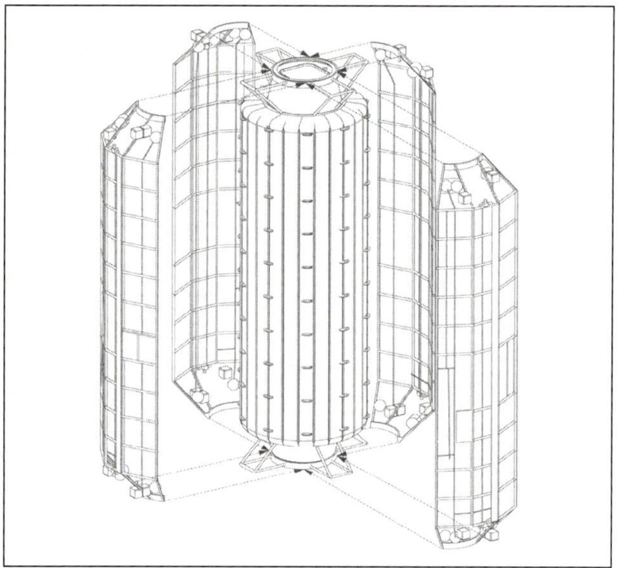

FIGURE 10 Exploded isometric of folded skin and exterior truss segments.

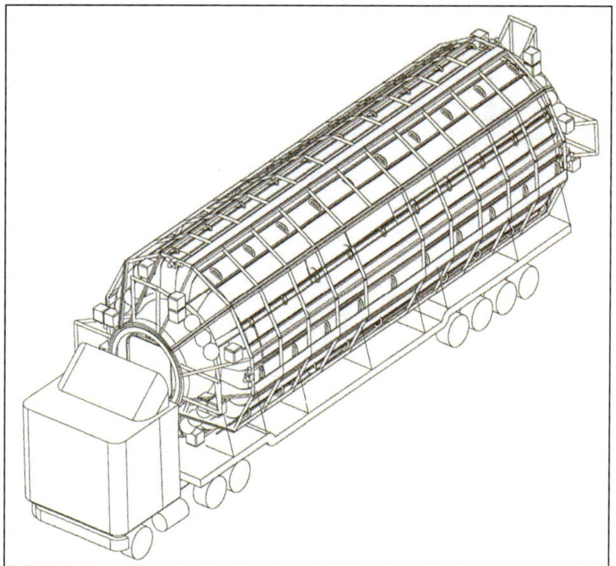

FIGURE 11 Truck transport of the XV system.

required for thrusters, radiators, solar arrays, antennas, EVA aids, and other externally mounted equipment and utilities identified as core system requirements. The external truss also aids ground transportation and launch-vehicle integration by providing hard points for lifting cranes and other ground-support equipment (Figure 11).

Although a weight penalty is paid for the exostructure, the requirements of a mission profile using a heavy-lift vehicle rather than the space shuttle presented the design team new challenges and opportunities not considered in the TransHab project. The exterior truss serves many functions, justifying its mass from an overall systems standpoint. Not

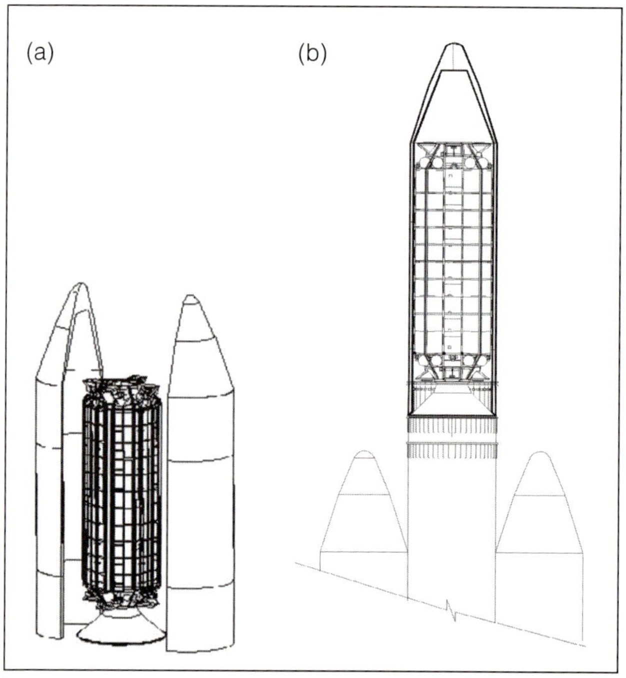

FIGURE 12 XV in folded configuration a) on payload attachment fairing (PAF) prior to fairing mating and b) on launch vehicle.

only does the exterior truss protect the flexible skin during ascent (Figure 12), it provides a rigid structural form and wider base for the payload attachment fitting. This helps the XV meet typical frequency requirements of the heavy-lift launch vehicles available during preliminary analysis.

Studying the Russian configurations and system requirements for manned space vehicles helped generate concepts for an independent spacecraft, with vehicle capabilities similar to the SM and FGB but retaining the enhanced volume of an inflatable, TransHab-like module. The resulting parti diagram could be described as an SM sliced into four sections parallel to the long (x) axis with an inflatable module packed inside. This SM-like shell breaks apart along the section lines and allows the inflatable module packed inside to expand out from the core.

Two concepts evolved using the exterior truss parti. Concept 2A split the exterior truss into three sections. After reaching orbit the three sections would separate and articulate 90 deg from the x axis, pivoting at the aft endcone (Figure 13). Two of the sections would then rotate around the y axis to face their solar arrays toward the sun. The third truss segment would align with the z axis, positioning its radiator panels edge on to the sun. Additional core systems equipment would be located toroidally on the exterior of the back endcone.

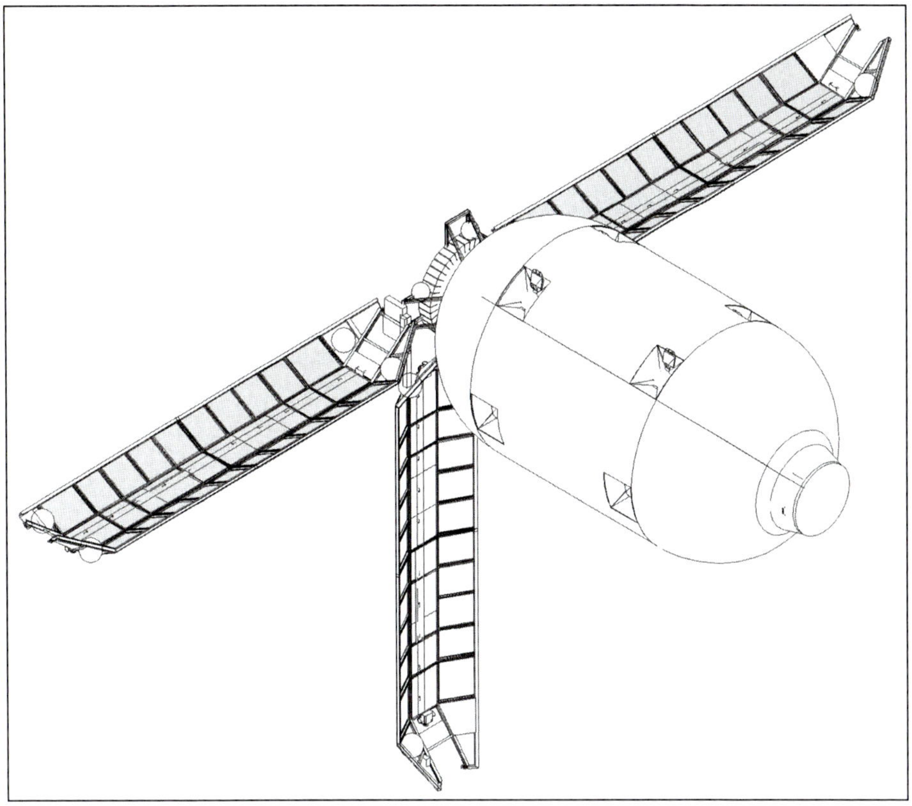

FIGURE 13 Concept 2A isometric view, deployed configuration. The two lateral exterior truss segments contain solar arrays, whereas the nadir segment contains the thermal radiator.

paths and utility-line attachment points from the endcone to the truss segments. Although concept 2B could fly as a solo vehicle, the designers envisioned 2B to be part of a larger station constructed as a series of similar additional vehicles berthed to it. The additional modules were assumed to have core systems identical to the first XV launched. Interior systems and crew accommodations would vary to augment the ECLSS and crew systems included on the first XV launch.

Concept 2B split the exterior truss into four sections that would expand in the y and z directions during inflation (Figure 14). The exterior truss segments would parallel the x axis at the outside diameter of the inflatable skin (Figure 15). This provides EVA translation paths and work sites along the exterior surfaces. GNC, EPS, TCS, communications equipment, and utility lines could be integrated on the truss segments as required by the client's mission (Figure 16). Scaffolding provides EVA translation

Evaluation of First-Launch Capabilities

To evaluate the overall robustness of each architectural configuration, certain assumptions were baselined for both:

1) Launch vehicle with a 22.7 mt (50,000-lb) lift capacity to 51.6-deg inclination LEO.
2) A successful inflation and checkout period for the first module is followed by a *Progress* logistics mission to bring additional supplies and reboost.
3) After successful docking of the *Progress*, the first three-person construction crew arrives for a short-duration commissioning phase.

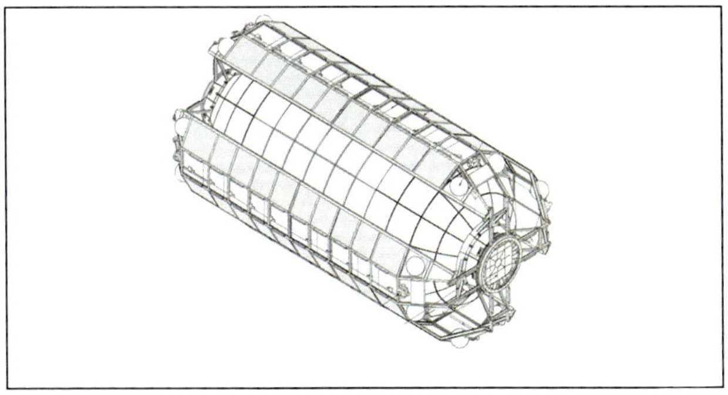

FIGURE 14 Concept 2B isometric view, deployed configuration. Shaded areas represent infill panels with solar arrays.

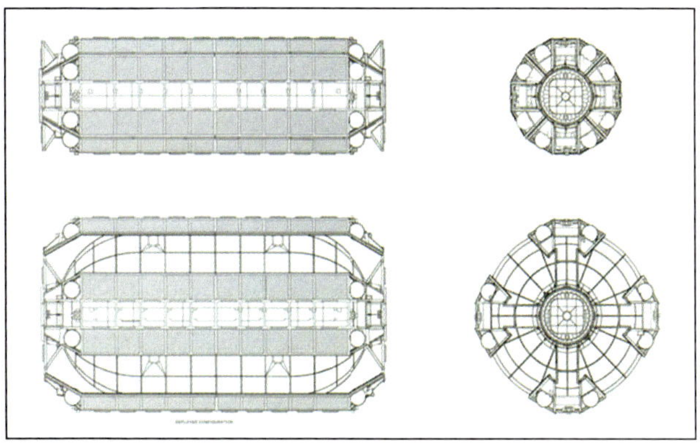

FIGURE 15 Side and front views of concept 2B: top—folded configuration with a diameter of 426.7 cm (14.00 ft); bottom—inflated configuration with a diameter of 670.6 cm (22.00 ft).

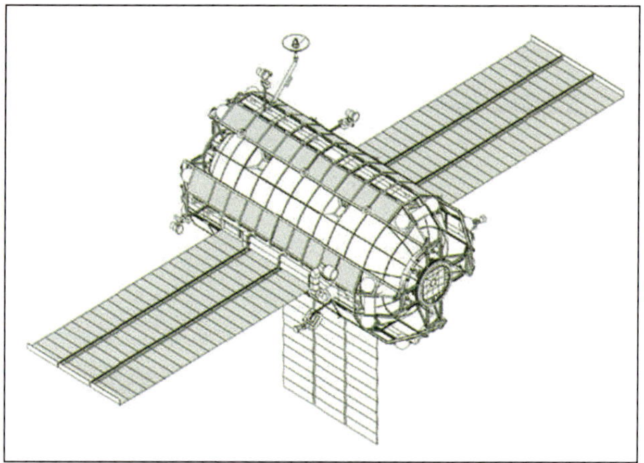

FIGURE 16 Concept 2B with deployed solar array and full exterior truss outfitting; additional equipment could be added as needed.

4) Common preliminary XV mass estimates and assumptions for core systems (Table 1).

Using these assumptions, members of the design team conducted a series of architectural studies to evaluate the strengths and weaknesses of each concept. Although individual scores varied between the concepts, the total scores for both concepts were equal for first-launch and initial-crew operations. Equipment for both concepts is located on exterior truss segments having EVA accessibility from the integrated air lock located in the aft endcone. Members of the review panel were concerned about the need for actuators and rotating joints to swing out the exterior truss segments prior to inflation in concept 2A. Additional comments arose from the proposal to use fabric handholds sewn along the flexible skin with a wire pulley system for EVA translation. This was an issue for EVA operations because the only hard point on the flexible skin between the fore and aft endcones is at the window locations (Figure 17).

Concept 2B uses the inflation process of the flexible skin itself to expand the four exterior truss segments after the restraints are released. Although this eliminates the need for mechanical actuators to pivot the exterior truss segments as in concept 2A, there was some concern about stress and thermal conditions on the structural restraint layer at the points where the exterior truss segments attach to this layer of the flexible skin (Figure 18). However, the rigid translation path and work area for EVA operations were both significant advantages. Deployable solar arrays could be mounted on the port and starboard truss segments to provide adequate power for manned operations, and solar cells mounted directly to the in-fill panels of the truss segments could provide keep-alive power prior to inflation. The number of these panels could be increased as needed.

TABLE 1 Mass budget common to concept alternatives

CORE SYSTEMS	ESTIMATE[a] (FIRST LAUNCH ONLY)
5.0 Data systems (hubs, routers, multiplexers, bus, and distribution lines, etc.)	300 kg
6.0 GNC (star trackers, GPS[b] antenna, CMG)	680 kg
7.0 Propulsion (reaction control thrusters, fuel tanks, lines, etc.)	450 kg
8.0 Comm Systems (S band, Ku-Band UHF)	250 kg
9.0 EPS (solar arrays, energy storage, primary and secondary distribution, etc.)	1700 kg
10.0 TCS (active and passive heat collection exchange and rejection)	1800 kg
11.0 ECLSS (ACS, THC, AR, CO_2 removal, filtration, ventilation, WRM)	1100 kg
12.0 Structures	13,500 kg total
Flexible skin	5000 kg
Longerons and core	1800 kg
Endcones	2700 kg
Exterior truss	4000 kg
13.0 Payloads	TBD per client
14.0 Robotics (grapple fixtures only)	130 kg
15.0 EVA (Camera and light group, hand rails, translation aids, and work site equipment	360 kg
16.0 Habitability systems (PBA, PFE, tools, exercise equip.) (additional equipment launched on Progress)	230 kg
TOTAL	20,500 kg 2100 kg (float)

[a]Mass estimates are for preliminary design purposes only. Estimates are based on preliminary information provided to the design team by manufacturers. Mass estimates are based on existing designs for space qualified hardware.

[b]GPS = global positioning system.

Evaluation of Station Growth Capabilities

The design study then looked at station growth patterns by including a second module consisting of identical core systems (Figures 19 and 20). Interior systems in the second module were assumed to augment first-module capabilities. This study revealed the strengths of concept 2B over 2A. No robotic arm is baselined for the core XV system, so

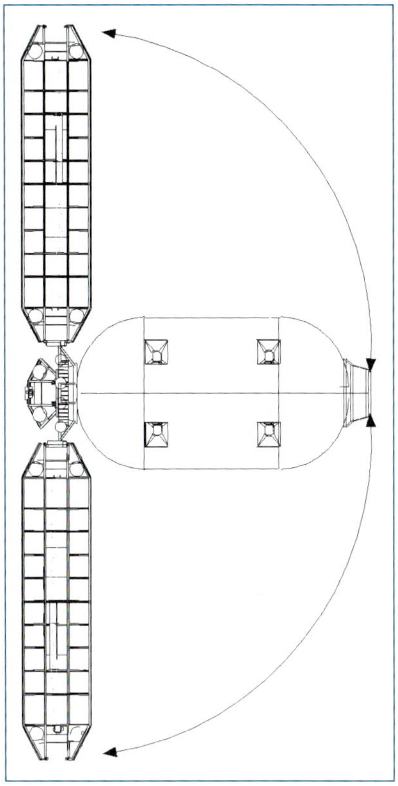

FIGURE 17 Top view of concept 2A. The exterior truss must open out and pivot before the inflatable skin can expand.

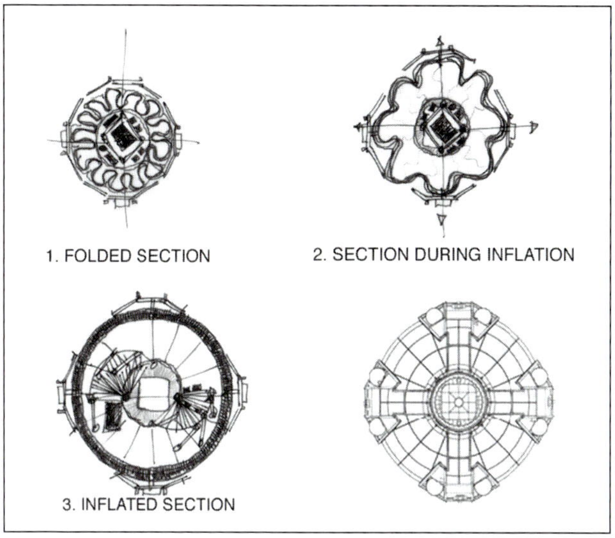

FIGURE 18 Inflation sequence of concept 2B in section, showing unfolding of the flexible skin and separation of exterior truss segments.

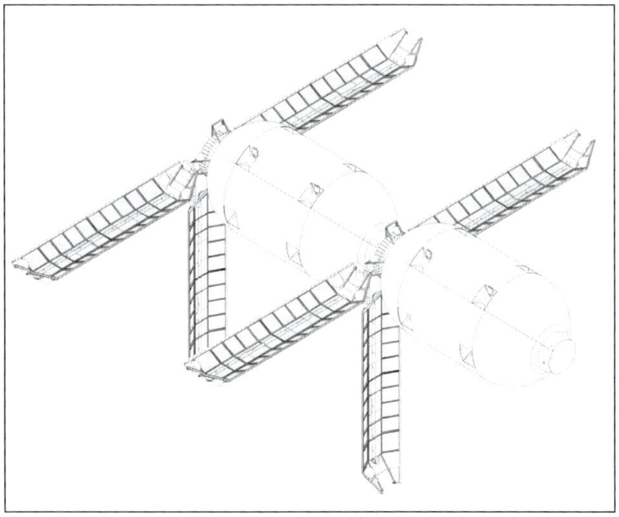

FIGURE 19 Concept 2A station growth configuration; isometric view of two XV modules berthed along the *x* axis.

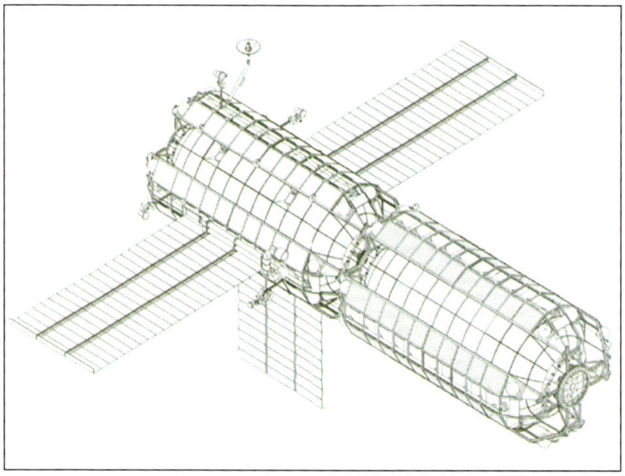

FIGURE 20 Concept 2B station growth configuration; isometric view of two XV modules berthed along the *x* axis.

that EVA operations and accessibility are necessary for contingency operations. When two 2B modules with the same basic architectural configuration are berthed, their exterior truss segments form four continuous, rigid translation paths for EVA and robotic operations. With the aid of work site stanchions and portable foot restraints, the crew could access a majority of the exterior surface (Figure 21).

The concept 2B exterior truss configuration was especially advantageous:

- It is located near window locations for deploying sunshades and for maintenance.
- Orbital replacement unit (ORU) locations and work sites could be placed as needed along any of the four exterior truss segments.
- A crew and equipment translation aid (CETA)-like cart could run from end to end on a rail system mounted to the exterior truss.

FIGURE 21 Multiple views of EVA operations on the rigid exterior of the ISS (Courtesy of NASA).

- Additional solar arrays and deployable radiators can be added or subtracted to the configuration depending on the client's mission. This could be done prior to launch or on orbit.
- It reduces or eliminates the need for a robotic arm while still providing mounting points for grapple fixtures.

The design review and comparative evaluation provided a vision for the XV architectural configuration based on concept 2B. The XV architecture takes advantage of the increased habitable volume enabled by an inflatable pressure vessel, yet retains advantages of a rigid exostructure through the use of the exterior truss segments. The XV is intended to offer a unique combination of features: commercial availability; design for commercial launch vehicles; increased interior volume compared to rigid modules; large rigid exterior surface area for mounting equipment and exterior payloads; rigid EVA translation paths along entire vehicle as station grows; adaptability to meet individual client needs without major redesign costs for the primary structure; and exterior truss structure that serves many functions before, during, and after launch, adding efficiency to the design by making maximum use of the structure weight.

FUTURE APPLICATIONS

Following the design review, an architectural study was conducted to investigate future options using the basic configuration provided by the exterior truss and core systems of concept 2B. During this study, use of an inflatable skin was seen as advantageous mostly for habitability issues. Several advanced concepts were sketched showing vehicles that used the core systems, endcones, core, and exterior truss of the XV without the flexible skin. These sketches show the value of using a design method that identifies desirable characteristics of existing vehicles and integrates them with new requirements, driven by the need for flexibility, repeatability, and robust capability.

The first concept is for a large automated transfer vehicle (ATV) (Figure 22). This concept eliminates the inflatable skin and replaces the core with a 3.7-m (12-ft)-diam pressurized aluminum cylinder between the fore and aft endcones. A docking mechanism is located in one endcone. The in-fill panel of the exterior truss serves as micrometeoroid and debris (MMOD) shielding with photovoltaic

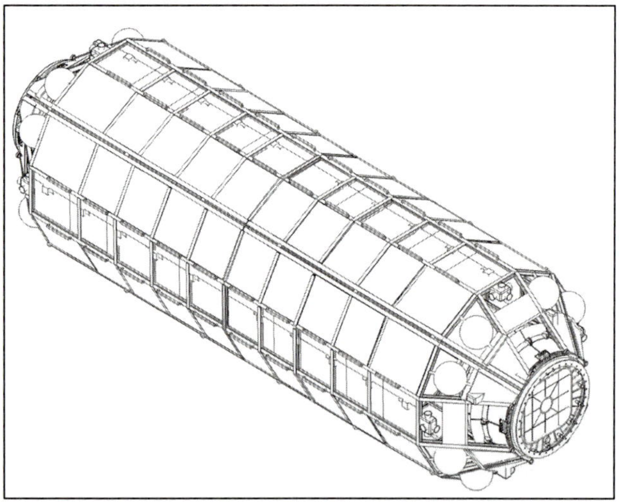

FIGURE 22 XV-based concept for an automated transfer vehicle, using core systems with a rigid pressure shell.

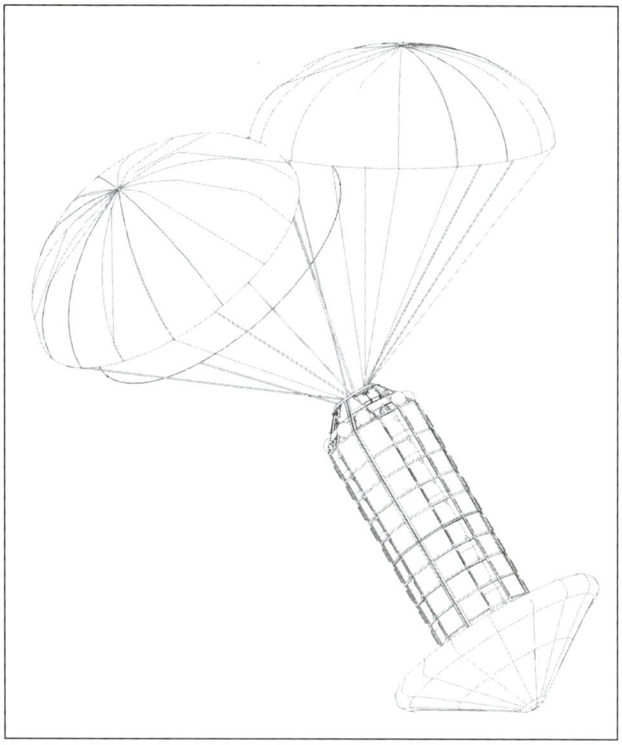

FIGURE 23 XV-based concept for a cargo return vehicle, using core systems with heat shielding panels over exterior truss, and shuttlecock-shaped inflatable heat shield deployed prior to reentry.

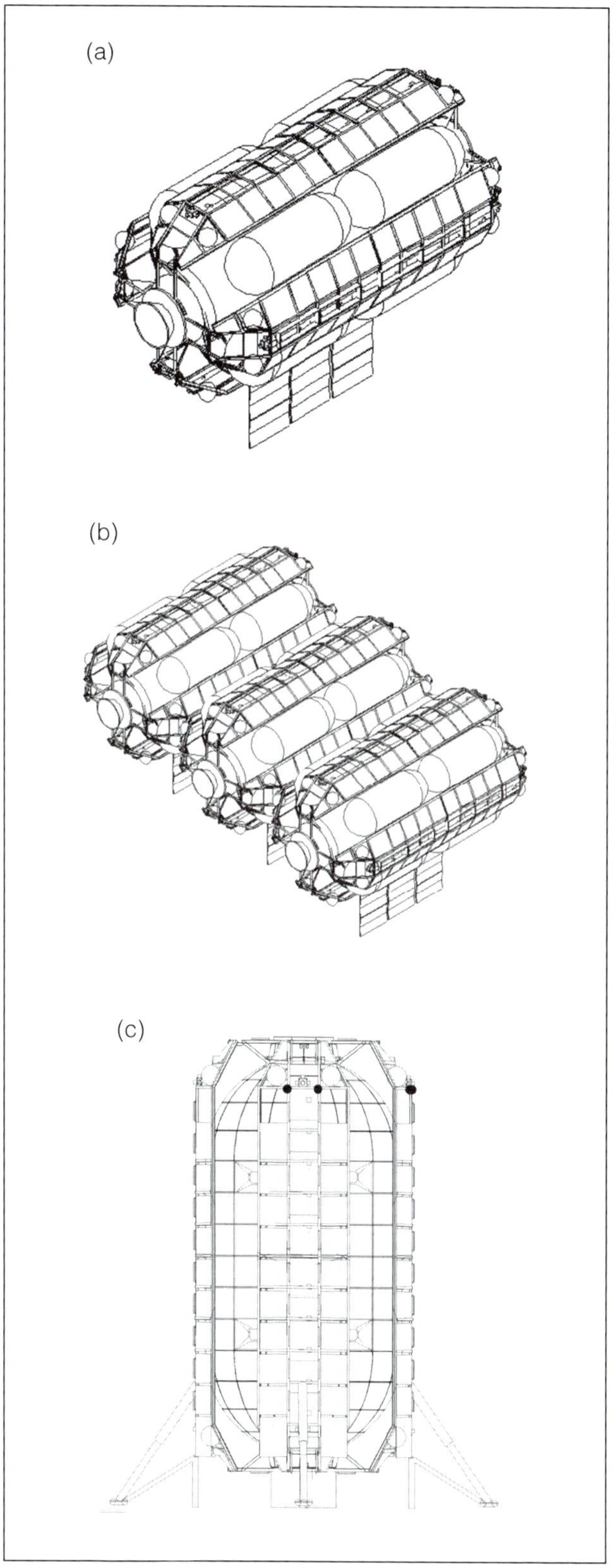

FIGURE 24 XV-based concept studies for a propulsion module with removable fuel tanks: a) single module; b) ganged propulsion stage; and c) lander stage core.

arrays and radiator panels mounted on top. Core systems for GNC and command and control are assumed similar to the XV.

The ATV study led to a cargo-return-vehicle concept that covers the exterior truss and one endcone of the ATV with reentry heat shielding (Figure 23). The exterior truss segments and heat shield would be removable for servicing after landing, and the aluminum cylinder cargo container could be reused.

These studies drove additional concept studies that would retain the XV's core structure and systems architecture, but explore additional functional roles. One proposal adapts the XV core systems for a propulsion module (Figure 24). In this concept, the rigid cylindrical core provides compartmentalized shirt-sleeve access to propulsion components for maintenance operations, if safe. This concept also provides MMOD shielding for propulsion components. Replaceable external fuel tanks are mounted to the external truss segments. Multiple modules could be arranged in a raft-like configuration. The XV might also be adapted as the core of a lander stage.

SYSTEM DEVELOPMENT PLAN

Profitable large-scale commercial human space operations remain unrealized. Although steps are being taken to promote manned commercial operations, cost is still a limiting factor. In response to these realities, the XV project outlined a staged development process to spread cost over time and reduce overall project risk.

Stage 1

Design and development of quarter-scale prototypes lead to full-scale prototypes of structural core, endcones, and flexible skin for ground testing. Ground-support equipment and facility needs are identified. Manufacturing procedures are defined, developed, and tested. Full-scale mock-ups and training facilities are constructed to prove full-scale concepts and layouts for crew systems (Figure 25).

Stage 2

Design, development, and qualification of first flight article. The first flight article serves as an orbital test. This mission is a crew-tended module with limited functionality. The crew-tended module is part of, or in proximity to, a space station. Mission requirements are based on proving performance data relating to inflatable technologies for crewed operations, rather than a crewed vehicle with full systems capabilities.

Stage 3

A crew habitat vehicle is the final precursor to the XV. This vehicle tests the core systems required for the XV but lacks some of the systems necessary for an independent vehicle. The habitat vehicle is launched on a heavy-lift vehicle and berthed to other modules that provide long-term stationkeeping functions and life support (Figure 26).

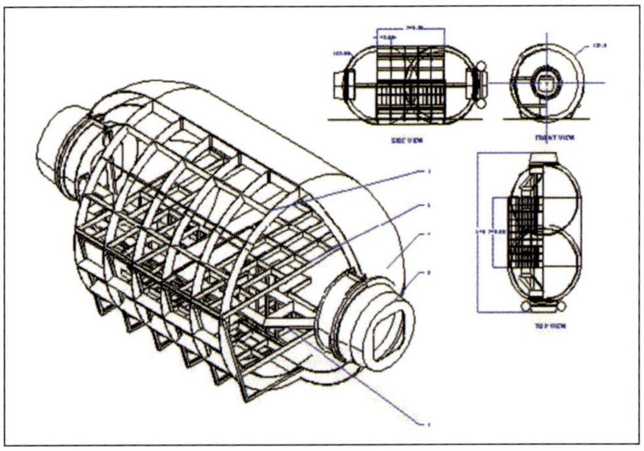

FIGURE 25 Construction drawings for full-scale mock-up.

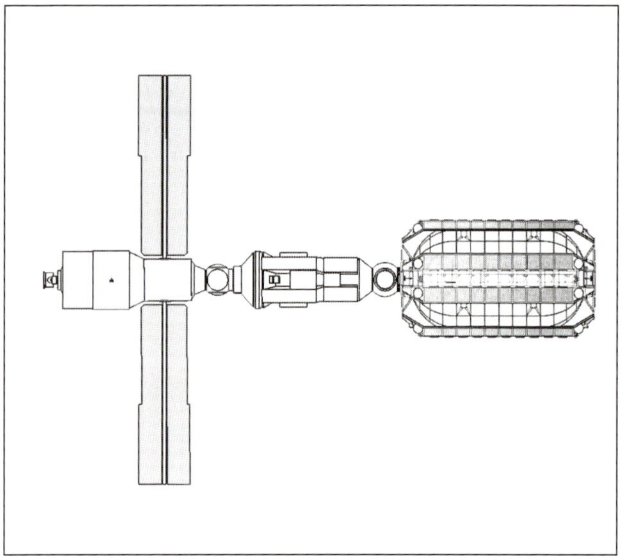

FIGURE 26 Top view of stage 3 developmental vehicle with support modules for housekeeping functions.

Stage 4

The final development phase is an independent vehicle capable of supporting commercial crew operations in one launch supported by logistics flights. This stage of development marks the refinement of inflatable technologies for space operations. The designers anticipate that the XV program will provide the core systems and capabilities for a variety of commercial space operations and play a vital role in increasing manned orbital infrastructure.

Stage 5

Refinement and adaptation of the XV core systems for future space exploration and commercial applications.

ROLE OF THE ARCHITECT

The design process used for XV concept development included the architect in design, development, and configuration of a commercial space vehicle. Architects worked with a team of engineers from diverse fields on preliminary design proposals that addressed technical, functional, and investor client needs. By conducting a series of architectural configuration studies, the design team envisioned an end goal for the system design process during early phases of the project.

Because of the uncertain requirements of potential user groups for the XV, the design process was structured in a flexible manner. This fluid design process adapted a methodology and structure similar to that used by architecture firms to fit within a more traditional aerospace research and development context as documented in various government-funded human space flight projects. This adaptation led to a flexible and robust design proposal that addressed economic realities and limitations, multifunctional client needs, future configurations, and growth patterns.

The architects provided valuable contributions to the design team by integrating information from previous space-vehicle design proposals into a new configuration that addressed unique design challenges related to an inflatable vehicle. Analysis of precedents and general understanding of the many phases of the design process allowed the architects to assist the project manager in communicating the project's goals, requirements, solutions, and historical context to the entire design team and to the investor client. |

References

Boeing Co. (1998), *Space Station Data Handbook.*

Cadogan, Stein, Grahne (1998), "Inflatable Composite Habitat Structures for Lunar and Mars Exploration," *Proceedings of the 49th International Astronautical Congress,* International Astronautical Federation, Paris.

Cohen, M. M. (1996), "Design of a Planetary Habitat Versus an Interplanetary Habitiat," Society of Automotive Engineers, Paper 961466, July.

Cohen, M. M. (1995), "The Suitport's Progress," AIAA Paper 95-1062, April.

ESA-ESTEC (2002), *European Cooperation for Space Standardization,* ESA Publications Div. Noordwijk, The Netherlands.

Hyde, Ishikawa, and Wood (1990a), "The Great Exploration Plan for the Human Exploration Initiative," LLNL Doc. No. PHYS.BRIEF 90_402.

Hyde, Ishika, and Wood (1990b), "NASA Assessment of the LLNL Space Exploration Proposal and LLNL Responses," LLNL Doc. No. SS 90-9.

Kennedy, K. J. (1999), "ISS TransHab: Architecture Description," Society of Automotive Engineers, Paper 1999-01-2143.

Lopper, C. A. (2000), "EVA Concept of Operation for International Space Station," Society of Automotive Engineers.

Messerschmid, E., and Bertrand, R. (1999), *Space Stations, Systems and Utilization,* Springer-Verlag Berlin.

NASA Engineering Directorate (1998), *JSC Design and Procedural Standards Manual, A Manual of Standards and Criteria for Design of Equipment Developed for Operation in Space.* JPG 8080.5, NASA Lyndon B. Johnson Space Center, Houston, TX.

NASA EVA and Crew Equipment Projects Office (1995), "Extravehicular Activity (EVA) Hardware Generic Design Requirements Document," JSC 26626A, NASA Lyndon B. Johnson Space Center, Houston, TX.

NASA Mission Operations Directorate, (1998), "International Space Station Familiarization," Space Flight Training Div., NASA Lyndon B. Johnson Space Center, Houston, TX.

NASA Office of the Director (1998), "NASA's Human Rating Requirements for Space Transportation Architecture Study," JSC 28354, NASA Lyndon B. Johnson Space Center, Houston, TX.

NASA Space Station Program Office (1999), "International Space Station Crew Integration Standard, Rev. C," SSP 50005, NASA Lyndon B. Johnson Space Center, Houston, TX.

NASA (1996), "NASA Space Flight Health Requirements Document," JSC-26882, Lyndon B. Johnson Space Center, Houston, TX.

NASA (2000), "NASA TransHab Program," ISS/TranHab Technical Interchange Meeting, Lyndon B. Johnson Space Center, Houston, TX.

Nixon, D., Miller, C. R., and Fauquet, R. (1998), "Space Station Wardroom Habitability and Equipment Study," NASA CR 4246, NASA Scientific and Technical Information Div. Washington, D.C.

Osberg, Uhl, Messerschmid, E. (2000), "An Interdisciplinary Engineering/Architectural Approach to the Conceptual Design of Space Stations," Society of Automotive Engineers, Paper 2000-01-2330.

RCS Energia (2002), "Preliminary Meetings and Conversations," www.energia.ru.

Volger, A. A., and Arch ETH (2000), "Micro-G-Architecture-A Transdisiplinary Education, Research and Product Development Project for Engineers and Architects," Society of Automotive Engineers, Paper 2000-01-2328.

Wieland, P. O. (1998). "Living Together in Space: The Design and Operation of the Life Support Systems on the International Space Station," NASA/TM-1998-206956/VOL1.

10

space hotel based on the TransHab

PAOLA FAVATA

INTRODUCTION

MODERN ADVENTURE TRAVEL started in the 16th century during the age of colonialism and from then until now has constantly increased. The first adventure travelers were explorers looking for new lands, cultures, and resources. Many of today's travelers are people involved in sedentary office work who desire vacations where they can be physically active. For this reason many countries are investing resources in "adventure travel" as an important industry in their economies.

Today, thousands of travel agents promote many types of adventure travel, and impressive amounts of money are spent annually on this sort of vacation and on its required equipment. For example, from 1990 to 1993, the U.S. tourist sector grew more than 50% and has overtaken agriculture to become the first source of profit from foreign capital (Spencer 1996).

Future space tourism will take advantage of unique and pioneering amusement activities in microgravity conditions. Based on public opinion polls, hundreds of thousands of people dream of having an experience in space; tens of thousands of people are ready to pay more than $100,000; and a few extremely rich people are already paying roughly $20 million each to visit the International Space Station (Space Travel and Tourism Div. 2000).

Considering mass space tourism as a long-term target, we can realistically predict development of the necessary means to carry out this dream. To do it, we must have safe and efficient human-rated launch, commercial tourist space platforms designed to high standards, and advanced ground facilities for training travelers.

SPACE HOTEL: WHY, HOW, AND WHEN

A realistic look at the space business landscape reveals that space tourism as presented in the literature is still not feasible. Enormous platforms remain utopian, and the design features of government space stations are not attractive for expensive vacation purposes. Fervent interest in this new market sector arises from young and brave companies working to commercialize the sector.

Space access by nonspecialist travelers will not grow through the scientific or government sectors, nor will it be based on philosophical, sociological, or humanitarian grounds. Very rich people will realize the first tourist platforms because they will invest their own money and reputations in then-available technologies for the pure passion, pride, and enjoyment of being pioneers in this "adventurist" sector. Potential profit will be a key motivation.

Given this reality, the most productive efforts space architects can dedicate to this ambitious objective consist in focusing their knowledge and competencies on developing fascinating and unique designs for high-end tourist habitats. Such designs cannot be accommodated on giant platforms, but at first must be hosted in small, feasible, and realistic modules based on already tested systems. This turns out to be easily feasible with current technology, as long as specific attention is devoted to optimizing and improving the confined environment for an adventure-travel lifestyle.

Such a practical approach can help make the new market sector real. At the beginning only very few people would have access to an exclusive vacation conceived in this simple way, within small habitats and hosting quite normal activities. Later,

Copyright © 2008 by Paola Favata. Published by the American Institute of Aeronautics and Astronautics, Inc., with permission.

however, larger platforms will be developed based on continuing technology advancements, and new types of space activities will begin to be planned and supported by completely different habitat types. This process will eventually open affordable space tourism access to less-rich people. The super-rich will start to look for new targets for their leisure time.

DESIGN DESCRIPTION

Anticipating a futuristic scenario of hotels, thermal baths, and sport centers in space, this chapter describes the design of an early space leisure environment to host 12 vacationers and their amusement activities in weightlessness.

The design conceit involves two principal elements. The first derives inspiration from the language of the toy world of childhood as a solution to many space interior design issues. Rounded furniture shapes avoid dangerous corners. Simplicity of the interiors—and arrangement and placement of functions—facilitates recognition, orientation, and identification. Eccentric use of material, fabric, and especially color breaks free of the traditionally rigid approach to space habitat design (Figure 1). Finally, use of the simplest available and safe technologies (mainly inflatable furniture) demonstrates feasibility of a completely new space environment despite inexpensive development on a near-term schedule.

FIGURE 1 Bar-fun-zone interior: Top—projection of the external space view on e-ink covering; middle—two windows allow restraints for comfortable viewing; and bottom—curved floor is equipped with handrails and a lighting system provides reference cue. Spherical deployable chairs offer mobile seats for use of computers.

The design project focuses on three main goals:

1) The first goal is freedom of creativity with regard to shapes, colors, materials, functional arrangement layouts, direction of furniture use, and relation of every design detail to the enjoyment of a new, artistic, and charming amusement environment (Figure 2).

The second design element is related to amusement activities in microgravity conditions, which are emphasized often when people talk about future space tourism. This project aims to demonstrate near-term feasibility by focusing on common human activities and terrestrial habits, which become absolutely new, completely different, complex, and fun to carry out in space.

This approach to communicating about space habitation demonstrates that it is not necessary to amaze lay people with huge, pioneering, utopian platforms. Rather, it is better to use available, already tested and safe technology, thereby reducing the scope of our future vision a little bit but making it possible now.

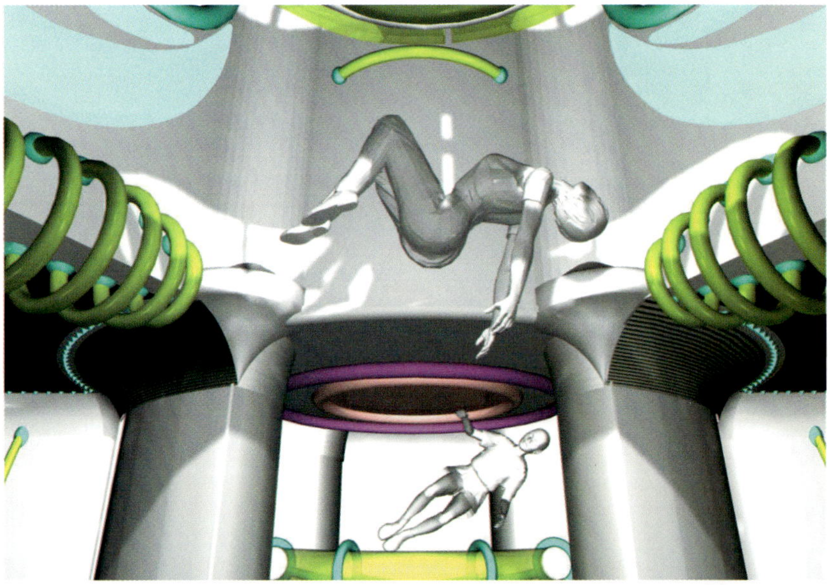

FIGURE 2 Restaurant interior. Inverted, T-shaped room hosts double-level restaurant connected directly to hotel entrance (central in the image). Inflated soft restraints preclude dangerous corners and offer recognizable color cues.

2) The next goal looks for a rational approach in choosing appropriate, effective, and especially safe technologies. This results in designing the whole interior using inflated and reconfigurable units to take advantage of minimum storage volume and weight (Figure 3).

3) Simplicity and spontaneity providing for both psychological and physiological comfort of the guests is the final goal. Much attention is paid to sociological factors (number of guests, positive subgroup arrangements, ages, activities, volumes) while pursuing the best ergonomics for furniture design (Figure 4).

FIGURE 3 Health care room interior. Restyled treadmill and bicycle arranged in dedicated area furnished with soft, inflated containers acting as restraints for movements and as storage spheres for towels, drinks, and personal items. E-ink textile screen follows wall curves and offers pleasant views during exercise.

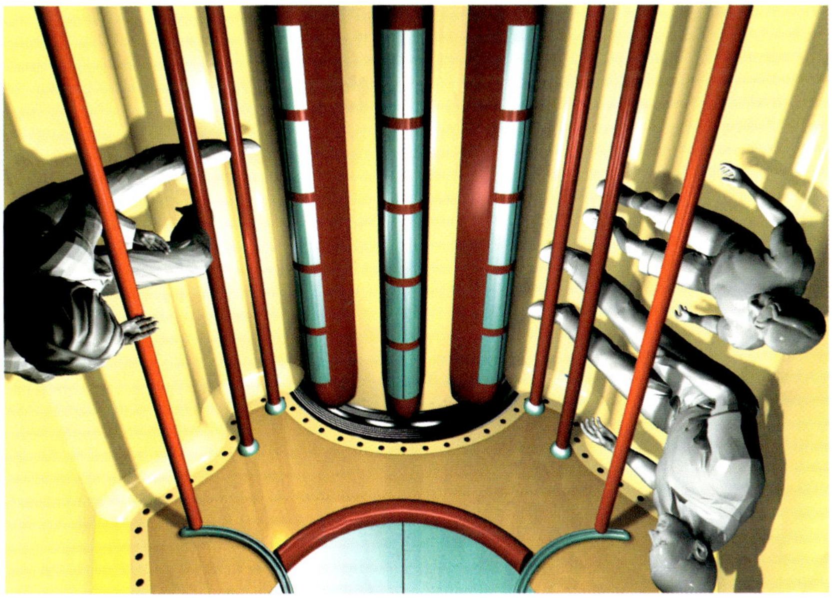

FIGURE 4 Mars room interior. Entire room is configured horizontally. Sliding inflatable restraints run along positioning rails that accommodate different body sizes. Fixed, soft pipes ensure correct lower-body restraint. Soft containers hold drinks and snacks (middle).

Layout

The hotel is based on the inflatable TransHab module because of its larger dimensions compared to traditional, aluminum, hardshell pressure vessel modules. Such a hotel module could be hosted by the ISS or a private space platform and provide accommodations for 12 to 16 people, which is probably the minimum number necessary for meaningful hotel social life.

TransHab's interior layout consisted of four levels arranged vertically (see Chapter 8). The hotel proposal maintains TransHab's vertical zone distribution but plans a lifestyle that follows a different outfitting logic, purposefully exploiting all directions in the microgravity environment. In the original TransHab concepts, a "terrestrial" way of using the interiors arises from the desire to maintain Earth living habits and perceptual references without complicating psychophysiological conditions for the crew.

The hotel concept instead suggests using local references rather than total-habitat references because many astronauts and cosmonauts have confirmed their adaptability to nonuniform interior orientations. Because of the free flight of weightlessness, and the use of equipment throughout space habitats, astronauts and cosmonauts have already demonstrated a nonterrestrial living style requiring only local orientation cues for well-being (e.g., sleeping comfortably attached to the "ceiling"). A direct benefit of a layout optimized for local orientation references is that it allows a far more efficient use of volume. This in turn allows packing a great diversity of environment experiences into the TransHab-size module.

The hotel has six private rooms equipped with private bathrooms and varied in size, each capable of hosting up to four people. The largest spaces distributed on each level are allocated for social activities: entrance, restaurant, bar-fun-zone-disco, health-care room, gymnasium, and two lounges for relaxing and communicating with Earth. These common areas fill almost two-thirds of the entire volume and intercommunicate with each other.

The design pays special attention to human factors, shapes, materials, and colors to facilitate living, activities, and spatial recognition. Physiological conditions in microgravity, color theory, and perception of the interiors during private and common activities were all studied to complete the design. Space-compatible industrial fabrics and materials were investigated to understand how various fabric textures and material densities could realize new shapes and new functions. This led to the decision to use predominantly rounded and soft shapes. Diverse functions are fitted into multi-use areas but differentiated by different cues, and colors are used according to the respective functions and expected human mood during their fulfillment. Social and amusement activities were investigated to achieve an innovative space vacation cruise with a futuristic design, characterized by an enjoyable vacation environment, different from those used now for scientific and engineering missions, which would not be attractive to travelers for holiday purposes.

Finally, with the exception of technical equipment and a few furnishings, the whole interior is designed with inflatable furniture, which can be compacted for launch. Once in orbit, it would be necessary only to deploy and inflate the TransHab module, configure the structural frames, and finally inflate the interior furniture. Only two shuttle-equivalent launches would be needed to assemble the whole hotel, thus completing its activation in the shortest time and with the lowest cost possible.

The hotel offers accommodation for 12 to 16 people. A shuttle-equivalent launch system could deliver this number of people in a *Spacehab* double module outfitted with passenger seating; *Spacehab* originally conceived this as one of its mission types.

A comfortable sojourn for this small community is possible as a result of the bigger dimension of the inflatable TransHab module (342 m^3 of total volume) compared to traditional aluminum modules (average of 42.5 m^3). Here, a short-term voyage is foreseen (one to two weeks). This would minimize negative physiological impacts of microgravity on the guests (longer stays would require serious physical activity and monitoring) as well as provide time in space beyond the first few days (the most acute period of space adaptation). For a two-week mission physical activity can be enjoyed as merely a source of amusement rather than as a physiological need.

The original TransHab interior layout consisted of four functional levels stacked inside the pressurized volume (Figure 5). The first contained living room and light storage; the second contained system equipment and, inside a safe haven in the rigid central core, six crew quarters; the third contained a health-care room for physical exercise and light storage; and the fourth was the entry tunnel. The spaces partially intercommunicated, and the living room was a double-height space.

The hotel design maintains the four levels, but its interior outfitting uses a completely different logic. It exploits local orientation to accommodate the functions. This approach allows mobile furnishings to be used in different and unusual orientations (e.g., the "upside-down" table shown in Figure 6). At the same time the furnishings are designed to be always unmistakably recognized and approached because they are neither symmetrical nor uniform, and they present diverse physical characteristics in each direction (Figure 7).

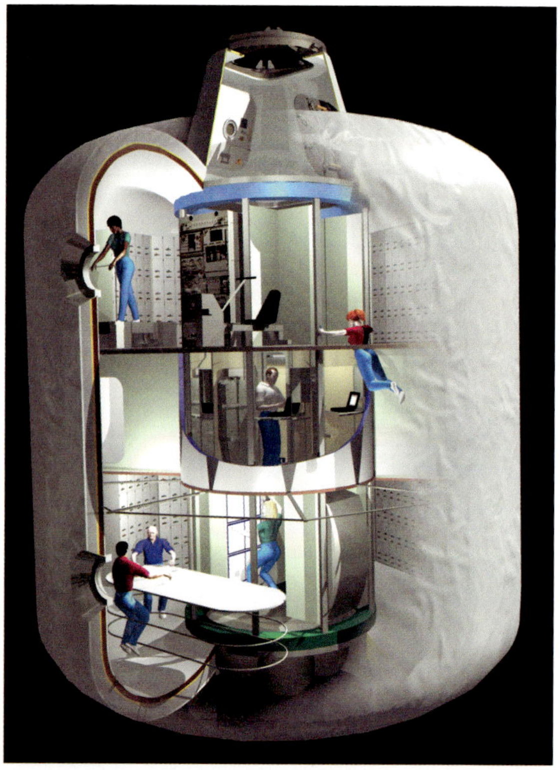

FIGURE 5 Original TransHab inflatable module interior design. (Courtesy of http://www.astronautix.com/ © Mark Wade, http://www.nasatech.com/Briefs/Jan01/MSC22900.html, and http://www.alespazio.it/program/infr/iss/transhab/transhab.htm.)

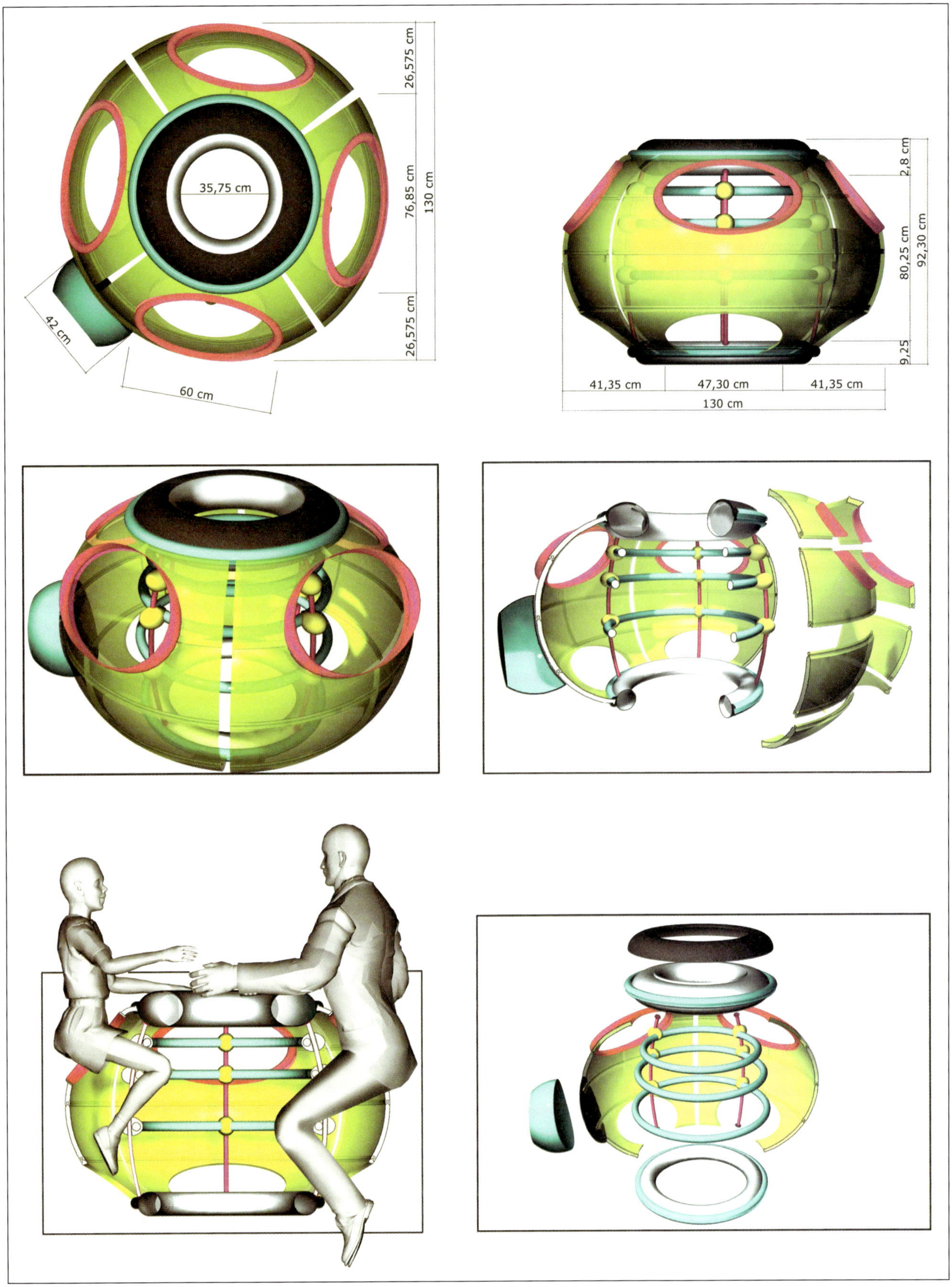

FIGURE 6 Restaurant table. Composed of four main elements made of laminated aerogel, it can be linked to the walls via Velcro® joints and arranged and turned in any direction.

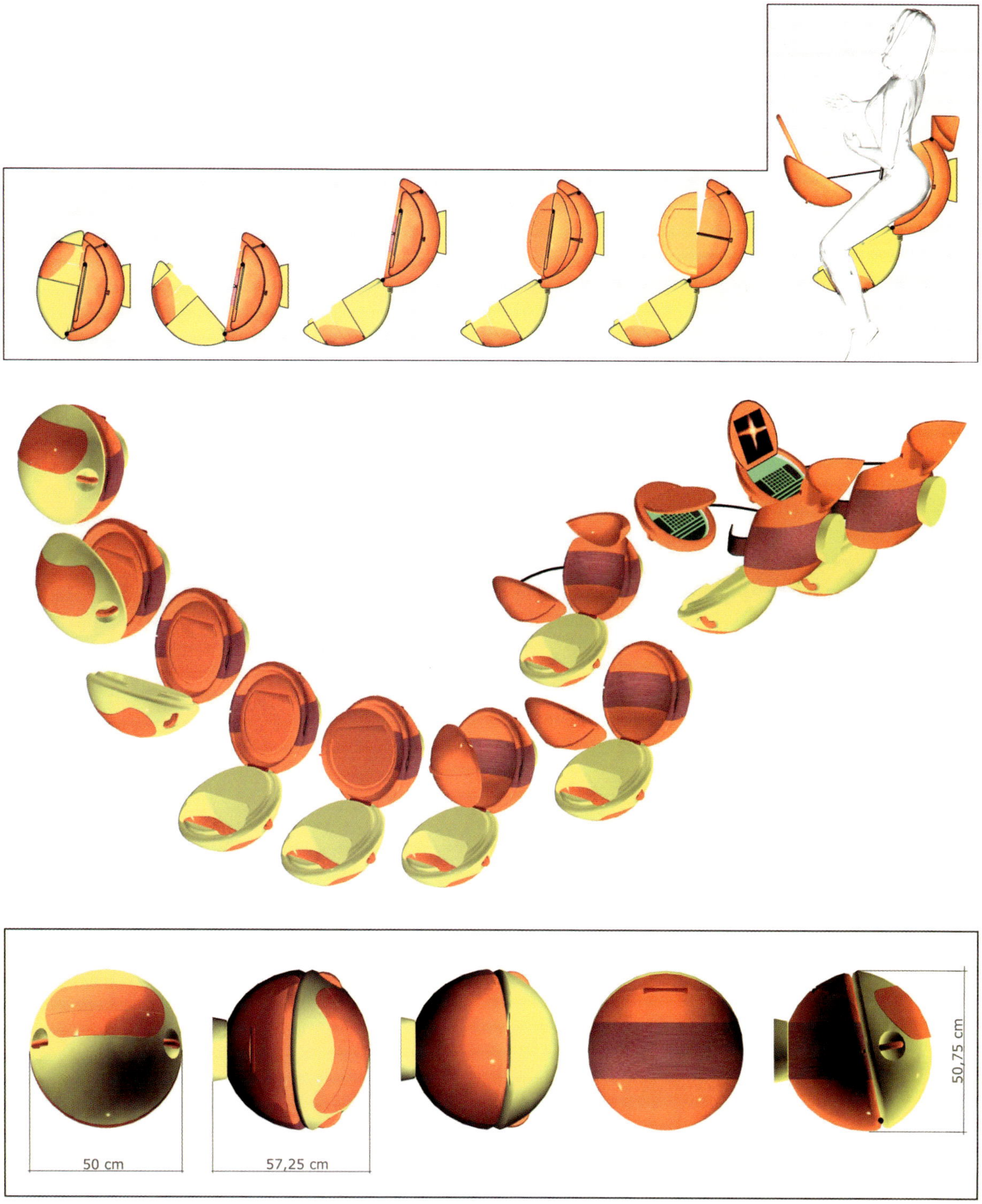

FIGURE 7 Equipped inflatable sphere: front, top, bottom, back, and side. Made of both soft and rigid materials, semi-inflatable, and mountable to the walls via a Velcro® joint, it is deployable as a chair equipped with hardware and soft body restraints.

It is interesting to compare the internal free volume of the TransHab with the rearrangement shown here for the hotel program. While the TransHab would provide about 18 m^3 of free volume per person with a capacity of 12 crew, the hotel would offer about 22.5 m^3 per person. This is even more striking because the basic TransHab already increases free volume per person by 27% compared to aluminum ISS modules.

From a structural point of view, nothing is changed except the water-wrapped safe haven inside the rigid core. Instead of having only the second level dedicated to the safe haven with six private cabins, in the hotel the safe haven is doubled to arrange two common lounges able to host eight people each. This function, of course, is not needed for a low-Earth-orbit hotel, but it is important to understand how it could be accommodated within a hotel design. This modification was considered with the support of Alenia Spazio engineers who suggested it to reinforce central core compression strength for launch loading and structure stiffness for launch vibration. These two common rooms are placed on the first and second levels within the central core. They intercommunicate but can be closed and separated for security reasons in emergency situations (Figures 8–10). The structural panels inside the core, used for strength during launch, are removed once in orbit.

For anchoring racks, current technology is used; the racks are rigidly linked to the structural central core, close to the utility chases. The only two exceptions are the health-care rack, which in any case has been kept close to the central plant and anchored to a rigid partition wall (Figure 11), and the galley rack (oven-refrigerator-freezer facility) in the restaurant room, where two of the four units have been positioned next to the shell (Figures 11–13).

In this design the racks are a variation of current International Space Station (ISS) racks; they follow a common, serial production approach but with half the volume of ISS racks. They can be vertically paired to form the traditional rack volume and stored without any change in the structural core.

Internal ventilation and air-handling are provided as usual by life support systems (ECLSS) equipment to ensure a safe and comfortable

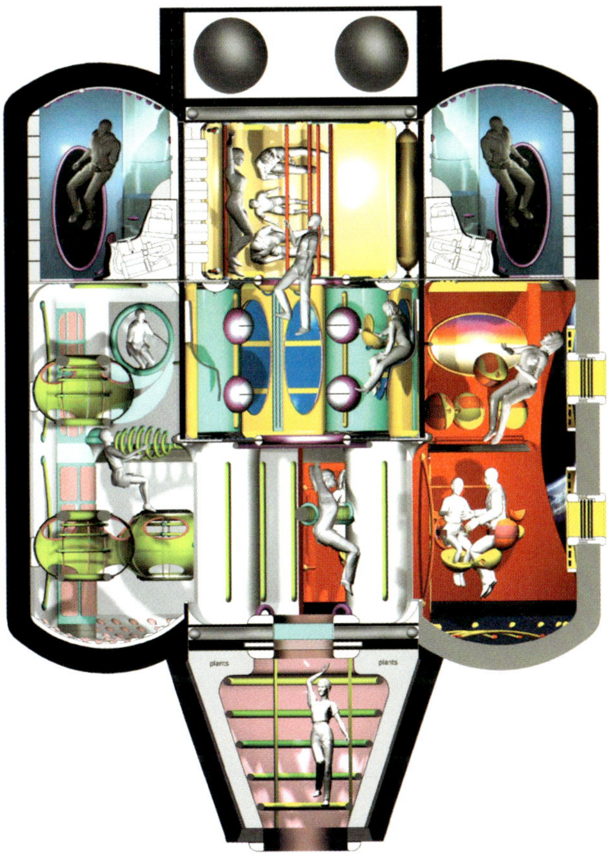

FIGURE 8 Vertical section through common rooms: left—restaurant; and right—bar.

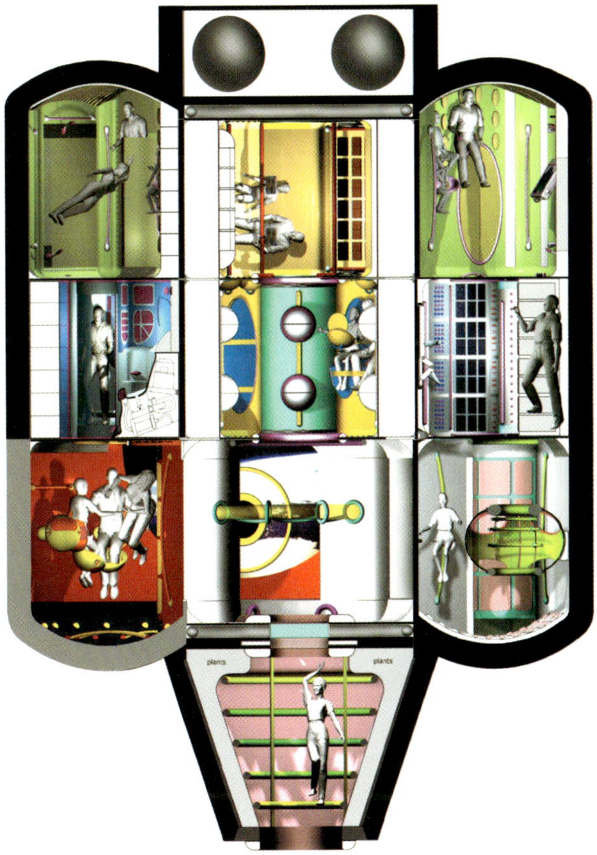

FIGURE 9 Hotel vertical section through common and private rooms: clockwise from top right—large private room, health-care room, restaurant, hotel entrance, bar-fun zone, and bathroom; central core, starting from top—Mars room, Moon room, and central hall.

FIGURE 10 Hotel vertical section through common and private rooms: clockwise from top right—large private room, bathroom, bar-fun zone, hotel entrance, restaurant, and gymnasium; central core, starting from top—Mars room, Moon room, and central hall.

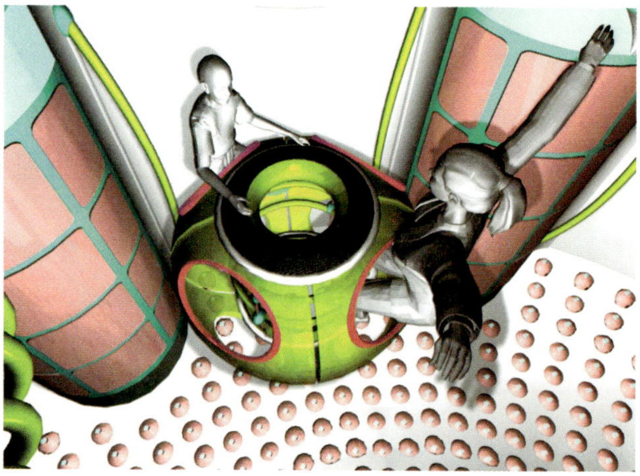

FIGURE 11 Restaurant table between two galley racks.

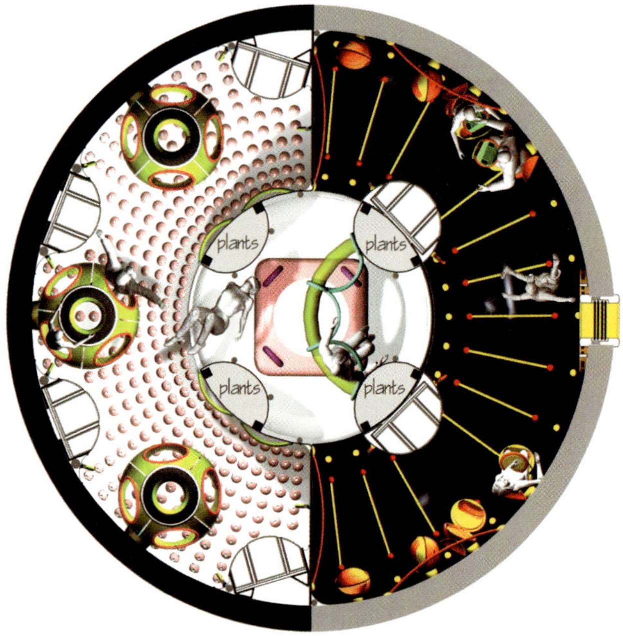

FIGURE 12 Third-level floor plan: left—restaurant; and right—bar-fun zone with window.

FIGURE 13 Third-level ceiling plan: left—bar-fun zone with window; and right—restaurant.

breathing, humidity, and temperature environment throughout the whole hotel. Circulation is aided by spatial communication between the common rooms.

The windows unfortunately remain the weakest design feature of the project by being limited to the TransHab configuration: two small, nadir-oriented portholes (Figures 14–17). The reason is structural: to increase the number or size of the windows would mean departing from TransHab design heritage, requiring redesign of the shell and reducing the technical credibility of the hotel concept. The only modification assumes locating both windows in the

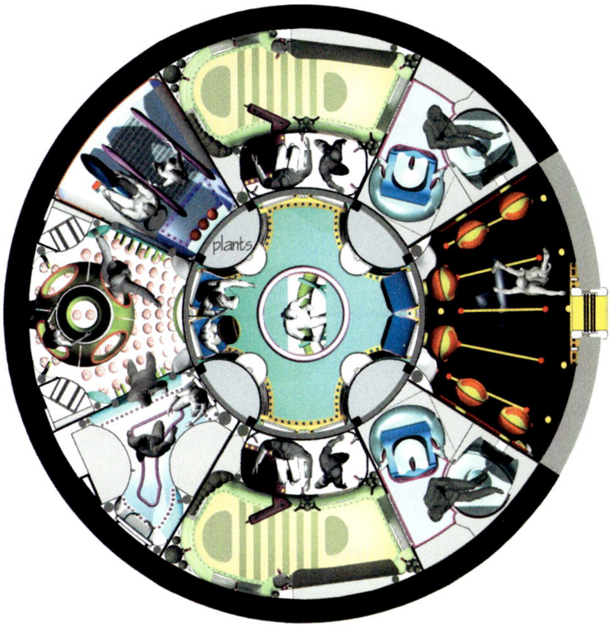

FIGURE 14 Second-level floor plan: clockwise from top—private room, bathroom, bar upper level, second bathroom and small private room, health-care room, restaurant upper level, and gymnasium.

FIGURE 16 First-level floor plan: four large private rooms furnished with bathroom.

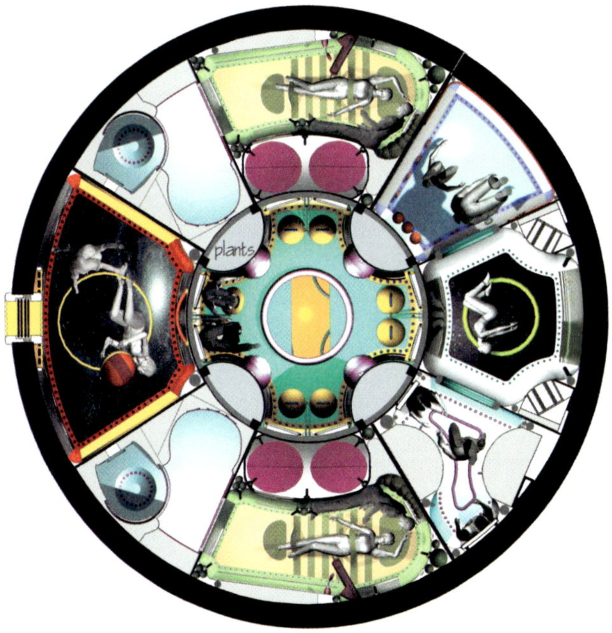

FIGURE 15 Second-level ceiling plan: clockwise from top—private room with bathroom, gymnasium, restaurant upper level, health-care room, second small room and bathroom, and bar upper level.

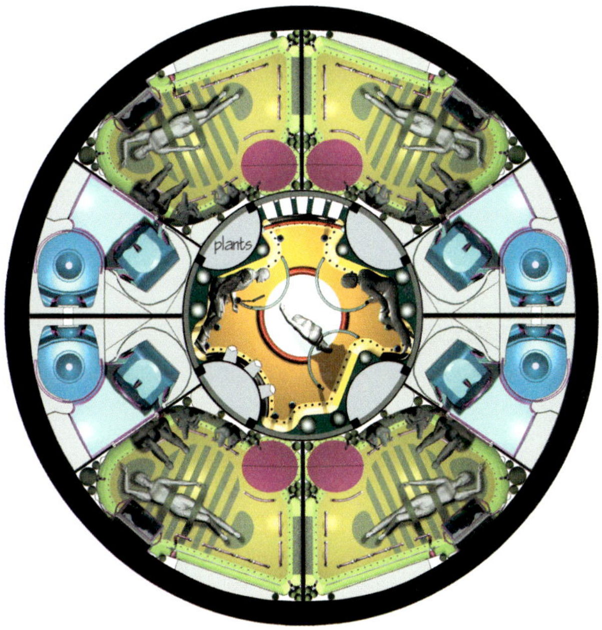

FIGURE 17 First-level floor plan, alternate view: four large private rooms furnished with bathroom.

bar-fun-zone-disco, so as to arrange two unusual areas for external viewing.

The hotel is conceived as a dependent module; as many technical functions as possible are located external to the module, in an adjacent node, for noninterference access by the station's technical crew. This relocation reduces noise and vibration in the interior environment. The few system racks located within the hotel module are acoustically and vibrationally isolated. The private-room areas are entirely wrapped with a soundproofing resin that has performed well also for vibration absorption, as suggested by automobile and other industrial applications.

Design Drivers

The idea underlying this project is an intention to change our mental image of space habitats, moving away from what we are used to seeing, and to utilize currently available, space-compatible technologies to 1) make this creative dream a reality with minimum economic investment and 2) ensure a safe and high-quality lifestyle. What better circumstance than space tourism to explore these new languages?

The first step of the project was focused on analysis of basic human needs and then on common human activities related to leisure time. Many new microgravity-environment functions and possibilities have been raised in the literature, but the final focus for this project was investigating how common private and social human activities could be carried out in space: sleeping or watching TV (Figure 18), taking care of body hygiene (Figure 19), circulating (Figure 20), relaxing with friends (Figure 21), floating in weightlessness (Figure 22), and moving between private and common areas (Figure 23). In addition, a few activities were included as a result of obligations caused by the extreme environment, such as the gymnasium for experiencing how to exercise the body in weightlessness (Figures 3 and 15), and the health-care room, which is indispensable in orbiting habitats as well as in vacation facilities.

FIGURE 18 Large private room: mobile soft restraints allow hanging seats; elastic strips allow restraints for unusual beds on floors and ceilings.

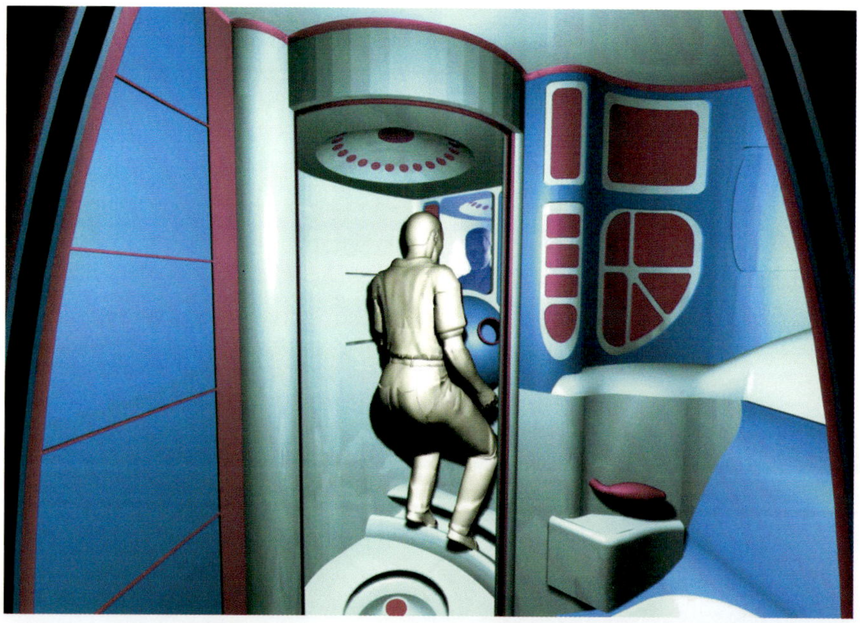

FIGURE 19 Bathroom: view from pass-through window toward shower.

FIGURE 20 Small private room: guest passing through elastic soft-layer door from bathroom.

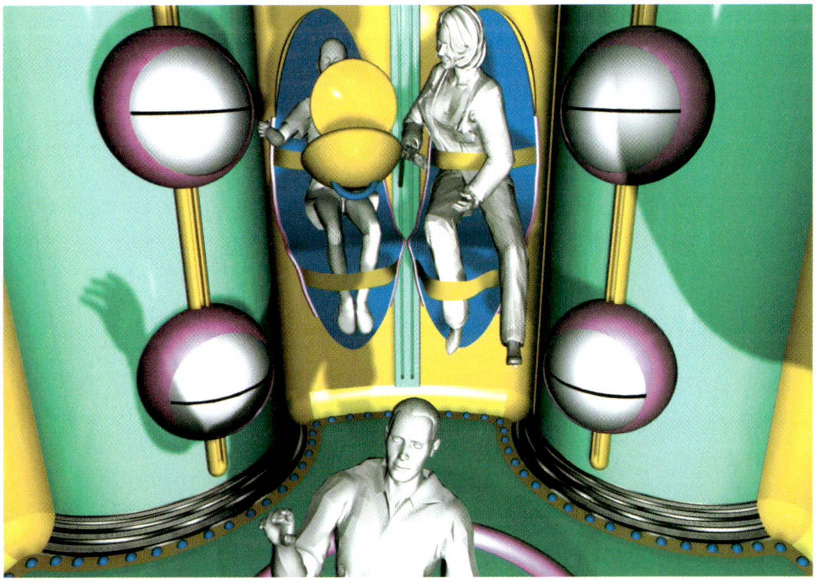

FIGURE 21 Moon room: inflatable adaptable bed-chairs offer lounges equipped with computers.

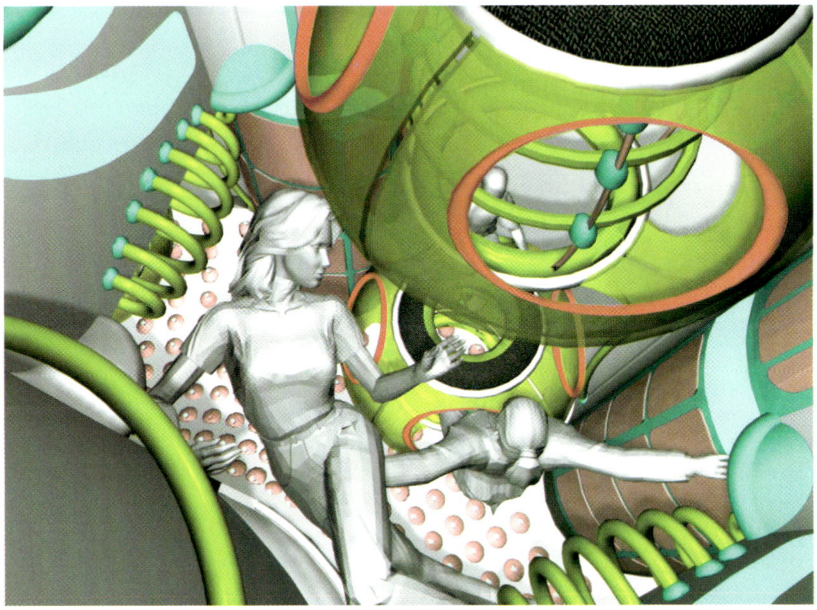

FIGURE 22 Restaurant: view through suspended rotating tables.

FIGURE 23 Bar-fun-zone: view of pass-through between common and private areas.

At the beginning, the possibility of utilizing fewer, bigger common rooms for space amusement, instead of a dense subdivision into very small spaces, was considered. Even if microgravity vertigo is not an issue (especially in the case of a "tower" typological configuration), in the end the choice converged onto two criteria: 1) traditional typology of public buildings for leisure and 2) typology of luxury cruise ships. Obviously, differences would be manifest because of the different environment. Therefore, efforts were concentrated on how to arrange private and social areas in the best way to facilitate circulation, technical equipment distribution (which remains one of the most difficult problems to resolve—where to find the space to put boxes, cables, ducts, etc.), and human orientation. It was essential also to optimize space with the specific challenge of "maximizing" such small rooms using the resources at hand: shapes, furniture, light, colors, materials, textures, body restraints, and other details.

Fundamental was the investigation of human behavior and preference, both in private and social occasions, with specific focus on proximity issues critical for small, confined habitats. Much effort centered on how to create a very flexible environment characterized by mobile, personalizable, and interchangeable furniture that is rich in variety, multi-use in function, full of diverse sensory inputs, and easily recognizable in coding and comfortable orientation cues to aid circulation.

In the end, the usable interior volume of the TransHab module is allocated according to anticipated intervals of time for typical behaviors: approximately one-third dedicated to private rooms and two-thirds to larger common rooms used for social interaction. In this way, both types of activities are comfortably accommodated.

PRIVATE ROOMS The four large private rooms (Figures 16 and 17) and the two small ones (Figures 14 and 15) are equipped with private bathrooms whose novel design incorporates a shower unit. A polymeric multilayer film offers a large, safe mirror that is also the bathroom entrance portal. In the rooms, inclined (not horizontal) beds and lounges are arranged to provide a relaxing environment with soft colors and materials, such as elastomers, nylons, etc. Inflated Velcro® accessory restraints provide easy circulation and anchorage for the guests. Communications, computer, and video and stereo equipment complete these interiors.

COMMON ROOMS (Figures 12–15). The social activities areas begin with the entrance, linked to a rounded central lounge that is itself connected with the bar-fun-zone-disco. Here a few racks provide drinks and light restaurant food; spheric-chairs can freely rotate to face any direction, attached with Velcro® to the walls. Two windows equipped with circular soft restraints offer outer space views, and the whole interior is covered with electronic paper, which can reproduce exterior or other images, turning this interior space into a virtual space-outdoors. The walls could also display video projections or whatever material, texture, or color is desired by the guests.

The restaurant is designed with light colors, and racks provide gourmet food and beverages. Very light, aerogel-core spheric tables are "Velcroed" to

FIGURE 24 Moon room for quiet entertainment in safe haven.

the walls and, once again, can be used in any direction, hosting up to four guests each. Soft inflated toruses provide ergonomic restraints for different body sizes, and a "Velcroed" upper ring ensures anchorage of food trays.

From this room it is possible to reach the gymnasium and the health-care room, which are separated from the restaurant for hygiene reasons but remain partially in communication with it.

Finally, two lounges inside the central core (safe haven areas) offer quiet rooms equipped with electronic libraries, video projections, and other entertainment outlets including foods and beverages. The Moon room is furnished with inflated bedchairs made with parallel faces and zig-zag sewing. The Mars room, by contrast, is arranged in the opposite horizontal direction, inspired by amusement park facilities and equipped using the same logic as the Moon room (Figures 24 and 25).

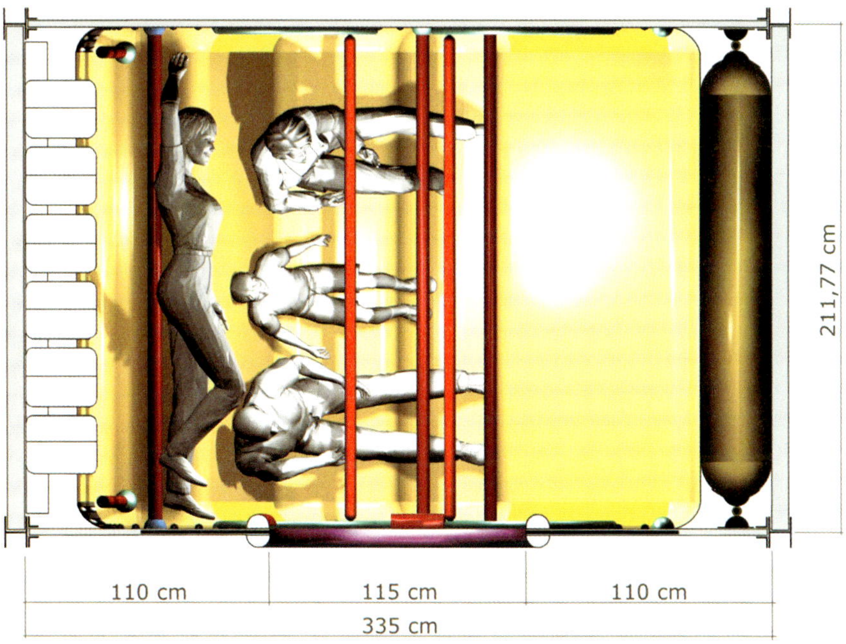

FIGURE 25 Mars room for communal entertainment in safe haven: inflated hinged column visible on the right; these units link furnishings to the walls.

INFLATABLE TECHNOLOGY, MATERIALS, AND DEPLOYMENT

One of the main features of the project is the use of inflatable technologies to equip and characterize the habitation functions with minimum cost, weight, and deployment time.

Module commissioning occurs in three stages. The first follows the same procedure designed by NASA to deploy the TransHab module: the external shell is inflated; vertical floor struts hinged to the structural core rings (parallel to the longerons in the launch phase) are deployed, the upper four toward the bottom and the lower four toward the top; honeycomb vertical wall panels are attached to the rigid frame; these panels in turn host soft floors through metal eyelets; and wall panels and soft floors are covered with resin layers using Velcro® strips to provide sound and vibration conditioning.

The second stage, outfitting the basic hotel interior, involves hooking two hinges for inflatable cylindrical columns to ceiling and floor joints within each room. The columns are inflated, pulling with them the entire continuous cover of the room. The cover is then attached to the walls using colored eyelets. The materials are flame-resistant, low-outgassing fluoroelastomers already tested and approved for space applications, which coat the aramid fibers before being vulcanized and varnished with silicone elastomers.

FIGURE 26 Hotel interior design uses well-understood technologies to create a novel environment for space tourists.

The third stage comprises two manual operations inside the configured rooms: 1) attachment, hookup, and functional verification of the racks; and 2) configuring the furniture, that is, assembling the rigid unit puzzle pieces and inflating the soft units.

All of the inflatable structural units and furniture are conceived by adapting traditional marine boat and raft technologies to the various functions of the furniture shapes and the mechanical characteristics they require. Three inflatable shapes are used: 1) tubular with minimum diameter of 10 cm; 2) complex shapes rigidized with internal diaphragms; and 3) parallel-plane boards with zig-zag sewing. Manual deposition in molds produces the more complex shapes; pultrusion produces the constant-section beams and units; and spray-up produces serializable elements. Welding (radiofrequency, hot air or hot wedge systems), gluing (epoxy resins or other glues) and sewing techniques can be used to manufacture the furnishings, depending on their specific shapes and requirements.

Predominant use of inflatable furnishings allows the equipment to be compacted into a very small envelope for launch inside the structural core. Various technologies were considered for this delicate packing operation, and complex approaches (including vacuum packaging that causes embrittlement or talcum powder that causes outgassing) do not appear necessary.

CONCLUSION

This design case study shows how available technologies can be combined using a novel design sensibility to create a habitable space environment that is attractive for tourist purposes but also safe, realistic, and comfortable (Figure 26). By setting our sights realistically and applying skilled architecture design, near-term orbiting amusement facilities could be readily achievable even using already tested, safe, and available technologies.

The character of the design is inspired by the toy world of childhood. Apart from being appropriate for conceiving safely rounded space equipment and furnishings, this aesthetic is a bridge between the sophisticated style inevitably requested by the few rich people who can afford such space voyages and the funny, colorful style dreamed by all children, young and old. In the very end, isn't space tourism the dream of the child inside each one of us?

References

Space Travel and Tourism Div. (2000), "Going Public 2000: Moving Toward the Development of Large Space Travel and Tourism Business," Space Transportation Association, Arlington, VA, June 26.

Spencer, J. (1996), "Wealth from Space Tourism," Space Tourism Workshop, NASA and the Space Transportation Association, Paper.

Bibliography

Alpert, M. (1999), "Making Money in Space," *Scientific American,* Vol. 10, No. 1, May, pp. 92–95.

Anderton, D. A. (1985), *Space Station,* U.S. Government Printing Office, Washington, D.C.

Bell, L., and Hua, L. (1988), "Inflatable Space Structures," *SICSA Outreach,* Vol. 1, No. 7, May–June.

Bell, L., and Trotti, G. (1986), *Hab Module Outfitting Design Concepts*, Bell and Trotti, Inc., Houston, TX.

Caprara, G. (1998), *Abitare lo Spazio,* Mondadori, Milan.

Collins, P. (1998), "Tourism in Low Earth Orbit: the Trigger for Commercial Lunar Development?" http://www.spacefuture.com/archive/tourism_in_low_earth_orbit_the_trigger_for_commercial_development.shtm [retrieved 5 June 1999].

Collins, P. (1999), "Design and Construction of Zero-Gravity Gymnasium," http://www.spacefuture.com/archive/design_and_construction_of_Zero_Gravity_Gymnasium.shtm [retrieved 21 April 1999].

Collins, P., Akiyama, T., Shiraishi, I., and Nagase, T. (1996), "Service Expected for the First Phase of Space Tourism," *Space Energy and Transportation*, Vol. 1, No. 1.

Collins, P., Iwasaki, Y., Kanayama, H., and Ohnuki, M. (1994), "Commercial Implications of Market Research on Space Tourism," *Journal of Space Technology and Science*, Vol. 10, No. 2.

Computer Sciences Corp. (2000), "International Space Station Operations Architecture Study: Final Report," NASA Contract GS-23F-8029H, Aug., http://spaceflight.nasa.gov/station/reference/iss_oas.pdf [retrieved 23 July 2002].

De Grandis, L. (1984), *Teoria e Uso del Colore,* Mondadori, Milan.

Dunn, M. (1999), "Launch. Inflate. Insert Crew," *Air and Space Magazine*, April–May, pp. 20–27.

Gaubatz, W. A. (1995), *Space is a Place*, McDonnell Douglas, Huntington Beach, CA, Sept.

Goethe, J. W. (1999), *La Teoria dei Colori,* Il Saggiatore, Milan.

Hall, T. W. (1997), "Artificial Gravity and the Architecture of Orbital Habitats," http://www.spacefuture.com/archive/artificial_gravity_and_the_architecture_of_orbital_habitats.shtml [retrieved 5 June 1999].

Hall, T. W. (1995), "The Architecture of Artificial Gravity: Theory, Form, and Function in the High Frontier," http://www.spacefuture.com/archive/the_architecture_of_artificial_gravity_theory_form_and_function_in_the_high_frontier.shtml [retrieved 21 April 1999].

Haltermann, R. L. (1996). "Evolution of the Modern Cruise Trade and Its Application to Space Tourism," 1997 Space Tourism Association Rep., Space Tourism Association, Arlington, VA, Nov.

Harris, P. R. (1984), *Living and Working in Space: Human Behavior, Culture and Organization*, Ellis Horwood Library of Space Science and Space Technology, Crystal City, VA.

Hilton, B. (1967), "Hotels in Space," (American Astronomical Society Paper 67_126), May, http://www.panix.com/~kingdon/space/hilton.html [retrieved 5 Sept. 2002].

Itten, J. (1965), *Arte del Colore,* Il Saggiatore, Milan.

Kennedy, K. J. (1999), "ISS TransHab: Architecture Description," Society of Automotive Engineers, Paper 1999-01-2143, July.

Mullane, M. R. (1997), *Do Your Ears Pop in Space?* Wiley, Hoboken, NJ.

NASA (1995), "Man-System Integration Standards," NASA STD 3000, Rev. B, July.

NASA, and STA (1997), "General Public Space Travel and Tourism," *Workshop Proceedings*, Vol. 2, Space Transportation Association, Arlington, VA.

NASA (1998), "International Space Station Familiarization," NASA TD9702A, July 31, http://spaceflight.nasa.gov/spacenews/factsheets/pdfs/td9702.pdf [retrieved 23 July 2002].

NASA (2000), *International Space Station User's Guide*, Oct. 31, pp. 1–11, http://spaceflight.nasa.gov/station/reference/issug/ISSUG1–11.pdf [retrieved 23 July 2002].

NASA (2000), *International Space Station User's Guide*, Oct. 31, pp. 12–34, http://spaceflight.nasa.gov/station/reference/issug/ISSUG12–34.pdf [retrieved 23 July 2002].

NASA (2000), *International Space Station User's Guide*, Oct. 31, pp. 35–50, http://spaceflight.nasa.gov/station/reference/issug/ISSUG35–50.pdf [retrieved 23 July 2002].

NASA (2001), *A Key to Discovery: The International Space Station Fact Book,* Aug. 13, http://spaceflight.nasa.gov/spacenews/factsheets/pdfs/factbook/issfactbook2001.pdf [retrieved 23 July 2002].

NASA (2002a), "NASA Human Spaceflight" July 13, http://spaceflight.nasa.gov/spacenews/factsheets/ [retrieved 23 July 2002].

NASA (2002b), *NASA Human Spaceflight, International Space Station Fact Book*, July 13, http://spaceflight.nasa.gov/station/reference/factbook/ [retrieved 23 July 2002].

NASA (2002c), "NASA Human Spaceflight, Space Station Crew," July 13, http://spaceflight.nasa.gov/station/crew/ [retrieved 23 July 2002].

NASA (2002d), "NASA Human Spaceflight, Technology," July 13, http://spaceflight.nasa.gov/mars/technology/ [retrieved 23 July 2002].

Pine, II, J. and Gilmore, J. H., (1998), "Welcome to the Experience Economy," *Harvard Business Review*, July–Aug., pp. 97–105.

Portal, F. (1997), Sui Colori Simbolici nell'Antichit_, nel Medioevo e nell'Et_ Moderna, Luni Editrice, Milan.

Sadeh, W. Z., and Criswell, M. E. (1993), "Inflatable Structures - A Concept for Lunar and Martian Structures," AIAA Paper 93-0995, Feb.

Sadeh, W. Z., and Criswell, M. E. (1993), "Inflatable Structures for a Lunar Base," AIAA Paper 93-4177, Sept. 1993.

Sandy, C. R. (1995), "Development of the Mars Pathfinder Inflatable Airbag Subsystem," AIAA Paper 95-3796), Sept.

Satter, C. M., and Freeland, R. E. (1995). "Inflatable Structures Technology Applications and Requirements," AIAA Paper 95-3737), Sept.

Stine, H. G. (1997), *Living in space,* Co., New York.

Welch, J. F. (1990), *Van Sickle's Modern Airmanship*, 6th ed., Tab Books, Blue Ridge Summit, PA.

White, R. J. (1998), "Vivere in Assenza di Gravit," *Le Scienze*, No. 363, Nov.

Widmann, C. (2000), *Il Simbolismo dei Colori,* Edizioni Scientifiche Ma. Gi. Srl., Rome.

Wolff, H. J. (1997), "Trends Shaping the Design of Future Destinations," NASA and the Space Transportation Association, Paper, Public Space Travel Workshop, Feb.

Zubrin, R. (1999), "Sending Humans to Mars," *Scientific American*, Vol. 10, No. 1, May, pp. 46–51.

11

habitats for long-duration missions

BRENT SHERWOOD
AND
STEPHEN D. CAPPS

INTRODUCTION

HABITAT CONCEPTS naturally attract great interest for a human space-exploration program. Habitats are sophisticated pressure vessels that contain and protect the ultimate payload—people. They are complex, heavy, expensive elements around which other spacecraft systems are functionally arrayed, both for transportation vehicles and for permanent facilities like planetary bases. Primary requirements for any space habitation element are reliable structural integrity with adequate safety margins, graceful failure modes (leak before rupture), ability to be fully tested before being put into service, long life, straightforward outfitting and servicing, and successful accommodation of users' activities.

Additionally, long-duration habitats must address effectively the psychological needs of confined, high-performance crews. A final requirement is commonality, defined as the ability to use an element in multiple settings throughout the program architecture. The extremely high cost of developing space hardware mediates strongly in favor of multi-use elements.

This chapter reports the methodology and results used during the first Space Exploration Initiative (SEI), in 1990, to investigate preliminary design concepts for long-duration, rigid habitation elements. As Class I systems, these would be manufactured, integrated, and fully tested on Earth prior to launch to service both kinds of long-duration mission: transfer between Earth and Mars and surface bases for the Moon and Mars. Although the purpose of the trade study was to select the reference design to be incorporated into NASA Mars program architecture plans of the time, its comprehensive examination of architectural issues remains widely applicable and is thus a valid conceptual template for future work in the area. To remind readers of the study vintage, the figures are original, and references to the International Space Station are made using its name at the time, Space Station *Freedom* (SSF).

PROBLEM STATEMENT AND SCOPE

Long-duration surface and transfer missions pose a different problem regime than other kinds of exploration missions. Multiyear stay times represent an order-of-magnitude longer mission duration than the SSF expedition baseline. Direct escape from such a remote, long-duration mission is impractical, and resupply and crew rotation schedules are sparse. The mission duration and "distance from home" compound problems of crew isolation and confinement, ensuring the ascendancy of psychological needs.

The second decade of the SEI was planned to include an early, permanently manned lunar base (Figure 1). Although a host of applications for direct derivatives of SSF hardware are apparent, the study targeted mission regimes suspected of requiring enhanced capabilities. Our hypothesis suggested that the long-duration requirement of both surface-base and interplanetary-transfer applications would drive their needs to be more similar than different, and together quite different from shorter-duration needs.

For rigid, Earth-integrated habitation modules, the fundamental design parameters available are number and diameter of the modules and overall size of the habitat system. Space habitats have traditionally taken advantage of the maximum "throw" diameter of their launch vehicles; we studied diameters corresponding to commonly discussed SEI launch shroud sizes

Copyright © 2008 by the authors. Published by the American Institute of Aeronautics and Astronautics, Inc., with permission.

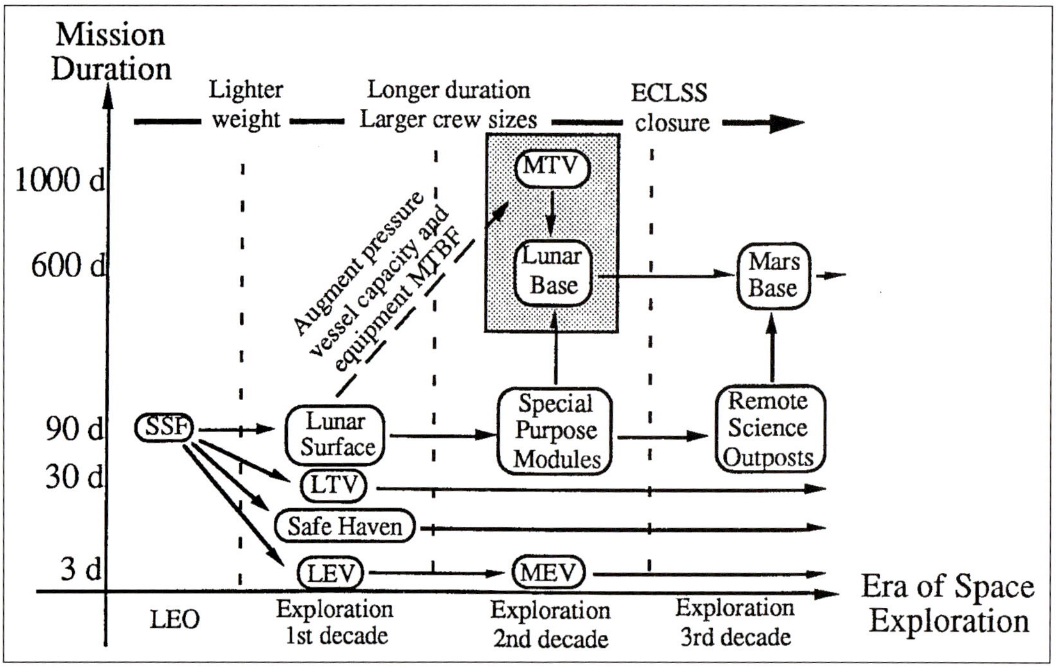

FIGURE 1 Second decade of SEI would require long-duration habitats for multiple uses (gray box): LEO = low Earth orbit, LTV/MTV = lunar/Mars transfer vehicles, LEV/MEV = lunar/Mars excursion vehicles, ECLSS = environmental control and life support system, and MTBF = mean time between failures.

(shuttle-compatible 4.4 m, shuttle-C block II-compatible 7.6 m, and advanced launch system-compatible 10 m). We limited the medium- and large-diameter options to unitary modules with internal pressure bulkheads for redundancy; the smaller-diameter options included clusters of up to six individual modules.

METHOD

The basic method was to 1) generate a broad set of options to span a useful trade space; 2) compare these options, successively narrowing the field; and 3) synthesize an optimum reference concept. We generated 1480 distinct options by varying crew size, the presence or absence of weightlessness, module size, critical structural dimensions, cluster topology, and module orientation for outfitting. Early on, we concentrated only on options for "gravity" applications. This subsumes microgravity requirements because even artificial-*g* vehicles must accommodate microgravity conditions some of the time, and enhances potential for commonality between test, planet surface, and transportation system applications. That is, a gravity-based design can be used also for microgravity conditions, whereas the reverse is not true. We applied a series of screens to reduce the remaining option domain another order of magnitude by eliminating many nonrevealing and poorly trading candidates. Figure 2 shows one-fifth of our reduced initial trade tree. None of several reviews of the work invalidated any of the specific cuts made. Detailed analysis was thus reserved for the 30 most representative concept types, which still spanned the trade space. (With crew size factored in as a parameter, this captured 150 distinct designs.) Geometrical merit was assessed by developing, applying, and interpreting several quantitative metrics described below. Structure mass was calculated and compared, and candidate manufacturing techniques were evaluated. The final reference concept for use by SEI was synthesized from the best features of the highest-scoring options.

A noteworthy feature of the method is its transparency. All rejection rationales were recorded. Also, the geometry metrics were developed to be rigorously measurable, repeatable, and generally applicable even to other habitat options not investigated by our study. They are assessed independent of internal outfitting design and so are properties of the pressure vessel and primary structure only. The perceptual implications of these metrics remain open to interpretation and alternative interpretations as needed for other applications.

Four categories of trade discriminators were used. *Functionality discriminators* are issues of user accommodation, such as access convenience, safety during contingencies, and sensory interference. *Integration discriminators* relate to fitting the habitat system into the rest of a space system or program architecture and

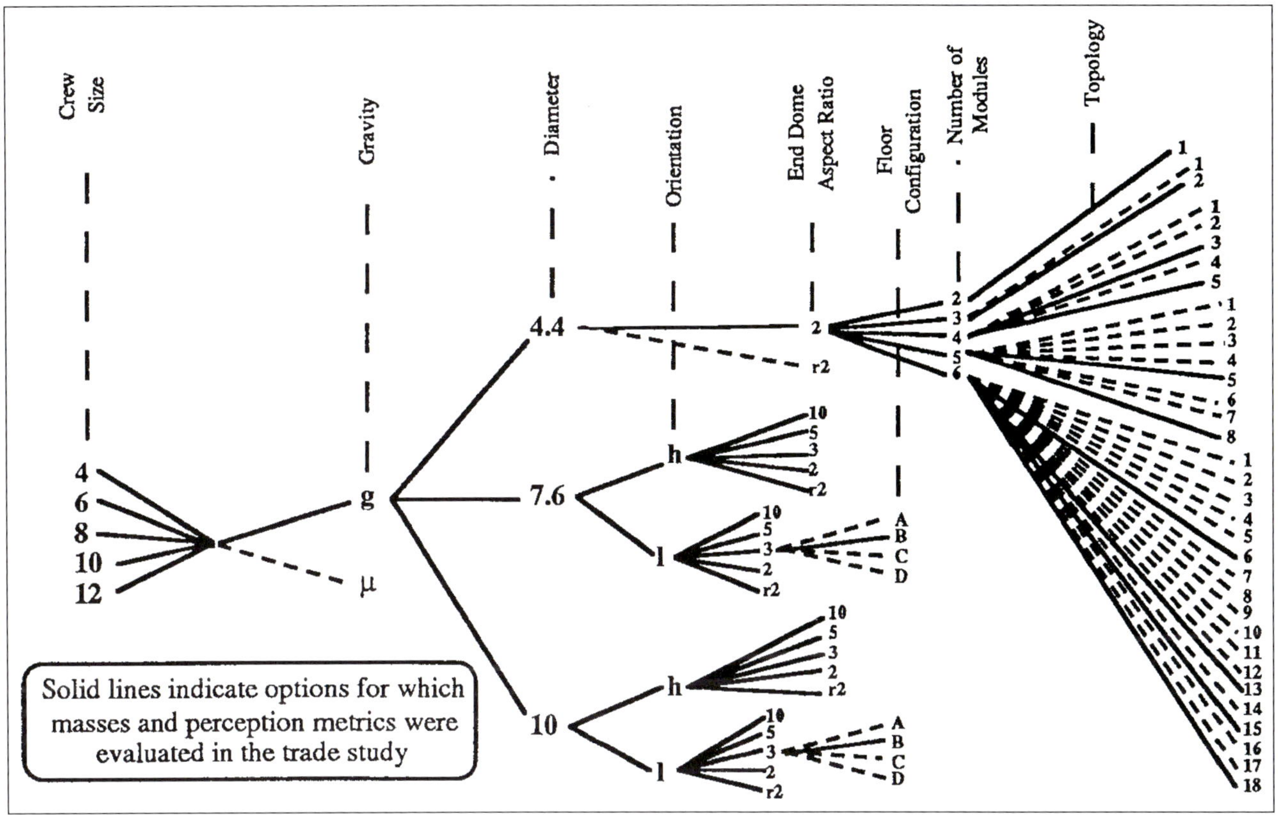

FIGURE 2 Eight trade parameters generated 1480 distinct concept alternatives. Gravity levels are μ for microgravity and *g* for partial-weight. Diameter values are in meters. Orientation designator is *l* for modules partitioned longitudinally and *h* for modules partitioned crosswise. Aspect ratio of end-domes ranges from square root 2 (r2) to 10. Floor configuration and topology designators refer to specific layouts compared in the trade study.

include considerations of assembly, packaging behind aerobrakes, and growth potential. *Cost discriminators* include commonality, manufacturability, processing, and mass. *Perceptual discriminators* involve crew responses to proportion and scale (specific and total volume, volume shape, internal views, pathway options for moving around in the habitat, and spatial variety).

Internal layouts, selection of science and environmental control (ECLSS) equipment, materials and finishes, furnishings, and hatches and windows required for extravehicular activity (EVA) were excluded from evaluation. To first order, these components are nondiscriminators for a pressure vessel trade study because the effect of varying them largely cancels across the trade options.

The overall study assumptions were as follows:

- Volumes based on extrapolating historical specific-volume values
- SSF habitability standards used as a minimum point of departure
- "Stacked" module partitioning valid only for diameters > 4.4 m
- Ceiling height minimum standardized at 2.3 m
- Floor thickness standardized at 0.5 m
- All hull penetrations occurred in barrel sections (not end domes)
- Cluster topologies based on direct connection (no "node" modules)
- Cluster topologies based on all same-length modules
- Galley/storm shelter integrated into floor structures
- All configurations optimized for gravity configuration

A reference pressurized-specific volume of 112 m^3/person was determined by extrapolating historical data for space habitats (Figure 3), after removing data for aeroentry vehicle cabins. Volumes were thus held constant for each crew size studied, across the module concept options. Of note on this graph is that the *Skylab* volume was not as anomalous as is often claimed. SSF is comparable when all nodes and international modules are accounted for; SSF seems less roomy than *Skylab* because it has much more internal system equipment. The graph says nothing about the psychological differences between acceptable specific volume and acceptable total volume; incidents of open conflict occurred on *Salyut*, where

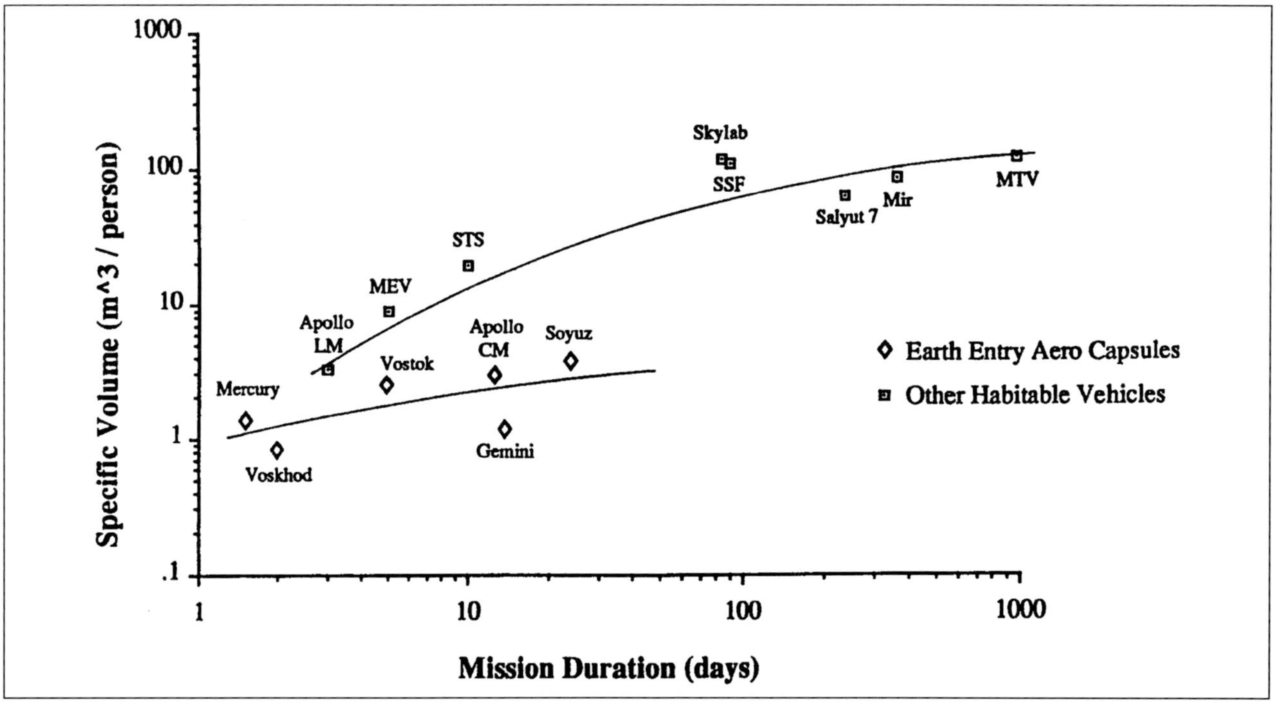

FIGURE 3 Specific volume was based on historical trends for space vehicles excluding those constrained by an atmospheric-entry outer mold line.

total volume was comparatively small and mission durations were long.

Structure concept assumptions for all module sizes in the trade study are shown in Table 1. This concept departs from baseline SSF module construction in two basic ways. First, the wall uses a monocoque rather than waffle-grid skin; the structural precedent for this hull design is the SSF node dock-port adapter barrels. Such a structure can be stiffened for launch in an unmanned vehicle by being overpressurized for that mission phase. Second, the module end geometry comprises simple, unpenetrated ellipsoidal domes rather than heavy cones sized for docking loads. Such modules would be soft-berthed, and an exoskeletal structure would react intermodule bending, torsion, and thermal expansion. These two conceptual alterations can reduce total structure mass substantially over the SSF baseline, critical for interplanetary transportation.

TABLE 1 Transportation mass penalties argue for a lighter, SSF-evolved structure system.

COMPONENT	SPACE-STATION STRUCTURE DESIGN	REFERENCE STUDY STRUCTURE CONCEPT	COMMENT
Material	2219-T8 Al (as welded)	Same	Long experience. Ultimate strength = 38 ksi (262 MPa)
Cylinder	45-deg waffle grid	Monocoque	SSF uses man-rated, side-mounted launch without overpressure rigidization
Cylinder cap	25-deg conical, with flat pressure bulkhead	Ellipsoidal, with no penetrations	Docking loads, assemblage stiffness, and axial penetrations drive ISS design
Support structure	Longitudinal support beams for launch loads; cylinder support rings	Cylinder support rings; intermodule support structure (4.4 m diam)	Overpressure for stiffness during unmanned launch; intermodule support for uneven bending loads on multimodule system
Pressure bulkhead	Monolithic, integrated into endcones	Al/Al honeycomb (10 m and 7.6 m diam); SSF-derived (4.4 m diam)	Honeycomb lighter than monolithic bulkhead, with acceptable volume penalty
Module connection	Pressurized nodes	Parallel tunnels with pressure bulkheads between modules (4.4 m diam)	Mass critical for planetary missions; no requirement for reconfiguration flexibility

Locating connection tunnels in the barrel sections of a nodeless design limits mass further. The launch load path for outfitting equipment is through mounting standoffs to the floor structure and then to girth rings for the vehicle interface.

GEOMETRY ANALYSES

We performed a topology comparison study for 34 cluster configurations of small-diameter (4.4-m) modules by developing and applying metrics to assess the following seven factors: 1) mass penalty of topologies requiring extra ECLSS strings to achieve two-fault-tolerance; 2) compactness for ease of integration behind as small an aerobrake as possible; 3) fraction of the original habitat volume left available for safe haven in the event of loss of a "keystone" module; 4) number of distinct spatial units; 5) parts count incurred by additional modules, tunnels, and connecting exostructure; 6) number of modules separating the starting point and destination for crew pathways inside the system, summed over all possibilities; and 7) proportion of hardware elements devoted to circulation among modules.

All metrics were formulated to yield values between 0 and unity, with higher numbers being better. For example, the safe-haven-split factor $F_{ss} = n_m/n$, where n_m is the number of modules left accessible after loss of the worst-case keystone module, given the best possible distribution of ECLSS equipment, and n is the total number of modules. The calculated value represents how much of the initial habitat system is still usable. Figure 4 shows the result for all possible cluster topologies (see Figure 5 for diagrams of these options). Arbitrarily setting a limit of $F_{ss} \geq 0.5$ eliminates many possible topologies, particularly "star-like" clusters.

Simultaneous evaluation of all of the metrics allowed a dramatic reduction of the cluster topologies to be considered further in the trade study. Figure 5 shows all of the topologies and highlights those admitted into later stages of analysis. Because of SEI priorities at the time, our work emphasized the aerobrake integration factor (which favors compact, conformal clusters over linear clusters, a result that would change if planet surface bases became the only use case), and the safe-haven-split factor.

Whereas small-diameter options use multiple modules as spatial units, unitary medium- and large-diameter module options use multiple floors instead. We performed a sectional-properties analysis of the tunnel-oriented unitary options (floors running parallel to the module axis, the *l* configuration). We compared usable floor area, volume that is accessible but unusable for human occupation, maximum ceiling height, and out-of-reach overhead volume, for several different floor arrangements. Reasonable

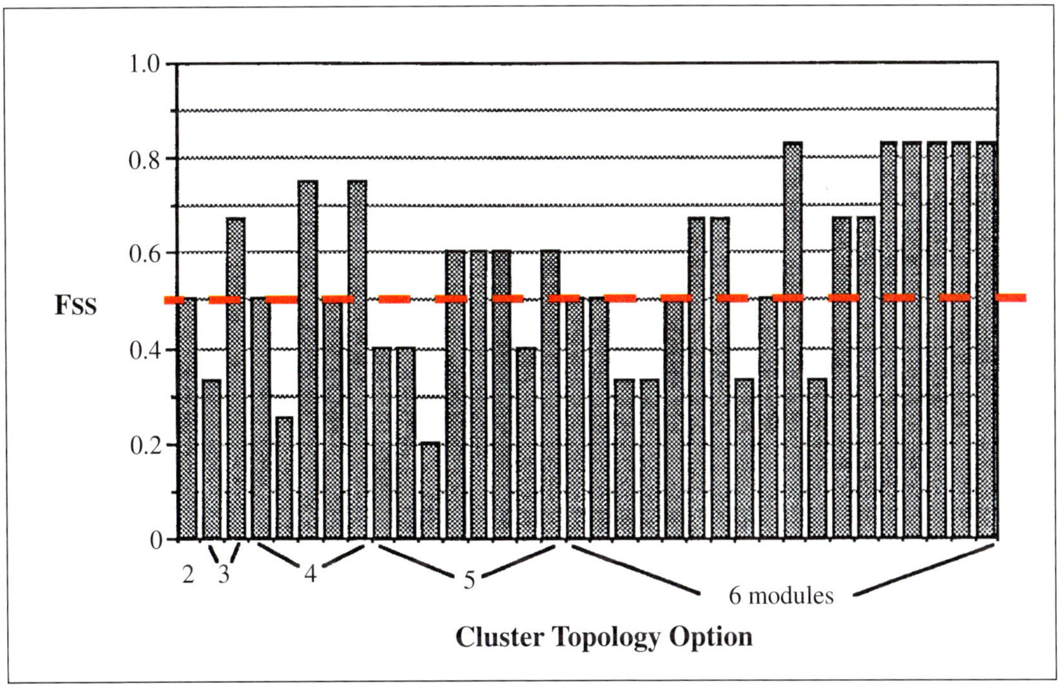

FIGURE 4 Limiting overall loss of habitat system usage to 50% (---) after a worst-case loss of a single unit eliminates many possible cluster topologies. Abscissa indicates topologies for increasing cluster sizes.

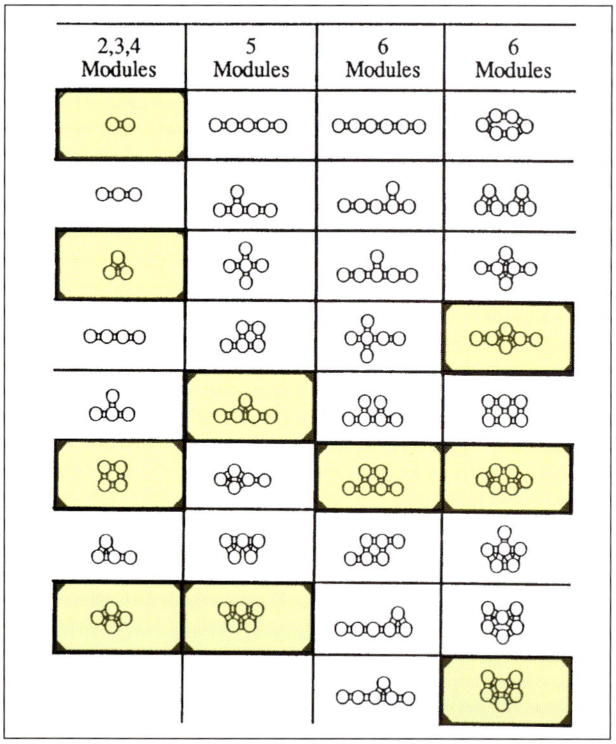

FIGURE 5 Topology analyses narrowed the field of possible small-module habitat configurations. to 10 cluster concepts appropriate for planetary aerocapture.

concepts for each module size were then selected and used in subsequent geometry analyses.

Ten geometry metrics were devised to compare the surviving small-diameter cluster options with the unitary medium- and large-diameter options, to assess: 1) usable floor area, normalized to total pressurized volume; 2) out-of-reach sectional area, normalized to total sectional area in the largest spatial unit; 3) longest end-to-end internal path, normalized to total pressurized volume; 4) aspect ratio of the floor plan; 5) volumetric aspect ratio; 6) number of spatial units, normalized to total pressurized volume; 7) number of floors, normalized to total usable floor area; 8) total useful wall perimeter in plan, normalized to total usable floor area; 9) distinct interior pathway options through spatial units, summed over all possibilities; and 10) ratio of largest single volume to volume of a crew cabin. Taken together, these geometry metrics permit quantitative comparison of crew functional and perceptual response to habitat configuration options. We discovered that options having extreme values (either large or small) of some metrics tended to appear less optimal over all metrics than those with consistently moderate values.

Two of the most revealing metrics are the pathway factor and the scale factor. The pathway factor F_{path} measures the number of different ways to get from one spatial unit to another in the habitat, summed over all combinations of starting and destination points (Figure 6). Over the study domain of crew sizes and concept options, F_{path} ranged from 2 to over 2000. Discontinuities in the curves represent break-points in assumptions about the number of separate circulation paths required for larger crews to avoid circulation congestion. Interconnected clusters provide many pathway options, which might be advantageous mitigating boredom

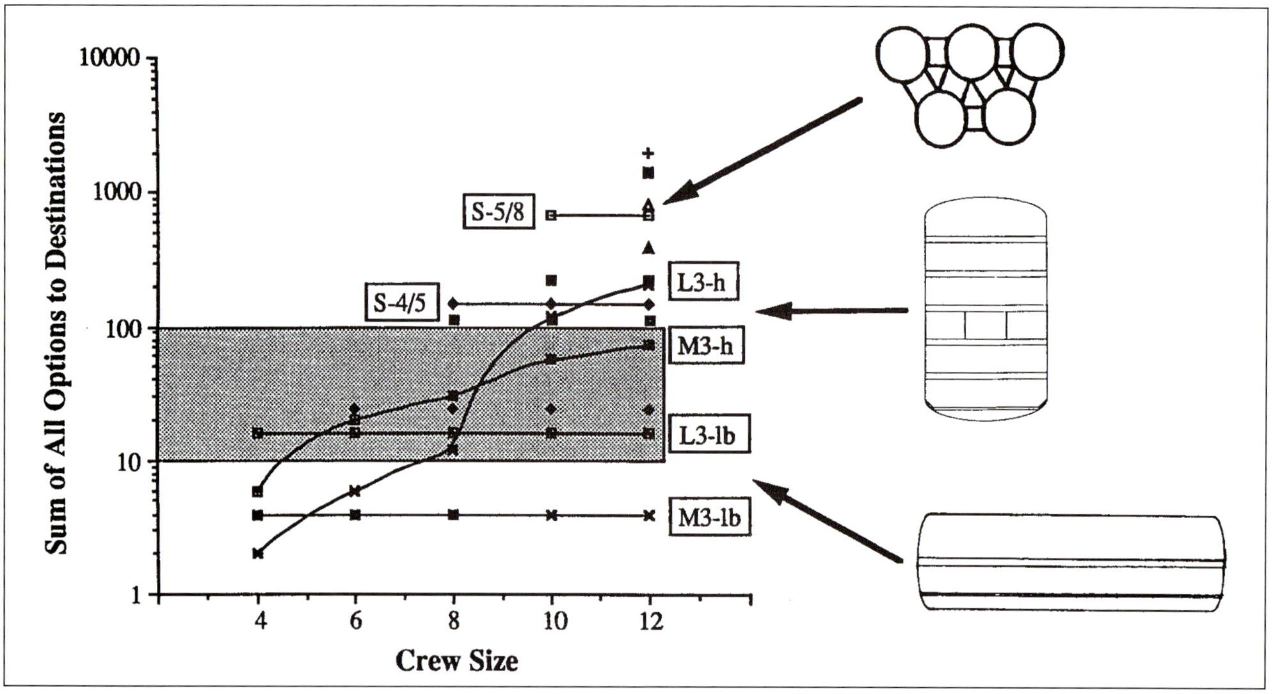

FIGURE 6 Multiple pathway options could mitigate crew boredom on long-duration missions. Gray highlighting shows a range selected to achieve this benefit without adding undue system complexity.

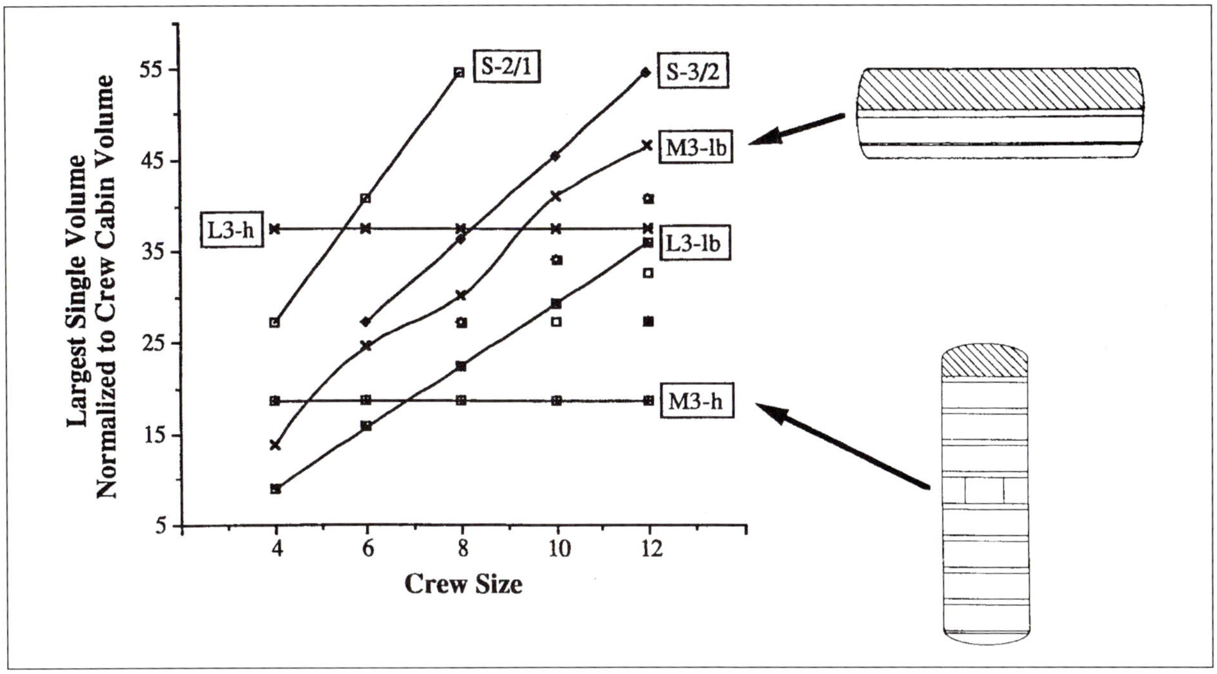

FIGURE 7 With increasing crew size, tunnel-oriented options allow more spatial variety than stacked options.

over long durations and in alleviating unavoidable social concentration.

Perception of the habitat's inherent ability to accommodate privacy might be enhanced with many pathway options, although internal partitions can also add environmental complexity to compensate for poor "pathway" performance. A suggested practical range of pathways options (10–100) is indicated on the figure.

The scale-factor ratios the largest single volume available to the volume of an individual crew cabin as an indication of the range of spatial scales intrinsically available to interior designers of the habitat system (Figure 7). Greater potential spatial variety could be important for mitigating perceptual boredom over long durations. For small crew sizes, the top-floor dome of large-diameter, vertically stacked options is dramatic; for large crew sizes, however, tunnel-oriented options are more favorable because their generous vault space grows with module length. The seemingly good performance of some cluster options for large crew sizes simply shows that their modules get very long (up to 27 m); this must be weighed against the high aspect ratio of their plan (the hallway factor).

In addition to quantitative geometry analyses, we generated representative interior outfitting sketches to determine if unforeseen complications might arise from the principal habitat module options being studied. Using prior spacecraft design studies as starting points, we assigned representative functional area allocations, presuming the presence of gravity. From these, proximity relationships (Figure 8) were

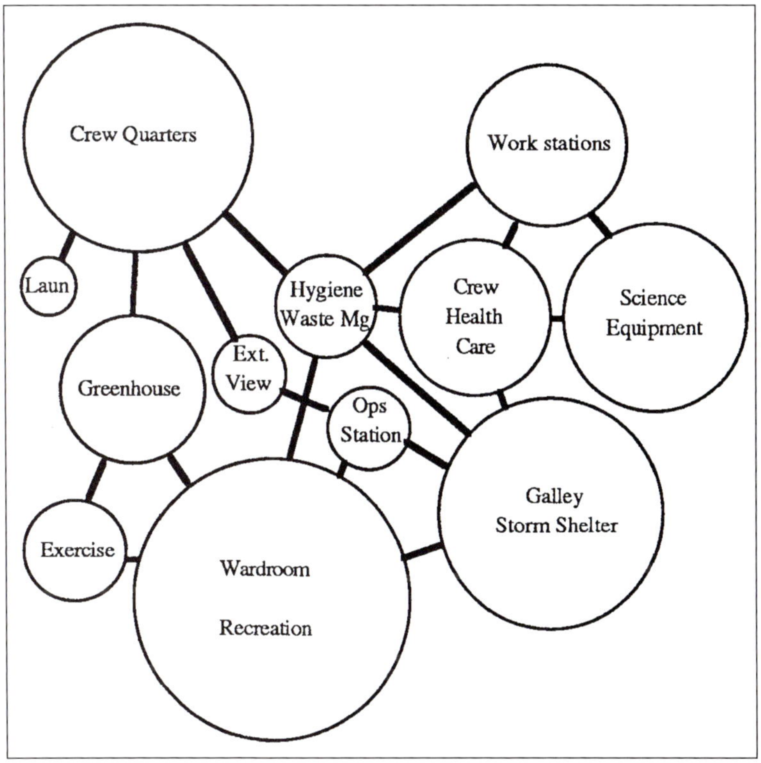

FIGURE 8 Scaled proximity analysis is the basis for reference interior configurations.

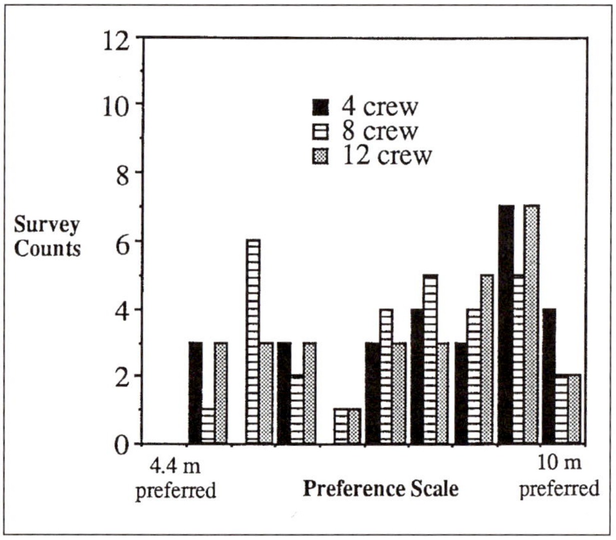

FIGURE 9 Thirty surveyed engineers tended to prefer larger-diameter module options when shown outfitted design examples.

developed into internal configurations for crews of four, eight, and twelve, using two module options: clusters of small numbers of 4.4-m modules and vertically stacked 10-m modules. It became apparent that efficient, and even quite interesting, habitats could be made out of all of the options. The configuration sketches were then used to conduct a preference survey among several dozen technical and nontechnical people. The results (Figure 9) suggested that engineers (perhaps an appropriate analog for mission crews) tend to prefer the larger diameter, generally citing presumed enhanced efficiency and perception of greater spaciousness.

MASS ANALYSIS

We compared pressure-vessel mass for the 150 options pursued through the study and estimated their total outfitted mass. Included in structure mass were primary structure (pressure vessel, structural rings and ribs, and pressure bulkheads) and some secondary structure (intermodule tunnels and pressure hatches; intermodule integrating structure; and meteoroid, debris, and thermal protection). Not included were items required identically by all the habitat concepts (such as airlocks and hatches associated with them). Also excluded were outfitting standoffs and floors, which take different forms in different concepts but do not weigh differently to first order. Figure 10 shows mass sensitivity to crew size for the best concepts. Cluster options with large numbers of separate modules trade poorly. (The high mass shown as reference for SSF is caused by its heavier primary structure design as well as its clustered configuration). For crew sizes over six, unitary module concepts enjoy an increasing mass advantage over small-diameter, cluster concepts. However, the linear relationship between skin stress and diameter argues for smaller diameters, so that the medium-diameter option trades best of all. Even the cross-sectional safe-haven pressure bulkhead required for unitary-module designs does not eliminate their mass advantage.

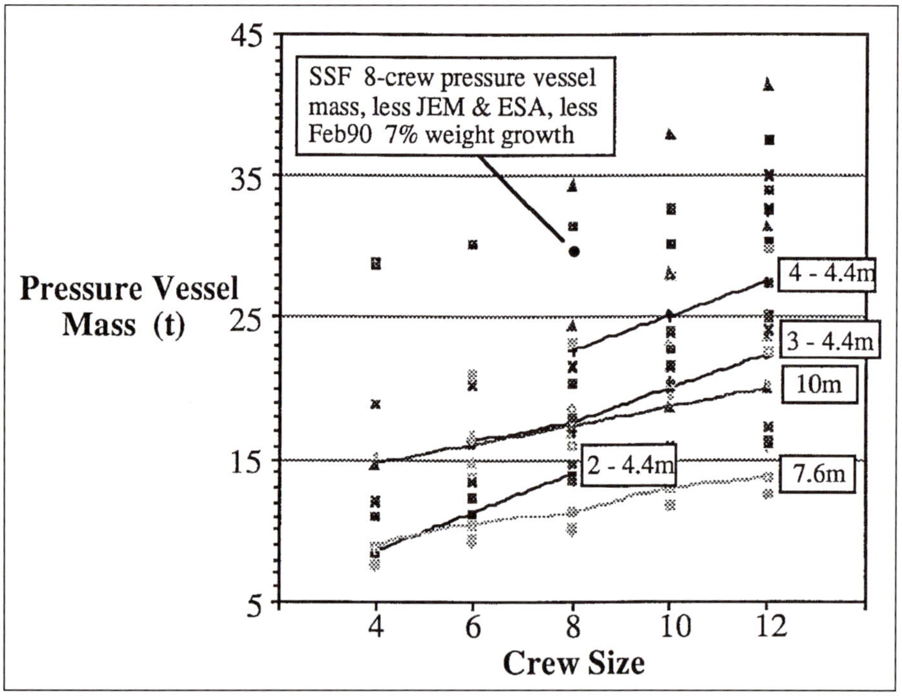

FIGURE 10 With increasing crew size, cluster options rapidly become less mass competitive.

A parametric algorithm, based on SSF designs adjusted for long-duration missions, was developed to estimate outfitting mass. The calculation did not include crew, effects, consumables, or spares, all of which are easily stowed on orbit or even *in situ*. Figure 11 shows the total structure plus outfitting mass for 30 selected

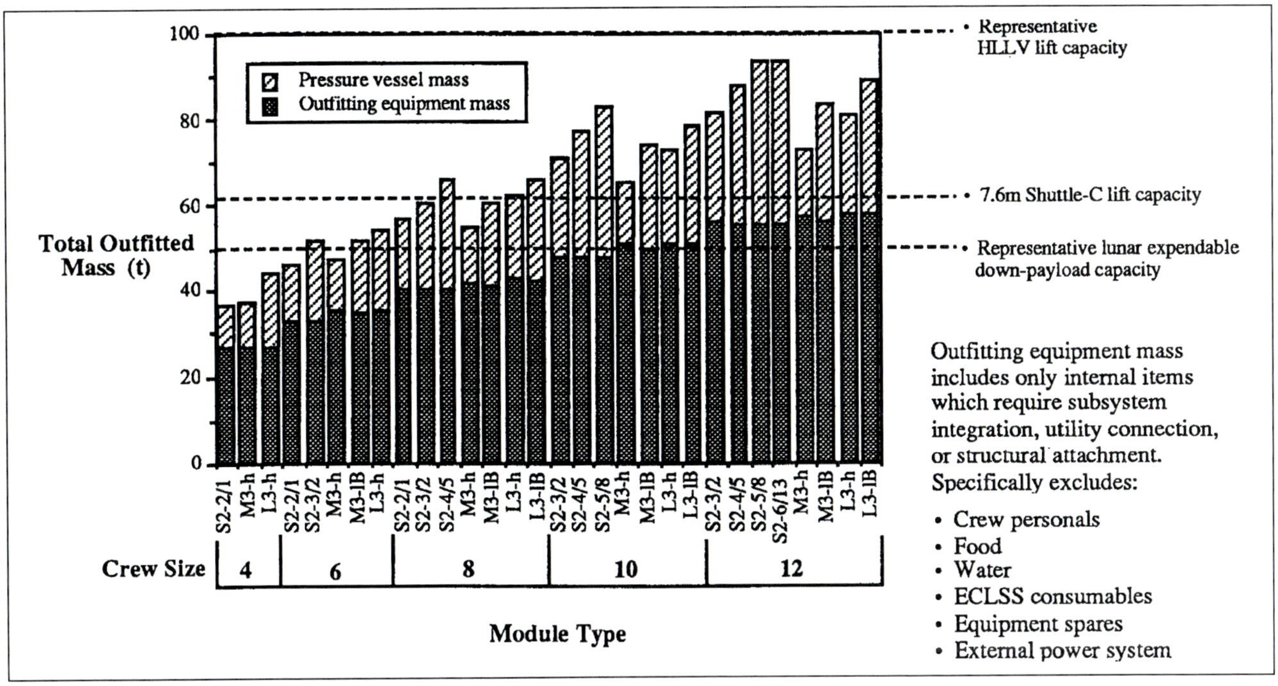

FIGURE 11 Trade study concluded that a nine-crew module made of advanced materials could be landed, fully outfitted, on the Moon by SEI-concept vehicles. (Note that today lunar lander capability is presumed to be less than half of what was being planned during SEI.) Alphanumeric codes on the abscissa designate specific configuration options studied. Outfitting equipment mass includes only internal items that require subsystem integration, utility connection, or structural attachment. This specifically excludes crew personals, food, water, ECLSS consumables, equipment spares, and external power system.

options and indicates 1990-vintage conceptual payload capacity for launch and one-way lunar landing. Assuming very advanced pressure-vessel materials and fabrication methods (e.g., fiber-wound SiC/Al composite with plasma-sprayed matrix), fully integrated unitary module options for crews of up to about nine could be transported intact. This could accommodate adequate solar-flare radiation protection if consumables are configured as part of a storm shelter.

REFERENCE MODULE CONCEPT SYNTHESIS

Figure 12 shows the configuration of a six-crew concept synthesizing the best features of the studied options. Although the medium-diameter unitary option with cross bulkhead was lightest, tunnel orientations traded better for functionality and perception, particularly at larger crew sizes. What resulted was therefore a 7.6-m-diam module, with unpenetrated ellipsoidal end domes, laterally bisecting pressure bulkhead of sandwich construction, and two floors configured for use in a gravity field. Locating the upper floor at the diameter enables it to also act as a structural tension-tie (analogous to large airplanes) to further reduce mass.

Functionally, this concept minimizes connection leakage and parts count, permits great flexibility of internal arrangement, maximizes useful floor area, and minimizes access and translation time in emergencies. The design provides a sense of spaciousness and convenience, although it depends on interior partitioning to mitigate short sightlines, few pathway options, and a compact inhabited domain. Its orientation is efficient for use in gravity fields and minimizes height and areal coverage of regolith shielding when required. It integrates well within a launch shroud, a Mars transfer vehicle, an aerobrake, and a planetary lander, and can be deployed to a planet surface outfitted for immediate use (Figure 13). With simple clustering, it can integrate with special-purpose modules or accommodate very large (several dozen) crews. Cost effectiveness is maintained through superior mass performance, ability to evolve from conventional welded-metal technology to high-performance materials and fabrication processes, straightforward access for integration and ground processing, and high potential commonality across a whole mission architecture.

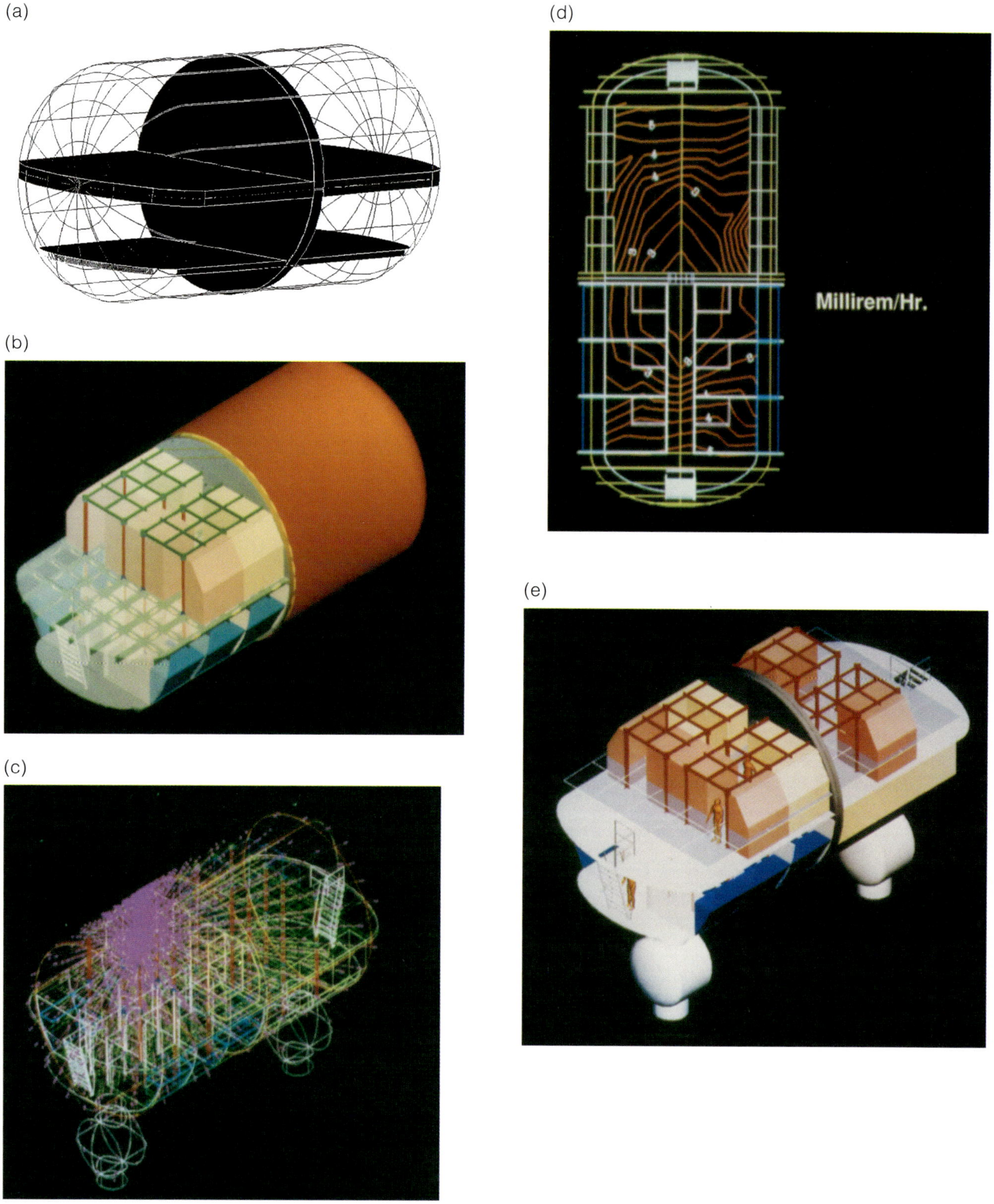

FIGURE 12 Synthesized concept, a 7.6-m-diam module arranged longitudinally, is shown sized for six crew: a) bare concept; b) layout of floor structure and primary outfitting; c) example ray-trace calculation for radiation exposure; d) integrated radiation dose map showing safe-haven in module core; and e) indication of outfitting and airlock locations for Mars transfer application (see Figure 13) (courtesy of NASA and Boeing Co.).

FIGURE 13 In addition to its interior advantages, the final concept integrates well with transportation vehicles: a) a nuclear-thermal Mars transfer ship (arrow indicates habitat module nested among hydrogen tanks for radiation protection); b) a roving Mars lander carrying a surface module; and c) a Mars entry vehicle to decelerate and target the roving lander (courtesy of NASA and Boeing Co.).

Acknowledgments

This chapter is derived from B. Sherwood and S. Capps, "Early Surface Habitation Elements for Planetary Exploration Missions, "AIAA Paper 90-3737, presented at the AIAA Space Programs and Technologies Conference, Huntsville Alabama, Sept. 25–28, 1990. Figures are reprinted courtesy of the Boeing Company.

The work is reported in full in Woodcock, G.R., Buddington P., Donahue B., Finley R., Henshaw E., McGhee J., Ryan P., Sherwood B., Appleby M., Burress J., Capps S., Cothran B., Cupples M., Eder D., Fisher E., Fouche M., Fowler R., LeDoux S., Nordwall J., Ramsey P., Rao N., Stanley K., Tanner R., Thrasher D., Tillotson B., Vas I., Wallace B., *Space Transfer Concepts and Analysis for Exploration Missions, Final Technical Report, Phase 1*. Boeing Company D615-10030-2, 30 March, 1991, Huntsville, Alabama, and supporting interim study reports performed under contract NAS8-37857 for NASA Marshall Space Flight Center.

12

artificial gravity

THEODORE W. HALL

INTRODUCTION

BEGINNING WITH THE ENGINEER Konstantin Tsiolkovsky more than 100 years ago, space-station designs have featured artificial gravity as one of the most important determiners of form. Early visionaries regarded microgravity as a dangerous or inconvenient curiosity unrelated to the station's mission: Earth observation, astronomy, logistics support for interplanetary excursions, or defense. Only after more than a decade of human spaceflights of increasing duration did microgravity finally supplant artificial gravity in space-station conceptual designs—and not for the sake of crew safety or convenience, but primarily as a complexity- and cost-saving measure. Although access to microgravity is now a principal research mission of the International Space Station (ISS), much ISS life-sciences research would be unnecessary if spacecraft provided artificial gravity.

Artificial gravity remains potentially important for Mars-class interplanetary travel, driven primarily by lack of confidence that astronauts could function well upon arrival at Mars after being deconditioned by microgravity for several months during transit. For such missions, long exposure to microgravity *is* dangerous and inconvenient, is not a mission objective, and might not be necessary.

Artificial gravity is well understood in abstract mathematical terms, but not in architectural terms, that is, as a pervasive life condition the same we know, feel, and can design for Earth's natural gravity environment. Terrestrial architects routinely design for life in 1*g*, using common sense developed from direct experience and backed up by design precedents evolved through millennia of trial and error. Whether explicit or not, considerations of gravity pervade nearly every aspect of terrestrial design. Most architects designing for nonterrestrial gravity regimes have neither experience nor many precedents to draw from. This is especially true for artificial gravity, which no human has ever experienced in its pure form, uncoupled from Earth's 1-*g* environment. There is neither common sense nor built precedent.

Historically, NASA governing standards for the design of human space systems, such as NASA-STD-3000, have been rich in prescriptive design requirements as well as performance requirements. However, reliance on design rules does not substitute for a deep, innate understanding of the environment; architects do not learn to design merely by reading building codes. In any case, NASA-STD-3000 has very little to say about artificial gravity (NASA 1995).

Before architects can design comfortable, functional artificial gravity habitats, they must develop empathy for the people who will inhabit them. Perhaps this empathy can arise only from direct experience of similar conditions. That leaves us with a "chicken-or-egg" dilemma designing the first artificial-gravity habitats. The only available option is partial experience and through simulation and visualization. Gravity and acceleration are experienced primarily through the vestibular and haptic senses and only secondarily through vision. However, visual sensations are much easier to generate, record, broadcast, and reproduce in portable media, so this is the place to start.

Copyright © 2008 by Theodore W, Hall. Published by the American Institute of Aeronautics and Astronautics, Inc., with permission.

MOTIVATION

Flaws in published design concepts based on artificial gravity are evidence of inadequate visualizations. This chapter presents visualizations that reveal those flaws to promote general guidelines for better design of artificial-gravity habitats. These visualizations reveal that conformance to the "comfort zone" for artificial gravity does not guarantee an Earth-like gravitational environment. They aim to help architects recognize important design constraints beyond the basic rotational parameters. In particular, the Coriolis effects associated with movement in a rotating environment impose specific north–south and east–west axes, in addition to the up-down axis. Axially symmetric plans that ignore the orientation of Coriolis effects are not well adapted to life in artificial gravity, especially at small radii.

COMFORT CHARTS AND CALCULATORS

This section reviews "comfort charts," well-known and oft-cited visualizations of the numerical parameters of artificial gravity. Comfort charts have been important references in the design of such environments, but are too abstract to impart empathy for living in them. The section also describes an artificial-gravity calculator that assists finding a set of parameter values within an agreed comfort zone.

Early government-funded space-station planning in the 1960s included research into the comfort conditions for human habitation of artificial gravity. This began as an outgrowth of centrifuge studies on military pilots who needed to endure high-*g* loads during combat maneuvers. Whereas those military studies focused on high intensity and short duration, space-station studies focused on low intensity and long duration. Much of the U.S. research occurred at the Naval Aviation Medical Acceleration Laboratory (U.S. Naval Air Development Center, Johnsville, Pennsylvania), the Naval Aerospace Medical Research Laboratory (Pensacola, Florida), and the Rotating Space Station Simulator at NASA Langley Research Center (Hampton, Virginia).

In 1962, Hill and Schnitzer published one of the earliest and best-known visualizations of the comfort zone for artificial gravity, shown in Figure 1 (Hill and Schnitzer

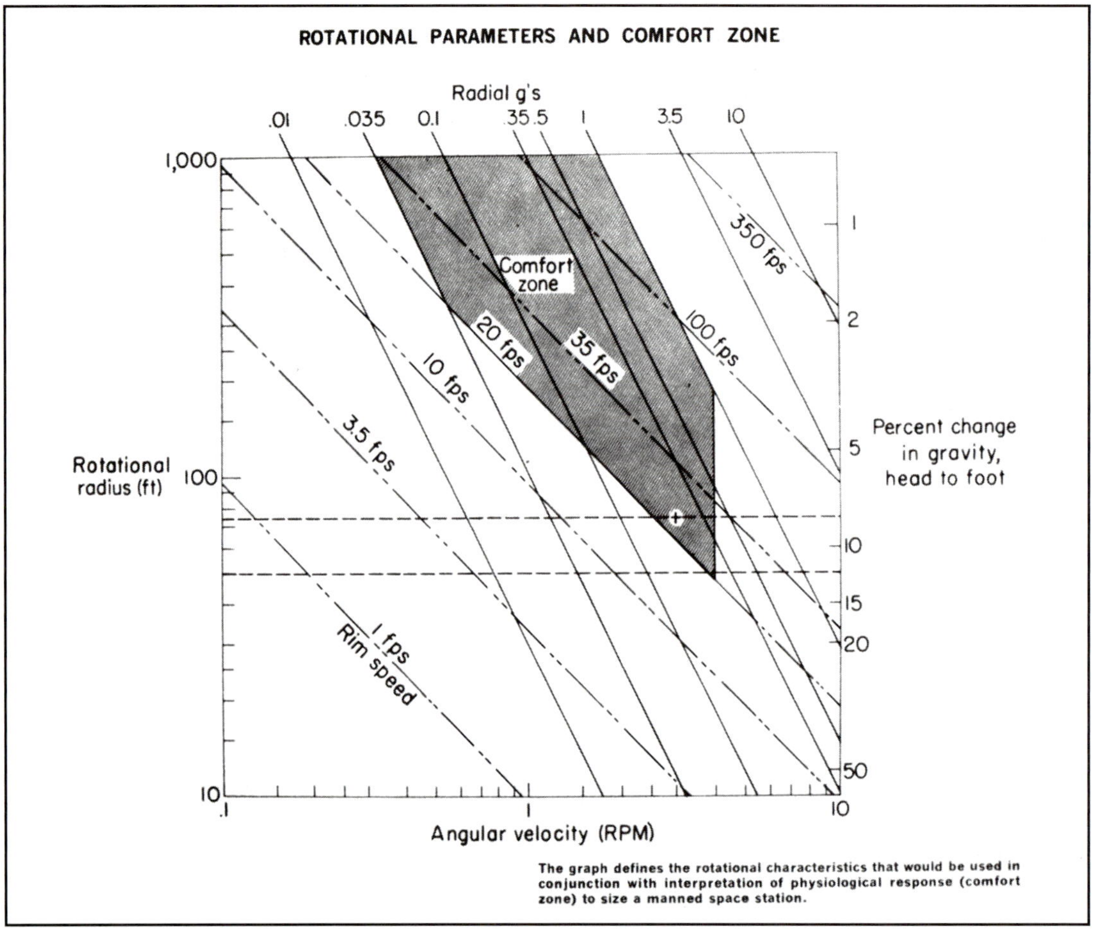

FIGURE 1 Comfort chart (Hill and Schnitzer 1962, courtesy of AIAA).

1962). This chart circulated widely in the pages of the journal *Astronautics* and was still reappearing up to 25 years later in at least three separate papers (Schultz et al. 1989; Staehle 1989; Tillman 1987).

A little essential algebra determines the features of this chart. Tangential velocity (rim speed) of habitat V and centripetal acceleration (nominal artificial gravity) A_{cent} depend on radius of rotation R and angular velocity of habitat Ω according to the following formulas:

$$V = \Omega \cdot R$$
$$A_{cent} = \Omega^2 \cdot R$$

(For numerical equality, Ω must be expressed as radians per time, V as distance per time, and A_{cent} as distance per time squared.) In logarithmic form, these become

$$\log(V) = \log(\Omega) + \log(R)$$
$$\log(A_{cent}) = 2 \cdot \log(\Omega) + \log(R)$$

Hill and Schnitzer measure $\log(R)$ on the vertical axis and $\log(\Omega)$ on the horizontal axis. Rewriting the last two relations in the slope-intercept form ($y = m \cdot x + b$), they become

$$\log(R) = -\log(\Omega) + \log(V)$$
$$= -2 \cdot \log(\Omega) + \log(A_{cent})$$

Thus, contours of tangential velocity V have a slope of -1, whereas contours of centripetal acceleration A_{cent} have a slope of -2. Using logarithmic rather than linear scales renders these contours as straight lines rather than curves.

Hill and Schnitzer's hypothetical comfort zone is sandwiched between minimum and maximum centripetal accelerations of 0.035 and 1.0 g (0.343 and 9.81 m/s^2). The lower bound seems to be an arbitrary value on their logarithmic scale. The zone is further bounded on the right by a maximum angular velocity of 4 rotations per minute (rpm) [0.419 radians per second (radians/s)]. The value is documented in their paper but not clearly labeled in their chart. This is to limit dizziness from cross-coupled head rotations. Another boundary, on the bottom left, is a minimum tangential velocity of 20 feet per second (6.1 m/s) so that a person's relative velocity v within the rotating habitat does not affect the net apparent gravity too much (the ratio of v/V should be small). A final, potential boundary at the bottom is minimum radius. If the head-to-foot gravity gradient for a six-foot person is limited to 12%, then $6/R \leq 0.12$, and $R \geq 50$ feet (15.2 m). The chart indicates this with a horizontal dashed line just above the bottom tip of the shaded area. A greater tolerance for gravity gradient, allowing a smaller radius, is superseded by the intersection of the limits on maximum angular velocity and minimum tangential velocity. There are no explicit limits for maximum radius or minimum angular velocity: the bigger and slower the rotation, the better.

This is not really a visualization of artificial gravity per se, but rather of mathematical relationships among numeric parameters. Charts such as this could just as well represent countless physical concepts unrelated to artificial gravity. For example, if the axes represented mass m and velocity v, then the sloping contours would represent momentum $m \cdot v$ and kinetic energy $m \cdot v^2/2$. If the axes represented linear load density w and beam length l, then the sloping contours would represent shear $w \cdot l/2$ and bending moment $w \cdot l^2/8$. Any chart that can represent all of these concepts cannot represent any of them very specifically. The chart does not adequately visualize a feel for artificial gravity. It visualizes the math, but not the physical phenomenon.

Hill and Schnitzer's comfort chart is neither the latest nor necessarily the best of these representations. Figures 2–5 show alternatives published by Gilruth (1969), Gordon and Gervais (1969), Stone (1973), and Cramer (1985). The first thing we notice trying to compare these charts is how incomparable they are; no two are formatted quite the same. The inconsistent formats veil significant disagreements regarding the comfort-zone limits. In fact, no two of the charts agree on all of the limits.

Gilruth's chart (Figure 2) uses linear rather than logarithmic axes, with centripetal acceleration on the vertical axis. Contours of radius appear as a set of parabolas with steeper curves corresponding to larger radii. A maximum head-to-foot gravity gradient of 15% implies a minimum radius of 40 ft (12.2 m), which clips across the lower-right corner of the comfort zone. The minimum and maximum centripetal accelerations are 0.3 and 0.9 g (2.94 and 8.83 m/s^2). Setting the upper limit less than 1 g could allow for some inevitable Coriolis acceleration without pushing net acceleration above 1 g. The chart distinguishes between "comfort" and "optimum comfort," limited by maximum angular velocities of 6 and 2 rpm (0.628 and 0.209 radian/s), respectively. There is no indication of any limit on tangential velocity.

Gordon and Gervais's chart (Figure 3) is the most similar to Hill and Schnitzer. It adopts essentially the same logarithmic format, but swaps the horizontal and vertical axes. It also indicates higher limits for nearly every parameter. Minimum

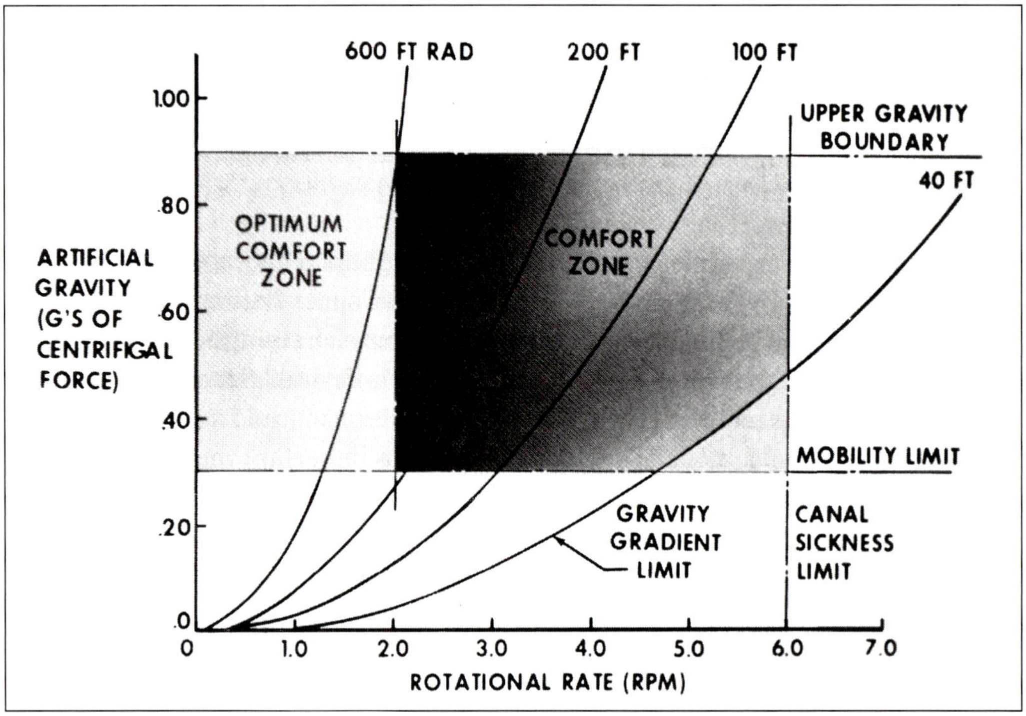

FIGURE 2 Comfort chart (Gilruth 1969, courtesy of Springer-Verlag).

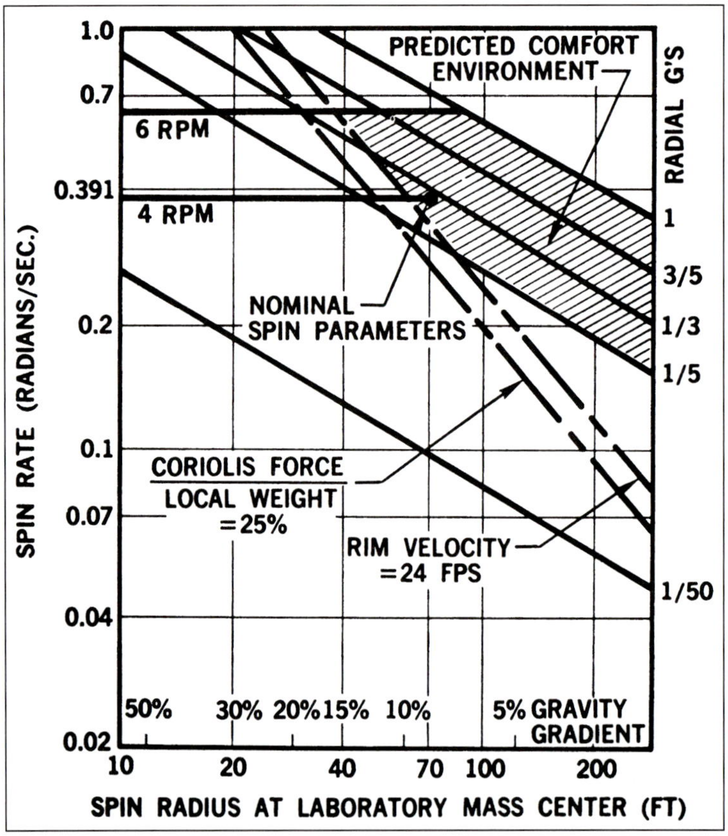

FIGURE 3 Comfort chart (Gordon and Gervais 1969, courtesy of Springer-Verlag).

centripetal acceleration is 0.2 g (1.96 m/s^2). Maximum angular velocity is 6 rpm (0.628 radian/s). Minimum tangential velocity is 24 ft/s (7.3 m/s). There is no indication of any limit on gravity gradient or radius.

Stone's chart (Figure 4) uses the same axis variables as Gordon and Gervais, but with linear rather than logarithmic scales; it measures Ω on the vertical axis and R on the horizontal axis. Because of the nonlogarithmic scales, contours of tangential velocity and centripetal acceleration render as curves. The velocity contours are steeper than the acceleration contours, especially at the left. The curve that delimits the bottom left of the comfort zone corresponds to a tangential velocity of about 10.2 m/s. A horizontal line marks a maximum angular velocity of 6 rpm. Vertical lines mark limits on radius based on the gravity gradient and the deflection of dropped objects. But again, these limits are superseded by the intersection of the limits on maximum angular velocity and minimum tangential velocity.

Cramer's chart (Figure 5) again uses logarithmic scales, but with $\log(\Omega)$ on the vertical axis and $\log(A_{cent})$ on the horizontal axis. The comfort zone is bounded on the left and right by minimum and maximum centripetal

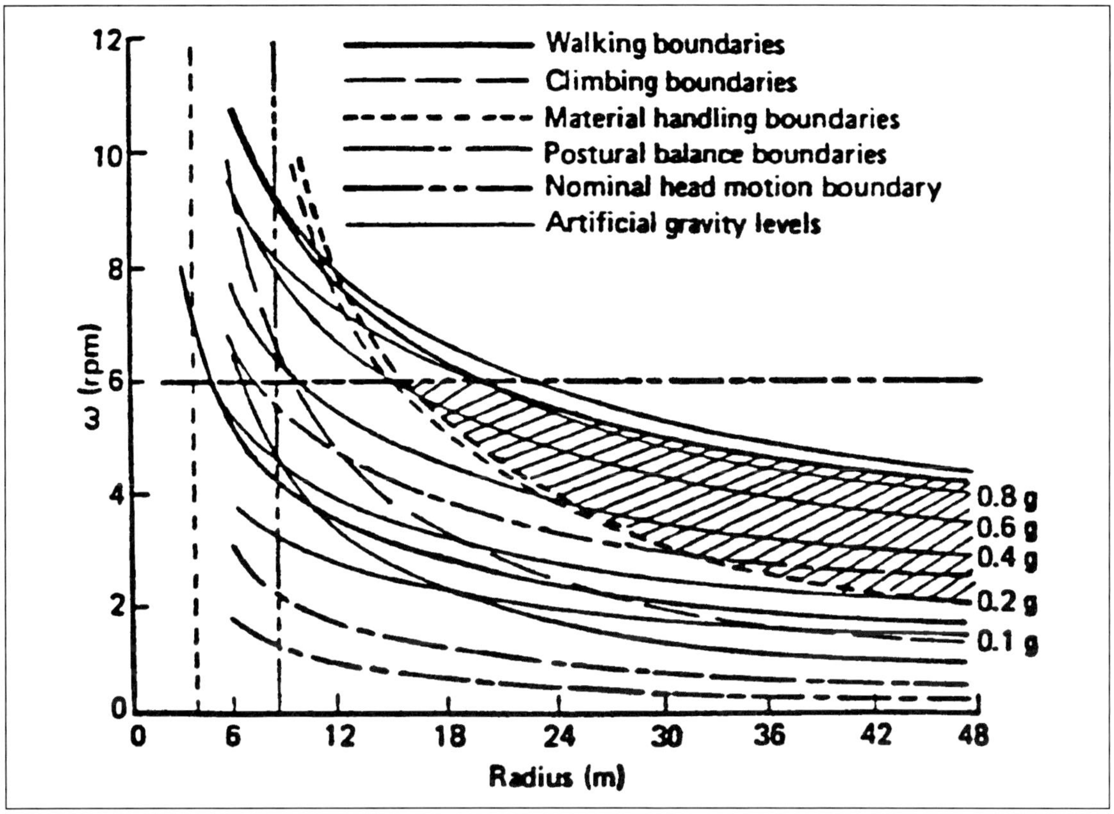

FIGURE 4 Comfort chart (Stone 1973, courtesy of AIAA).

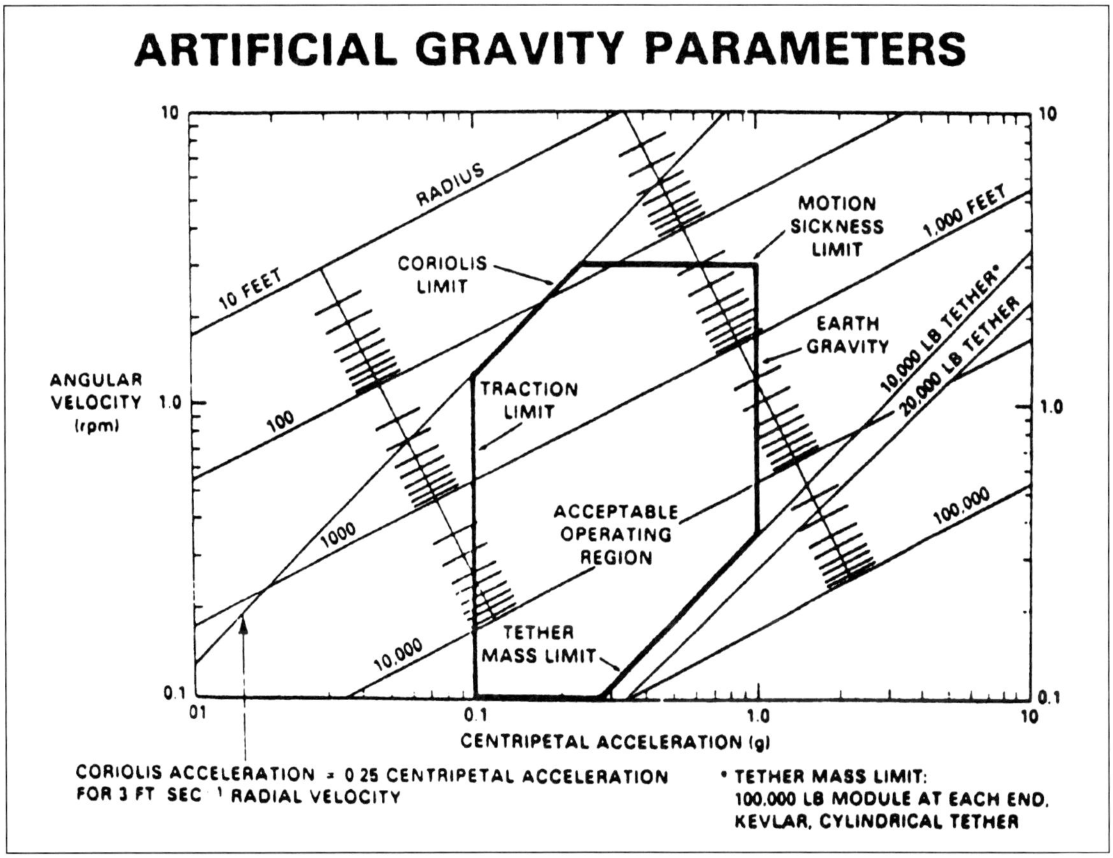

FIGURE 5 Comfort chart (Cramer 1985, courtesy of AIAA).

accelerations of 0.1 and 1 g (0.981 and 9.81 m/s^2). On top, the zone is bounded by a maximum angular velocity of 3 rpm (0.314 radian/s). Contours of radius appear as lines with a slope of one-half while contours of tangential velocity appear as lines with a slope of 1. A boundary for minimum tangential velocity of 24 ft/s (7.3 m/s) clips across the top-left corner of the comfort zone. This boundary ensures that the Coriolis/centripetal ratio will not exceed 0.25 for a radial velocity of 3 ft/s (0.91 m/s). A unique feature of this chart is a boundary for maximum tangential velocity, imposed not for human comfort but for the load capacity of tethers. (The self-weight of a cylindrical tether in artificial gravity is proportional to its centripetal acceleration and length, and $A_{\text{cent}}R = \Omega^2 \cdot R^2 = V^2$.) This limit is about 900 ft/s (274 m/s) and clips across the bottom-right corner of the comfort zone. The chart shows many contour values for radius, but none contributes to the comfort boundaries.

All of these charts suffer from the same fundamental weaknesses:

- They represent abstract mathematical relationships not unique to artificial gravity.
- They portray comfort boundaries as precise algebraic relations, ignoring the underlying statistics of means and deviations. Only when the charts are considered together—with considerable difficulty because of their disparate formats—do the uncertainties in the boundaries become apparent.
- They do not facilitate reading parameter values within the comfort zone, but instead seem to encourage the use of edge values on the frontiers of comfort.

Figure 6 begins to address the second weakness just described by compositing all of the alternative limits cited by these authors onto a single chart. The green zone depicts conditions that all agree are comfortable, requiring little adaptation. The red zone depicts conditions that all agree are not comfortable, even after a period of adaptation. The intermediate zones ranging through shades of yellow and orange depict areas of disagreement. Moving through these zones, away from the green and toward the red, demands greater adaptability of inhabitants to the peculiar conditions of life in constant rotation.

Figure 7 illustrates the user interface of SpinCalc, an artificial-gravity calculator that addresses the third weakness of comfort charts by helping the user find a set of parameter values conforming to all of the comfort boundaries (Hall 2000; 2003). The four parameters—R, Ω, V, and A—are interdependent. The user can input any two of them in any of several units selected from drop-down menus. The calculator computes the other two from the input. The text beneath each value indicates how the value was determined. The colored icon to the left of each value indicates its relationship to the presumed comfort boundaries for artificial gravity. Green indicates that all of the cited authors agree the value is comfortable. Red

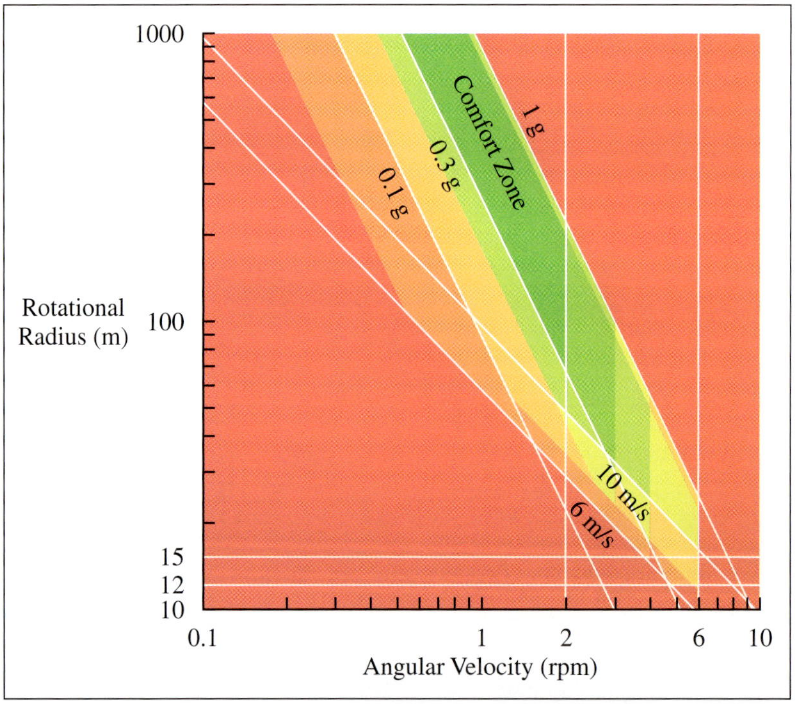

FIGURE 6 Comfort chart composited from Figures 1–5. The green zone depicts conditions that the cited authors agree are comfortable. The red zone depicts conditions that they agree are not comfortable. The intermediate zones ranging through shades of yellow and orange depict areas of disagreement.

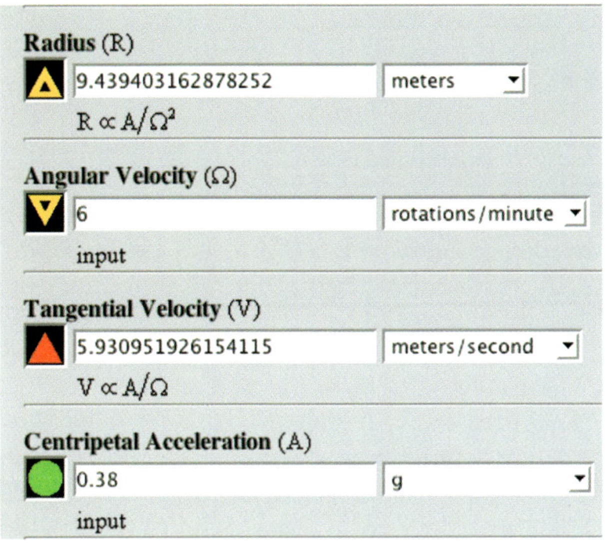

FIGURE 7 SpinCalc (courtesy of Hall 2000; 2003).

indicates agreement that it is not comfortable. Yellow indicates disagreement. Up arrows indicate that the value should be greater; down arrows indicate that it should be less.

In this example, the user is checking design parameters for a hypothetical Earth-to-Mars transit habitat, having guessed 6 rpm for Ω and 0.38 g (Mars surface gravity) for A and entered these as input. The calculator has determined that R must be 9.44 m and V must be 5.93 m/s. Although the input values for both Ω and A meet at least one set of comfort criteria when considered independently, when taken together they result in a V that is definitely too low. If the user now inputs a greater value for V, the calculator will recompute R and either Ω or A, whichever of the two was input least recently. Working back and forth among the parameters, the user can quickly arrive at a set of acceptable values.

FREE-FALLING PARTICLES AND INVOLUTE CURVES

This section begins to address the first weakness of comfort charts as artificial-gravity visualizations: their mathematical abstraction. Gravity, whether natural or artificial, is manifest through its effect on free-falling particles. The section describes a simple simulation of free-falling particles in artificial gravity and illustrates how still frames from such a simulation can contribute to the visual evaluation of these environments.

Figure 8 illustrates SpinDoctor, an interactive, dynamic graphical simulation of free-falling particles in artificial gravity (Hall 1987; 1999). The bottom-right quadrant is a control panel for specifying the initial height and velocity of a particle (in units of meters, meters per second, and degrees of elevation from east), and the rotational parameters R, Ω, and A (in meters, rotations per minute, and gravity). As in SpinCalc, the rotational parameters are interdependent. Specifying any two of them determines the third. Whenever the user changes one, the program recalculates another one—the one least recently specified by the user—so that the remaining parameter remains constant.

The top-right quadrant shows an inertial view of the situation. The top-left quadrant shows a rotating view, as experienced by an inhabitant of the rotating frame. The bottom-left quadrant shows the behavior of an identical particle thrown with the same initial height and velocity on Earth.

In this example, the habitat has a floor radius of 50 m and is rotating at 3 rpm for a centripetal acceleration of 0.5 g. Although all three parameters conform to most renditions of the comfort zone for artificial gravity, the particle's behavior under these conditions is profoundly different from its behavior on Earth. By adjusting the parameters and seeing the particle's motion in real time, the user can obtain some sense of the shape of the apparent artificial-gravity field and the meaning and significance of the comfort criteria.

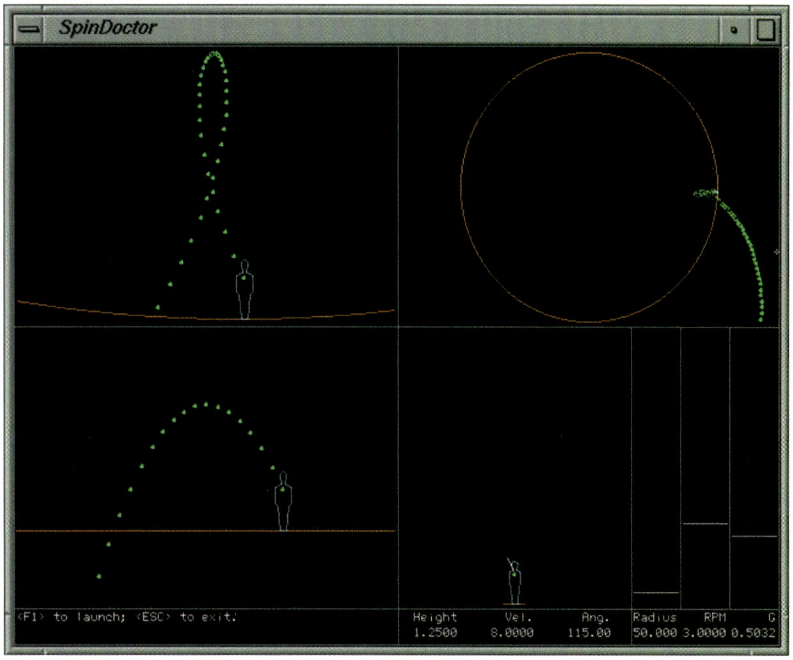

FIGURE 8 SpinDoctor (courtesy of Hall 1987; 1999).

Observing the dynamic behavior of falling objects in real time might be the best way to gain an understanding of artificial gravity, short of actually living in it and feeling it through the vestibular and haptic senses. As much as static printed books (and their printable electronic equivalents) remain important media for communicating design information, still frames derived from such an animation have a role to play. Although also abstractions of the actual phenomenon, they are perhaps less abstract than comfort charts and easier to relate to physical experience.

Figure 9 shows the situation for a particle dropping from a height h above a floor with a radius R_f in an artificial-gravity environment. Contrary to popular misconception, there is neither any centrifugal force pushing the particle toward the floor, nor any Coriolis force diverting it. Rather, in the absence of any impinging force, the particle's momentum carries it at constant velocity through some distance S until it strikes the floor. If the particle had not been released, it would have traveled the same distance S along an arc. The falling particle subtends an angle θ_p while the observer, constrained to circular motion, subtends an angle θ_o:

$$R_h = R_f - h$$

$$S = \sqrt{R_f^2 - R_h^2}$$

$$\theta_p = \arctan(S/R_h)$$

$$\theta_o = S/R_h$$

We can see in Figure 9 that the distances and angles are independent of the speed. One can imagine the animation running faster or slower, but the paths of the particle and the observer remain the same. This means the relative path of the particle is independent of the angular velocity, tangential velocity, and centripetal acceleration. The shape of the path depends only on the ratio h/R_f.

If h is some defined proportion of human height, then one can see the importance of R_f as a design parameter: h is given by anthropometry and ergonomics; the designer's choice of R_f determines the relative deflection of particles falling from that height. The apparent gravity is most natural when the ratio h/R_f is small.

Figure 10 shows the relative paths of free-falling particles dropped from a certain initial height h, in artificial-gravity environments with various floor radii R_f. The straighter paths correspond to the smaller ratios of h/R_f. The curves are involutes: the particle follows a path like the endpoint of a thread unwinding from a spool. The "thread" is the inertial path of the particle, which Figure 10 renders for one case as a set of dashed lines. In the observer's frame of reference, the inertial path rotates. The path itself is invisible except for the particle at its endpoint.

The apparent deflection of the particle, as seen by the rotating observer, is not caused by Coriolis force, but rather caused by the *absence* of the Coriolis force that would be required to constrain it to a straight line in the rotating frame of reference.

Figure 11 shows the situation for a particle hopping vertically from the floor with an initial relative velocity v. The inertial path cuts a chord across the circle with a slope equal to the ratio of the particle's relative hop velocity to the environment's tangential velocity: v/V. The proportions of the path

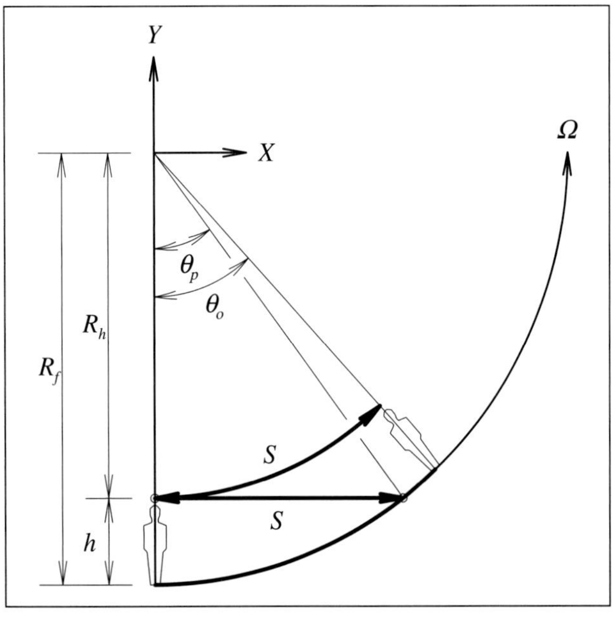

FIGURE 9 Dropping particle in artificial gravity.

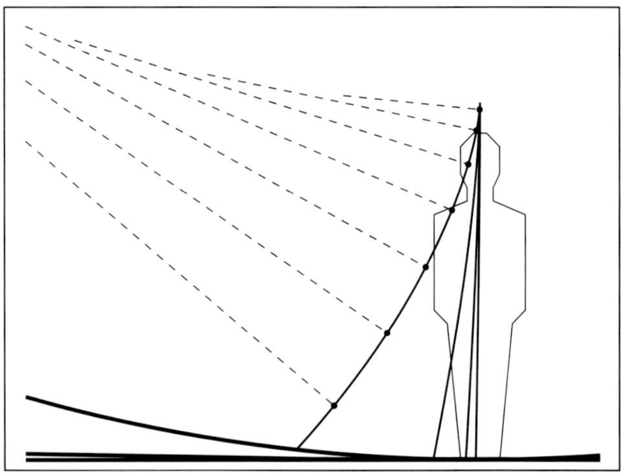

$h = 2$ m

$R_f =$ 10 m, 100 m, 1000 m, 10000 m

FIGURE 10 Dropping particles from an initial height h of 2 m. The shape of the path depends on the ratio h/R_f.

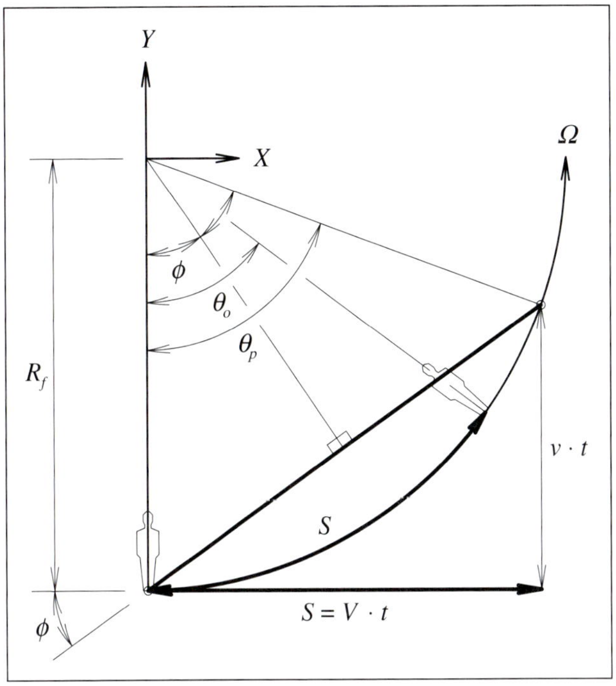

FIGURE 11 Hopping particle in artificial gravity.

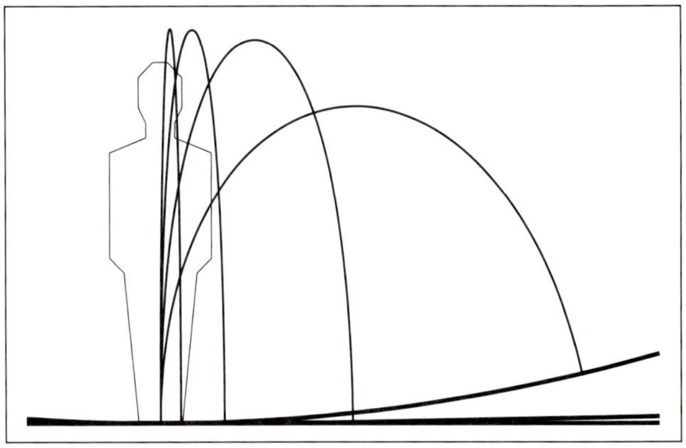

$A_{cent} = 1\ \text{m/s}^2$

$v = 2\ \text{m/s}$

$V = 100\ \text{m/s}, 32\ \text{m/s}, 10\ \text{m/s}, 3.2\ \text{m/s}$

$R_f = 10{,}000, 1000\ \text{m}, 100\ \text{m}, 10\ \text{m}$

FIGURE 12 Hopping particles with an initial relative velocity *v* of 2 m/s and centripetal acceleration A_{cent} of 1 m/s² (about 0.1 *g*). The proportions of the paths depend on the ratio *v/V*, whereas the size depends on A_{cent} and R_f.

depend only on that ratio, whereas its size depends on the circle's radius R_f. The velocity ratio v/V is also directly proportional to the ratio of Coriolis to centripetal acceleration, and so the shape of the path is indicative of that acceleration ratio:

$$\frac{A_{Cor}}{A_{cent}} = \frac{2 \cdot \Omega \cdot v}{\Omega^2 \cdot R_f} = 2\frac{v}{\Omega \cdot R_f} = 2\frac{v}{V}$$

Here, A_{Cor} is the Coriolis acceleration that would be required to maintain a constant relative velocity v on a straight path in the rotating frame of reference. Without that acceleration, the particle appears to deviate. A_{Cor} applies to any motion in the rotating frame perpendicular to the axis of rotation. Vertical "hopping" motion is just one example.

If v is some defined proportion of human speed, then one can see the importance of V as a design parameter: v is given by anthropometry and ergonomics; the designer's choice of V determines the ratio of Coriolis to centripetal acceleration and the relative deflection of particles launched with velocity v. The apparent gravity is most natural when the ratio v/V is small.

Figure 12 shows the relative paths of particles hopping with a certain initial relative velocity v in artificial-gravity environments with a certain centripetal acceleration A_{cent} and a range of tangential velocities V. Holding A_{cent} constant while varying V requires varying R_f as well. The path heights are fairly consistent, as one might expect from the consistent A_{cent} value, but the proportions vary with the ratio v/V. The straighter, narrower proportions correspond to the smaller ratios.

Figure 13 is similar, but holds V constant while varying A_{cent} and R_f. The consistent ratio v/V results in consistent path proportions, but the size varies with A_{cent} and R_f.

The point of Figures 9–13 is to visualize the apparent artificial-gravity field and its sensitivity to variations in the parameters A_{cent}, R_f, and V. As iron filings delineate a magnetic field, free-falling particles delineate a gravitational field. Contrary to the comfort charts, these figures reveal that there are no hard-edged boundary values where artificial gravity suddenly becomes Earth-like. Rather, deviations from Earth normalcy decrease as the radius and tangential velocity of the habitat increase, relative to the size and speed of the inhabitants. Designers can choose the dynamic characteristics they see as tolerable or desirable for the environment they seek to create.

A specific examination of the comfort chart boundaries is illuminating. First, we should establish a picture of normalcy as a basis for comparison. Figure 14 represents Earth-normal gravity. The dots represent particles dropping from an initial height of 2 m, spaced at time intervals of 0.1 s. The short

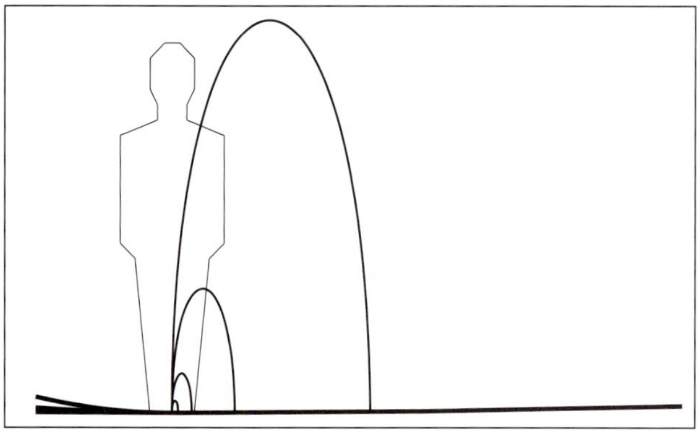

$v = 2$ m/s

$V = 10$ m/s

$A_{cent} = 32$ m/s^2, 10 m/s^2, 3.2 m/s^2, 1 m/s^2

$R_f = 3.2$ m, 10 m, 32 m, 100 m

FIGURE 13 Hopping particles with an initial relative velocity *v* of 2 m/s and tangential velocity *V* of 10 m/s.

horizontal line about 0.2 m above the floor marks the maximum height attained by a particle hopping from the floor with an initial velocity of 2 m/s. These values of $h = 2$ m and $v = 2$ m/s are within the envelope of common human activity and thus serve as standards for visualizing and comparing different gravity environments.

There is nothing remarkable about Figure 14 because it represents the most fundamental human experience: Earth-normal gravity. Figure 14 is the "gold standard" for gravity against which alternatives can be evaluated.

Figure 15 reconstructs Hill and Schnitzer's comfort chart (chosen merely because it is the most well known). The five boundary points of the proposed comfort zone are labeled as a, b, c, d, and e. For each

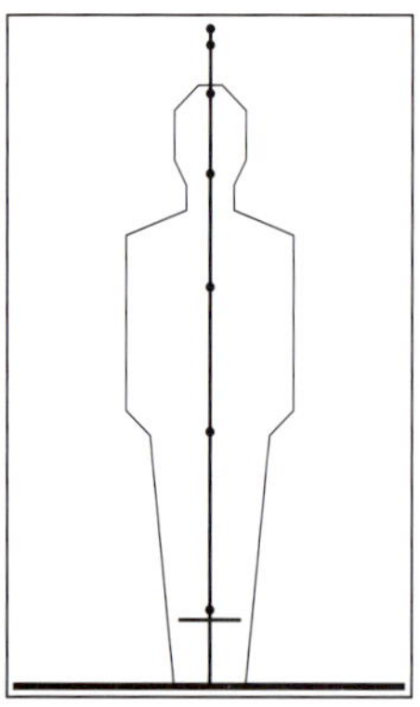

FIGURE 14 Earth-normal gravity benchmark: dropping a particle from 2-m height, and a particle hopping from the floor with initial velocity 2 m/s.

of those points, Figure 15 includes a "hop-and-drop" diagram of artificial gravity under those conditions using the two benchmark tests: dropping from an initial height of 2 m and hopping up from the floor with initial velocity 2 m/s. Compared to Figure 14, the deviations from normalcy show clearly that conformance to the comfort zone does not guarantee an Earth-like gravitational environment.

Point a is the most like Earth. Besides 1-*g* acceleration, it provides a large radius and fast tangential velocity, yielding small ratios of h/R_f and v/V. This environment comes at a systems price: this point demands the greatest investment in mass and kinetic energy for any particular material and structural configuration. (Mass is proportional to volume; kinetic energy is proportional to mass and tangential velocity squared.)

Point b also offers 1-*g* acceleration, so both the hop and the drop are comparable to their counterparts at point a with regard to speed and height. (The dots represent time intervals of 0.1 s.) However, this point pushes angular velocity to the maximum acceptable value in order to reduce the radius and tangential velocity, yielding larger ratios of h/R_f and v/V. Consequently, both the hop and the drop are more deviant.

Point c further reduces radius and tangential velocity to their minimum acceptable values, yielding still larger ratios of h/R_f and v/V and more deviant paths for both the hop and the drop. Because angular velocity is already at its maximum allowed value, reducing the radius also reduces centripetal acceleration. Consequently the hop is higher, and the drop is slower. In contrast to point a, point c represents the minimum investment in system mass and kinetic energy. The economic incentive to minimize these costs might contribute to the disparity of opinion (reference Figure 6) regarding the location of this comfort-zone point. There is pressure to push the envelope in this region of the parameter space, but also there is disagreement regarding how far it can be pushed while maintaining some semblance of comfort.

Point d maintains the same minimum tangential velocity while also reducing acceleration to its minimum comfortable (or useful) value. This reduces angular velocity and the dizziness it induces and increases radius. Because of the lower acceleration but equal ratio of v/V, the hop at d is higher but has the same proportions as at c. Because of the lower acceleration but also smaller ratio of h/R_f, the drop at d is slower but straighter than at c.

Point e maintains the same minimum acceleration while increasing radius to a large value. This further reduces angular velocity while it increases tangential

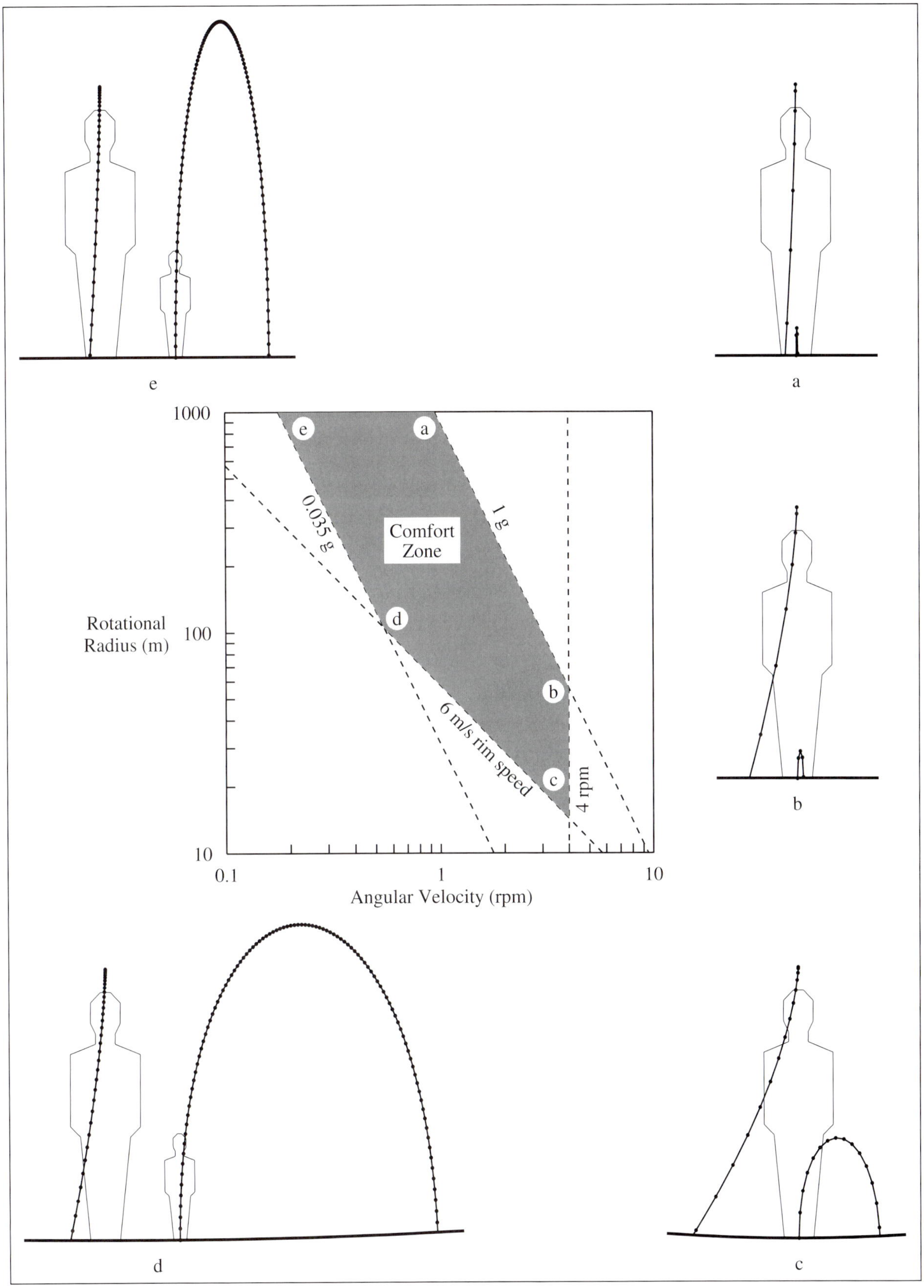

FIGURE 15 Artificial gravity at the boundary points of the Hill and Schnitzer comfort zone.

velocity. The hops at d and e are comparable in height because of the equal acceleration. The proportions at e are narrower than d, but wider than a, because of the intermediate tangential velocity and v/V ratio. The drop at e is as slow as d because of the equal acceleration, but as straight as a because of the equal radius and h/R_f ratio.

Of these five boundary points, point c demands the closest scrutiny because it seeks to minimize obvious costs in radius, tangential velocity, mass, and kinetic energy, but incurs less obvious costs in design, crew selection, training, and acclimatization to accommodate the peculiar gravitational environment. Moreover, as illustrated in Figure 6, there is considerable disagreement regarding where this point should be established.

Figure 16 explores the significance of this disagreement. For each of the five comfort charts just cited, Figure 16 overlays a hop-and-drop diagram for the boundary point at the proposed minimum radius and tangential velocity. Such comparisons help reveal how far the various authors push the comfort-zone boundary. Ultimately, designers need to decide which set of criteria to follow. Even the most conservative criteria allow considerable deviation from Earth-like gravity.

Standard tests such as the "2-m drop and 2-m/s hop" are useful for comparing gravitational environments in a generic way. Nevertheless, these or other tests that represent the envelope of common human activities are only partially revealing. Visualizing specific activities important to a particular situation is also necessary. Figure 17 uses basketball as an example. This simple simulation abstracts the ball to a particle at the ball's center, ignoring effects of its spin, drag, elasticity, etc. Similarly, it abstracts the player to a mannequin included for scale, ignoring the complexities of human kinematics. The dots represent a successful shot in Earth gravity. The curves represent the paths of particles thrown in precisely the same manner in 1-g rotating environments with radii of 10, 100, 1000, and 10,000 m. Artificial gravity in the 10-m environment is so distorted that the ball turns backward and falls

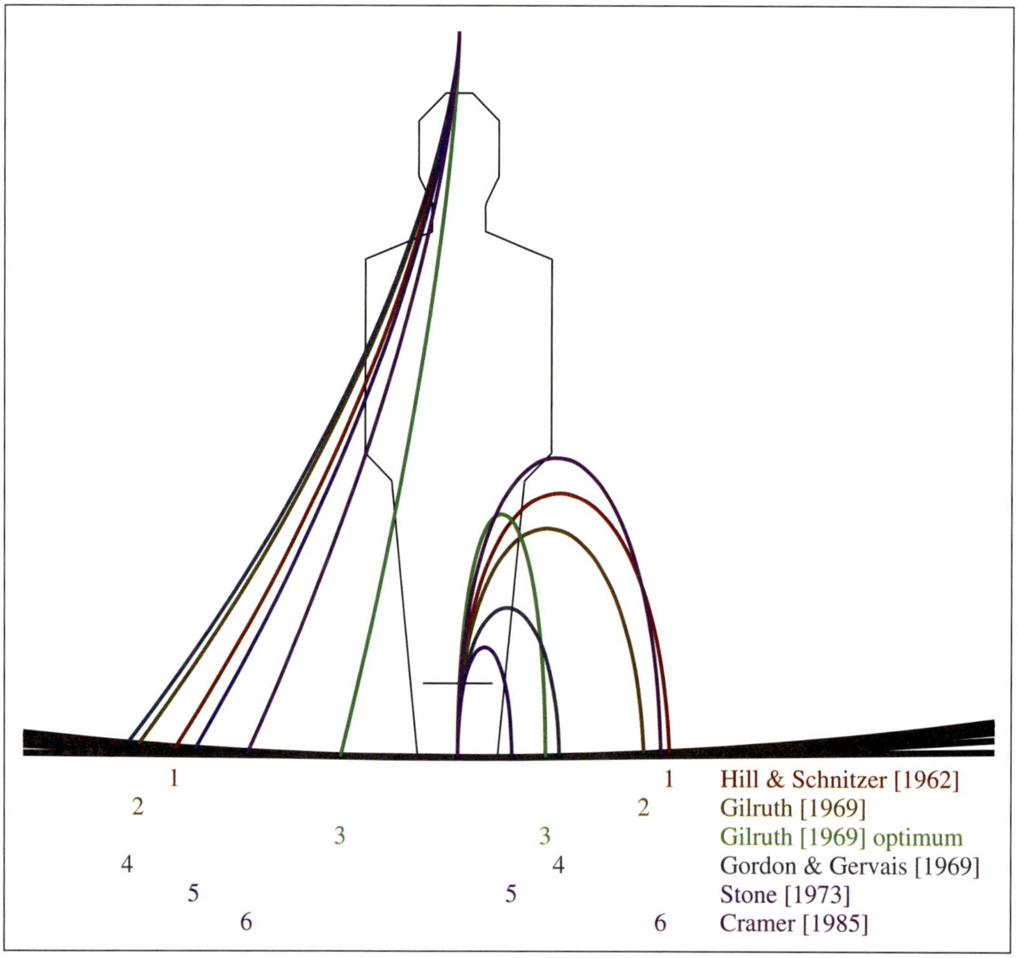

FIGURE 16 Artificial gravity at the minimum comfortable radius and tangential velocity proposed by Hill and Schnitzer (1962), Gilruth (1969), Gordon and Gervais (1969), Stone (1973), and Cramer (1985).

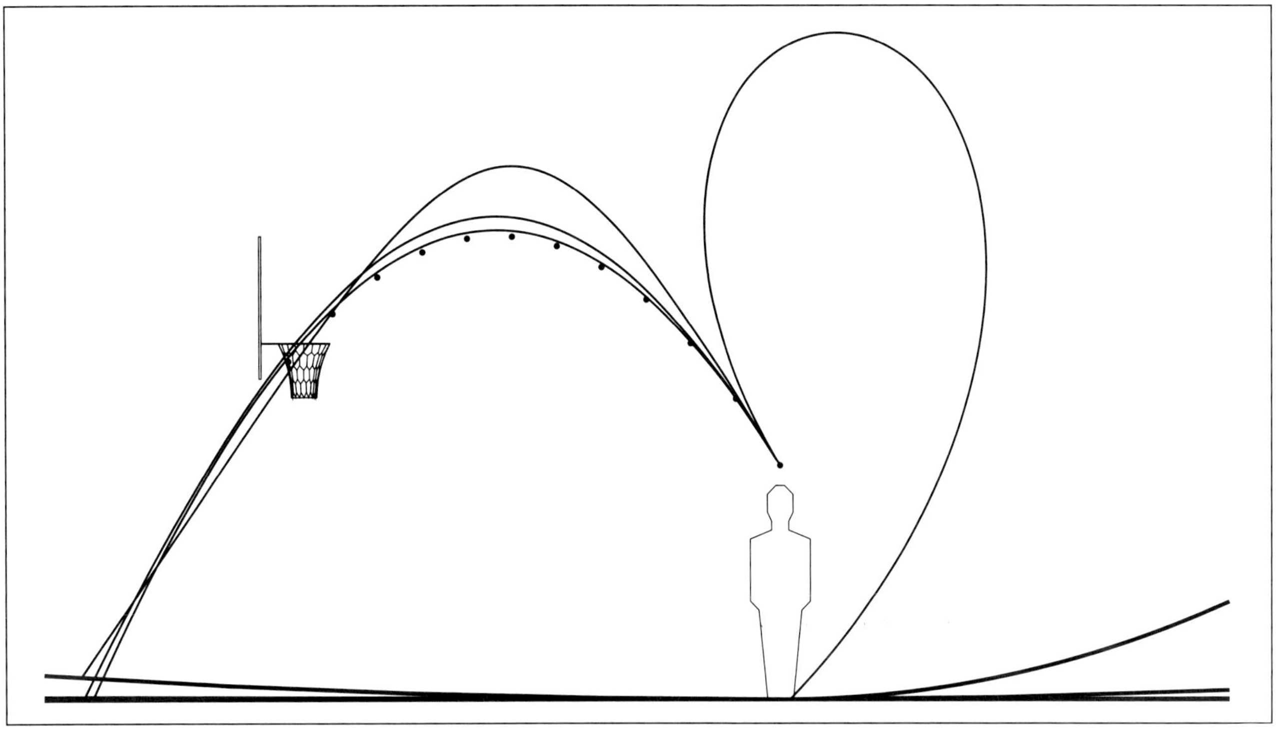

(a) Shooting from the free-throw line

$h = 2$ m
$v = 7.348$ m/s @ 122.4 deg
$R_f = 100$ m, 1000 m, 10,000 m, 10 m

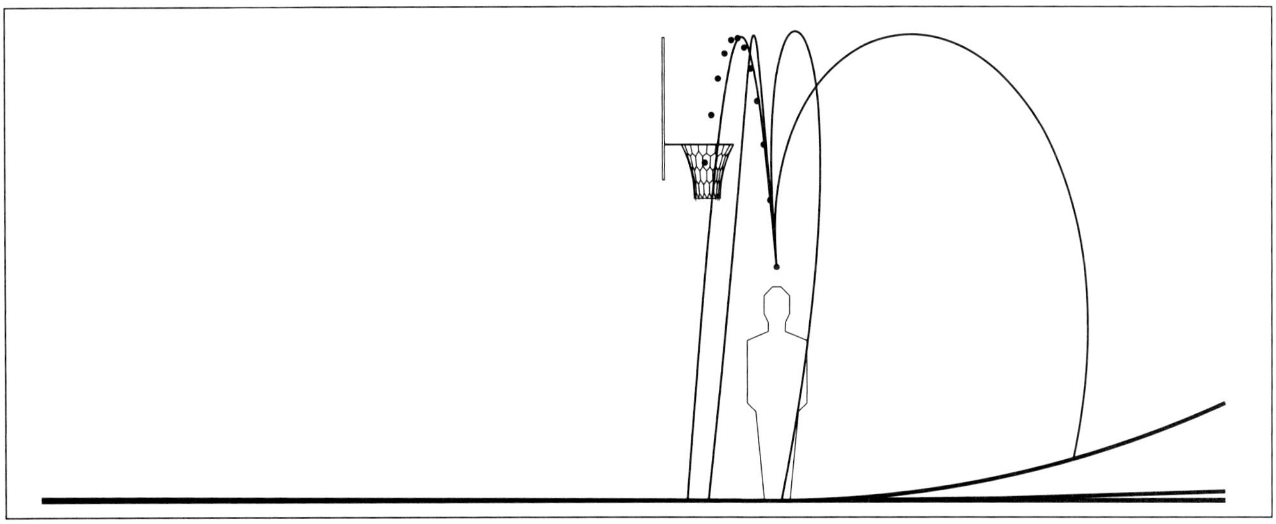

(b) Shooting from close under the hoop

$h = 2$ m
$v = 6.230$ m/s @ 95.3 deg
$R_f = 10{,}000$ m, 1000 m, 100 m, 10 m

FIGURE 17 Basketball in artificial gravity. The dots represent a successful shot in Earth gravity. The curves represent identical shots in 1-*g* rotating environments with radii of 10, 100, 1000, and 10,000 m.

behind the shooter. None of the cited authors regard this environment as comfortable. From the free-throw line, shots in the 100-, 1000-, and 10,000-m environments appear to hit. (The hoop needs to be rotated up slightly to account for the floor curvature.) But closer to the net, a shot in the 100-m environment also turns backward, and even in the 1000-m environment the particle misses the goal. This implies terrestrial players visiting space colonies as large as the Stanford Torus or Bernal Sphere (Johnson and Holbrow 1977) would need to adjust their technique to complete against the "home team." In smaller space hotels, the rules of the game itself would likely need substantial revision.

Figure 17 shows only the "west" goal, with the player shooting up and *against* the environment's tangential velocity. The effects would be different at the "east" goal, that is, shooting *with* the tangential velocity. Switching goals at the start of each period might cancel any advantage of one goal over the other. Turning the court's orientation from east–west to north–south—parallel to the axis of rotation—might also eliminate any such advantage. It would not eliminate the gravitational deviations, of course, but merely orient them in the plane of the backboard instead of perpendicular to it. One goal would be biased left and the other biased right.

APPARENT SLOPES AND CATENARY CURVES

This section describes the use of catenary curves to visualize the apparent slopes of straight chords in artificial gravity. This is especially important in the design of ladders and segmented floors. The orientation of a ladder with respect to the rotation is a critical design detail often ignored in renditions of artificial-gravity habitats.

As Figures 9–13 illustrate, a particle moving at constant velocity in an inertial frame of reference appears to change velocity in a rotating frame of reference. Conversely, a particle that appears to move at constant velocity in the rotating frame must undergo a change of velocity in the inertial frame. Any change in either its speed or direction constitutes acceleration. This is the Coriolis acceleration. Its magnitude and direction are determined by the vector cross product of the environment's angular velocity $\mathbf{\Omega}$ (radians per time) and the particle's relative linear velocity v (distance per time) in the rotating frame:

$$\mathrm{A}_{\mathrm{Cor}} = 2\,\Omega \times \mathrm{v}$$

As a cross product, the Coriolis acceleration is perpendicular to both the axis of rotation and the relative velocity, with magnitude proportional to the sine of the angle between them. For velocity parallel to the axis, the Coriolis acceleration is zero. For velocity perpendicular to the axis (in the plane of rotation), the magnitude is simply

$$A_{\mathrm{Cor}} = 2 \cdot \Omega \cdot v$$

Constant relative velocity in the rotating frame requires Coriolis acceleration in the inertial frame, and this contributes, along with the centripetal acceleration, to the total apparent gravity. This is especially important to walking on flat floors and climbing straight ladders.

Figure 18 shows the situation for an observer walking east (prograde) on a flat floor in a rotating environment. The floor is a chord of a circle, with its midpoint at radius R_c from the center of rotation. The observer's centripetal acceleration is always toward the center; its magnitude and direction depend only on his position, independent of his relative speed. On the other hand, his Coriolis acceleration depends only on his relative speed, independent of his position. The total acceleration $\mathbf{A} = \mathbf{A}_{\mathrm{cent}} + \mathbf{A}_{\mathrm{cor}}$ determines his net apparent gravity and sense of "up." At the center of the chord, its magnitude is

$$\begin{aligned} A &= A_{\mathrm{cent}} + A_{\mathrm{Cor}} \\ &= \Omega^2 \cdot R_c + 2 \cdot \Omega \cdot v \\ &= \Omega^2 \cdot \left(R_c + 2 \cdot \frac{v}{\Omega} \right) \end{aligned}$$

So, the Coriolis component effectively raises the convergence point for the net acceleration by $2 \cdot v/\Omega$ from the center of rotation.

Figure 19 shows what happens when the observer reverses direction and walks back west (antigrade). The Coriolis acceleration also reverses direction, thereby lowering the convergence point.

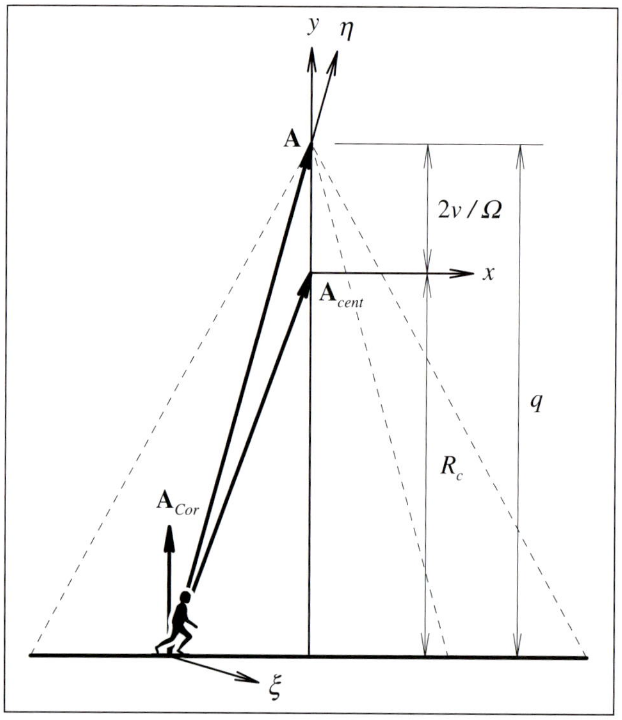

FIGURE 18 Walking east on a flat floor.

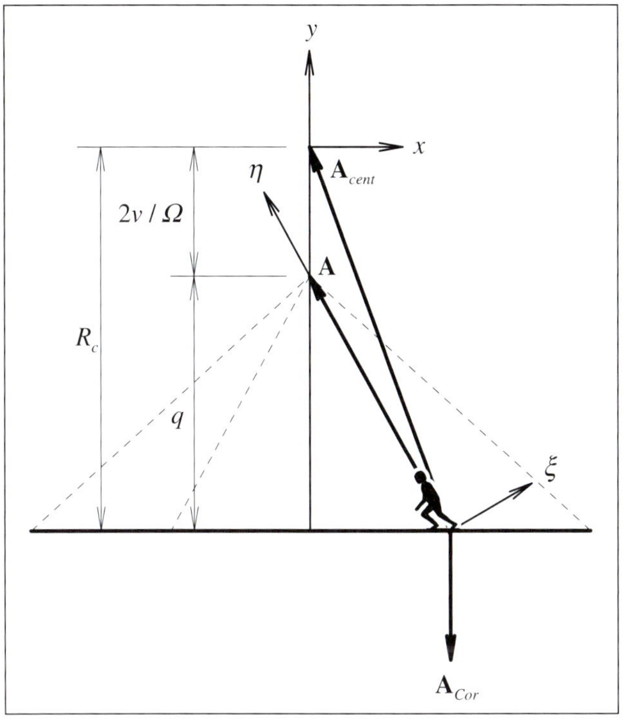

FIGURE 19 Walking west on a flat floor.

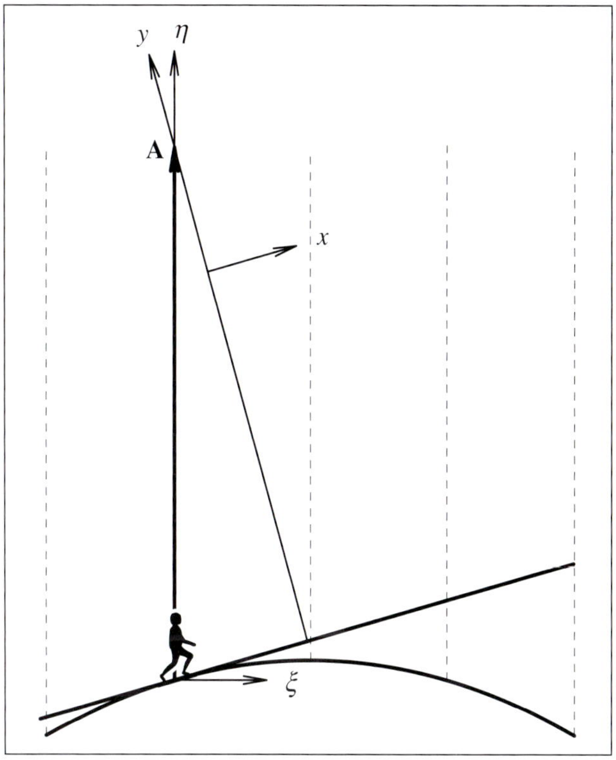

FIGURE 20 Apparent slope while walking east.

In Figures 18–21, the x, y coordinate axes are tied to the rotating environment. The observer carries his own coordinate axes with him, tied to his sense of horizontal and vertical, determined by his net acceleration. These axes are labeled ξ, η. The apparent slope of the floor changes with respect to these axes as the observer walks across it.

Figures 20 and 21 show the situation from the observer's point of view. Though the floor looks straight, when being walked it feels as though it teeter-totters over a hill. Both the shape of the apparent hill and the magnitude of the apparent gravity follow catenary curves (Hall 1999):

$$\eta / q = -\cosh\left(\frac{\xi}{q}\right)$$

$$A / q = \Omega^2 \cdot \cosh\left(\frac{\xi}{q}\right)$$

$$q = R_c + 2 \cdot \frac{v}{\Omega}$$

The dotted lines in Figures 18–21 represent the magnitude and direction of the apparent gravity. Compared to eastward motion, westward motion yields a steeper hill but with weaker gravity.

Figures 22 and 23 illustrate ascending a "vertical" ladder that passes through the center of rotation. In this case, R_c is zero. Although the centripetal acceleration is zero at the center point, the Coriolis acceleration remains constant and nonzero as long as the observer continues to ascend at constant velocity.

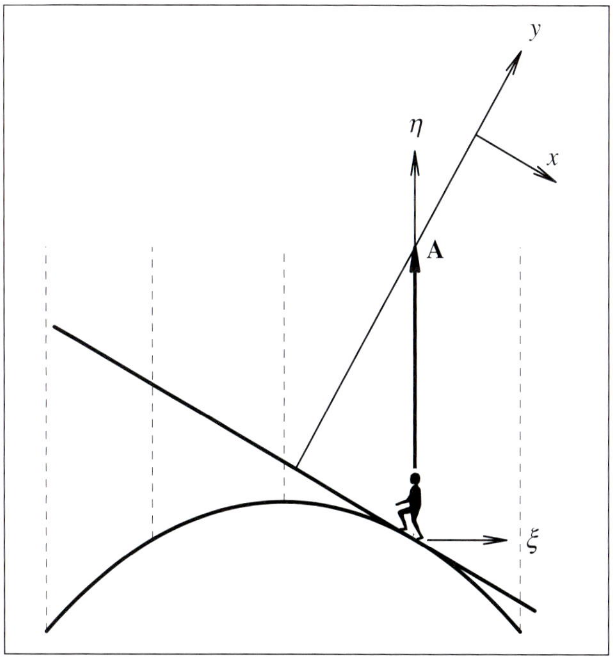

FIGURE 21 Apparent slope while walking west.

The stiffness of the ladder provides the observer's Coriolis acceleration by constraining him to straight-line motion in the rotating frame. As he ascends toward the center, the ladder decreases his tangential

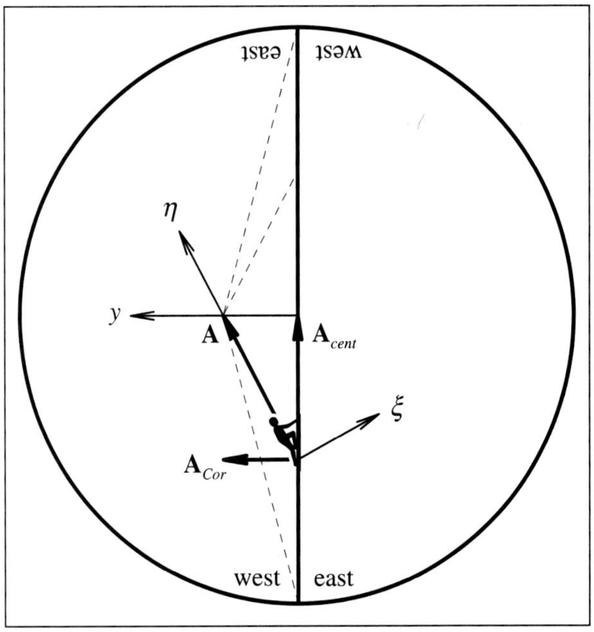

FIGURE 22 Ascending a ladder.

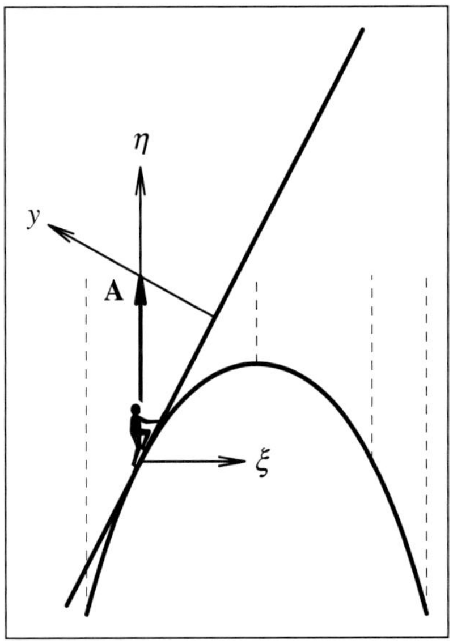

FIGURE 23 Apparent slope while ascending.

velocity; as he descends toward the perimeter, the ladder increases his tangential velocity. To stay above the curve, he must ascend on the west side, as shown, and descend on the east side, so that the ladder presses toward him. If he attempts to climb on the wrong side, the ladder will pull away from him, as if he were climbing on the underside of the curve. It is difficult enough to climb a truly vertical ladder. A ladder that leans backward is not only uncomfortable, but unsafe. For the designer, this raises several key points:

- The plane of the ladder should be parallel to the axis of rotation, perpendicular to the plane of rotation and to the Coriolis acceleration.
- Either the ladder should be accessible from both sides, or separate ladders (accessible from opposite sides) should be provided for ascending and descending. Figure 24 illustrates three possibilities.
- It might be best to avoid ladders entirely, by leaning the line of motion far enough away from the center of rotation so that it functions as a stair.

Will these floor and ladder slopes be "apparent" in any real implementation, or are they only theoretical? Figures 25 and 26 address this question by considering a multistory module similar to TransHab

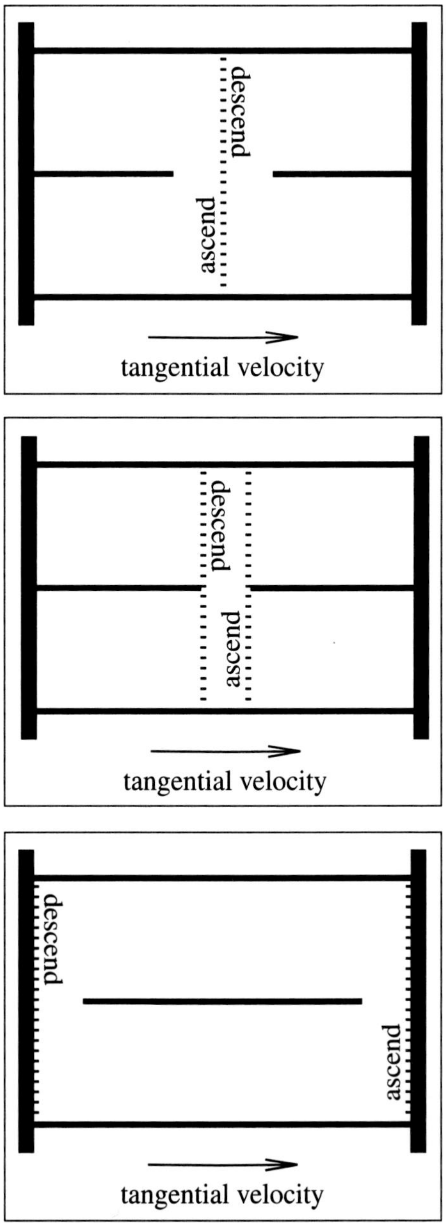

FIGURE 24 Three recommended options for orienting ladders in artificial gravity.

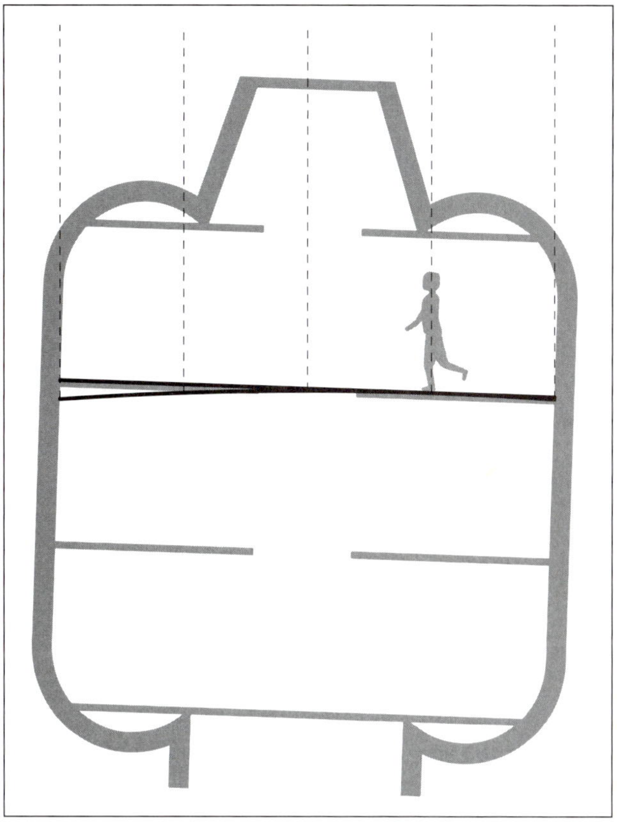

FIGURE 25 Walking west at 1 m/s on a flat floor rotating at radius 67.1 m and tangential velocity 14.0 m/s. These are the minimum radius and tangential velocity that satisfy all cited comfort criteria. The apparent floor slope at the wall is nearly 4 deg (about a 7% grade, or 1:15).

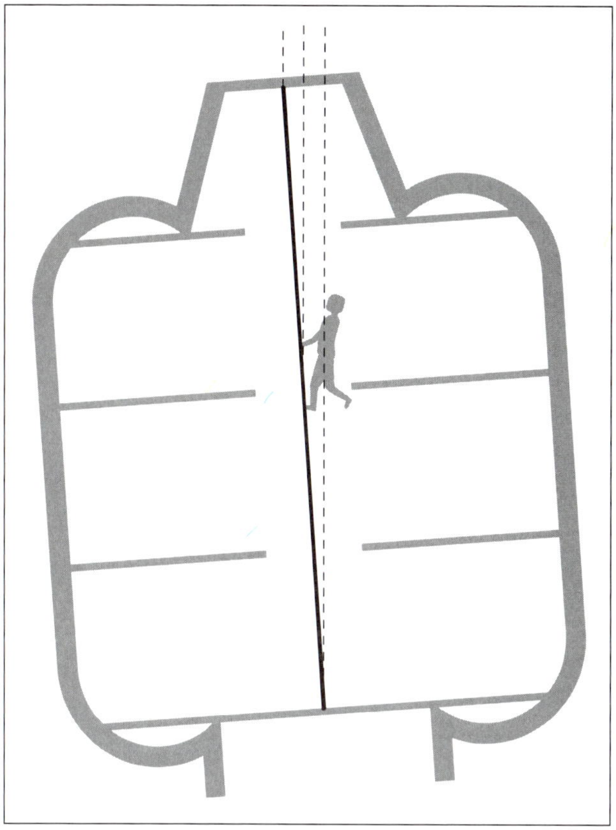

FIGURE 26 Descending a ladder at 0.5 m/s on a "vertical" ladder in the same rotation conditions as Figure 25. The apparent lean of the ladder is 4 deg from vertical.

(Kennedy 1999) in an artificial-gravity application. The top floor is at the minimum radius and tangential velocity that satisfies all of the comfort criteria cited earlier—about 67.1 m and 14.0 m/s, respectively. Figure 25 shows an observer walking west on the top level at a modest speed of about 1 m/s, whereas Figure 26 shows him descending from that level at 0.5 m/s. Once again, these figures show that even the most conservative comfort-zone criteria yield considerable deviation from Earth-like gravity. The floor's maximum apparent slope is nearly 4 degrees (about a 7% grade, or 1:15). The ladder's apparent lean from vertical is also about 4 deg. These deviations would be even greater with faster walking and climbing speeds, a wider floor, a shorter radius, or a slower tangential velocity, which most renditions of the comfort zone allow.

Published illustrations of artificial-gravity habitats often fail to account for these effects. For example, Figure 27 shows two scenes from a computer animation of a TransHab module in an artificial-gravity crew transfer vehicle (Borowski et al. 2006). To provide 0.38 *g* at 4 rpm, the top level would have a rotational radius of only about 21.2 m and a tangential velocity of 8.9 m/s. Although these values are within most renditions of the comfort zone, they are much less than the values assumed in Figures 25 and 26. Consequently, the floor would seem considerably more sloped—as much as 13 deg (23% grade, 1:4.33) for westward motion at 1 m/s and 3.8 m from the centerline. The ladder at this level would seem to lean more than 6 deg from vertical for descent at 0.5 m/s. From this image, the ladder's orientation with respect to the rotation is not clear, but it is clear that there is only one ladder that cannot function well for both ascending and descending.

Figure 28 is an iconic scene from the movie *2001: A Space Odyssey* (Kubrick and Clarke 1968), showing a crewman descending a ladder set into the side wall of the rotating section of a spacecraft. The floor radius is about 5.33 m, and it rotates at about 5.3 rpm to simulate lunar gravity. Under these conditions, if he descends at 0.5 m/s,

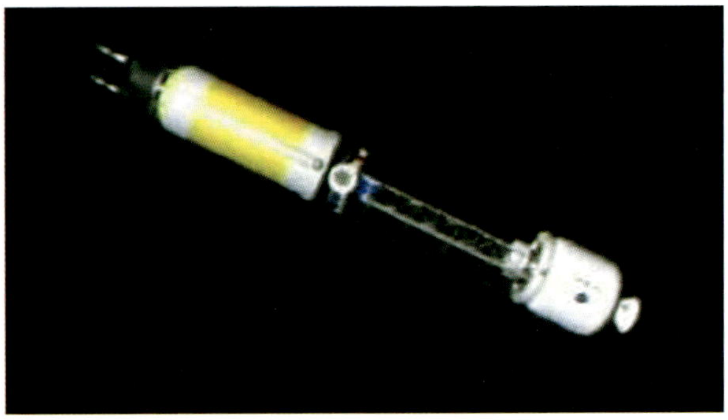

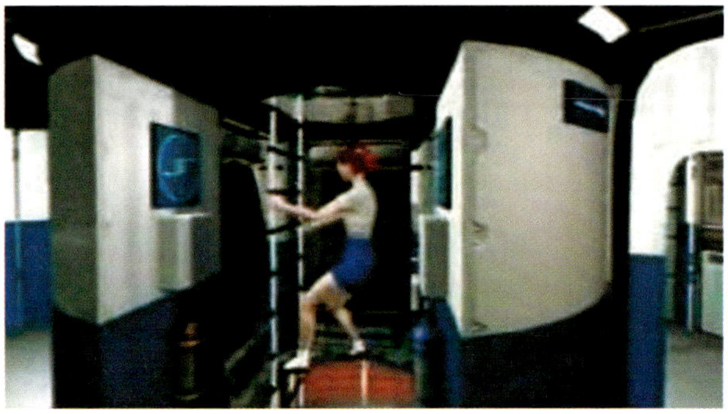

FIGURE 27 Scenes from a computer animation of a TransHab module in an artificial-gravity crew transfer vehicle (Borowski et al. 2006). With 0.38 *g* at 4 rpm, the top floor radius would be 21.2 m with a tangential velocity of 8.9 m/s. The apparent floor slope when walking west at 1 m/s would be as much as 13 deg (23% grade, 1:4.33). The apparent lean of the ladder when descending at 0.5 m/s would be more than 6 deg from vertical. Because of the reversal of Coriolis acceleration from ascending to descending, a one-sided ladder cannot provide safe transit in both directions.

FIGURE 28 Descending a ladder in the side wall of a rotating spacecraft section [from the movie *2001: A Space Odyssey* Copyright © Turner Entertainment Co. A Warner Bros. Entertainment Company. All rights reserved. (Kubrick and Clarke 1968)].

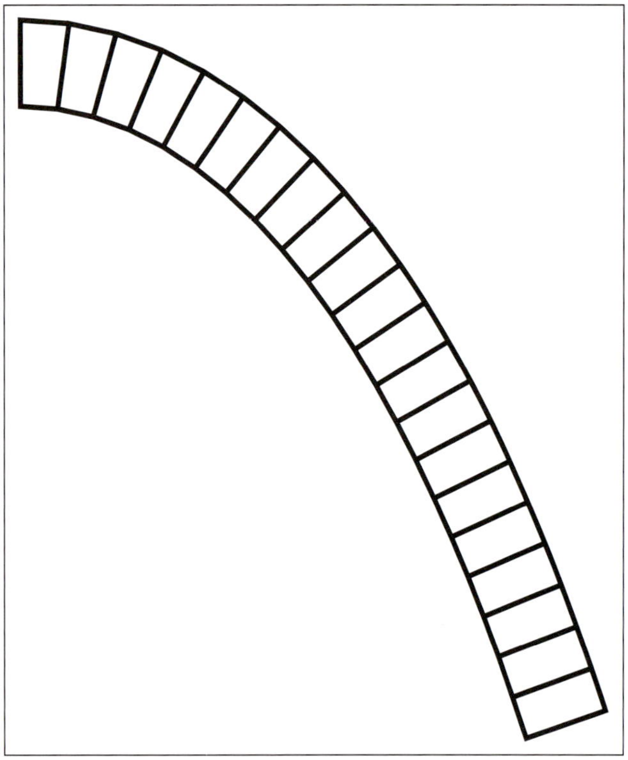

FIGURE 29 Apparent lean of the ladder in Figure 28 caused by the lateral Coriolis acceleration.

the apparent lean of the ladder will be as depicted in Figure 29. The Coriolis acceleration acts in the plane of the ladder, instead of perpendicular to it, and near the center is proportionally much greater than the centripetal acceleration. The effect is as if the ladder were bent over to the side. To avoid this perception the ladder should be turned perpendicular to the side wall so that it feels more like a ramp.

Figure 30 is a detail from an illustration of a "variable gravity research station" published in the proceedings of the third Case for Mars Conference (Staehle 1989). It shows a ladder coplanar with large external parabolic collectors that seem to be aimed at a fixed target, thus making both the dishes and the ladder coplanar with the rotation. This ladder, similar to the one in Figures 28 and 29, will seem to lean sideways in relation to the Coriolis acceleration. The danger can be exacerbated by the flow of blood away from the brain as a person descends into stronger centripetal acceleration. A series of stairs would be considerably safer.

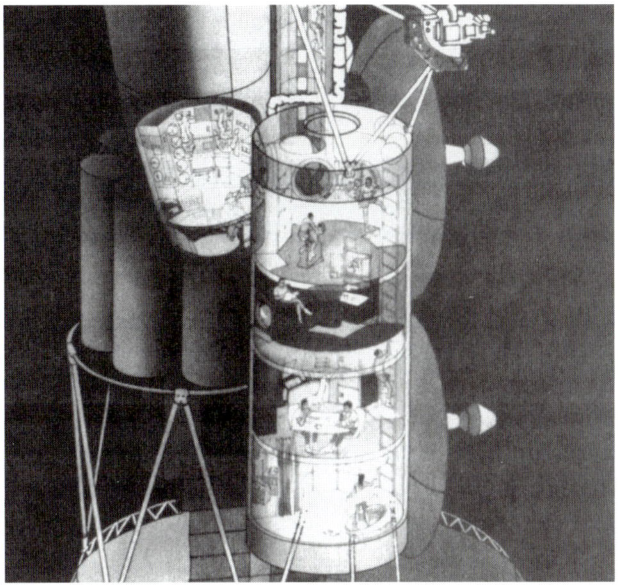

FIGURE 30 Climbing a ladder in the side wall of a variable-gravity research station. Detail from an illustration by Carter Emmart for the third Case for Mars Conference (Staehle 1989). Similar to Figures 28 and 29, this ladder will seem to lean sideways.

ADVANCED VISUALIZATIONS

Computer aided design (CAD) can be a powerful tool for simulating and testing environments before they are built. However, as with all tools, the craftsman must use them skillfully: the best hammers and saws cannot compensate for an unskilled or careless carpenter. Several common CAD systems provide the means for computing walk-throughs and other kinds of animations. Some of these systems are capable of rendering particle physics, with realistic rules for force, mass, acceleration, velocity, and momentum. The onus is on the designer and renderer to use these tools with understanding and skill.

Simple key-frame animations, with erroneous built-in assumptions about the gravitational environment, cannot be used to visualize or validate a design for artificial gravity. An animation done with realistic rendering of light and material and articulated human figures, but without regard for the physics of rotating systems, is actually detrimental. To the extent that such animations mislead viewers into unrealistic expectations about mobility in artificial gravity, they are worse than no visualization at all.

A human body is a complicated mechanism, and realistic simulation of its articulations is difficult to achieve. A common approach to life-like animation is "motion capture": recording body segment positions and rotations of an actor performing various tasks in a studio, then applying these data to a geometric model of a human or some other creature. Because the effects of Earth gravity are implicit in the recorded data, such motion-capture data are not valid for predicting mobility in any other gravitational environment.

Human mobility is a complex interaction of muscular force, body segment mass and distribution, inertia, gravitation, and acceleration of the frame of reference. A robust, predictive simulation, adequate for validating a new gravitational environment, must account for all of these factors by applying the laws of physics to an accurate and detailed anthropometric model. This is the realm of "digital human modeling," and there is a considerable body of research, data, and software to support it. Designers seeking to validate an artificial-gravity environment via computed human animation must use these sophisticated tools. If lack of resources precludes it, then design integrity is better served by simple, yet physically accurate, particle simulations rather than visually stunning but physically flawed cartoons.

Immersive, high-fidelity, virtual-reality simulations could be very powerful for exploring certain aspects of artificial-gravity design. Unfortunately, the most sophisticated simulations tend to be least accessible. And the designer must be a cautious and knowledgeable user in any case; virtual reality systems can be superb at visualizing imaginary environments that defy the laws of physics. Although simple, the visualizations presented in this chapter are correct and intended for easy and immediate accessibility.

DESIGNING FOR ARTIFICIAL GRAVITY: BASIC PRINCIPLES

Minimizing radius and tangential velocity averts obvious costs in mass and kinetic energy of rotating habitat systems. Less obvious, though, are potential costs implicit in accommodating a distorted gravitational environment. At the outset, these ought to include a greater investment in design—including research, testing, and scrupulous review of assumptions carried over from terrestrial experience. Other cost increases might include more stringent crew selection, longer periods of training and acclimatization, and reduced crew productivity. Highly skilled candidates for an interplanetary expedition using artificial gravity might be disqualified by physiological intolerance and inability to adapt to conditions at the edge of the rotation comfort zone.

The habitat pressure shell does not shield its interior from physics and mechanical dynamics.

The same concerns that govern overall configuration for a rotating spacecraft must be brought to bear on its interior architecture. In an artificial-gravity environment, the direction of rotation should be visually obvious throughout the design, to establish a connection between visual and vestibular cues to rotation so that crew can orient their movements advantageously with regard to Coriolis accelerations. Floors that are wide with respect to the rotational radius should not be flat; they should be cylindrical arcs so that centripetal acceleration remains perpendicular to the surface, thus avoiding unwanted apparent slopes.

Circular plans without obvious orientation to the habitat's rotation should be avoided. If other design considerations mandate a vertically oriented, cylindrical habitat, the interior should nevertheless be laid out on a grid aligned with the axis of rotation (north–south) and tangential velocity (east–west). The ISS's cylindrical modules are inscribed with a square living volume presenting identifiable wall, floor, and ceiling surfaces. Similarly, an artificial-gravity module should be inscribed with axis-aligned partitions presenting distinct east, west, north, and south walls regardless of the module's exterior shape. Bending the floor into an arc, as suggested in the preceding paragraph, is fundamental to establishing this orientation.

Color and pattern can further distinguish east (prograde) from west (antigrade). This can help keep inhabitants visually oriented with respect to the rotation, so that they can anticipate the direction of Coriolis accelerations that will accompany various actions such as standing up and sitting down.

Multistory designs should be avoided if possible. If not, then vertical circulation must be planned with Coriolis acceleration as a principal consideration. Keeping the plane of a ladder parallel to the axis of rotation will keep the Coriolis acceleration perpendicular to that plane. Moreover, upward and downward motion should be accommodated on opposite faces of such a ladder, or on opposite ladders, so that the acceleration always presses the ladder against the climber rather than pulling it away.

When designing a space habitat for artificial gravity, question all assumptions related to the perception of gravity and the motion of objects and inhabitants. Use dynamically correct simulations to aid design visualization. |

References

Borowski, S., Dudzinski, L., Sauls, B. and Minsaas, L. (2003), "Bimodal NTR Artificial Gravity Mars Mission," NASA Glenn Research Center and John Frassanito & Associates, http://www.lunar-rocket.com/flash.html.

Cramer, D. B. (1985), "Physiological Considerations of Artificial Gravity," *Applications of Tethers in Space* edited by A. C. Cron, Vol. 1, NASA, Washington, D.C., pp. 3.95–3.107.

Gilruth, R. R. (1969), "Manned Space Stations – Gateway to Our Future in Space," *Manned Laboratories in Space*, edited by S. F. Singer, Springer-Verlag, New York, pp. 1–10.

Gordon, T. J., and Gervais, R. L. (1969), "Critical Engineering Problems of Space Stations," *Manned Laboratories in Space*, edited by S. F. Singer, Springer-Verlag, New York, pp. 11–32.

Hall, T. W. (1987; 1999), "SpinDoctor Artificial Gravity Simulator," original coding 1987/02/14, last revision 1999/06/22, http://www.artificial-gravity.com/ag/sw/SpinDoctor/.

Hall, T. W. (1999), "Inhabiting Artificial Gravity, AIAA Paper 99-4524, Sept. 1999.

Hall, T. W. (2000; 2003), "SpinCalc Artificial Gravity Calculator," original coding 2000/01/27, last revision 2003/05/05, http://www.artificial-gravity.com/ag/sw/SpinCalc/.

Hill, P. R., and Schnitzer, E. (1962), "Rotating Manned Space Stations," *Astronautics*, Vol. 7, No. 9, pp. 14–18.

Johnson, R. D., and Holbrow, C. (eds.) (1977), *Space Settlements: A Design Study*, NASA SP-413, NASA, Washington, D.C., http://www.nas.nasa.gov/About/Education/SpaceSettlement/75SummerStudy/Design.html.

Kennedy, K. J. (1999), "ISS TransHab: Architecture Description," Society of Automotive Engineers, Paper 1999-01-2143, July.

Kubrick, S., and Clarke, A. C. (1968), *2001: A Space Odyssey*, Metro-Goldwyn-Mayer, Hollywood.

NASA (1995), *Man-Systems Integration Standards, Revision B*, NASA STD-3000, Vol. 1, NASA, Washington, D.C., Sec. 5.3.2.3, http://msis.jsc.nasa.gov/sections/section05.htm#_5.3_ACCELERATION.

Schultz, D. N., Rupp, C. C., Hajos, G. A., and Butler, J. M. (1989), "A Manned Mars Artificial Gravity Vehicle," *The Case for Mars III: Strategies for Exploration — General Interest and Overview*, edited by C. Stoker, Science and Technology Series, Vol. 74, American Astronautical Society, pp. 325–352.

Staehle, R. L. (1989), "Earth Orbital Preparations for Mars Expeditions," *The Case for Mars III: Strategies for Exploration – General Interest and Overview*, edited by C. Stoker, Science and Technology Series, Vol. 74, American Astronautical Society, Univelt, SanDiego, pp. 373–396.

Stone, R. W. (1973), "An Overview of Artificial Gravity," *Fifth Symposium on the Role of the Vestibular Organs in Space Exploration*, edited by A. Graybiel, NASA, Washington, D.C., pp. 23–33.

Tillman, B. (1987), "Human Factors in the Design of an Artificial Gravity Research Facility," unpublished report prepared for Lockheed Missiles and Space Co., Sunnyvale, CA.

13

orbital cities

BRENT SHERWOOD

INTRODUCTION

EVENTUALLY HUMANS will extensively develop Earth orbital space. Over the next several hundred years, we will continue to expand our habitable zone by applying technology, as we have throughout tens of millennia of human history. Fundamental principles governing large-scale architecture and urban planning in Earth orbital space—some of which derive from unique constraints and some of which are invariant despite them—are discernible now. Resolving a viable parti even now—decades before it is manifest—is useful because it can help focus development of strategic options for design and technology, leading to the emergence of an organic architecture. Such a parti, for an integrated infrastructure configuration, must respond to a combination of constraints: low-Earth-orbit (LEO) operations, capabilities of anticipated technology, predictable utilization requirements and human drivers, and long-term growth.

EVOLUTION OF A VIABLE LEO PARTI

Infrastructure principal axes (i.e., mass properties axes) are gravity-gradient stabilized and torque balanced and therefore are aligned with the orthogonal, cardinal directions of LEO: zenith-nadir, port-starboard, and forward-aft (Figure 1). As constructions become larger, they approach actual geometrical symmetry about these cardinal axes. The basic parti is a spine, stretched along the orbit track, serving as the principal access conduit—"Main Street."

Cantilevered laterally from this spine are ribs—"side streets"—spaced at intervals separated by gaps of free space consistent with their vacuum-access or viewing requirements (Figure 2). The ribs are dendritic and taper away from the root. This accommodates the natural bundling of all types of service feeds closer to the spine (Figure 3). As the facility grows, its longitudinal aspect ratio becomes very high: the spine is many times longer than the lateral dimension of the ribs (Figure 4). Closest to the spine are functions requiring the greatest concentration of utilities services (e.g., environmental control equipment, laboratories, manufacturing) and uses supported by the greatest human traffic (e.g., public gatherings, retail, workplaces). More naturally remote functions like habitation are deployed farther out in the ribs.

Internal traffic is conveyed through a continuous pressurized tunnel along the spine. Early constructions use concatenated modules for this tunnel; mature construction techniques include *in situ* extrusion assembly analogous to subterranean tunnel construction. Conveyance in the large-scale implementation is carried in redundant, adjacent tunnels, mechanically assisted by railed linear induction. Exterior conveyance rails are located on the zenith ridge of the spine, with redundant external mechanical systems distributed along each side of, and therefore accessible from, the rail corridor. The belly of the spine is open for nadir Earth views as noted next (Figure 5).

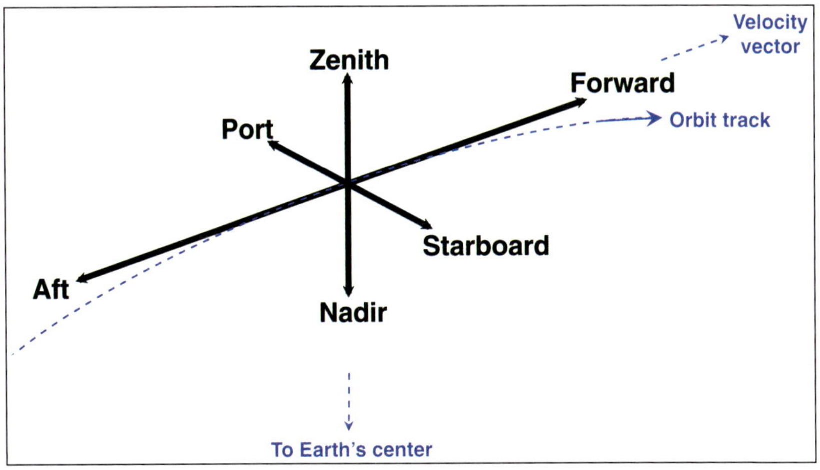

FIGURE 1 Infrastructure principal axes align with LEO cardinal directions.

Copyright © 2008 by Brent Sherwood. Published by the American Institute of Aeronautics and Astronautics, Inc., with permission. This chapter is based on material presented in AIAA Paper 90-3737, reprinted with permission of Boeing Company.

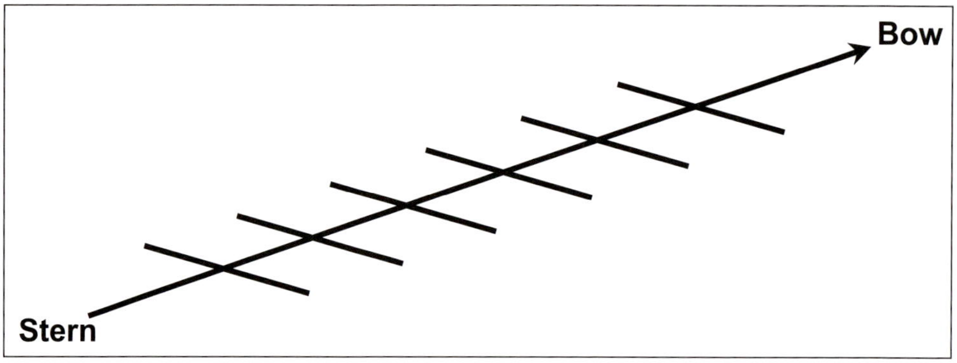

FIGURE 2 Basic spine-and-rib parti enables growth.

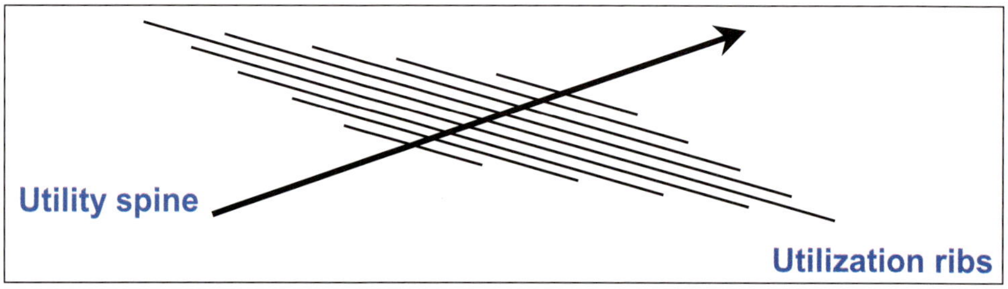

FIGURE 3 Utility trunks bundle near the spine.

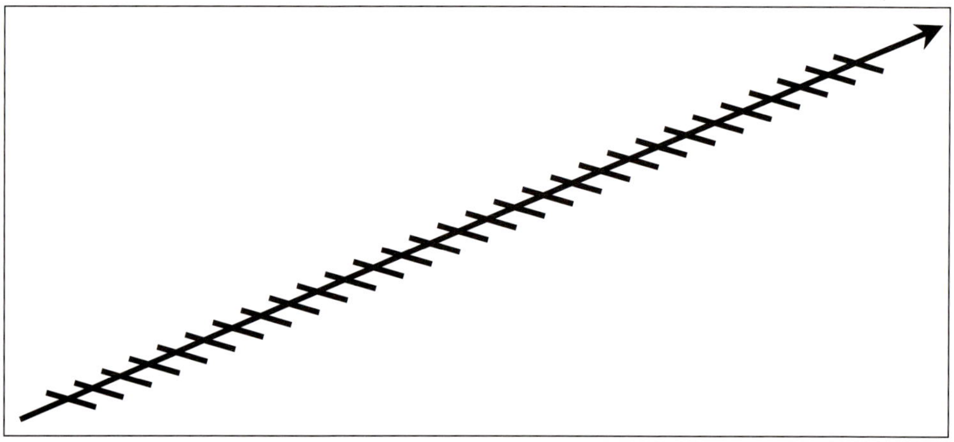

FIGURE 4 High aspect ratio emerges with growth.

Hung below the spine on tethers are facilities (e.g., habitats) that benefit from genuine isolation, near-4π steradian viewing, and very slight partial weight. For urban constructions, such facilities are clustered in "neighborhoods" widely spaced along the tethers (Figure 6). Wide spacing yields minimum mutual view obscuration between adjacent clusters. Because the tether threads the mass centers of each cluster, elevator cabs crawling up and down the tether pass through the geometrical center of generally symmetrical constructions. The elevator cabs berth laterally to the clusters with extensible mating adapters (Figure 7).

Because the best views in LEO are Earthward, vistas of Earth substitute for the vistas of open sky people have been accustomed to throughout human history—alternating every 45 minutes between the bright blue, variegated, day-lit hemisphere and the city-light and lightning-spangled dark hemisphere. The blue Earth becomes LEO inhabitants' blue sky. Habitable microgravity spaces, including private apartments, the spine tunnel, and places of assembly,

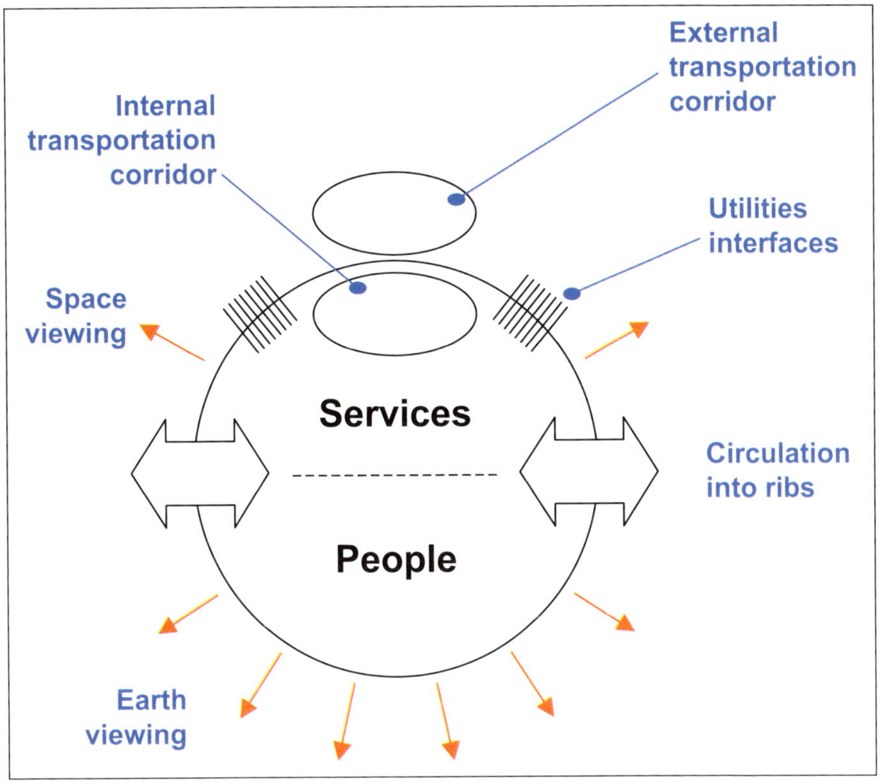

FIGURE 5 Spine cross-section functional zoning responds to orientation within the environment.

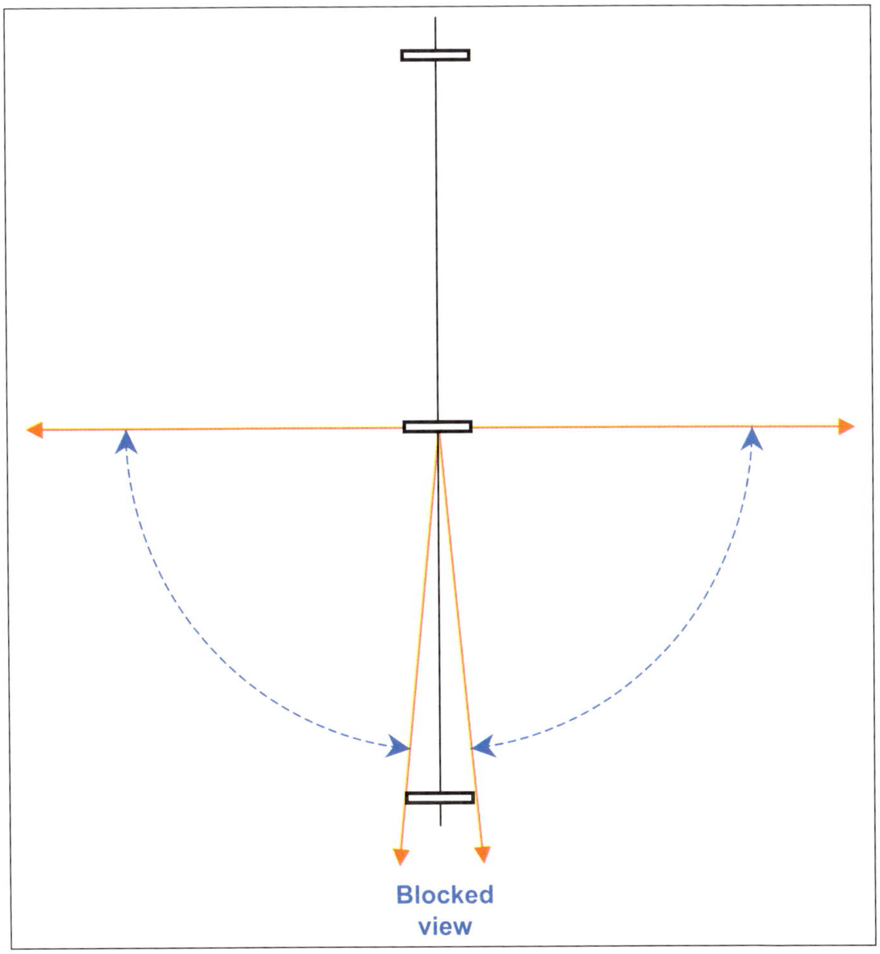

FIGURE 6 Separation along tether preserves view factor for each habitable cluster.

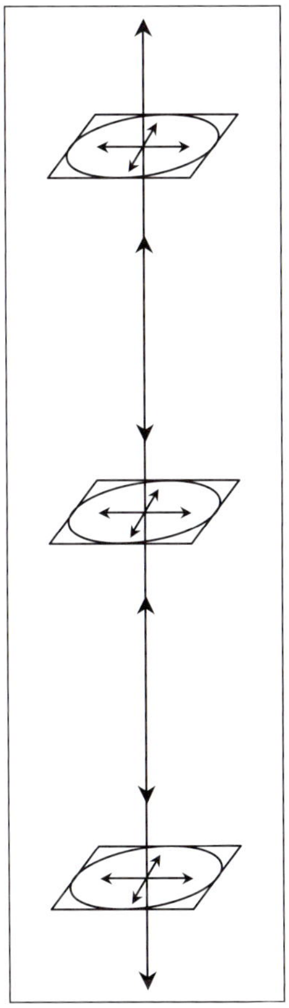

FIGURE 7 Pressurized elevator cabs berth laterally for shirt-sleeve transfer among separated habitable clusters.

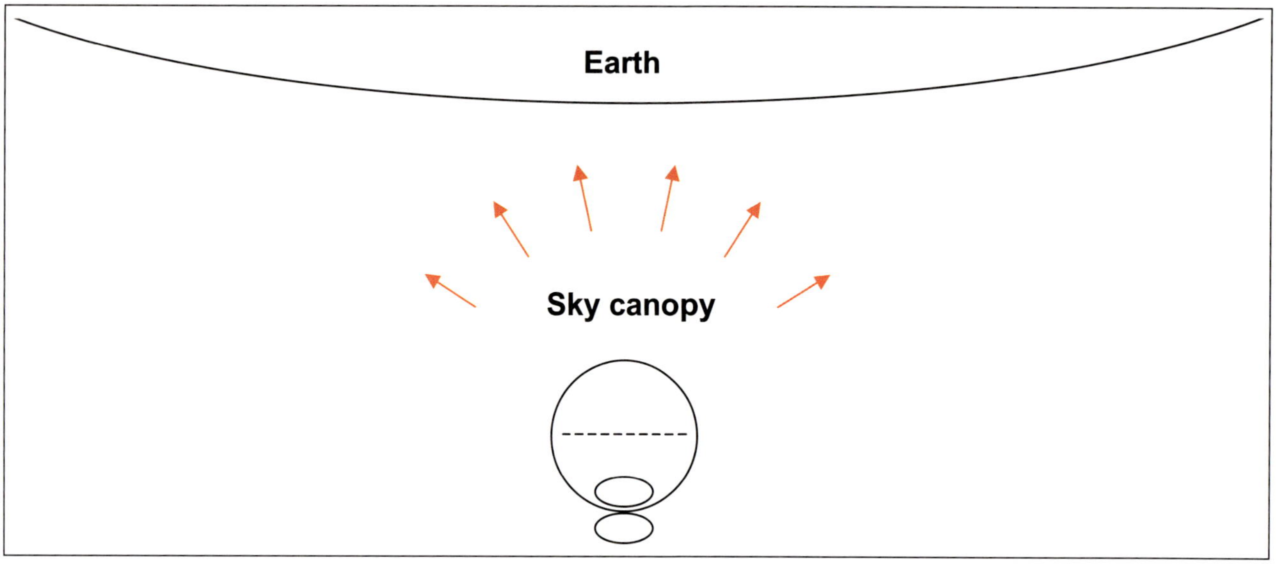

FIGURE 8 Blue vista of planet Earth provides the sky canopy in LEO architecture.

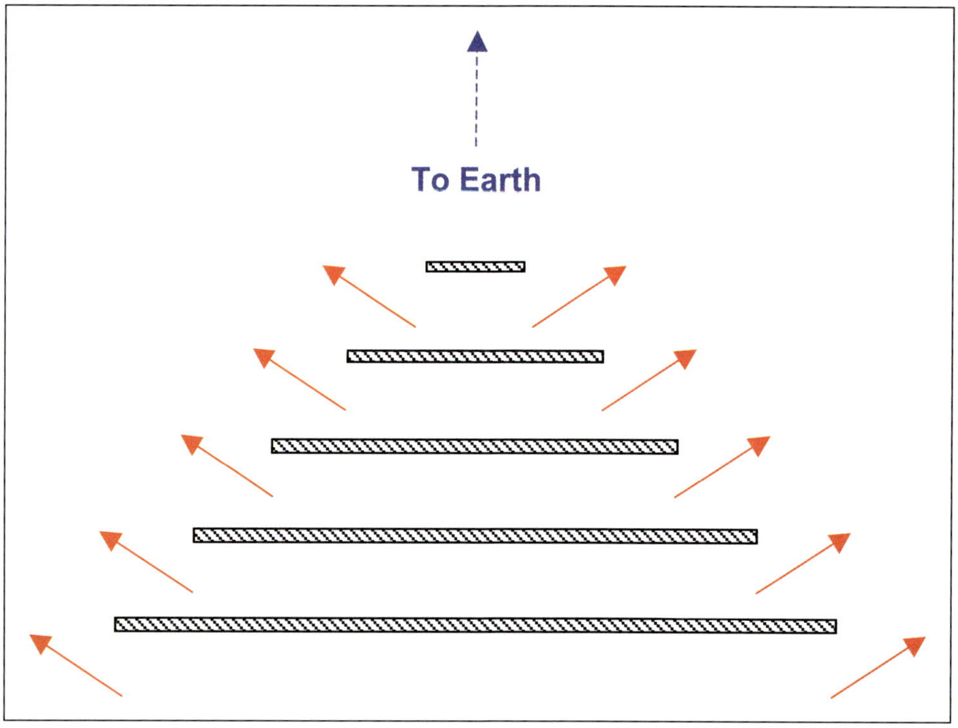

FIGURE 9 Ziggurat topology maximizes access to Earth view.

are principally oriented "upside down," the local "up" is directed toward the nadir, and ceiling windows enable the Earthscape to become the sky (Figure 8). For the general case where development density outweighs optimal viewing, a ziggurat configuration maximizes both (Figure 9).

The key Earth-view orientation just described is reversed in partial-weight habitats hung below the spine because in these areas the nadir feels "down" and objects dropped will collect on the "floor." So whereas in microgravity facilities the "sky" is overhead, in suspended facilities it is underfoot. Here, the sensation is palpably of flying high above the Earth. The two primary window configurations here are oblique downward views through lateral windows and viewing galleries surrounding windows in the floor (Figure 10). This in turn means that the primary architectural parti for suspended

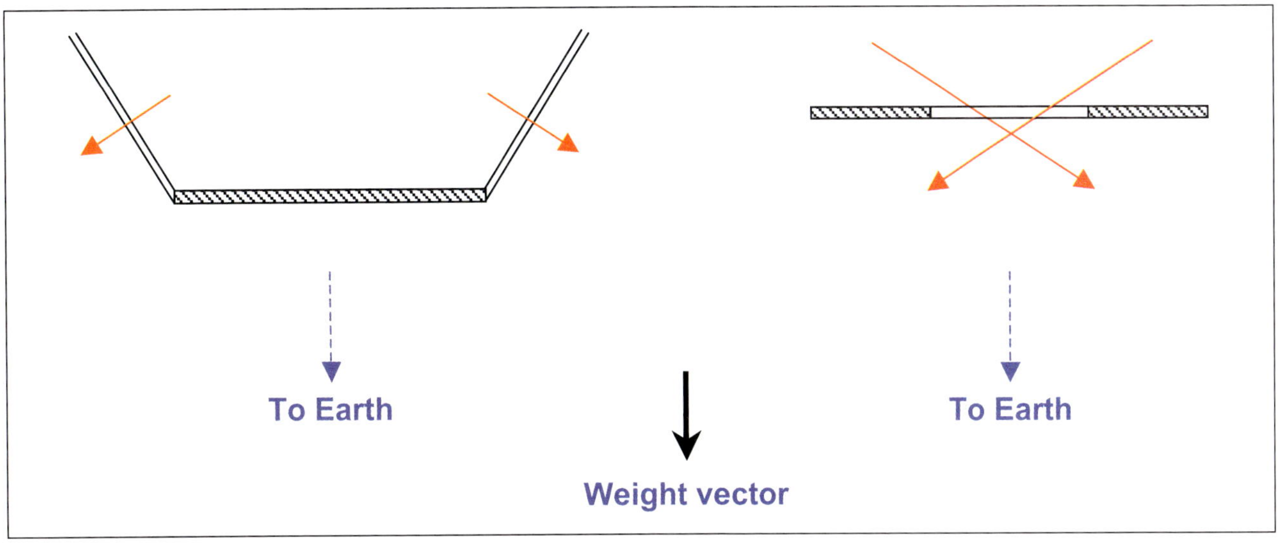

FIGURE 10 Earth view is downward rather than overhead in tethered partial-weight zones.

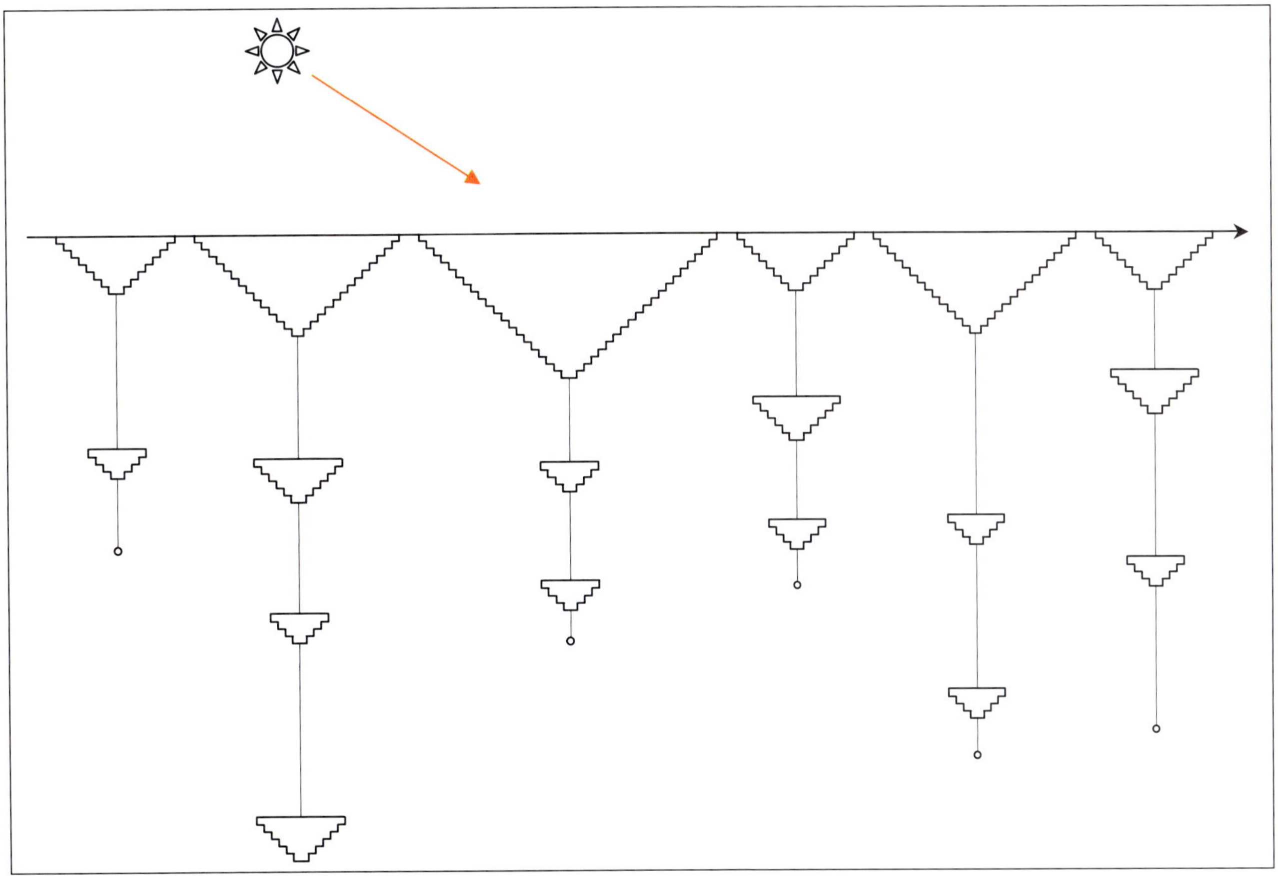

FIGURE 11 Suspended parti marries maximizing Earth view with the downward tug of partial weight.

module clusters is inverted ziggurats to maximize the opportunities for unobstructed, oblique-downward viewing (Figure 11).

Above the spine are the energy-exchange systems that require "up-and-out" viewing for power collection and thermal radiation. Either direct insolation or beamed power enters the complex here for electricity to be generated. Active-loop radiators are also gimbaled, but oriented normal to the instantaneous sun vector to view cold space. Large collector-radiator arrays extend fore and aft from lateral cantilevers parallel to the ribs beneath them on both sides of the spine (Figure 12). A typical average ratio of module volume supported per unit array area can be visualized from the International Space Station (ISS) configuration (solar-dynamic plants and beamed-power collectors are more area efficient).

Powerplants are also tethered up in the zenith direction to serve three purposes: provide mass balance for the nadir-tethered habitable facilities, to keep the center of mass of the overall complex within the envelope of the central spine, maintaining its microgravity condition; collect the power needed to support the nadir-tethered facilities without obstructing the Earth view from the main spine and ribs; provide nonpropulsive, electrodynamic orbit reboost for the entire facility. Tethered powerplants can be stacked up on tethers; their mutual view-factor obscuration decreases with increasing separation along the tether, and the general pointing angle is generally oblique. The parti prescribes zenith-tethered powerplants approximately in proportion to the amount of nadir-tethered facilities (Figure 13).

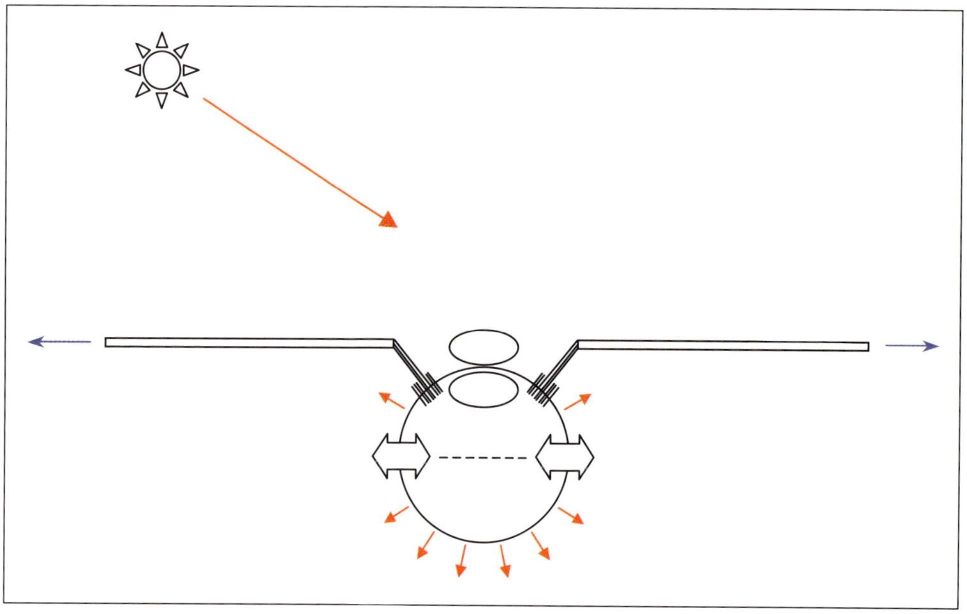

FIGURE 12 Dual energy-exchange crests surmount the spine to provide power in and heat out.

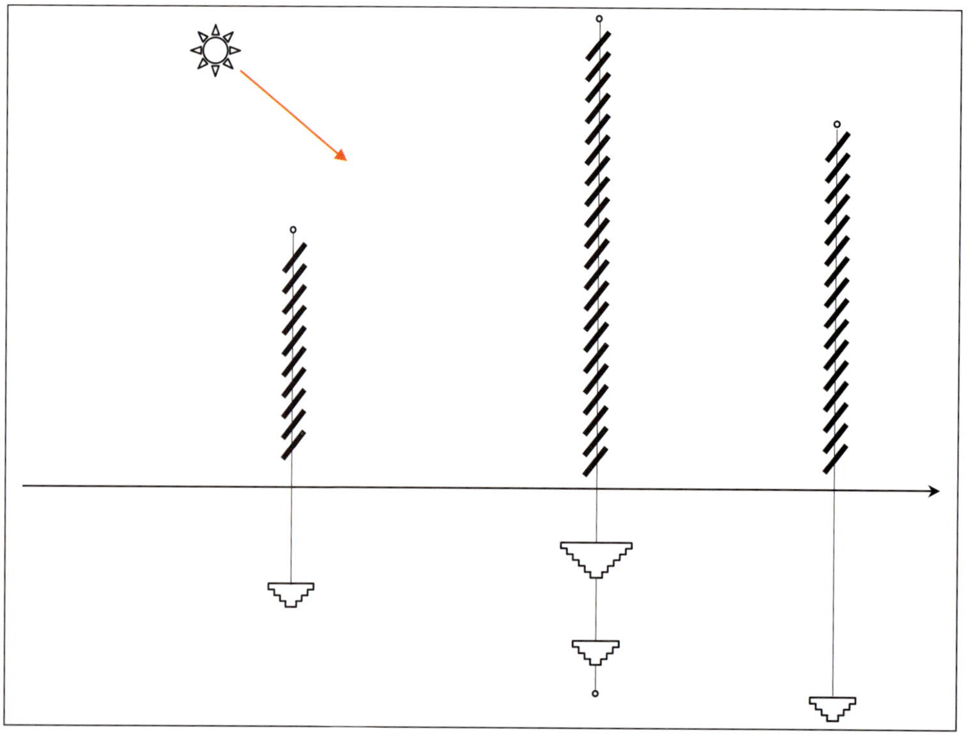

FIGURE 13 Tethered-up powerplants proportionately balance and service tethered-down facilities.

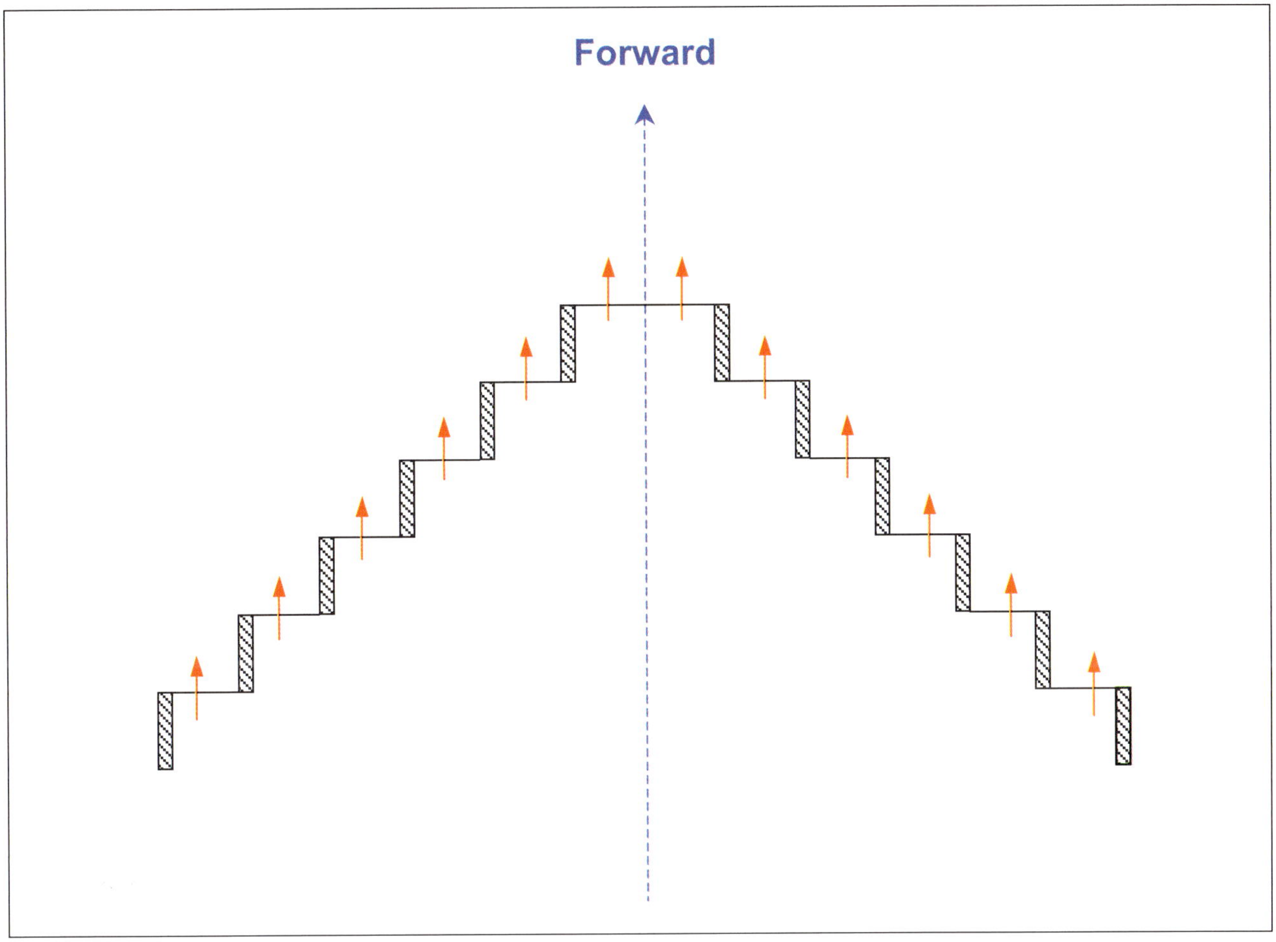

FIGURE 14 Staggered step-backs maximize the unique forward view.

Bow and stern are prime real estate. The view off the prow is the most exhilarating, for here the full effect of orbital velocity, with Earth's landscape continuously approaching over the horizon, can be experienced without any foreground for scale or distraction. This experience was first described by *Skylab* astronauts changing film canisters on the station's solar telescope, as feeling like being on the front end of a surreal locomotive: view and motion, but without foreground, wind, or sound. Forward-facing views are maximized in the LEO parti by staggered setbacks, just as seaside resort hotels are stepped back in plan to maximize the number of ocean-view rooms (Figure 14).

The stern provides the best opportunities for wake-shielded access to clean vacuum for industrial activities (Figure 15). Both bow and stern are also essential transportation approach corridors. V-bar rendezvous (approach and departure parallel to the velocity vector) is preferable to R-bar rendezvous (approach and departure parallel to the orbit radius vector) because the zenith and nadir approaches are complicated by tethered facilities described earlier.

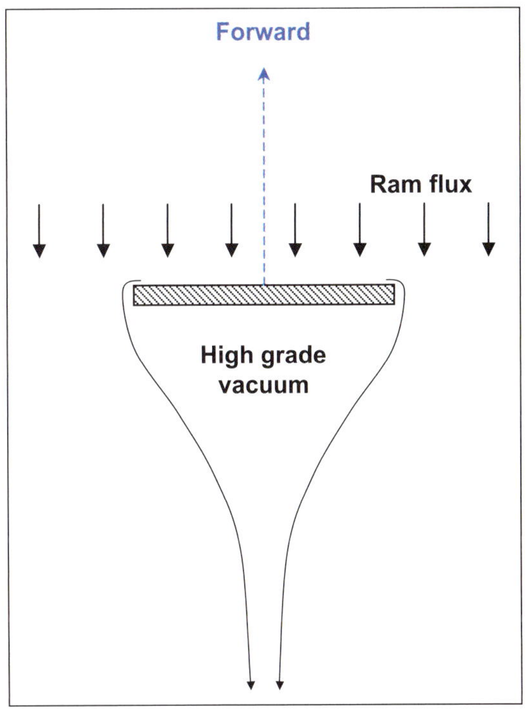

FIGURE 15 High-grade industrial vacuum is achieved behind wake shields.

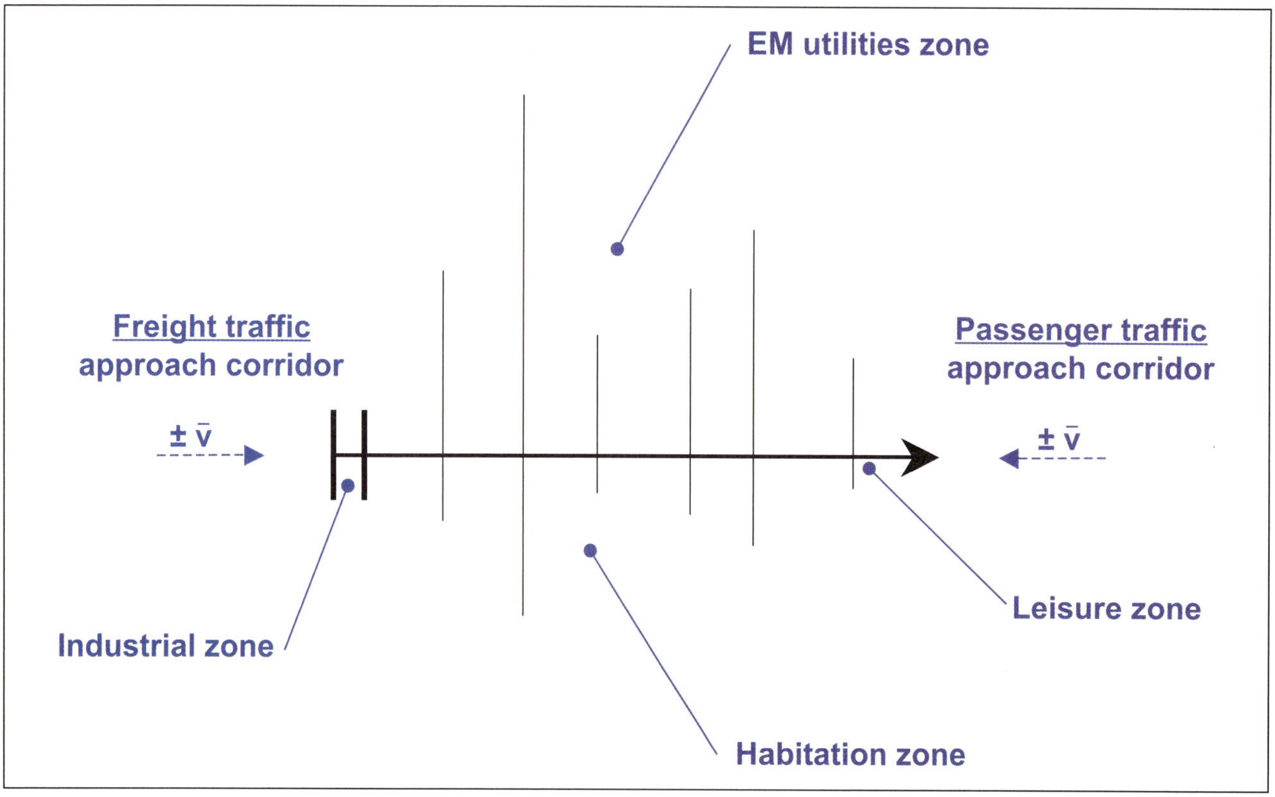

FIGURE 16 Bow and stern are zoned respectively for passenger and industrial traffic (EM = electromagnetic radiation).

What results is the use of the bow approach for passenger travel and the stern approach for industrial freight. This functional separation facilitates the desirable proximity of passenger traffic to the populated bow and of industrial traffic to the high-grade-vacuum industrial stern (Figure 16).

So far then, the midscale parti is a long spine hosting utilities and facility transportation, its sides bristling with short ribs hosting user functions, its ridge crested by a dual canopy of energy-exchange arrays, its belly open for viewing Earth, its best real estate in the prow, and its industry in the stern, with traffic approaching from the forward and aft directions and with utilization clusters tethered below and powerplants tethered above (Figure 17).

LARGE-SCALE GROWTH

For large-scale urbanization, proximity constraints extend the two-dimensional parti laterally into a "raft." The linear spine is replicated, yielding parallel spine "avenues" bridged by rib "streets" (Figure 18). Trains provide mass transit along the avenues; conveyors provide microgravity transit along the streets. The energy-exchange crest becomes a broad canopy above the raft. Complexes requiring Earth view are hung below, and powerplants are flown above, in tethered three-dimensional arrays serviced by elevators (Figure 19). In the extreme, the core of this assemblage becomes dense packed: a fully three-dimensional urban matrix with pressurized volume beneath the utility canopy, rich neighborhood texture, local exterior urban views, and vista views along the edges (Figure 20).

For very large constructions (of the same order as over-the-horizon surface distances on Earth, e.g., the urbanized Boston–Washington corridor), the along-track spines actually curve gently concave-downward, parallel to the Earth's surface (Figure 21). At local neighborhood and urban scales, this warping of the fore-aft cardinal axis is imperceptible. Also, the nadir tethers angle slightly toward each other, and the zenith tethers splay slightly, although this is not apparent either because at LEO altitudes they cannot be much longer than approximately 50 km anyway. The high-Earth-orbit (HEO) three-dimensional parti can be thicker, and geosynchronous-orbit (GEO) architecture can be much thicker in height than in lateral extent, depending on its function.

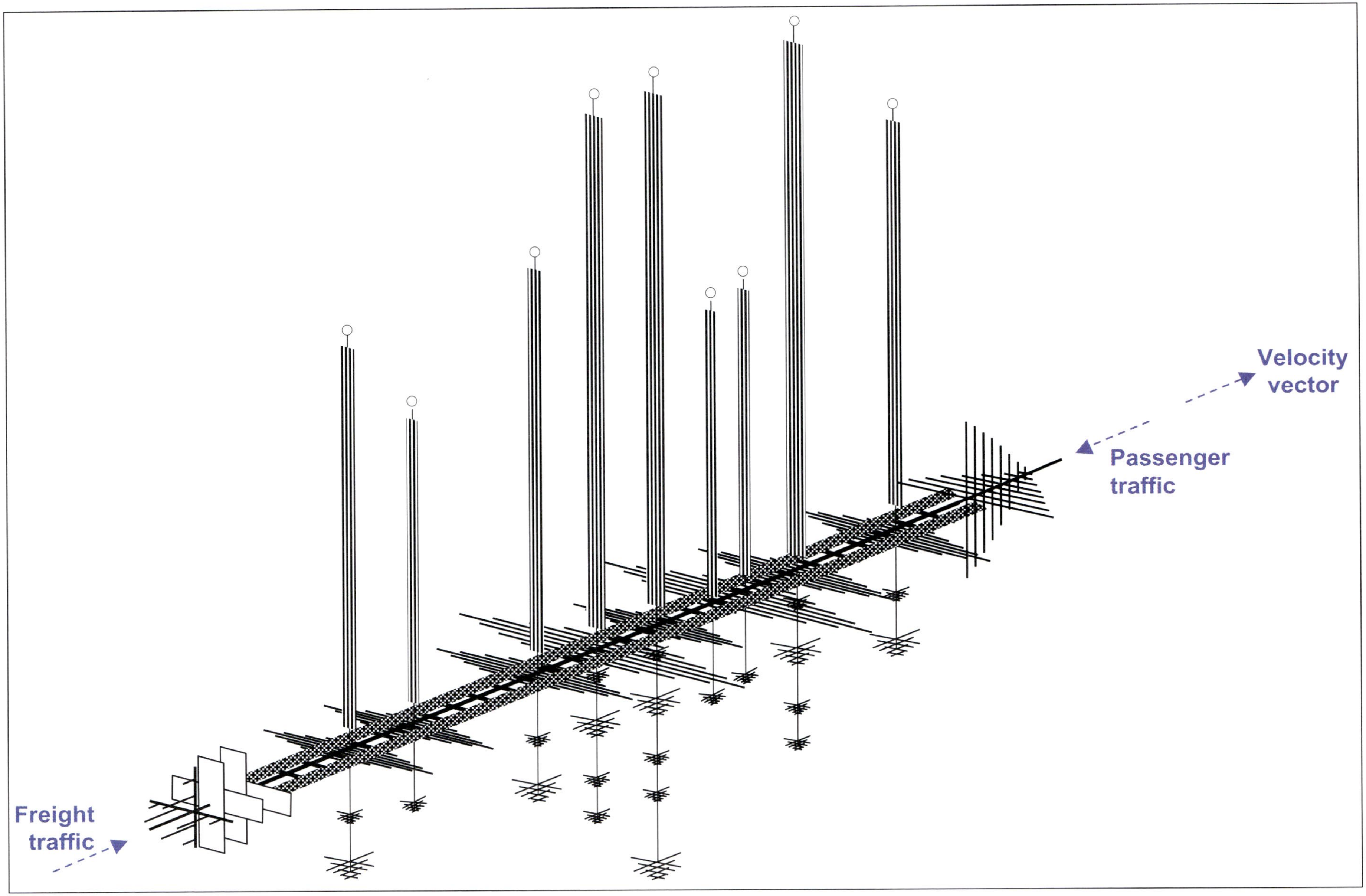

FIGURE 17 Two-dimensional LEO parti integrates the primary midscale constraints.

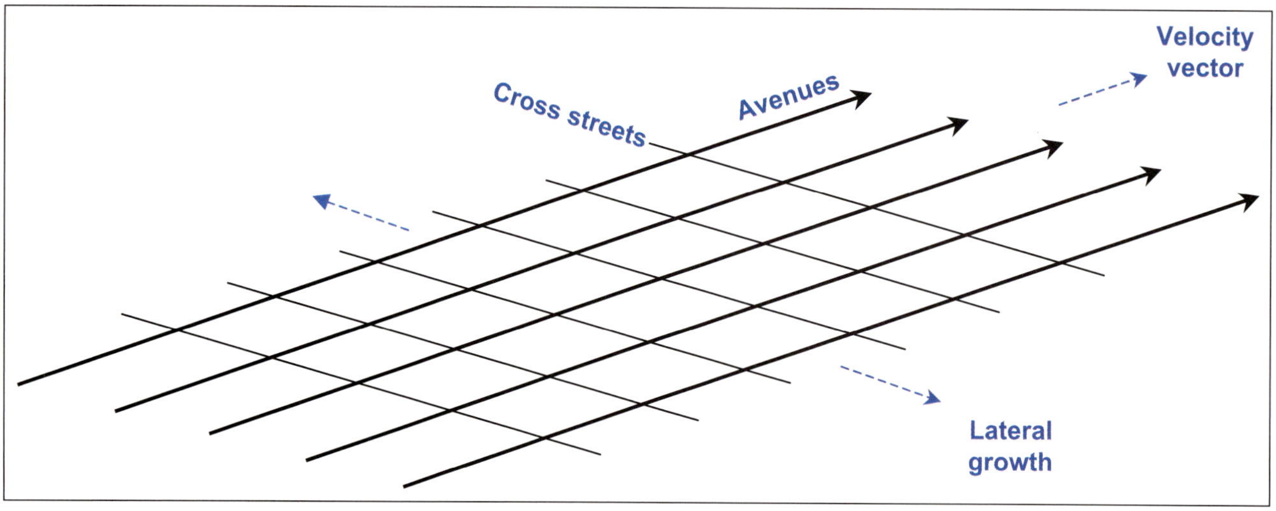

FIGURE 18 Multiple spines become avenues; connecting ribs become cross streets.

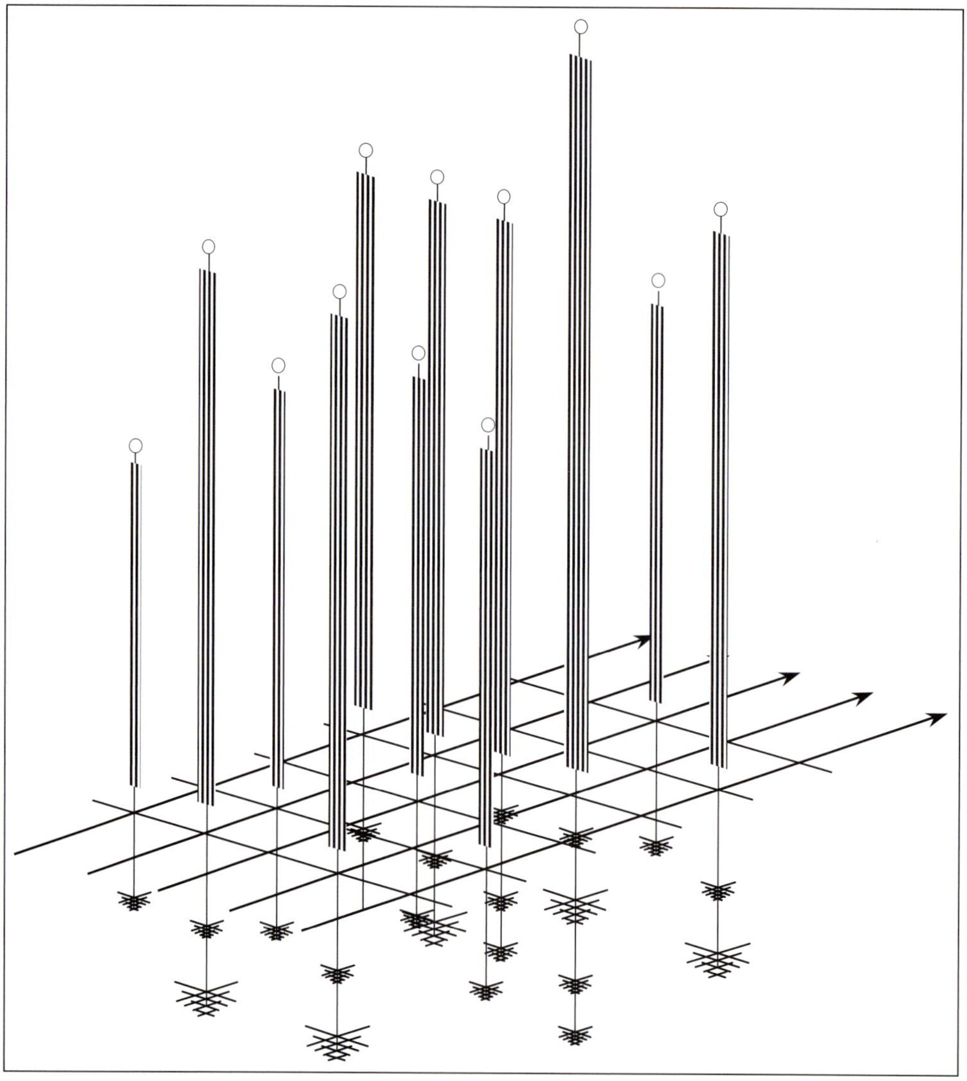

FIGURE 19 Tethered facilities grow above and below major intersections.

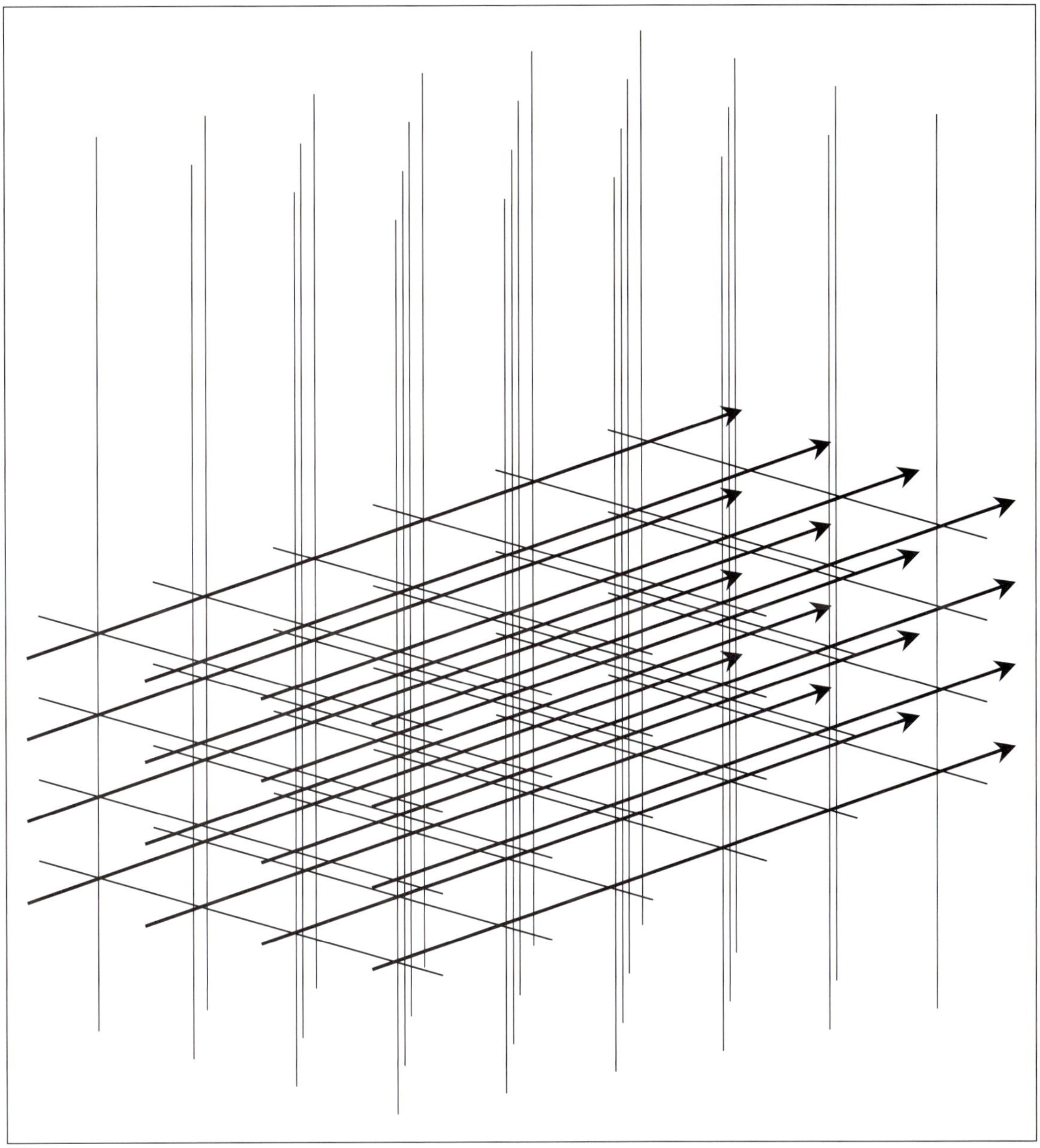

FIGURE 20 High density enables a fully three-dimensional microgravity and partial-gravity urbanism.

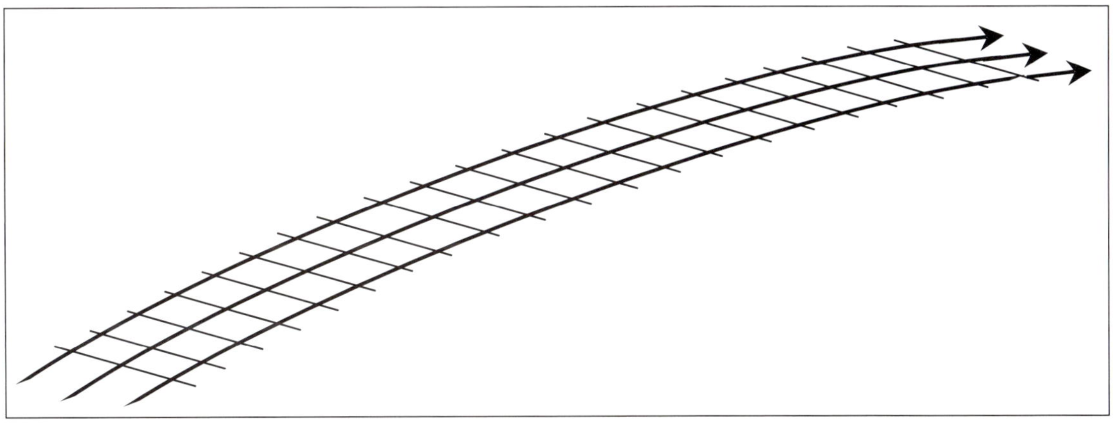

FIGURE 21 Megalopolis follows Earth's curvature as it grows in length.

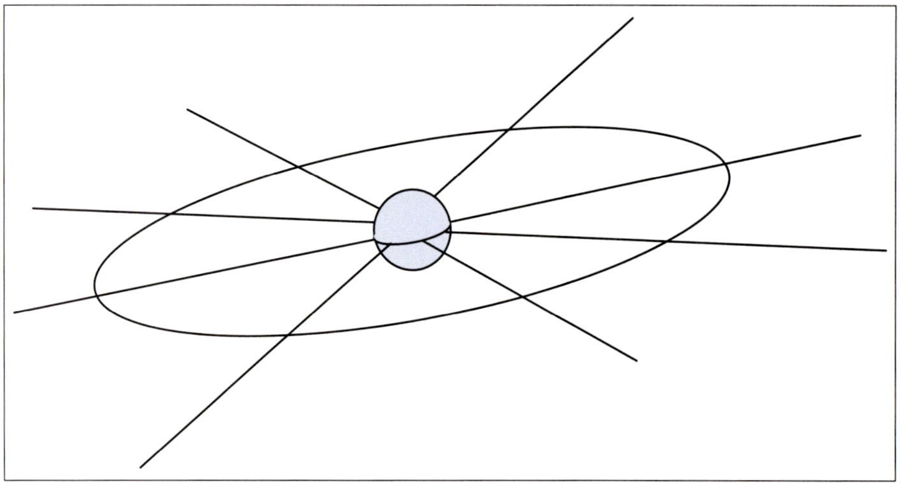

FIGURE 22 Ultimate planetary parti uses elevators from equator to ring at GEO.

Indeed, development of space-elevator tether technology would enable the ultimate planetary architectural parti: 72,000-km-long axes radiating from Earth's equator up through and beyond GEO, with an equatorial urban ring at GEO (Figure 22). The tethers provide spacelift, and high-speed trains connect points along the ring. The GEO ring is built out laterally for growth, built down for Earth-looking uses (communications, imagery, recreation), and provides a way station for orbit-transfer systems to other destinations. The extent of facility build-out at other altitudes is a function only of economic demand, tether material strength, and number of tethers used to hang them. Similarly, extensive cross-track build-out would enable nonequatorial spacelift tethers.

ARTIFICIAL GRAVITY

Apart from industrial needs, general mixed-use LEO architecture includes microgravity and partial-weight and artificial-weight zones. Internal ballistic motions in rotating environments are counterintuitive and therefore provide interesting opportunities for living, sport, and art (Chapter 12). However, 4π steradian motions in microgravity environments, and the settling bias in partial-weight environments, provide equally interesting yet very different opportunities for living, working, sport, and art.

The LEO urban assemblage rotates stern-over-bow (pitching down) at 1 revolution per 90 minutes because it is gravity-gradient oriented. Faster-rotating, artificial-weight complexes rotate the same way along the orbit track (i.e., spin axis parallel to the cross-track direction) because the gyroscopic moment from any other arrangement is incompatible with the gravity-gradient rotation behavior of the whole complex (Figure 23). Large-scale constructions overcome this constraint by counter-rotating multiple rotors with matched angular momentum. When the total angular momentum vector cancels, all gyroscopic torques vanish, and no complication ensues (Figure 24).

The simplest configuration for artificial weight is side-mounted rotors, located at the end of the lateral ribs (Figure 25). If artificial weight is the predominant

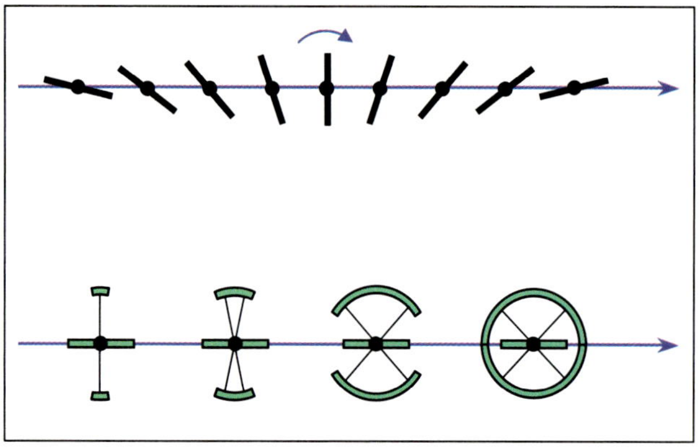

FIGURE 23 Basic artificial weight systems roll along the orbit track.

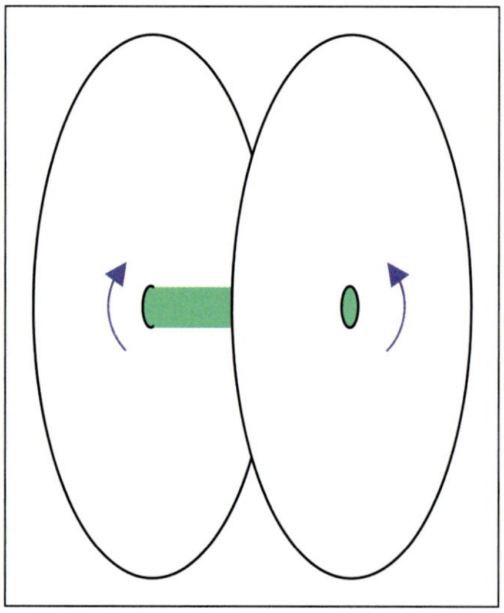

FIGURE 24 Counter-rotors neutralize gyroscopic moments for large complexes.

condition, the rotor diameter can exceed the length of the nonrotating spine; a single rotor would use a split spine, whereas a single spine would require a double rotor (Figure 26). Many variations are possible, depending on the relative need for microgravity vs artificial weight conditions. Electromagnetic utility systems (communications and energy exchange), at the least, require despun platforms in order to be independently pointed. Density is achieved by staggering the rotors so that they overlap (Figure 27).

Habitable spin interfaces are required. Even early implementations are operationally complex: radial elevator cabs using extensible berthing mechanisms to make pressurized connections at both ends of travel. As such a cab nears the despun hub, it releases the radial track and grapples onto the hub, prior to berthing (Figure 28). Mature, operationally simpler implementations are far more technologically complex: large-diameter, human-rated, pressure-containing despin joints that literally connect modules rotating with respect to each other. The best means to mitigate negative perceptual effects of Coriolis acceleration as passengers travel radially are unresolved.

Rotors have large radii and slow rim speeds on the order of 1 revolution per minute. Primary viewing is lateral (i.e., not up toward the hub or down through the floor). Remote vistas are less disorienting than nearby objects because parallax reduces their apparent motion (recall Dr. Haywood Floyd phoning home from the Orbiter Hilton in *2001: A Space Odyssey*—the distant Moon lazily rotates in his window, but the nearer Earth limb is nowhere in sight.). Therefore, circulation and housekeeping functions are concentrated in the inboard side of the rotor, which

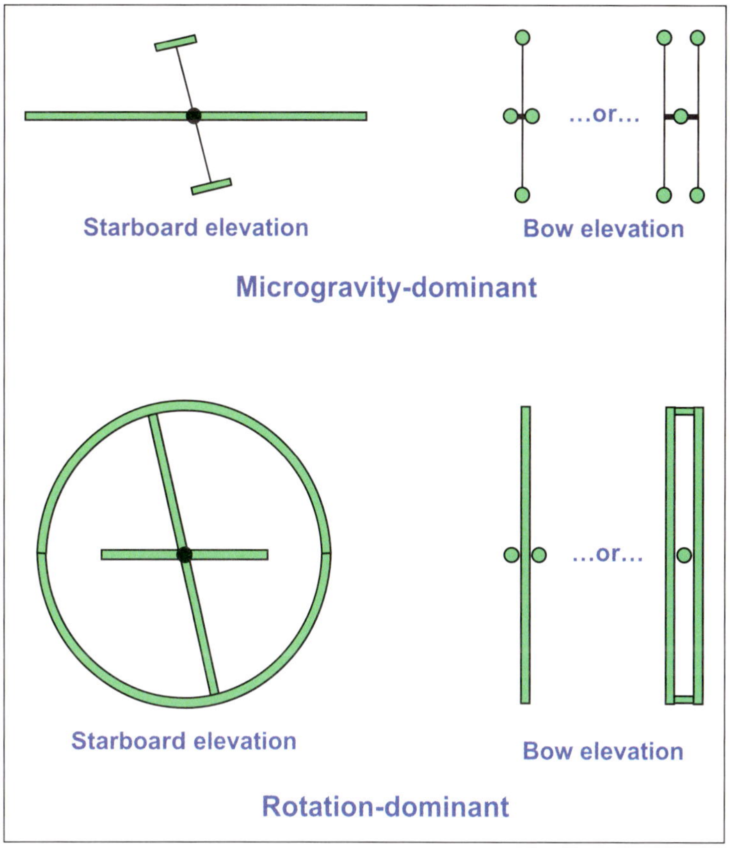

FIGURE 26 Relative emphasis of microgravity vs weighted functional uses determines the ratio of spun vs despun volumes.

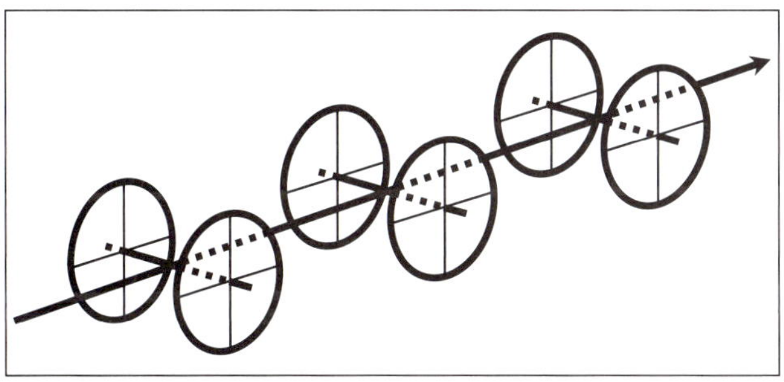

FIGURE 25 Side-mounted rotors add conditions for artificial weight to the basic LEO parti.

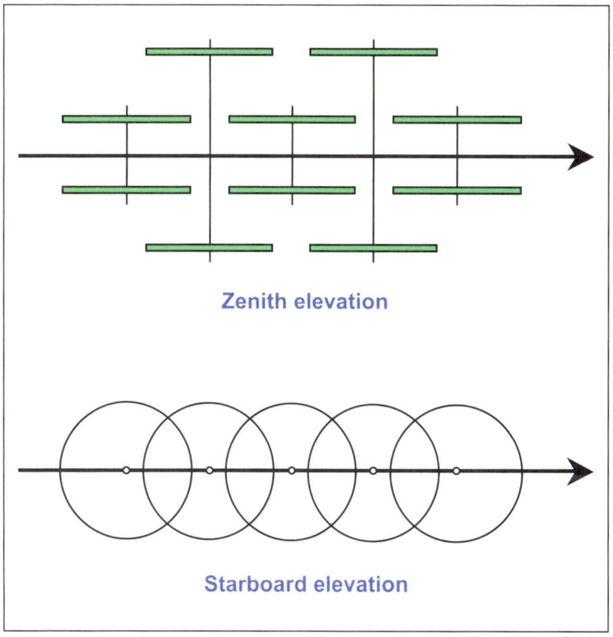

FIGURE 27 Staggered rotors enable a denser total complex.

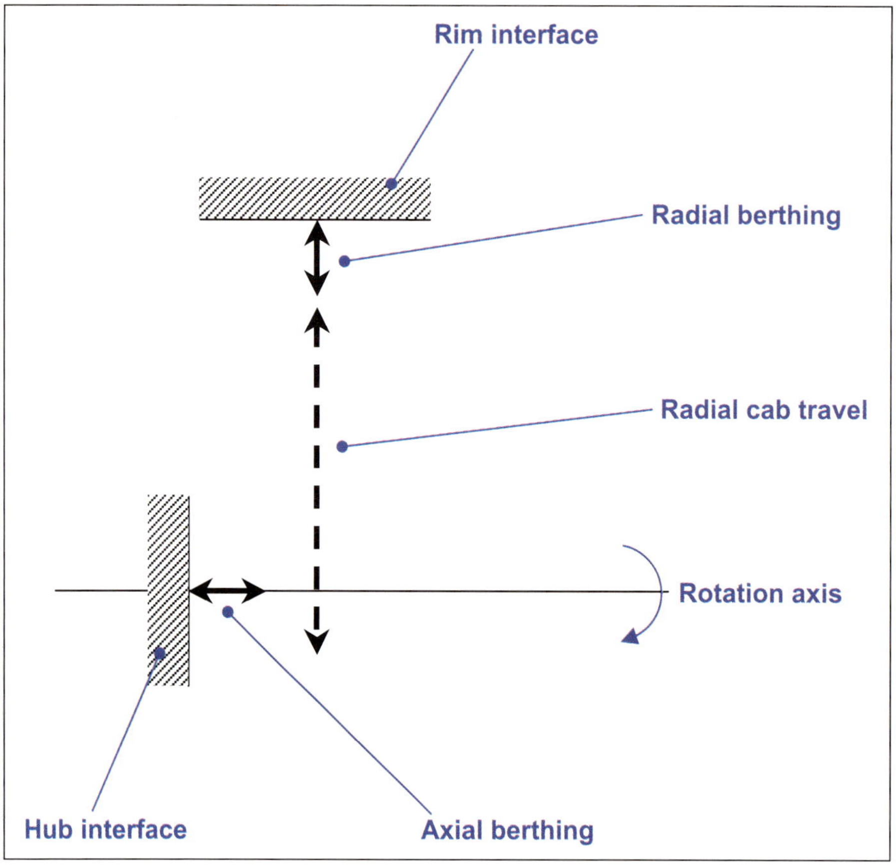

FIGURE 28 Pressurized interfaces for spinning artificial gravity facilities require complex mechanisms.

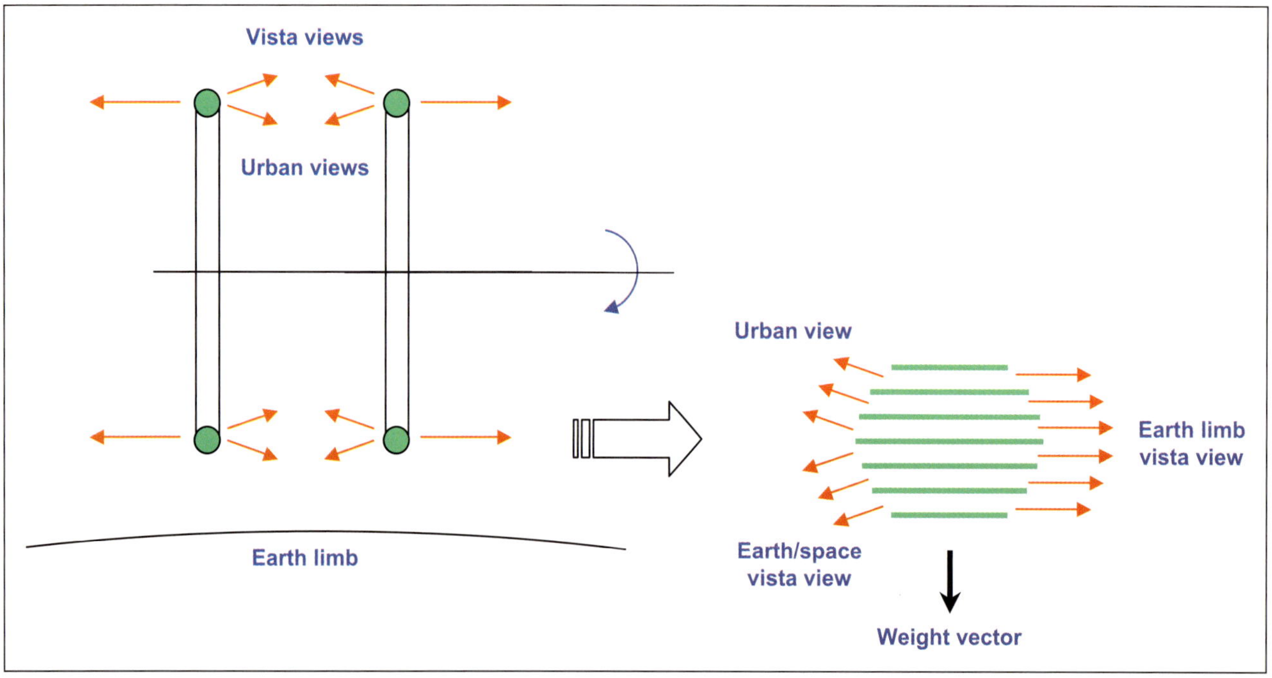

FIGURE 29 Lateral-configuration Earth-view lateral is least disorienting from a rotating system.

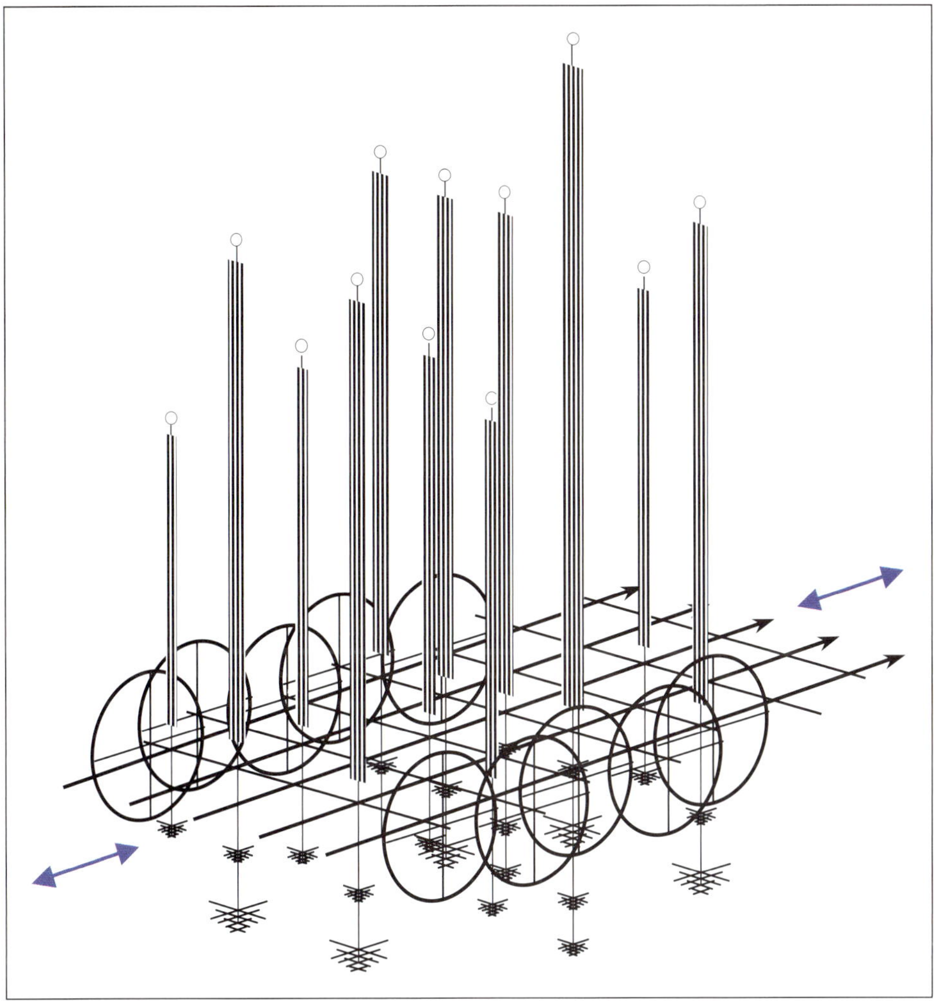

FIGURE 30 Complete three-dimensional urban parti integrates all basic large-scale constraints.

faces the rest of the complex nearby, while occupied spaces are concentrated on the outboard side, with rotating views of the Earth's limb to port or starboard (Figure 29). This preferential orientation is consistent with the one described earlier, required when only a single rotor is used. The addition of multiple, staggered rotors mounted along the port and starboard edges of the city, much as wheels rolling along the orbit track, completes the mixed-use urban-scale parti (Figure 30).

CONCLUSION

Predictable, unavoidable constraints enable us to develop an integrated parti for LEO urbanism. This parti is consistent with infrastructure ranging in scale from ISS all of the way to a future urban megalopolis and the evolutionary growth to get from one to the other. It accommodates realistic design capabilities ranging from current state of practice to reasonable projections of future technology. It embodies design patterns that respond to ancient human needs like activity zoning, vista viewing, and growth. It uses the inherent properties of LEO to introduce novel experiences and anticipate needs based on them. It provides a conceptual framework within which specific design ideas can be addressed, compared, and integrated. Foreknowledge of viable end-state configurations can help guide development of design concepts and might even facilitate strategic selection among options.

PART 3
Planet Surface Architecture

Planet surfaces are the complement to orbital locations. Images of planetary bases have been with us practically since the dawn of the space age. But what will planet surface architecture really be and how will it evolve?

Part 3 opens as did Part 2, by summarizing the basics. In "Design Constraints for Planet Surface Architecture" (*Chapter 14*), Brent Sherwood and Larry Toups catalog the governing environmental conditions on the moon and Mars." Marc Cohen and Haym Benaroya then survey the history of proposed approaches to building planet surface structures intended to contain atmospheric pressure and support human activity, in *Chapter 15*, "Lunar Base Structures." Three sections follow, which discuss 1) issues and concepts for early outposts; 2) concepts and methods for extending outposts into permanent bases; and 3) architecture of eventual planetary settlements.

OUTPOSTS

Planetary outposts will be the first human beachheads on other worlds, including the Moon and Mars. The outpost phase takes us from initial landings through short-term sortie missions to repeated visits to the same site with a gradual buildup of re-used assets. The earliest outposts might be "campsites" near lunar landers, and various schemes have been proposed for incorporating landers themselves into outpost architectures.

Larry Toups and Kriss Kennedy, in "Lunar Habitat Concepts" (*Chapter 16*), describe contemporary NASA concepts and analysis aimed at determining the very first outpost buildup scenario. These approaches highlight compactness, pre-integration, and the lunar polar environment as design drivers.

Chapter 17, "Lunar Surface Airlocks," by Brand Griffin, delves into the design problem of enabling routine access between the dusty lunar surface and a clean habitat interior. Balancing issues like suit design, airlock volume, routine and contingency operations, dust control, maintenance access, and system integrity in the harsh lunar environment makes this one of the toughest challenges facing planet surface architects.

NASA planning includes mobile options as strong candidates for post-*Apollo* exploration. In *Chapter 18*, "Habot Concept," Marc Cohen and Ross Tisdale describe a revolutionary lunar outpost approach in which mobile, reconfigurable architecture enables flexible exploration.

This section concludes with "Roving Laboratory" (*Chapter 19*), where Andreas Vogler et al. propose a design for a compact, mobile laboratory for initial planet surface exploration. Optimized for fast coverage of large distances, this large, live-in rover extends the geographical reach of an outpost crew.

EXPANDABLE BASES

How do outposts grow into larger permanent bases, and what architectural elements enable this? The ideas in this section offer space architects clues for new directions in planet surface architecture.

In *Chapter 20*, "Flat-Floor Inflatable Structures," Jim Lowe presents an initial investigation of pressurized structure geometries better optimized than cylinders and spheres for habitat applications in vacuum on planet surfaces. Scott Howe and Ian Gibson, in "Trigon Modular Robotic Construction System" (*Chapter 21*), analyze principles of modular, reconfigurable robotic construction systems. They propose a mass-producible product line of functional panel elements to enable flexible, automated expansion and reconfiguration of planetary base stationary and mobile elements. Brent Sherwood describes the quantitative method used to develop a site plan for an expandable, oxygen-producing lunar base in "Lunar-Base Site Design" (*Chapter 22*).

LARGE-SCALE SETTLEMENTS

Crossing over from the planetary base stage to genuine settlement takes space architecture to a new level. It requires the use of local materials and development of techniques for *in situ* fabrication, construction, verification, and outfitting. It also introduces societal issues and requirements far beyond those needed for small bases.

In "Mars Habitat Using Locally Produced Materials" (*Chapter 23*) Bas Lansdorp and Kristian von Bengtson propose a hybrid Mars architecture that combines pre-integrated elements brought from Earth with large, cast-glass pressure vessels made onsite. Such concepts provoke space architects to think about all of the practical details of making architecture in alien environments from what we find there. Brent Sherwood explores what it means for architects to take on the challenge of designing the infrastructure for off-world civilization in *Chapter 24*, "Lunar Architecture and Urbanism," and he outlines basic principles likely to shape what lunar architecture becomes as it approaches the scale of towns and cities.

14

design constraints for planet surface architecture

BRENT SHERWOOD
AND
LARRY TOUPS

INTRODUCTION

A first-generation lunar outpost is a major space project likely to require sponsorship by one or more governments. Such an outpost differs in two basic ways from large terrestrial construction endeavors. First, there is no large, preexisting experience base of directly relevant, successful design solutions. Second, despite that lack of knowledge an early base can only embody a very few approaches. Vast monetary resources, diverse expertise, and years of development and testing ultimately converge in just a few hundred metric tons of hardware sent to the moon. So a limited set of solutions, all drawn from an unproven set of alternatives, gets implemented in a remote environment under intense, critical public scrutiny. The uncertainties, programmatic risks, and visibility are unprecedented in terrestrial civil engineering and constitute a highly selective filter for candidate approaches to lunar base structures.

Space architects need a common understanding of the many relevant issues confronting planet surface system design. This chapter, companion to Chapter 3, summarizes the major issues governing what architecture must do and can be on the Moon and Mars. It begins with the Moon and divides these issues into three equally important categories: programmatic, environmental, and operational. It then examines salient differences for Mars.

PROGRAMMATIC ISSUES

The premier programmatic issue is the base's function. Early lunar bases will likely be small-scale installations driven by geological and astronomical science investigations and the development of viable methods of *in situ* resource utilization (ISRU). Economic reality dictates that any substantial growth of lunar facilities be enabled by large-scale ISRU industrialization. The Moon's material resources and location make it a potential catalyst for space industrialization, as many scenarios have proposed.

There are four key program parameters: 1) scope, what the program includes; 2) scale, how big it is; 3) timing, how fast it proceeds; and 4) technology, how much sophistication is needed. Political, economic, and technical challenges imply lunar development will most likely start small and proceed slowly, using technology already in hand or introduced conservatively a step at a time, for the purpose of enabling lunar-based science (Augustine 1990; Synthesis Group 1991) and learning how to operate in planetary environments (NASA 2005). These objectives are far less ambitious than many published scenarios, but far more likely. Regarding potential commercial investment, no business plans have emerged that are capable of driving lunar architecture growth without also depending on government funds. Lunar tourism, energy production for Earth, or material mining for large-scale space industrialization are all far-horizon visions.

Flexibility and *multiuse design* are two important programmatic drivers for realistic space engineering problems including lunar base architecture. Flexibility has three components: *adaptability*, *resiliency*, and *evolution*. Adaptability means accommodating off-nominal conditions. A structure system that requires exquisitely prepared substrate conditions is less flexible than one that can rest on uneven rocks or regolith (or permafrost in the case of Mars). Resiliency is the ability to accommodate failure. A habitat system that becomes unusable if one critical part fails to perform as predicted is poorer than one that absorbs the loss gracefully. Evolution means adapting gradually over time to changing requirements. Because large space projects are developed over decadal timescales, the state of knowledge and art changes along the way and continues to change over the service life of the systems. What planetary exploration will become decades from now is indiscernible, yet the systems we design must remain usable for decades.

Multiuse design has two components: *commonality* and *reuse*. Commonality means using systems

Copyright © 2008 by the authors. Published by the American Institute of Aeronautics and Astronautics, Inc., with permission.

from the same production line for more than one architectural application. An example is Class I habitable modules: a family of small, hard-wall pressure vessels can be widely applicable for transportation uses (e.g., lander crew cabs and short-duration lunar-class transfer habs) as well as surface uses (e.g., biological life-support, special-purpose science, and safe-haven outpost modules). Reuse means using the same hardware more than once. A modular thermal system (radiators, conduits, heat exchangers, secondary structure) designed to be broken down, refurbished, and integrated into other applications can outlive the habitat system it was originally part of. Although reuse would be operationally challenging on the moon, it would also minimize delay and expense in solving emergent design needs, as well as wasteful accumulation of defunct components. Learning how to recycle equipment (and therefore designing it for this) can be enabling for deep-space missions where logistics support is impossible.

The dual programmatic principles of flexibility and multiuse design are important constraints because they can lead to technological solutions capable of robust integration in many ways, in various settings, for a long time. They affect performance, safety, variety, and cost effectiveness. With these constraints, infrastructure elements designed for early, modest applications can provide the basis for more elaborate bases later in subsequent stages of an evolutionary architecture.

PHYSICAL ENVIRONMENTAL ISSUES

Details of the lunar physical environment that constrain architecture there are fundamental and unique. Specific data presented here are collected principally from three sources (Heiken et al. 1991; Hartmann 1983; Beatty et al. 1990).

Pressure Differential

The Moon's atmosphere is so tenuous that for purposes of engineering structures it can be completely neglected. The resulting hard vacuum means that lunar base subsystems tend to be more like spacecraft subsystems than like terrestrial ones. Habitation systems containing Earth-normal atmosphere will experience 70–100 kPa (10–15 psi) across their entire enclosing surfaces. And some materials processing plants and storage systems might require the containment of much higher pressures (of order 10 atmospheres) in large vessels. The structural challenge of accomplishing this containment reliably constrains both the size of Class I structures that foreseeable transportation systems can deliver to the moon and the complexity with which practical Class II structures can be designed.

One potential benefit of the hard lunar vacuum is that the desiccated environment can allow *in situ* manufactured glass to approach its theoretical strength because anhydrous melts are subject to fewer surface defects upon forming as they cool. Many other manufacturing advantages might accrue in a place where hard vacuum is free; although they might have importance for structural concepts, such processes are outside the scope of this chapter.

Diurnal Cycle/Temperature Extremes

The length of a lunar diurnal cycle (the synodic period) is 29.53 Earth days; at the lunar equator, this leads to almost 15 days of daylight and 15 days of darkness, although this value can vary by up to 13 hours because of orbital eccentricity. Without an atmosphere to scatter it, sunlight is unattenuated and unidirectional (except where reflected by local surfaces) on short timescales; contrast is extreme. The near side is illuminated at night by Earthshine; the full Earth is roughly 50 times brighter than is a full Moon viewed from Earth. Lunar surface temperatures change drastically from high noon to predawn, posing thermal expansion and thermal cycling challenges for lunar structures. Temperatures measured on the midlatitude lunar surface during *Apollo 17* ranged between 111°C and –171°C. Temperatures decrease rapidly as sunset approaches, falling about 5°C per hour.

Process plant and storage system temperatures exceed even this range by quite a wide margin. Basalt-melting techniques require temperatures on the order of 1500 K, and cryogen (propellant) depots and delivery systems must handle temperatures as low as 20 K. Specialized structural systems are required for these lunar-base components.

Radiation

The two major sources of deep-space radiation are 1) high-energy galactic cosmic rays (GCR), an isotropic flux of protons, alpha particles, and heavier nuclei; and 2) solar flares [solar proton events (SPE)], a wind of high-energy protons with fluxes up to 100/cm^2/s resulting from explosions in the sun's chromosphere. Because of the Moon's very small magnetic field and nearly absent atmosphere, space radiation bombards lunar-base structures directly (albeit only from the overhead hemisphere), with negative consequences for both biological and materials systems. Allowable dose limits for astronauts are periodically revised by the National Council on Radiation Protection. There are few data yet on materials degradation from deep-space radiation. Lunar-base structures associated with nuclear power systems will also require hardening against neutron bombardment and gamma rays, as well as disposal scenarios for materials that become activated.

Micrometeoroids

The lack of significant atmosphere on the Moon allows even the tiniest particles to strike the Moon's surface with their full encounter velocities, as high as 20 km/s. Almost all lunar rock samples contain numerous microcraters on exposed surfaces. Lunar-base structures, especially those not already shielded against radiation, will face possible material degradation and even structural failure. Sensitive surfaces, such as optical elements and tailored thermal coatings, have to be shielded when not in use.

Lunar Gravity

At 1.62 m/s^2, lunar surface gravitational acceleration is about one-sixth that at Earth's surface. It varies slightly over the surface, but not significantly for surface operations. The presence of lunar gravity means that, unlike in orbit, gravity stabilization is available to help during construction and operations; its reduced level relative to Earth's means that lighter structural members can be used for comparable spans and that unprecedented spans are possible, but that gravity anchoring is less effective.

Substrate Mechanics/Dynamics

Lunar regolith (soil) has formed over billions of years, as meteoroid bombardment has pulverized the native lunar rocks, and mixed and settled the resulting fragments. Regolith depth ranges generally from 2–30 m and mostly from 5–10 m in mare regions. Local chemical composition primarily reflects the composition of underlying bedrock. The bulk density of regolith is very low (0.8–1.0 mt/m^3) in the upper few millimeters, but increases to 1.4–2.2 mt/m^3 at depths down to and below 3 m. Seismic data suggest the density in each successive brecciated layer below the regolith increases with increasing depth toward values in excess of 3 mt/m^3 at depths over 20 km. Seismic experiments (Cooper et al. 1974) imply the upper few hundred meters of the lunar surface are rubble. A few localities might have intact bedrock, however, as evidenced by visible layers of mare basalt flows in crater walls and sinuous rilles at the *Apollo 15* landing site.

Except for the uppermost few millimeters, lunar regolith is more cohesive (0.1–1.0 kN/m^2) than most terrestrial soils and has a 30–50-deg angle of internal friction. Long-term vibrational settling of the regolith caused by meteoroid impacts has led to extremely high relative densities, approaching 100% just a couple of meters down. This means that 1) excavation of native material is difficult; 2) undisturbed subsurface layers of material can provide excellent foundation substrate; and 3) once disturbed, no amount of compaction can regain the native relative density.

The Moon's core is presumed to be rather cool because of its small size, but it is not yet known whether a liquid remnant core exists; measured moonquakes are extremely weak. Based on current knowledge, natural seismic disturbances can be neglected for structural engineering purposes.

Dust Composition/Behavior

Fully 50% of lunar regolith consists of particles finer than 70 micrometers (μm) (too fine to be resolved by the unaided eye). This extremely penetrating dust sticks electrostatically to objects that touch it because of the anhydrous vacuum environment, and much of the dust is abrasive (high-hardness particles like alumina). Macroscopically, the regolith clumps together like damp beach sand because of the interlocking action of its agglutinate particles. The *Apollo* missions verified that regolith attraction and clumping can be an operational nuisance. In particular, pressure seals and life-support equipment are subject to contamination. Performance of sensitive equipment such as photovoltaics, optics, and thermal radiators is prone to degradation by dust layers. Dust-removal technologies are under development but not yet tested in the lunar environment.

Lunar Tribology

Tribology is the science of friction, lubrication, and wear. Mechanisms (such as hinges, latches, axles, pivots, and end effectors) suffer tribology challenges on the moon because of the hard vacuum, temperature extremes, and dust conditions just outlined.

Chalcogenide salts like MoS_2 are the best-known lubricant for the vacuum and temperature conditions but compromise mechanism lifetime because they are solid lubricants—relubricating means replacing the part. New approaches to prophylaxis and mitigation are required (Sherwood 1990).

Available Material Combinations (Class III)

The bulk (>98%) of all lunar rocks and regolith is made up of seven major elements: oxygen, silicon, magnesium, iron, calcium, aluminum, and titanium. Iron is abundant (9–14%) in the mare regions where ilmenite-rich basalts are found, such as at the *Apollo 11* and *Apollo 17* landing sites. Titanium is also available as a major element in the mare regions, where it occurs at the 1–6.5% level. Aluminum is found in the lunar highlands where it occurs at the 9–18% level. From these major elements many structural material combinations are possible; in addition, transporting additives from Earth for *in situ* fabrication of structural materials opens a great range of options. The presence of ores would substantially facilitate industrial processes, but their existence on the moon is speculative. (Nonhydrologic concentration processes have been proposed but not proved.) The lighter elements, including those required for organic materials and biomass (hydrogen, nitrogen, and carbon), are notably deficient on the moon, presumably having escaped its weak gravity when the planet was still young and hot. Albeit scarce, hydrogen is somewhat enriched in regolith compared to the other volatiles because of implantation of solar wind ions over billions of years. Glasses are found naturally and can be made from available CaO and SiO_2. Structural components can also be made directly from unbeneficiated native basalt, by casting and sintering.

Ground Processing Environment

Lunar systems brought from Earth must of course withstand 1-*g* gravitational loads on the Earth's surface, as well as the dynamic conditions encountered during ground transportation, testing operations, and launch-vehicle integration. These can be disabling; the *Galileo* Jupiter orbiter antenna-deployment failure was traced to loss of lubricant on release pins, as a result of extra, unplanned cross-country ground transportation. Coastal launch sites also introduce problems associated with humid, salt-rich atmospheres (Griffin and French 1991).

ETO Launch Environment

Ascent to Earth orbit is often the most strenuous time in a space system's life. Shuttle payloads are designed to withstand acceleration loads of 3 *g* and acoustic loads of up to 138 dB sustained for several minutes because of the noise created by two large solid rocket boosters adjacent to the payload. Future cargo launchers can mitigate the noise level, but some proposed cargo-launching schemes incur higher accelerations, particularly for small payload packages. Operational limitations imposed on payloads by launch safety requirements can sometimes dominate.

LEO and Orbit-Transfer Environment

The low-Earth-orbital (LEO) environment is dynamically benign compared to launch and in terms of radiation and temperature compared to the moon. However, it is by far the most hazardous for both collision and material erosion. Earth's hypervelocity orbital-debris halo is severe and growing and already surpasses the natural micrometeoroid hazard. The probability of collision increases dramatically with 1) increasing projected target area; 2) increasing residence time in LEO; and 3) advancing mission year. The probability varies also with orbital inclination and altitude and solar activity. (The upper atmosphere swells when heated, increasing drag and decreasing orbital lifetime of particles.) Impact energies can be higher than can be replicated on Earth for testing: a 1-*g* particle moving at 20-km/s relative velocity has the same kinetic energy as a 2-mt vehicle moving at over 50 km/hr. For sensitive payloads remaining in LEO long enough to incur a worrisome collision probability, debris shielding is the only solution. The approach vector distribution is heavily weighted at 40 deg to the right and left of the orbital-velocity vector, parallel with the Earth's surface below. Most collisions are noncoplanar because of the statistical distribution of nodal regression rates for particles with slightly different orbital elements.

Earth's tenuous outer atmosphere contains atomic oxygen, a potent reactive species. In LEO, the ram flux of atomic oxygen is high enough to completely erode thin vulnerable materials in a matter of months. For lunar-base structures specifically, this means that components made of polymeric materials or epoxy-matrix composites must be protected if warehoused or remanifested during any significant stopover in LEO. Payload integration modules or enclosed carriers should be sufficient, as should residence times less than 60 days. Most of the transfer environment from LEO to the moon is comparatively benign, with two exceptions: 1) a few hours' residence time crossing Earth's trapped proton and electron zones (the van Allen radiation belts); and 2) a deep-space radiation and micrometeoroid flux double what the payload will see on the lunar surface, due the full 2π steradian exposure geometry.

OPERATIONAL ISSUES

Taken together, lunar architecture's functions, adaptability, and manner of construction and maintenance are unprecedented. These operational issues comprise the third set of constraints.

Thermal Rejection

Vacuum is a superb insulator because convection and conduction are precluded. The presence of hard vacuum between grains of lunar regolith makes even that substrate a very poor thermal conductor. Thus, lunar systems of all kinds cannot reject heat in general except via radiation to space. For low-grade heat (like that generated by electronics, people, and lights), radiative rejection is challenging during the lunar day because of the high surrounding ground temperature and the sun overhead. For high-grade heat (like that generated by process plants and nuclear-power systems), high-temperature materials must be used for supporting structures. Most applications will require fluid-loop heat-rejection systems.

Radiation Safe-Haven Provision

Habitable lunar-base structures intended for long occupancy require heavy radiation shielding. Some scenarios use water jackets for this purpose, requiring the water to be brought from Earth. Using *in situ* materials inevitably means operating with lunar regolith. Regolith sheltering schemes and operations scenarios cannot reasonably be appended to a structural concept but must instead be designed in from the start (Sherwood 1990).

Human Mobility Envelopes

Human locomotion in lunar gravity is quite different from familiar Earth locomotion. Walking and running gaits, motion postures, and traction are all affected (Capps 1990). Indoors (without the complicating constrictions of a pressurized suit), a body inclination of 20–45 deg and walking/running speed of 1.25–4 m/s imply 2.5-m headroom clearance might be necessary for normal habitation areas. As speed increases, body inclination increases to compensate for reduced ground friction. Other aspects of architectural design affected include stair riser/tread ratios and dimensions, workstation design, and corridor heights.

Habitat Internal Outgassing and Health Environment

The materials used for hermetic space structures have serious implications for crew health. Great efforts are made to conserve atmosphere and reduce leakage; thus, "sick-building syndrome" created by this containment presents a real concern for lunar architects. Metal pressure vessels are largely inert biologically, as would be *in situ* vitreous and stone materials, but uncoated polymeric materials and membranes, secondary structure, and outfitting equipment might outgas over time. A long-term health risk issue not yet studied is radon decay daughters evolved from the uranium fraction of *in situ* materials.

Accommodation of Robotic Assembly, Operations, and Maintenance

The notion that lunar bases would be built manually by suited construction crews persists even today. But trends in machine and manufacturing technology, space experience to date, and analysis all indicate that lunar architecture will predominantly be built, operated, and maintained by machines. These machines will function with many kinds of and degrees of autonomy, ranging from operator driven to artificially intelligent. Lunar architecture operational requirements extend far beyond conventional human safety to include machine-environment accommodation. Analysis shows these considerations must be embedded in the design from the conceptual stage (Sherwood 1990).

Joining Techniques

The design of tolerant, robust, and repeatable *in situ* joining techniques is essential to lunar system expansion. In the case of highly modular Class II habitable structures, the joining method might be the essence of a successful assembly design. But even for Class I structures of all kinds, developing connection mechanisms and protocols that work in the dusty, weighty, hot and cold, hard-vacuum environment of the Moon is an enabling challenge. This is one of the highest-leverage (make or break) areas of technology development for lunar-base structures.

Retrofitting

Because space-system hardware is so expensive to design, manufacture, test, and transport to the Moon, a high premium will be placed on reusing available equipment to meet freshly derived requirements long into the future. Unlike most terrestrial architecture, lunar-base systems will be most useful, long lived, and cost effective if they are designed to be taken apart, reconfigured, re-outfitted, and reverified at least as simply as they went together the first time. Except for the earliest applications, or specialized systems produced and used in large numbers, single-use concepts will price themselves out of practicality.

Expansion

Over time, larger crew sizes and evolving mission objectives might demand substantial growth. Lunar architecture should be scarred for open-ended construction wherever sensible. This is especially difficult for heavily shielded habitation systems: how can they be expanded? One of the most interesting design challenges for lunar habitats is to develop shielding schemes that minimize initial construction and regolith moving as well as disruptive deconstruction and refurbishment to allow expansion.

Disposal

In time, all manufactured objects must be disposed of. On the Moon, given the tremendous cost of delivering or fashioning high-grade materials, defunct systems represent an important resource. We should expect that disposal will initially mean recycling substances easy to recover and stockpiling the rest for retrieval later, as more sophisticated recycling techniques arise.

Eventually the industrial capacity to recycle exotic materials from this stockpile and fabricate new components can occur on the Moon. An exception would be materials activated by nuclear power systems, which truly require permanent disposal. Options include permanent landfill or launching into space, but responsibility lies with the designers of such systems (including irradiated support structures, casings, and plumbing) to design appropriate disposal schemes as well.

MARS CONSTRAINTS

Through robotic exploration, the surface of Mars is becoming almost a familiar place. Yet photographs do not convey the full reality of this alien environment, which shares some characteristics with the moon but is different and unique in other ways. Especially given long-range plans for lunar activities to prepare for human Mars exploration, space architects need to understand how Mars is different.

Weather/Windblown Dust

Mars does have an atmosphere (mostly CO_2), albeit a tenuous one; surface pressure averages only about 0.7% of Earth's sea-level pressure and varies greatly regionally with surface elevation and with the change in seasons. Wind speeds get quite high (of order 100 m/s), but direct wind loading by the thin atmosphere is not a dominant problem. However, surface fines are lofted and therefore can achieve relatively high kinetic energy (hence erosive and penetrating capability). Local dust storms, and occasional global, long-lived dust storms are unavoidable. So unlike on the airless Moon, contamination avoidance cannot be achieved simply by locating components up off the ground. Local microtopography is transient, and sand-dune mobility has been observed on the surface. Thus, simple foundation pads (as are possible on the Moon) might be impractical. Experience with solar-powered rovers has demonstrated that occasional, random dust devils often are sufficient to clear accumulated dust from photovoltaic and radiator surfaces.

Temperature

Mars' surface temperature varies mostly between about −120°C and −25°C daily (Mars' diurnal cycle, the sol, is 1.03 Earth days long) and rarely exceeds the freezing point of water. Mars receives less than half the specific solar flux that the Moon (or Earth) does because of its greater distance from the Sun; its elliptical orbit causes a 39% annual insolation variation. Together, the shorter diurnal cycle and moderating atmosphere make Mars a more hospitable place for materials systems than the Moon.

Gravity Level

Mars' gravitational acceleration is roughly three-eighths Earth normal. Dead loading is thus 2.25 times the lunar case for the same equipment. Another consequence of Mars' greater gravity is its ability to retain an atmosphere. Mars landers inevitably use aerobraking for entry, which limits allowable dimensions of delivered payloads compared to eventual lunar delivery systems.

Substrate Properties

Soil mechanical properties have been observed from concrete-hard and sticky-clumpy with admixed ice (*Phoenix* northern-latitude site), to exposed-bedrock, rocky sand, and soft deposits that can trap wheeled vehicles (*Spirit* and *Opportunity* rovers). *In situ* chemistry experiments indicate the regolith has pH 8–9 and contains water-soluble elements and inorganic compounds (Cowen 2008) including perchlorate. Speculation that chemical reactions in Martian soil would be detrimental to some materials, including polymers, did not prevent the *Spirit* and *Opportunity* rovers from far outliving their design lives. Surface frost is common at dawn. Permafrost can occur throughout large regions; if so, foundations would likely require special thermal-management attention. The Martian poles are particularly active regions seasonally, with massive alternating freezing and sublimation of CO_2 ice; it is difficult to posit permanent structures being located there (see Chapter 27).

Forward Contamination

Mars might have once harbored native life; it is conceivable that in protected microclimates it still does. A conclusion of barrenness can only follow detailed, global exploration, most likely involving human field work. Thus paradoxically, we will probably remain uncertain about the possibility of contaminating a Martian ecology with Earth organisms until the human operations that might contaminate it are occurring. Precluding such contamination is one of the most significant design challenges for Mars missions (Sherwood 2004).

Material Combinations

Raw materials available for fabricating Class III structures on Mars largely comprise silicate rocks and regolith, as on the Moon (see Chapter 23). Mars obviously has iron in the form of FeO_2, and the availability of water makes conventional concrete a more promising option than on the Moon. No information exists yet about ore bodies on Mars; however, the planet has an extremely diverse geology, with abundant evidence of past hydrologic and volcanic processes that might have concentrated useful compounds.

The presence of lighter elements (notably deficient on the Moon) means that organic polymers and related materials could ultimately be made there. Based on what we know so far, Mars' supply of elements is well balanced for human uses. It should become possible eventually to make anything there that can be made on Earth, including not only Mars base structures but also outfitting equipment, consumables, and biomass. In environment and resources, Mars is the most hospitable planet after Earth.

CONCLUSION

Lunar architecture is likely to consist of government-sponsored exploration projects for many years, once such missions begin. The most immediate infrastructure elements will support simple habitation and modest scientific activities. Although it is premature to contemplate writing a lunar building code, extant information can begin constraining productively our concepts for such lunar architecture as roadways, foundations, shelters, supporting systems, and perhaps second-generation habitation envelopes.

What the lunar architecture community lacks is an experience base of prototypes, pilot experiments, attempts, failures, revisions, and successes in the lunar environment itself, upon which to base recommendation of proper techniques or standards. There remains plenty of room for innovation in this nascent field. Laboratory development of relevant early construction techniques, followed by field testing calibrated analytically to match lunar conditions, can help advance the field. Of greatest value might be options capable of being demonstration payloads on small, robotic lunar landers. Modest, solid steps appear to be the quickest and surest way to grander lunar architecture.

Acknowledgment

This chapter is derived from "Technical Issues for Lunar Base Construction," *Journal of Aerospace Engineering,* Vol. 5, No. 2, 1992.

References

Augustine, N. (1990), "Report of the Advisory Committee on the Future of the U.S. Space Program," U.S. Government Printing Office, Washington, D.C.

Beatty, J. K., O'Leary, B., and Chaikin, A. (1990), *The New Solar System*, Cambridge Univ. Press, Cambridge, England, U.K.

Capps, S. D. (1990), "Partial Gravity: Human Impacts on Facility Design," *Engineering, Construction and Operations in Space II*, edited by S. W. Johnson and J. P. Wetzel, American Society of Civil Engineers, New York.

Cooper, M. R., Kovach, R. L., and Watkins, J. S. (1974), "Lunar near-Surface Structure," *Rev. Geophys. & Space Phys.,* Vol. 12, Aug., pp. 291–308.

Cowen, R. (2008), "Lander Hints at Water, Nutrients on Red Planet," *Science News*, Vol. 174, No. 2, July 29, p. 11.

Griffin, M. D., and French, J. R. (1991), *Space Vehicle Design*, AIAA, Washington, D.C.

Hartmann, W. K. (1983), *Moons and Planets*, 2nd ed., Wadsworth, Belmont, CA.

Heiken, G., Vaniman, D., and French, B. M. (eds.) (1991), Lunar Source Book, Cambridge Univ. Press, Cambridge, England, U.K.

NASA (2005), "Vision for Space Exploration," NASA-TM-2005–214062.

Sherwood, B. (2004), "Progressive Protocol for Planetary Protection During Joint Human and Robotic Exploration of Mars," International Astronautical Congress, Paper IAC-04-IAA.3.7.2.10.

Sherwood, B. (1990), "Robotic Lunar Surface Operations: Engineering Analysis for the Design, Emplacement, Checkout and Performance of Robotic Lunar Surface Systems," NASA NAS2-12108, Boeing Co. D615-11901.

Synthesis Group (1991), *America at the Threshold: Report of the Synthesis Group on America's Space Exploration Initiative*, U.S. Government Printing Office, Washington, D.C.

INTRODUCTION

THIS CHAPTER SURVEYS representative concepts and structure systems proposed for lunar architecture over the first half-century of the space age. It applies the three-tier structure classification system described in Chapter 2 and explores how various concepts accommodate the environmental and operations constraints described in Chapter 14. It concludes with a checklist of lunar-base structures issues and an extensive bibliography.

15

lunar-base structures

MARC M. COHEN
AND
HAYM BENAROYA

HISTORICAL CONTEXT

The space age is now well past the date assumed for the vision of lunar science and commerce depicted in *2001: A Space Odyssey*. Progress toward sustained human lunar presence has been interrupted repeatedly; in the mid-1970s and again in the mid-1990s, the U.S. political climate shied away from such an expensive, challenging goal. The human spaceflight constituency is fragmented as well; a target-centered debate continues between factions favoring use of the Moon to develop planetary operations experience or a direct assault on the herculean challenge of getting humans to Mars.

Most design professionals concur that without extensive, routine human spaceflight and planetary infrastructure, leaping to Mars would pose unacceptably high risks and do no more to expand civilization into the solar system than did *Apollo* in its era. We do not yet have the technology and experience, let alone the resources, to send people to Mars on multiyear missions. The physiology and reliability issues alone are as yet unresolved for such a trip. Summing up the philosophy of gradualism, Krafft Ehricke (1985) famously said, "If God wanted man to go to Mars, He would have given him a Moon."* The Moon is our best first goal.

*In his address, Ehricke stated his aphorism as quoted. The written version is, "If God wanted man to become a space-faring species, He would have given him a Moon."

Lowman (1985) offers the following candidate purposes for a post-*Apollo* lunar base, which remain valid a quarter-century later:

- Science and astronomy
- Test bed for technologies required to send humans to Mars and beyond
- Establishment of a U.S. presence (on the "high ground" of another planet close to Earth)
- Stimulus to space technology
- Stimulus to science and engineering interest and education among young Americans
- Utilization of lunar resources
- Beginning a long-range program to ensure survival of the species

Each purpose introduces unique considerations. For example, the practicality of lunar-based optical astronomy is still highly debated, but it is feasible and might benefit from servicing by routine human presence. Several bold proposals for astronomy from the Moon have been made (Burns et al. 1990), for example, a 16-m-diameter optical reflector telescope. Nearly all such proposals involve use of advanced materials and structure concepts to erect large, long-life lunar astronomy facilities. Such facilities would challenge structural designers, constructors, and logistics planners (Johnson 1989; Johnson and Wetzel 1990b).

Selection of an appropriate site for a lunar astronomical facility involves many difficult decisions. Scientific advantages of a polar location for a lunar base

Copyright © 2008 by the authors. Published by the American Institute of Aeronautics and Astronautics, Inc., with permission.

FIGURE 1 Two versions of LESA modules placed on the Moon by Boeing (1963) (courtesy of Mendell 1985, p. 37).

(Burke 1985) are that half the sky would be continuously visible and that cryogenic instruments might be operated more readily in regions of perpetual darkness. Disadvantages arise also from the fact that the Sun essentially traces the horizon, leaving the outside workspace in extreme contrast and posing practical problems regarding solar power and communication with Earth. Van Susante (2002) studies the possibility of using the lunar south-polar region for an infrared telescope. Other wavelength regimes, for example, long-wave radio, might be the best candidates for lunar-based astronomy.

Some studies (Collins 2002) proffer space tourism as a viable, driving force behind sustained growth of human economic activity on the Moon that would in turn require increasingly elaborate lunar-base structures. More recently, numerous studies conclude space tourism will eventually be accessible to more than just the very wealthy (Collins 2006; Hempsell 2006; Parkinson 2006). Visions of elaborate lunar settlement are inspirational, deep rooted, and persistent. But the type of lunar civilization we can evolve will depend on the infrastructure we are capable of building, both technically and economically. Basic necessities such as shelter, water, waste disposal, communication, power, and transportation are the foundation of any viable settlement. Adequate infrastructure and resources are key to survival and growth of a society. At the core of infrastructure is physical structure.

Significant studies have occurred since *Apollo*, when it first appeared feasible that the Moon could become a second home for humankind (Figure 1). For a notably early example of research and development efforts, see the U.S. Army Corps of Engineers study (Office of the Chief of Engineers 1963). In the half decade from the late-1980s to mid-1990s, the U.S. Space Exploration Initiative intensified interest in lunar-base structures both within NASA and outside the government in industry and academe. Several references sample this thinking (Mendell 1985; Johnson and Wetzel 1988; Benaroya and Ettouney 1989; Benaroya and Ettouney 1990; Johnson and Wetzel 1990a; Johnson and Wetzel 1990c; Ettouney and Benaroya 1992; Sadeh et al. 1992; Duke and Benaroya 1993; Benaroya 1993a; Galloway and Lokaj 1994; Benaroya 1995; Johnson 1996; Galloway and Lokaj 1998; Benaroya et al. 2002a; Benaroya et al. 2002b). Sponsored progress resumed with the U.S. Vision for Space Exploration (Benaroya and Bernold 2008).

An extensive bibliography appears at the end of the chapter. The discussion draws in particular on two general references: "Habitats and Surface Construction Technology Development Roadmap" (Cohen and Kennedy 1997) and "An Overview of Lunar Base Structures: Past and Future" (Benaroya 2002).

CLASSIFICATION OF LUNAR STRUCTURES

In 1993, one of the authors edited a special issue of *Applied Mechanics Review* dedicated to lunar-base construction (Benaroya 1993a). In that issue, Smith proposes a three-phase evolutionary development for lunar-base construction: 1) prefabricated and pre-outfitted modules, 2) assembly of components fabricated on Earth with "some assembly required," and 3) building structures comprised substantially of indigenous materials (Smith 1993).

A version of this classification was codified in 1996–1997, when NASA created technology development roadmaps for technical and scientific areas critical to exploring the Moon, Mars, and beyond. The habitats and surface construction roadmap defined three major types of surface construction (Figure 2):

Class I (pre-integrated): The *Apollo* program intended to build to a 14-day base using enhanced lunar modules. In the 1960s, the U.S. Air Force was the first to propose pre-integrated cylindrical modules that landed on the lunar

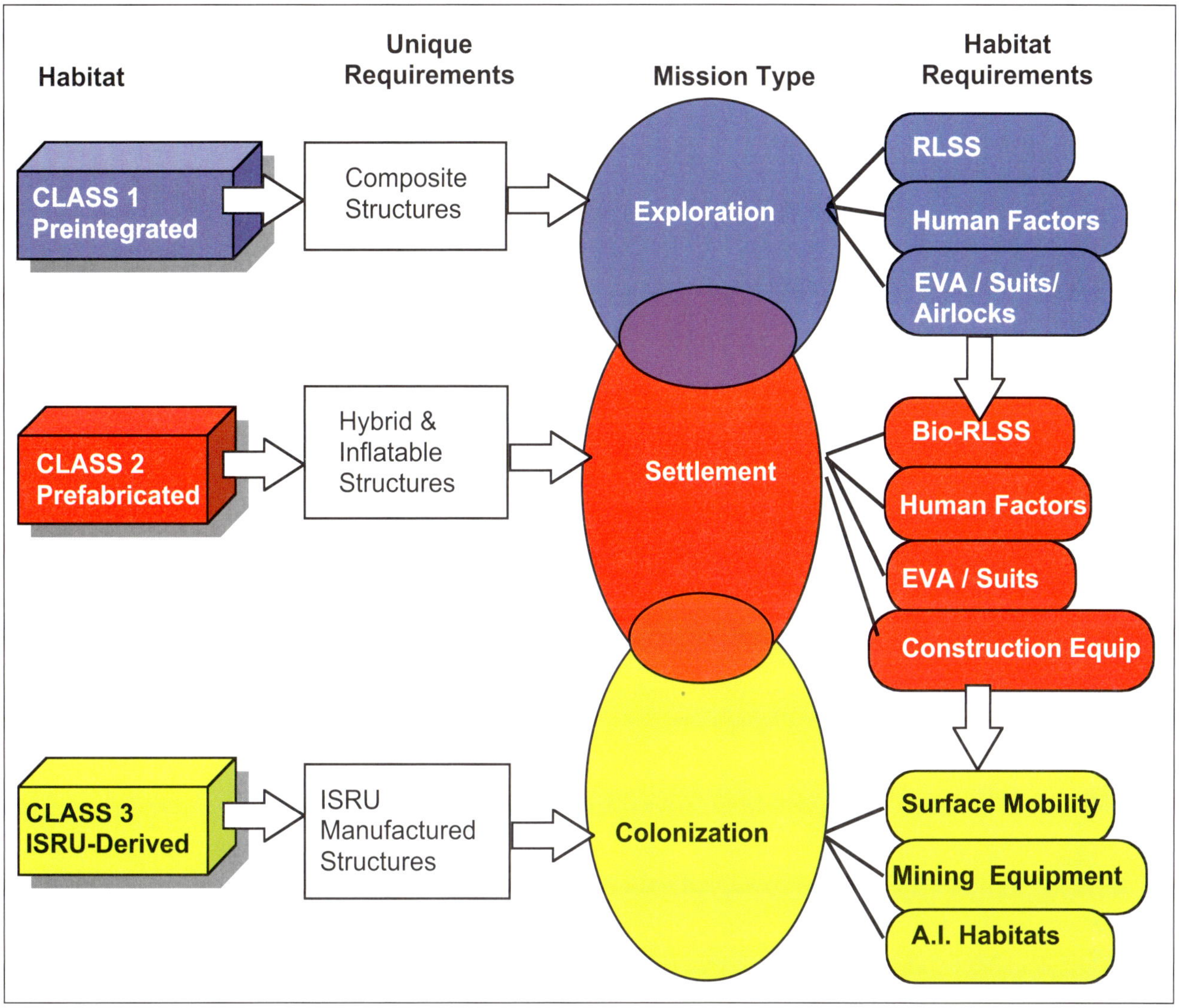

FIGURE 2 Roadmap indicates progressive stages of development through three classes of lunar/planet surface construction (RLS = regenerative life support system; EVA = extravehicular activity; A.I. = artificial intelligence).

surface (Richelson 2000), leading to many base concepts that would connect multiple modules. Thangavelu (1990) proposes pre-assembling a module-and-hub base in orbit and landing it on the Moon intact (Thangavelu 1990).

Class II (pre-fabricated but assembled, deployed, erected, or inflated): In the late 1980s, the NASA "90 Day Study" proposed a 20-m-diameter inflatable sphere for a lunar habitat (Alred et al. 1989). Abarbanel and Criswell (1997) propose rectangular inflatable habitats with lunar regolith ballast on the flat top.

Class III (use of *in situ* materials and site characteristics): Lunar masonry, lunar concrete, retrofitted caves, vitrified *in situ* structures, and other uses of natural lunar landforms have been proposed. Eichold proposes a base constructed inside a crater, using the rim as part of the structure (Eichold 1996).

The classification roadmap indicates how lunar-base development could eventually lead to large-scale settlements and colonies. Realistically, development of a lunar or planetary base will likely combine all three types of construction. Harper and Connell (1990) propose a "technosociological" development scenario consistent with such a hybrid roadmap. Their first step is construction of suitable artificial environments that provide life support and amenities to promote health, well-being, productivity, and crew acceptance of living in "a radically different environment for very long periods of time." The second step is to upgrade that artificial environment enough for crew members to be willing to "spend their lives and eventually bear and raise children" there. Third and finally, such a settlement achieves self-sufficiency and growth through the use of *in situ* resources to generate consumables.

DESIGNING FOR THE LUNAR ENVIRONMENT

Lunar architecture must overcome threats to life, health, equipment performance, and safety imposed by the alien and hostile environment described in the preceding chapter. These external environmental stressors critically shape lunar architecture just as environmental forces shape architecture on Earth (Fitch and Bobenhausen 1999), but perhaps even more notably given the unfamiliar nature of the lunar environment.

Lunar-base structures support, stabilize, or contain the materials, devices, processes, and people comprising a lunar base. A few examples that represent the large domain of lunar-base structures are process-plant frameworks and vessels, foundations and anchors, equipment masts, roads, environmental shelters, construction scaffolding, and habitation systems. This short list of applications requires all of the classical structure systems of foundations, pads, columns, beams, arches, frames, trusses, plates, shells, cables, and membranes. Gravity, pressure, and dynamics produce the same loading patterns on the Moon as everywhere: compression, tension, bending, and torsion. However, the unique conditions found on the Moon make the integrated problem of designing structure systems novel there. Lunar structure schemes are driven by hard vacuum, partial gravity, radiation, micrometeoroids, extreme thermal cycling, and other specific considerations.

Space architecture is about much more than just the structure, though. For example, while recognizing the key role of a habitat's structure in protecting against the natural lunar environment, space architects also recognize how each countermeasure affects the quality of life inside. Fujii et al. (1990) summarize key habitability needs: food, clothing, housing, communication, and mental and physical requirements. They correlate these needs across three dimensions: basal life, passive pursuits, and active pursuits. Design considerations shape the quality of life in a lunar base and inform the ensuing discussion of environmental threats and how structure systems, materials, and construction techniques address them.

Effects of Vacuum

The hard vacuum on the lunar surface is several orders of magnitude emptier than can be achieved on Earth for practical purposes. This vacuum precludes materials (e.g. many plastics) that are not chemically stable under such conditions.

Construction in vacuum poses multiple challenges. One is outgassing (loss to vacuum) of oil, vapors, and lubricants from mechanisms, interfering both with their operation and with optical and thermal performance of nearby systems as the gases condense and attract dust (Chua and Johnson 1991). Hydraulic systems using space-rated lubricants are in use today. Another challenge is vacuum tribology. Surface-to-surface contact becomes much more chemically active in vacuum because of the absence of thin oxide layers. Increased dynamic friction can cause fusion at the interface; for example, a drill bit might fuse with the metal, rock, or other substrate being drilled. This behavior is aggravated by the fact that in vacuum heat cannot be removed by convection. Increased abrasiveness at interfaces increases wear and tear on mechanisms, for example, pivots, hinges, latches, bearings, and axles.

Blasting in vacuum is also problematic. Blasts create gas overpressures that can exceed 100,000 atmospheres. It is difficult to predict how such an explosion and the resulting ejecta might affect the area around the blast. The Air Force Institute of Technology (Johnson et al. 1969) studied cratering at various gravities and/or in vacuum. Considering that particles blasted by a rocket plume could go into orbit around the Moon, the effects of construction blasting on the lunar surface must be carefully engineered. Discussion of tests involving explosives can be found in Watson (1988). Joachim (1988) discusses different candidate explosives for extraterrestrial use. Bernold (1991) presents experimental evidence from a study of blasting to loosen regolith for excavation.

Pressure Containment

For habitats, the immediate corollary to vacuum is the need to contain an atmosphere in hermetically enclosed volumes pressurized up to 15 pounds per square inch (psi) (103 kPa). This is the governing load for such structures, which must be fail-safe against catastrophic (explosive) decompression caused by accidents and natural impacts. A pressure vessel can be "hardshell" or "inflatable," although in either case the pressurization makes it rigid. Secondary structure and interior outfitting must accommodate dimensional changes caused by temperature and pressure, especially if the shell is inflatable. Note that in the TransHab prototype (Kennedy 1999) (Chapter 8), the secondary structure attaches to the rigid axial core; none of it makes a connection to the inflatable shell.

Habitats require life-support systems that provide and condition an atmosphere inside by continuously removing CO_2 and contaminants and replenishing O_2. The life-support discipline plays an all-pervasive role in determining quality of life at the lunar base. Ferrall et al. (1994) use the NASA Jet Propulsion Laboratory Life

Support Systems Analysis software tool to perform a comprehensive life-support and power economy modeling of a lunar habitat. Their model consists of three "baseline system" definitions: metabolic load basis, hygiene load basis, and configuration. The metabolic load increases linearly with the number of crew and the intensity of their activities and drives the air revitalization system sizing, almost independent of total atmospheric volume. Hygiene load is the water needed for washing, which is typically far larger than that needed for drinking water or rehydrating food. Clothes washing increases the hygiene load substantially. Configuration is the geometry, volume, and equipment complement.

Partial Gravity

After vacuum, the most evident architectural attribute of the lunar environment is the reduced, 0.18-*g* acceleration of gravity at the surface, about one-sixth of what we experience on Earth. However, if regolith overburden is used for radiation shielding (see what follows), its mass is considerably greater than conventional terrestrial roof deadloading. For Class I and II habitat structures, reduced gravity does not typically introduce a structural mass benefit for two reasons: 1) pressure containment dominates and 2) structures made on Earth and transported to the Moon must be sufficiently robust to resist ground-processing and launch loads.

Partial gravity does pose some advantages over microgravity for operations. Dust settles, outside and inside. Convection occurs in interior atmospheres. Gravity-fed flow can be used for sieving, filtration, and other routine operations. Gravity-stabilized mechanical needs like traction and leveling are particularly important for mobility, regolith moving, construction, and maintenance. However, the U.S. Bureau of Mines sounds a note of caution about the effect of lunar gravity on construction equipment and operations:

> The lunar gravity, being only one-sixth of Earth gravity, causes different dynamic conditions for equipment movements and operation. Stability of human and robot movements [is] impaired, and tall equipment [can] easily topple when lateral loads are applied. (Podnieks 1990)

The architectural effect of partial gravity on crew locomotion, both inside and outside a lunar habitat, remains not fully understood. Video of *Apollo* astronauts moving about is misleading for understanding ⅙-*g* gait because the suits of that era were pneumatically stiff. A Massachusetts Institute of Technology team investigating perambulation under varied gravity regimes in a neutral-buoyancy water tank found significant locomotion differences among the several regimes, despite complicating effects of water resistance in the tank (Newman et al. 1994).

Loading and Substrate

A given structure basically offers six times the weight-bearing capacity on the Moon as it would on Earth. Conversely, to support a certain loading condition, one-sixth the load-bearing strength is required on the Moon. Mass-based rather than weight-based criteria drive the approach of lunar structural engineers.

Regolith differs from Earth soil in several respects significant for construction. Although the top layers (10–20 cm) are loose and powdery, easily observable in *Apollo* video, lunar regolith quickly reaches a relative density of 90–100% below 30 cm. The lunar grain size distribution and high subsurface compaction are rare in the terrestrial environment (see Chapter 14). These conditions create unique problems for excavating, trenching, backfilling, and compacting lunar regolith (Goodings et al. 1992). Such operations are needed to create foundations, roadbeds, launchpads, buried utilities (e.g., for power, fluid, and data networks), shelters and covers, open-pit mining, and underground storage facilities.

The topic of lunar-regolith mechanics was exhaustively explored in the 1970s. Much of this work used classical soil mechanics. Typical analytical foundation design is primarily based on the limit-state condition, that is, the plastic limit of loading a wall or footing to the point of total collapse. Because many lunar structures require accurate positioning for connection or pointing, a settlement-based design method would be more useful. Chua et al. (1990) propose a nonlinear hyperbolic stress-strain model that can be used for lunar regolith in a finite element analysis and show how the finite element method can be used to predict settlement of a railway under a large telescope. Chua et al. (1992) show how a large, deformation-capable finite element program can predict the load-displacement characteristics of a circular spud-can footing designed to support a large lunar optical telescope. Newer work and development of nonlinear stress-strain models to describe the mechanics of lunar regolith appear in Johnson et al. (1995a) and Johnson and Chua (1993). Chua et al. (1994) show how structure-regolith simulations can be done using the finite element approach.

Chua warns against assuming that less gravity means a footing can support more load. If regolith is assumed to be linearly elastic, then the elastic modulus is not affected by gravity. However, the load-bearing capacity of any particulate substrate depends on the confining stress around it. If the regolith surrounding the point of interest were heavier because of stronger gravity, the confining stress would be higher, and the regolith could support a higher load without collapsing. Under reduced gravity artificially compacted regolith will be less consolidated and provide less containment than undisturbed, naturally comminuted lunar regolith.

Water-ice has been theorized to exist in some permanently shadowed craters near the poles of the Moon, deposited over the eons by water-laden comets. *Lunar Prospector* data imply ice content in the regolith at the bottom of such craters might be between 0.3 and 1%. Admixed ice would yield regolith mechanical properties unlike those found elsewhere on the Moon, as *Phoenix* lander excavation on Mars made evident.

Radiation Shielding

Radiation exposure by continuous galactic cosmic radiation (heavy nuclei) and episodic solar flares (high-energy protons) is one of the toughest design challenges for lunar architecture. It can be a "showstopper" for otherwise interesting surface system concepts. Accumulated exposures might reach several times those typically allowable on Earth, and acute solar-flare radiation events can be directly lethal if not shielded against.

Much U.S. research has determined damage to human beings and electronics resulting from nuclear weapon detonation, but little is known about long-term, sustained, low-level radiation effects such as encountered on the Moon. According to Silberberg et al. (1985), during times of low solar activity, the annual dose-equivalent to humans on the exposed lunar surface might be about 30 rem (30 cSv) and the dose-equivalent over an 11-year solar cycle is about 1000 rem (1000 cSv), with most of the solar proton dose arriving in one or two gigantic flares lasting one to two days each.

Radiation transport codes are used to simulate effects of cosmic radiation. Gallium arsenide-based electronic components (e.g., sensors, processors, and actuators embedded in "smart structures") are more radiation hardened than those typically made of silicon or germanium. Bulk shielding can also be used for components that can be enclosed.

For a Class I system supporting duty tours of a few months per crew member, it might be impractical to bring shield mass from Earth except for a small flare shelter. But for long stays, some sort of shielding must be applied. Many investigators over the years have considered how lunar regolith could be used for radiation shielding. So far, however, no one knows how much is really required, and the literature contains an unhelpfully wide range of assumptions. As much as 2.5 m of regolith cover might be required to keep the annual radiation dose at 5 rem (5 cSv), the current allowable level for radiation workers (0.5 rem for the general public). Thinner shields might be inadequate to protect against the primary radiation, and moderately thicker shields might cause excessive secondary radiation (Brehmsstrahlung and elementary particles resulting from the primary radiation being deflected by or fragmenting shield atoms). Only *in situ* measurements and shielding experiments can provide the data required for good dose and design models.

Thick regolith is a severe design driver for shield structures, construction operations, and equipment. NASA Langley Research Center researchers conclude a discussion of the difference between prefabricated and *in situ* radiation shielding thus:

> Lunar regolith still appears to be an attractive option for radiation protection for the habitat configurations considered in this analysis. However, if much smaller habitats are selected, then the mass of the regolith-moving equipment may approach the mass requirements of pre-fabricated shields launched from Earth. One of the major trade-offs will be the EVA time requirements, EVA risk, and the reliability of the regolith-moving equipment. If it is deemed necessary to provide a flare shelter while the habitat is being covered, a viable option appears to be polyethylene or water. (Simonson et al. 1992)

Depending on whether other regolith-moving activities are part of the outpost plan, Class I structures might need to include their own radiation shielding. Materials brought from Earth present significant advantages over regolith for this purpose. Water provides excellent shielding for solar flares and can be brought separately from the habitat to be pumped into internal shielding tanks (Cohen 1997). For Class II structures, radiation-shielding options tend to comprise regolith, water, or externally applied polyethylene panels. For Class III structures, use of *in situ* regolith for shielding would be consistent with major construction operations, although lunar concrete, masonry, or vitreous construction might obviate regolith burial.

Comprehensive radiation shielding directly affects crew well-being and quality of life. Adams (1998) discusses this issue with respect to a buried habitat:

> ...a Lunar or Mars habitat may be expected to be buried under a meter or more of soil for radiation protection. Designing Crew Quarters

> for a planetary habitat therefore involves taking into account the problem of how to mitigate claustrophobia without the benefit of windows. In the BIO-Plex HAB chamber [at NASA Johnson Space Center], strategies are being investigated for compensating for the lack of outside views which include the integration into partitions of flat-panel monitors which function as "Virtual Windows" as well as planning for upper-surface penetrations for optic-fiber light "straws" to draw external sunlight into light baffles designed to reflect that light into many interior spaces.

Criswell et al. (1996) discuss a scheme using a regolith-shield's overburden weight to "balance" 10–20% of the internal pressure force of inflatables, but designing for a depressurization contingency would eliminate the benefit of such offloading.

Micrometeoroid Bombardment

The lunar surface is exposed to a steady flux of micrometeoroid particles in a range of sizes and densities. Vanzani et al. (1997) extrapolated the lunar surface micrometeoroid flux from NASA Long Duration Exposure Facility results. Their findings were dramatic:

> As an example, a surface of about 150 m^2 located on the Moon is hit, on average, by one micrometeoroid larger than 0.5 mm in diameter per year: a projectile that size, impacting with an average velocity of about 13 km/s, excavates in aluminum alloy material of [a] hypothetical lunar basis structure a crater with diameter larger than about 1.8 mm and depth greater than about 1 mm….
>
> The actual risk to critical structures exposed on the Moon is difficult to estimate, but the flux of meteoroids represents a significant hazard [to] and requires proper protection [for] critical structures—habitats, base support facilities, processing plants or research instruments, especially optical systems and detector packages—that are expected to last on the lunar surface for many years.

Very large impacts that would cause major, catastrophic damage are virtually impossible to shield against, but their occurrence is vanishingly rare. Accepting this low-probability, high-consequence risk is a fact of life on the Moon. Shielding against micrometeoroid impacts is typically accomplished by a Whipple bumper (thin layers of foil separated by gaps, that absorb the kinetic energy as it disperses through the successive layers), but permanent structures could also use dense shields such as compacted regolith. Lunar rock might be more effective than regolith because of its fracture toughness, but might be more difficult to obtain and much more difficult to emplace.

One-Month Diurnal Cycle

The slow lunar diurnal (synodic) cycle, yielding over 14 days of intense sun followed by over 14 days of dark, cryogenic cold, has direct implications for the performance of systems and quality of life for the human crew. The most obvious human effect is disruption from our normal 24-hour day-night, wake-sleep, and work-leisure cycle. Systems effects include availability of power, feasibility of extravehicular activity (EVA), and performance and safety of other operations that may require or suffer from darkness.

The diurnal cycle also indirectly affects base site selection. Toward the poles the cycle becomes highly asymmetric (i.e., sunlit vs dark fractions of the 29-day period) and topography dependent (i.e., because the sun hugs the horizon) toward the poles. There have been numerous proposals to locate a lunar base on a hill or crater rim near one of the lunar poles so that it can receive near-constant solar energy and benefit from a more moderate and stable thermal environment. The possibility that water-ice might be found in permanently shadowed craters in these regions is a strong additional incentive for such site planning. The NASA Space Science Data Center (NASA 2002) reports the following:

> Much of the area around the South Pole is within the South Pole-Aitken Basin…a giant impact crater 2500 km in diameter and 12 km deep at its lowest point. Many smaller craters exist on the floor of this basin. Since they are down in this basin, the floors of many of these craters are never exposed to sunlight. Within these craters the temperatures would never rise above about 100K.

This long-lived cryogenic environment provides a theoretical mechanism for retaining water ice deposited by cometary impacts over geologic time and is therefore the basis for inferring the excess-hydrogen signature found by *Lunar Prospector* indicates ice. Operations in this locale would certainly be challenging: Sun on the horizon; high-contrast lighting environment; rugged terrain surrounding a landing zone and base perched on a crater rim or ridge Goldstone radar investigations confirmed (steep slopes in this area in 2008); extreme, starlit cold in the areas where ice might be found; and upslope and downslope mobility needed to commute between base and ice site. Despite the uncertain trafficability in these rugged areas, the U.S. Constellation program is considering planning for such polar exploration.

Thermal Cycling

The lunar thermal environment poses several challenges for structures. At a midlatitude site, habitat systems have to maintain tolerable and comfortable internal temperature ranges despite the extremely different conditions of lunar day and lunar night. NASA Marshall Space Flight Center engineers caution: "Lunar bases with a long-term human presence (90 days or more) present a very challenging thermal control problem. Typical concepts propose power levels of 12 to 30 kW, which means that a great deal of waste heat must be rejected" (Walker et al. 1995).

This is a problem at night as well as during the day because of the insulating nature of vacuum. But daytime is an especially challenging time, when radiators must avoid viewing the sun or the surrounding regolith. Radiator surfaces with tailored absorptivity and emissivity properties are vulnerable to obscuration by dust layers (Walker et al. 1995) and degradation by micrometeoroid erosion. Keller and Ewer (2000) modeled dust on horizontal and vertical radiator surfaces and on a "parabolic shade"; they recommend heat-rejection systems be located at least 1 km from landing zones because "lunar dust accumulations can drastically alter the thermal performance of those systems that have either specular, low-absorptivity or low-emissivity surfaces."

For those times during long lunar nights when heating would be required, NASA Johnson Space Center has proposed an *in situ* resource utilization (ISRU) based heat sink for energy storage using cast basalt blocks inside the habitat. The concept would use a fan to circulate air around blocks spaced to maximize the heat exchange surface area (Sullivan 1990).

Thermal control is also an issue for individual EVA-suited crew members, where uneven heating and cooling can occur. Where one side of a suit is in sunlight and the other in shadow, prolonged exposure can create stresses on both the portable life-support system and the crew member's metabolic thermoregulatory system (Koscheyev et al. 1996).

From the structural perspective, severe temperature swings pose the threat of material fatigue, especially for exposed structures (Benaroya et al. 2002a; 2002b). Very low-temperature effects include the possibility of material embrittlement and brittle fracture in structural members (Benaroya et al. 2002a; 2002b). This suggests base and habitat structures should use passive thermal treatments like multi-layer insulation and tailored paints to moderate the cycles seen by structural members.

Dust

Lunar dust is a ubiquitous design and operational challenge. Repeated bombardment by hypervelocity micrometeoroid impacts over the eons has pulverized lunar rocks into the regolith we find there today. Half the mixed regolith comprises dust finer than talcum powder; this dust is charged and easily disturbed and lofted. The charge might come from the fractured crystalline structure of the material, or it might be surficial because of charged particles from the solar wind adsorbing to the dust particles. Criswell (1972) reports that dust particles are levitated at the lunar terminator (line between lunar day and lunar night), and this might be because of a change in polarity of the surficial materials.

The particles cling to surfaces electrostatically, infiltrate everywhere and coat everything, and are highly abrasive, and so they interfere with construction equipment, mechanisms, hatch seals, fabrics, coatings, thermal control emissive surfaces, and all other exposed surfaces (Slane 1994). Johnson et al. (1995b) discuss the effects of lunar dust on operations. Halajian (1964) and Seiheimer and Johnson (1969) study the adhesive characteristics of regolith dust. "The *Apollo* experience indicated that in a low-gravity, vacuum environment dust traveled easily from the surface and adhered ferociously to equipment. Four of the 12 *Apollo* rock boxes did not seal adequately" (Allton and Lauer 1991).

Figure 3 shows a photo from *Apollo 11*—after the shortest total EVA time on the Moon so far—of Buzz Aldrin with lunar dust clinging thickly to his boots and knees. The design of a lunar habitat or base must present effective, proactive measures to mitigate dust intrusion, degradation of equipment, and other effects. Other *Apollo* astronauts returned from EVA on the lunar surface with the legs of their spacesuits completely coated in the dark gray dust. Physiological effects of inhaling the dust inside the habitat are unknown, but lunar dust includes grain sizes comparable to those that cause silicosis, making dust prophylaxis perhaps even vital.

Landing-Zone Safety

A landing zone (LZ) is a human-made environmental hazard. First, dust lofted by engine exhaust plumes travels long distances at unattenuated velocities because of the low gravity and absence of atmosphere and can cause erosion of surface coatings and optical elements at a base: "Keeping in mind that a particle set in motion by the firing from a rocket from a lander could theoretically travel halfway around the Moon, the effects of surface blasting on the Moon

FIGURE 3 *Apollo* astronaut Edwin "Buzz" Aldrin with gray lunar dust clinging thickly to his boots and knees (courtesy of NASA SP-350).

would be something to be concerned about" (Benaroya et al. 2002a; 2002b). Second, blast effects or shrapnel from an accidental crash or explosion could be far more severe. This suggests that base infrastructure should be located a considerable distance from the LZ and that debris blast shields would be prudent. A large separation between base and LZ highlights the design and operational challenge of moving payloads between them. See Chapter 22 for more discussion of these issues.

Structures Design

Lunar structural design should evaluate the total system life cycle, that is, from conception through retirement and disposition, including recycling of components. Many factors affecting system life are difficult to predict because of the unprecedented nature of the lunar environment and our inability to assess a lunar surface system realistically before it is built and put into use. Concurrent engineering approaches consider system design, manufacturing, construction, and use simultaneously, and anticipate potential problems or opportunities by modeling major items early in the design cycle. Given the extreme nature of the lunar environment, maintaining multiple design solutions could help ensure project completion in the face of unanticipated difficulties. A discussion of lunar design codes started early (Benaroya and Ettouney 1992a), including how lunar and Earth codes must diverge.

Standards for lunar structures design can take advantage of the unique conditions. Structures unsuitable for Earth construction might prove adequate for the reduced-gravity lunar environment (Chow and Lin 1989). In a light, flexible structural system in low gravity, light structural members such as composite cylinders with wall thickness of only a few mils (50–150 μm) can be designed to limit their load-carrying capacity by buckling first, so that their load is redistributed to other, less-loaded structural members. Such an approach offers possibilities for inflatable and other lunar surface structures where it would be simpler and less costly to include limit-state and sacrificial structural elements.

Chua et al. (1993) discuss guidelines and the developmental process for lunar-based structures. They present governing criteria for, and general misconceptions about, designing space structures. For example, devices that are simple and meet functional requirements with few moving parts are preferred over those having multiple degrees of freedom. Other common misconceptions are that lunar construction approaches can be simply scaled from similar operations on Earth; that computer-based, theoretical-predictive tools can accurately model the lunar circumstance; and that crew members will do extensive manual work on EVA.

Safety and Reliability

Lunar structures are designed with safety and reliability as prime considerations because of the high value of both space-faring crews and their equipment. Minimizing risk implies building in robustness, redundancy, and (for habitable structures) means of escape or safe haven. Acceptable risk is subjective, deeply rooted in pragmatic considerations. What is an acceptable level of safety and reliability for a lunar site, a place that by its nature is highly hazardous? Such questions go beyond engineering into program policy: what kind of failure can be afforded or allowed? See Cohen (1996) for a related discussion.

In particular, the issue of construction safety looms large. Even on Earth, the construction industry has one of the highest injury rates of any vocation (rivaled only by commercial fishing, logging, and slaughterhouse work) and the third highest death rate (15 per 100,000, after mining at 30 per 100,000 workers and agriculture/forestry/fishing at 19 per 100,000) (Centers for Disease Control 2001). Add to this the inherent risks of spaceflight and EVA, and crew safety emerges as a leading concern for lunar-base construction. Construction operations protocols and particularly machine-mediated methods for planet-surface construction, are essential.

Construction Operations

Lunar construction techniques are necessarily different from those on Earth: in addition to the environmental considerations just discussed, the on-site construction team is small and operates in pressure suits and via teleoperated machines. Lunar-base construction might benefit from capabilities developed by the U.S. Army Corps of Engineers (Simmerer 1988; Sargent and Hampson 1996).

Toups (1990) divides lunar construction methods into three categories: 1) require Earth support, 2) use natural surface and subsurface features, and 3) primarily use lunar resources. Published concepts and analyses are contingent on practical aspects of building on the Moon: machinery that can move equipment and astronauts about the surface; methods to construct in ⅙ *g* with fine, abrasive dust working its way into every interface; and determination of the appropriate layout of structures considering human safety and operations needs. For a structure optimized for the lunar environment, realistic prototyping cannot be performed on Earth or even in orbit. It is not currently possible, for example, to experimentally assess the effects of lofted (caused by ⅙ *g* and electrostatics) lunar-regolith fines on lunar machinery. *Apollo* experience can be extrapolated, but new information is necessary. Using harsh Earth environments like the Antarctic as testbeds for extraterrestrial operations might be helpful for some dimensions of the problem (Bell and Neubek 1988).

Drake and Richter (1990) qualitatively studied the design and construction of a lunar-outpost assembly facility. Such a facility would be used to construct structures too large for transportation to the Moon intact and support operations and maintenance during the functional life of the outpost. The remoteness of lunar construction sites and high cost of launch from Earth suggest lunar structures be designed for ease of construction to minimize the complexity of construction robots and EVA construction operations. Construction components must be practical and modular.

Sherwood (1990) presents an integrated base concept designed for machine construction, assembly, and shielding prior to crew arrival. Cohen and Kennedy (1997) provide another discussion of these issues including automated delivery and emplacement of habitats and surface facilities.

Local Resources

Use of planetary materials found or mined on site, referred to as ISRU, is enabling for long-term visions of extraterrestrial habitation. ISRU has been studied intensely because of the possibility of establishing human presence on the Moon, near-Earth objects (NEO), and Mars. Some of the earlier work is found in Johnson and Chua (1992). Many approaches have been theorized, and some techniques for oxygen extraction and regolith sintering have been tested on Earth, but practical ISRU methods require *in situ* testing and development after initial human presence has been reestablished on the Moon.

ISRU can play a role providing life-support consumables to produce water and air. Duke (1994) predicts "oxygen will probably be the first material produced on the Moon," primarily for rocket propellant. Although O_2 has obvious application in a life-support system, the 80% of sea-level-pressure makeup air that is N_2 cannot easily be derived from lunar sources anyway.

LUNAR STRUCTURE AND CONSTRUCTION CONCEPTS

The lunar exploration literature includes many discussions of structures concepts. In an early lunar structure design study, Johnson (1964) details what was then known about the lunar environment, discusses lunar soil from the perspective of foundation design, and reviews excavation concepts. Johnson and Leonard (1985) review the evolution of concepts for lunar bases up through the mid-1980s, and Johnson and Wetzel (1990c) update this discussion through the end of that decade. Richter and Drake (1990) compare extraterrestrial building system concepts including pneumatic, framed/rigid foam, prefabricated, and hybrid (inflatable/rigid) schemes. Hypes and Wright (1990) survey surface and subsurface concepts for lunar bases and recommend that preliminary designs focus on specific applications. Benaroya and Ettouney (1989; 1990) use decision-science and operations-research tools to assess overall efficiency of lunar structure concepts.

The complexities and costs of building on the Moon depend on the mission, for example, supporting scientific research, exploiting lunar resources for use in building space infrastructure, or attaining self-sufficiency in the lunar environment as a first step in planetary settlement. Hoffman and Niehoff (1985) use criteria such as scientific objectives and transport requirements for preliminary design of a permanently manned lunar surface research base. Surface system architectures, operations scenarios, and site plans (Sherwood 1990a; Sherwood 1990b; Pieniazek and Toups 1990) are iteratively developed with individual infrastructure element designs.

Taxonomy of Lunar-Base Structures

By "lunar-base structures," most people presume to mean habitats, and most of the discussion that follows is indeed about habitat structures. However, habitation systems comprise only about 10% of the delivered mass of an early, oxygen-producing lunar base (Sherwood 1990b), and only about 25% of that is the actual pressure-containing structure. (The rest is outfitting equipment and its mounting structures.) Such early habitat structures for the Moon will likely be adapted from spacecraft habitation systems already developed (Woodcock et al. 1991). Therefore the greatest variety of lunar-base structures is actually the nonhabitat infrastructure, including sitework. By far the dominant mass fraction of even a modest lunar base (about 90% of the total constructed mass, two orders of magnitude more than the habitation element mass) comprises foundations, roads, and other native material deposition. Especially for advanced scenarios with more delivered mass, it is most accurate to think of lunar bases as factories or laboratories supported by habitation systems.

The three structures classes provide a convenient, albeit not clean, schema for considering and comparing planet-surface base concepts. When applied to actual habitat concepts, division into three classes only works to first order. Table 1 indicates how even straightforward concepts for lunar structures are almost always a hybrid of multiple classes. In the matrix, hybridizing each class with secondary features from all three classes opens up rich, diverse possibilities.

Class I

Figure 4 shows an early concept for a temporary post-*Apollo* lunar base, featuring the multiple, near-coincident landing of several vehicles that serve as lunar surface habitats or "living stations." They are not linked together with pressurized connectors, precluding crew exchange without EVA. An unpressurized

TABLE 1 Matrix of lunar/planetary habitat structure classes reveals usefulness of hybrid concepts.

PRIMARY CLASS STRUCTURE	SECONDARY CLASS STRUCTURE		
	PRE-INTEGRATED	PREFABRICATED	*IN SITU*
PRE-INTEGRATED	"Tuna can"/node ISS long module ISRU fuel plant EVA access module Power system	"Tuna can" with node or flex-docking tunnel Inflatable expansion of a hard, pre-integrated module	Regolith packed on a hard module for radiation and micrometeoroid protection Location on lunar/Mars surface affords a measure of protection Location adjacent to high cliff face or crater wall can add protection
PREFABRICATED	Inflatable with pre-integrated safe haven Telescoping modules with pre-integrated nodes	Inflatable dome or sphere with flex tunnels or telescoping modules	Regolith packed on an inflatable or telescoping module for radiation and micrometeoroid protection Inflatable in a crater, surface depression, cave, or lava tube
IN SITU	Concrete from regolith Masonry vault or cave with a hard, pre-integrated module inside	Concrete shell Masonry vault or lava tube with an inflatable, pressurized habitat liner	*In situ* vitrification and pressure sealing of cave, bored tunnel, or lava tube with concrete or masonry pressure bulkheads

FIGURE 4 Early NASA concept for a post-*Apollo* temporary lunar base, using all Class I, pre-integrated landers with hard-shell modules (courtesy of NASA Marshall Space Flight Center).

lunar rover provides access among habitat modules and EVA stations for working and drilling. Complex issues like adjacent, precision landing of habitats, ingress up a long ladder for sick or injured crew members, and lack of pressurized emergency egress from the habitats are all unaddressed.

Figure 5 shows a design concept derived from NASA's First Lunar Outpost (FLO) study. The EVA crew uses a crane to unload a habitat module from the *Artemis* multipurpose cargo lander for flatbed transport from the LZ to the base area, a considerable distance away in the left background. To maintain this safe separation between base and LZ, this concept requires an infrastructure of unpressurized cranes and transporter vehicles. It is unclear how the crane was delivered to the Moon.

Figure 6 shows the MALEO concept (modular assembly in low Earth orbit) conceived to solve the problem of assembling a lunar base. This concept recognizes that landing a Class I base piecemeal on the lunar surface, then assembling, deploying, and integrating the modules and other elements, partially compromises the benefits of Class I pre-integration. The MALEO elements are assembled and verified in LEO, then boosted onto a translunar trajectory, and landed on the Moon intact by a modular propulsion system. Although

FIGURE 5 1992 joint study by McDonnell Douglas and Shimizu Corporation using the *Artemis* lander concept (courtesy of NASA Johnson Space Center).

FIGURE 6 Fully pre-integrated class I MALEO lunar base arriving at the Moon (courtesy of Madhu Thangavelu).

the scheme involves significant complexity in the propulsion system and risk in the landing operation, it offers advantages from a surface architecture standpoint:

- No need to move modules one at a time from the LZ
- No cranes, transporters, or other mobility equipment to move them
- Reduced time, expense, and labor of assembling components on the lunar surface
- No dust intrusion at module-to-module connections

Figure 7 shows a pair of pre-integrated "tuna can" modules that have landed and been moved into a side-by-side configuration. (The image shows the Mars surface, but the concept applies to the Moon as well.) Each lander has four wheels for towing by a rover-tractor from LZ to common base site. The mobility design for unknown surface conditions is critical for safely moving such large modules (in the 40–50-mt range). Note the low-hanging EVA airlocks, located for direct berthing by the pressurized rover shown. Inflatable airlock vestibules or an interconnect tunnel could be attached here, addressing a deficiency noted for the concept in Figure 4. IVA interaccess is a practical necessity for efficient routine operations and crew safety. Inflatable tunnels would be Class II elements.

Class II

The two principal types of Class II structures are inflatables and erectables.

INFLATABLES Broad (1989) reports on inflatable lunar-base structural concepts as a way to simplify and accelerate construction while limiting cost. Vanderbilt et al. (1988) propose a pillow-shaped structure for a permanent lunar base. The base consists of quilted, inflatable, pressurized, tensile structures using fiber composites. A regolith overburden

FIGURE 7 Pre-integrated "tuna can" hard modules, landed intact on the Mars surface, and then moved and assembled into a base (courtesy of John Frassanito and Associates, circa 1993).

provides shielding, with accommodation for sunlight ingress. Nowak et al. (1990; 1992) consider the foundation problem and reliability concerns. This concept marks a significant departure from numerous other inflatable concepts by showing an alternative to spheroidal inflatables that optimizes volume for habitation (see Chapter 20). An inflatable structure can be used as a generic testbed structure for a variety of lunar applications (Sadeh and Criswell 1994). Criswell et al. (1996) propose design criteria.

Chow and Lin (1988; 1989) proposes on a pressurized membrane structure constructed of a double-skin membrane filled with structural foam. A pressurized, torus-shaped substructure provides edge support. Shielding is provided by a regolith overburden. The construction procedure requires shaping the ground and spreading the uninflated structure upon it, then pressurizing the torus-shaped substructure. Structural foam is injected into the inflatable component, and the internal compartment is pressurinzing. The concept proposes the bottoms of both inflated structures be filled with compacted regolith to provide stability and a flat interior substrate, although health and maintenance issues from lunar dust would require such regolith to be fully sealed. Eichold (2000) presents a similar concept for a lunar base built into a crater. Kennedy (1992) proposes a detailed architectural master plan for a horizontal, inflatable habitat.

Finite element simulations of inflatable structures are needed to analyze hard-vacuum and low-gravity conditions. Such a model must be large-deformation capable and have membrane elements (beam and plate elements without bending) with shear stiffness and axial tensile stiffness but no axial compression stiffness because a membrane cannot resist compression. Ideally, the program should also be able to model regolith-structure interaction. GEOT2D (Chua et al. 1994) is one tool capable of analyzing interactions between inflatable structures and regolith.

Figure 8 shows the NASA Johnson Space Center Lunar Return Habitat (LRH), a simple inflatable concept. After landing, the bellows habitat pressure vessel is extended and fully inflated, and the round photovoltaic array, rectangular radiators, radio dish, and stair ladder are deployed. Key attributes of this kind of Class II habitat are that all components are manufactured and tested on Earth and simply launched in a stowed configuration. Deployment on the Moon—by a combination of unstowing, unfolding, assembling, erecting, or inflating the habitat—could be autonomous or remotely operated.

Figure 9 shows a much larger, more iconic inflatable concept: a sphere approximately 16 m in diameter proposed for the Lacus Veris site on the Moon. This sphere would have five interior levels, including the subsurface "level zero," and accommodate a lunar population of 12 crew members. The long inflatable module attached to the dome is a different

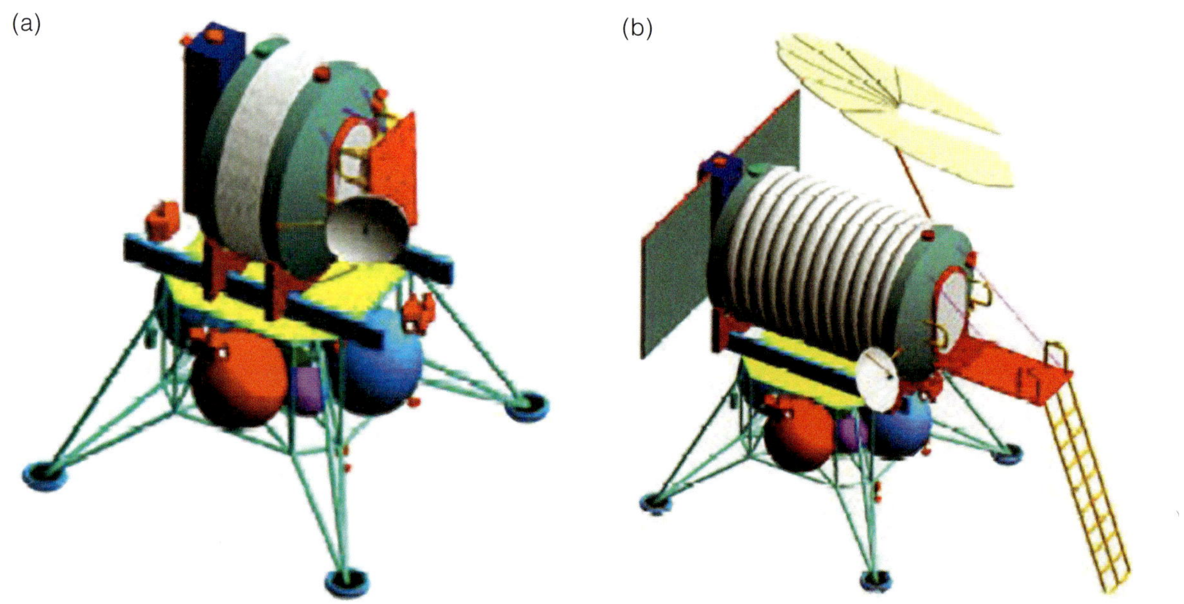

FIGURE 8 Lunar return habitat concept: a) landing configuration with inflatable habitat and equipment stowed and b) deployed and inflated for use (courtesy of Kriss Kennedy, NASA Johnson Space Center).

| **FIGURE 9** Lacus Veris lunar outpost (courtesy of NASA Johnson Space Center, 1989).

configuration, but still quintessentially Class II. The concentric rings stacked up from the lunar surface to the middle of the sphere are a radiation shield comprising regolith-filled fabric tubes. In the image, these shield-tubes are still under construction: a machine to the right of the dome deposits regolith into a fabric tube for emplacement. The oblong module to the right supports a similar configuration of regolith tubes and supports a photovoltaic array. Additional "solar-power farms" appear in the upper-left and upper-right corners of the image.

Running from behind the dome to the vanishing point at the horizon is an unusual feature for lunar concepts: a prepared road (see Chapter 22). On this road in the distance, a pressurized rover is driving. Another such pressurized rover appears at the left side of the inflatable sphere. This rover is docked to a pre-integrated airlock element. Although the inflatable habitat is basically a Class II structure, it includes components of all three classes: Class I airlock and rover elements, Class II inflatable tunnel connecting the two large habitat modules, and Class III regolith radiation protection.

Figure 10 shows a cutaway view of the Lacus Veris habitat's four levels of living and working areas. At lower left is an airlock. The subsurface level zero contains environmental control and life-support systems (ECLSS). Levels one, two, and three accommodate living-environment functions including crew quarters, wardroom, galley, hygiene facilities, science laboratories, base operations stations, EVA support, and significant allocations for food growth. Level four at the top of the dome houses more specialized crew-support functions, notably health and recreation facilities. This concept is primarily about construction technology, as the activity area allocations are not resolved (for comparison with concepts that are more fully integrated, see Chapters 8 and 16).

ERECTABLES Mangan (1988) proposes an expandable structure concept consisting of various geometrically configured, three-dimensional, trussed octet, or space-frame elements utilized both as building blocks and as a platform for expansion. Tetrahedral, hexahedral, and octahedral shapes are used.

King et al. (1989) propose a concept for adapting the space shuttle external tank's liquid-oxygen tank as a lunar habitat. Modifications of the tank, to take place in LEO, would include separation from the rest of the external tank structure and installation of living quarters, instrumentation, airlocks, life-support systems, and environmental control systems. The habitat would then be transported to the Moon for a soft landing.

Schroeder et al. (1994) propose a modular approach as a flexible way to develop a variety of structures. Schroeder and Richer (1994) propose a membrane approach for an open structure that can be utilized for assembly on the lunar surface. Benaroya (1993b) suggests a tensegrity structure concept for larger surface structures. A semi-quantitative

FIGURE 10 Cutaway view of the 1989 Lacus Veris inflatable habitat concept (courtesy of NASA S89-20084, July 1989).

approach to lunar-base structures is provided by Kelso et al. (1988). Some attention is given to economic considerations.

Class III

The two principal types of Class III structures use *in situ* materials or landforms. In situ lunar materials generally refer to products made from regolith such as lunar concrete, sintered or vitreous masonry, and bulk-regolith applications such as radiation shields, roads, and berms. Landforms include use of natural features such as a crater's floor, rim, and walls; cliffs; lava tubes; and caves.

Structure and architecture designs, along with manufacturing plants and construction methods, are presented for habitable lunar structures using concrete modules (Namba et al. 1988a) made of frames and panels. The interconnected modular construction withstands internal pressurization. Lin et al. (1989) provides a structural analysis and preliminary design of a precast, prestressed concrete lunar base. A floating foundation is proposed to maintain structural integrity, and thus air-tightness, should differential settlement occur. All raw materials for such a lunar concrete structure might be derivable from lunar resources, albeit at high cost. To sidestep the lack of water on the Moon, Gracia and Casanova (1998) suggest a concept for sulfur-based concrete. Sulfur is more readily available on the Moon than water, but still requires extensive processing of regolith to be obtained. Several research efforts have attempted to produce construction materials such as cement, concrete, and sulfur-based materials from elements available on the Moon (Agosto 1988; Leonard and Johnson 1988; Lin 1987; Namba et al. 1988b; Yong and Berger 1988; Strenski et al. 1990).

Utilizing unprocessed or minimally processed lunar materials for base structures as well as shielding might be possible by adopting and extending terrestrial techniques developed in antiquity for harsh environments (Khalili 1989). Khalili developed and tested a variety of materials and techniques that are candidates for unpressurized applications. Happel (1992a; 1992b) bases his design of a tied-arch

FIGURE 11 CalEarth Institute masonry dome, a possible model for vitrified *in situ* lunar masonry.

structure on indigenous materials. The study is extensive and detailed and includes an exposition on lunar materials.

Figure 11 shows a dome built of native masonry by the late architect Nader Khalili at the CalEarth Institute. In Khalili's approach, the postconstruction interior surface is coated with a glaze and then fired by an intensely hot incendiary source inside the dome to achieve a well-sealed surface. Such *in situ* vitrification technology might be applicable to hermetically sealed lunar masonry or concrete domes.

Construction using layered embankments of regolith and filmy materials (geotextiles) is an option using robotic construction (Okumura et al. 1994), as are fabric-confined soil structures (Harrison 1992). Some studies suggest using geosynthetics as soil reinforcement to construct regolith structures such as berms, walls, slopes, etc. Chua (in Benaroya et al. 2002a; 2002b) highlights several challenges for this to become practical: plastics are susceptible to degradation when subjected to radiation; the glass transition temperature of most terrestrial geosynthetics is well above the cold temperatures encountered at lunar sites, which would render them uselessly brittle as reinforcing. There is no experience on how geosynthetics fare in hard vacuum or when exposed to highly abrasive lunar regolith.

Figures 12 and 13 show lava tubes, a type of natural landform known to exist on the Moon and sometimes proposed as a candidate for habitable shelters. Lava tubes are underground conduits through which lava flows during volcanic eruptions.

FIGURE 12 Lava tube in Oregon (courtesy of Bryce Walden, Oregon L5 Society, http://www.oregonl5.org/lavatube/).

Often they are run horizontally for hundreds of meters, and leave flat floors when extinct. Remote-sensing photographic evidence indicates very long extinct lava tubes on the Moon, some partially collapsed and presumably therefore accessible from the surface. The technology to directly build into and line a lava tube using conventional construction

FIGURE 13 Hawaiian lava tube with active flow (U.S. Geological Survey; courtesy of R.D. "Gus" Fredricks, Oregon L5 Society, http://www.oregonl5.org/lavatube/).

materials is undeveloped. Another approach might be inserting a tube-shaped inflatable and then pressurizing it. If stable, the surrounding rock would provide excellent shielding against radiation, thermal swings, and meteoroids. However, such a base would be locationally constrained to accessible portions of pre-existing lava tubes.

Figure 14 shows a lunar construction concept proposed by the late space architect Alice Eichold (1996), using a crater as part of the support structure for a cable-suspended habitat shelter. This structure was conceived as a way of pressure sealing the natural bowl of a crater to enable construction of an expansive, shielded "outdoors" for inhabitants. The unconsolidated nature of crater-rim regolith, and the way even large craters are excavated in shock-fractured bedrock, would make the stabilization and sealing quite complex.

In some respects, Class III structures may seem more familiar than Class I or II structures because they are more closely related to terrestrial structures built of native materials. However, approaches for working with regolith, processing it, and making various Class III building materials out of it are not yet developed. One sobering paradox concerns the question of water on the Moon and extracting it for use by a human crew. The topic of lunar concrete offers a particularly illuminating calibration. ISRU advocates propose to mine and extract the approximately 1 to 2% concentration of putative water frozen into polar regolith. Compare this low-grade ore of ice-admixed soil to cured terrestrial concrete, which typically has residual moisture content of about 3%. If lunar concrete existed naturally on the Moon, it would make more sense to mine it for water than it would to mine the polar regolith. Recovering lunar water might not be impossible, but it will not likely be straightforward. This comparison might help temper enthusiasm for grand visions of lunar resource extraction industries and construction of urban-scale environments. In any case, it highlights both the overwhelming value of critical resources and how challenging lunar architecture implementation will be.

CONCLUSION

The environmental factors, concepts, and speculative ideas reviewed here indicate the state of the art for lunar-base structures design. Developing practical deployment scenarios for Class I or II concepts, or materials and production approaches for Class III concepts, would require a substantial technology program involving architects, engineers, manufacturers, roboticists, and builders. Any concept adopted for full-scale development would undergo rigorous testing in simulated environments including Earth-analogue locations (see Part 4 in this volume).

FIGURE 14 Transverse section through Class III lunar-crater-base concept.

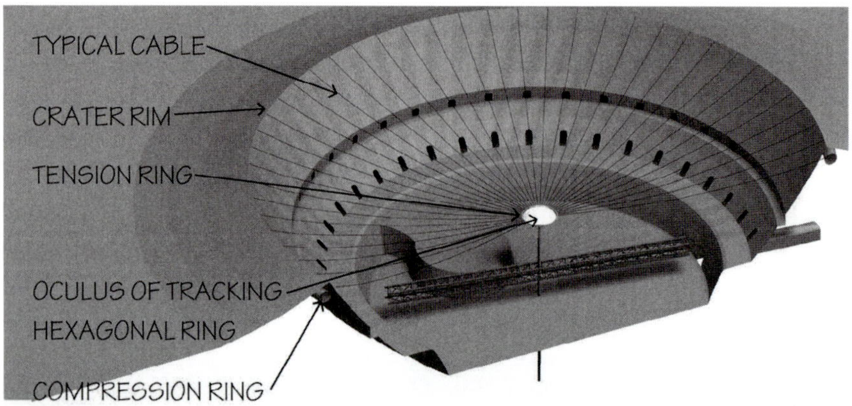

Appendix: Building Systems

Table A1 suggests topical directions for research into structure types, materials, tools and equipment, methods of operation and control, and construction techniques for the lunar environment. Because construction methods developed since the beginning of civilization both accommodate and use the terrestrial environment (e.g., 1-*g*, oxygen atmosphere, Earth substrates), technologies common on Earth either do not work at all on the Moon or do not work well because they are too costly, too inefficient, or impractical. The table can be used by space architects as a checklist of issues to consider when conceiving or specifying lunar surface structures.

TABLE A1 Checklist of issues for lunar surface structures.

STRUCTURES	ISSUES TO CONSIDER
APPLICATION TYPES	
Habitats	People (living and working) Agriculture Airlocks: ingress/egress Temporary storm shelters for emergencies and radiation Open volumes
Storage facilities/shelters	Cryogenic (propellants, life support, and science) Hazardous materials General supplies Surface equipment storage Servicing and maintenance Temporary protective structures
Supporting infrastructure	Foundations/roadbeds/launchpads Communication towers and antennas Waste management/life support Power generation, conditioning, and distribution Mobile systems Industrial processing facilities Conduits/pipes
APPLICATION REQUIREMENTS	
Habitats	Pressure containment Atmosphere composition/control Thermal control (active/passive) Acoustic control Radiation protection Meteoroid protection Integrated/natural lighting Local waste management/recycling Airlocks with dust-scrub areas Emergency systems Psychological/social factors

(Continued)

TABLE A1 Checklist of issues for lunar surface structures. (Continued)

STRUCTURES	ISSUES TO CONSIDER
APPLICATION REQUIREMENTS	
Storage facilities/shelters	Refrigeration/insulation/cryogenic systems Pressurization/atmospheric control Thermal control (active/passive) Radiation protection Meteoroid protection Hazardous material containment Maintenance equipment/tools
Supporting infrastructure	All of the above Regenerative life support (physical/chemical and biological) Industrial waste management
STRUCTURE TYPES	
Habitats	Landed self-contained structures (Class I) Rigid modules (prefabricated/*in situ*) Inflatable modules/membranes (prefabricated/*in situ*) Tunneling/coring Exploited caverns
Storage facilities/shelters	Open tensile (tents/awning) "Tinker toy" Modules (rigid/inflatable) Trenches/underground Ceramic/masonry (arches/tubes) Mobile Shells
Supporting infrastructure	Slabs (melts/compaction/additives) Trusses/frames All of the above
MATERIAL CONSIDERATIONS	
Habitats	Shelf life/life cycle Resistance to space environment (UV/thermal/radiation/abrasion/vacuum) Resistance to fatigue (acoustic and machine vibration/pressurization/thermal) Resistance to acute stresses (launch loads/pressurization/impact) Resistance to penetration (meteoroids/mechanical impacts) Biological/chemical inertness Reparability (process/materials)

TABLE A1 Checklist of issues for lunar surface structures. (Continued)

STRUCTURES	ISSUES TO CONSIDER
MATERIAL CONSIDERATIONS	
Operational suitability/economy	Availability (lunar/planetary sources) Ease of production and use (labor/equipment/power/automation and robotics) Versatility (materials and related processes/equipment) Radiation/thermal shielding characteristics Meteoroid/debris shielding characteristics Acoustic properties Launch weight/compactability (Earth sources) Transmission of visible light Pressurization leak resistance (permeability/bonding) Thermal and electrical properties (conductivity/specific heat)
Safety	Process operations (chemical/heat) Flammability/smoke/explosive potential Outgassing Toxicity
TECHNOLOGY DRIVERS	
Mission/application influences	Mission objectives and size Specific site-related conditions (resources/terrain features) Site preparation requirements (excavation/infrastructure) Available equipment/tools (construction/maintenance) Surface transportation/infrastructure Crew size/specialization Available power Priority given to use of lunar material and material processing Evolutionary growth/reconfiguration requirements Resupply vs reuse strategies
General planning/design considerations	Automation and robotics EVA time for assembly Ease and safety of assembly (handling/connections) Optimization of teleoperated/automated systems Influences of reduced gravity (anchorage/excavation/traction) Quality control and validation Reliability/risk analysis Optimization of *in situ* materials utilization Maintenance procedures/requirements Cost/availability of materials Flexibility for reconfiguration/expansion Utility interfaces (lines/structures) Emergency procedures/equipment Logistics (delivery of equipment/materials) Evolutionary system upgrades/changeouts Tribology

(Continued)

TABLE A1 Checklist of issues for lunar surface structures. (Continued)

STRUCTURES	ISSUES TO CONSIDER
REQUIREMENTS DEFINITION AND EVALUATION	
Requirements/options studies	Identify site implications (lunar soil/geologic models) Identify mission-driven requirements (function and purpose/staging of structures) Identify conceptual options (site preparation/construction) Identify evaluation criteria (costs/equipment/labor) Identify architectural program (human environmental needs)
Evaluation studies	Technology development requirements Cost/benefit models (early/long term) System design optimization/analysis

Designing a structure for the lunar surface is challenging in almost every way. Some important topics not discussed in this survey chapter are listed here for reference:

- Dead loads and live loads under lunar gravity
- Buckling, stiffening, and bracing requirements for lunar structures that are internally pressurized
- Relationship between severe, diurnal lunar temperature cycles and structural and material fatigue—a problem for exposed structures, seals, and hatches
- Structural sensitivity to temperature differentials between different sections of the same component
- Very low-temperature effects and brittle-fracture behavior
- Outgassing for exposed steels, and other effects of hard vacuum on steel, alloys, and advanced materials
- Reliability and risk (Benaroya 1994)
- Factors of safety originally developed to account for uncertainties in terrestrial design and construction processes undoubtedly need adjustment for the lunar environment—up or down depending on purpose, perspective, and risk tolerance
- Consideration of new failure modes, such as those caused by high-velocity micrometeorite impacts
- Nontechnical issues, for example, financing, and drivers on structures design arising from human physiology and psychology in space

References

Adams, C. M. (1998), "Four Legs in the Morning: Issues in Crew Quarter Design for Long-Duration Space Facilities," Society of Automotive Engineers, Paper 981794, July, p. 11.

Agosto, W. N., Wickman, J. H., and James, E. (1988), "Lunar Cement/Concrete for Orbital Structures," *Proceedings of SPACE 88; Engineering, Construction, and Operations in Space*, American Society of Civil Engineers, New York, pp. 157–168.

Allton, J. H., and Lauer, H. V., Jr. (1991), "Effects of Dust on Teflon Face Seals: Implications for Martian Soil Containers," *IDEEA ONE*, Univ. of Houston, Houston, TX, pp. 313–317.

Bell, L., and Neubek, D. J. (1988), "Antarctic Testbed for Extraterrestrial Operations," *Proceedings of SPACE 88, Engineering, Construction, and Operations in Space*, American Society of Civil Engineers, New York.

Benaroya, H. (ed.) (1993a), "Applied Mechanics of a Lunar Base," *Applied Mechanics Reviews*, Vol. 46, No. 6, pp. 265–358.

Benaroya, H. (1993b), "Tensile-Integrity Structures for the Moon," Applied Mechanics of a Lunar Base," *Applied Mechanics Reviews*, Vol. 46, No. 6, pp. 326–335.

Benaroya, H. (1994), "Reliability of Structures for the Moon," *Structural Safety*, Vol. 15, Nos. 1–2, pp. 67–84.

Benaroya, H. (ed.) (1995), "Lunar Structures," *Journal of the British Interplanetary Society*, Vol. 48, No. 1.

Benaroya, H. (2002), "An Overview of Lunar Base Structures: Past and Future," AIAA-Paper 6113, Oct.

Benaroya, H., and Bernold, L. (2008), "Engineering of Lunar Bases," *Acta Astronautica*, Vol. 62, Nos. 4–5, pp. 277–299.

Benaroya, H., Bernold, L., and Chua, K. M. (2002a), "Engineering, Design and Construction of Lunar Bases," *Journal of Aerospace Engineering*, Vol. 15, No. 2, pp. 33–45.

Benaroya, H., Bernold, L., and Chua, K. M. (2002b), "Engineering Design and Construction of Lunar Bases," *Journal of Aerospace Engineering*, Vol. 15, No. 2, April, pp. 1–13.

Benaroya, H., and Ettouney, M. (1989), "Framework for the Evaluation of Lunar Base Structural Concepts," *9th Biennial SSI/Princeton Conference on Space Manufacturing*, AIAA, Washington, DC, pp. 297–302.

Benaroya, H., and Ettouney, M. (1990), "A Preliminary Framework for the Comparison of Two Lunar Base Structural Concepts," *SPACE 90, Engineering, Construction, and Operations in Space*, American Society of Civil Engineers, New York, pp. 490–499.

Benaroya, H., and Ettouney, M. (1992a), "Design Codes for Lunar Structures," *SPACE 92, Engineering, Construction, and Operations in Space*, edited by W. Sadeh, American Society of Civil Engineers, New York, pp. 1–12.

Bernold, L. E. (1991), "Experimental Studies on Mechanics of Lunar Excavation," *Journal of Aerospace Engineering*, Vol. 4, No. 1, pp. 9–22.

Broad, W. J., (1989), "Lab Offers to Develop an Inflatable Space Base," *New York Times*, 14 Nov.

Burke, J. D. (1985), "Merits of a Lunar Polar Base Location," *Lunar Bases and Space Activities of the 21st Century, Proceedings of the Lunar and Planetary Institute*, Houston, TX, pp. 77–84.

Burns, J. O., Duric, N., Taylor, G. J., and Johnson, S. W. (1990), "Astronomy on the Moon," *Scientific American*.

Centers for Disease Control (2001), "Fatal Occupational Injuries—United States 1980–1997," *MMWR Weekly*, Vol. 50, No. 16, April 27, pp. 317–320, http://www.cdc.gov/mmwr/preview/mmwrht ml/mm5016a4.htm [retrieved Oct. 2002].

Chow, P. Y., and Lin, T. Y. (1988). "Structures for the Moon," *SPACE 88 Engineering, Construction, and Operations in Space, Proceedings of the ASCE*, edited by S. W. Johnson, New York, pp. 362–374.

Chow, P. Y., and Lin T. Y. (1989), "Structural Engineer's Concept of Lunar Structures," *Journal of Aerospace Engineering*, Vol. 2, No. 1, Jan., pp. 1–9.

Chua, K. M., and Johnson, S. W. (1991), "Foundation, Excavation and Radiation Shielding Emplacement Concepts for a 16-Meter Large Lunar Telescope," Society of Photo-Optical Instrumentation Engineers Proceedings, Vol. 1494.

Chua, K. M., and Johnson, S. W. (1998), "Martian and Lunar Cold Region Soil Mechanics Considerations," *Journal of Aerospace Engineering*, Vol. 11, No. 4, Oct., pp. 138–147.

Chua, K. M., Johnson, S. W., and Nein, M. E. (1993), "Structural Concepts for Lunar-Based Astronomy," *Applied Mechanics Review*, Vol. 46, No. 6, June 1993, pp. 336–357.

Chua, K. M., Johnson, S. W., and Sahu, R. (1992), "Design of a Support and Foundation for a Large Lunar Optical Telescope," *Proceedings of ASCE SPACE 92 Conference: Engineering, Construction and Operations in Space III*, edited by S. W. Johnson, ASCE, New York.

Chua, K. M., Xu, L., and Johnson, S. W. (1994), "Numerical Simulations of Structure-Regolith Interactions," *Computer Methods and Advances in Geomechanics*, edited by H. J. Siriwardane and M. M. Zaman, Vol. II, Balkema.

Chua, K. M., Yuan, Z., and Johnson, S. W. (1990), "Foundation Design of a Large Diameter Radio Telescope on the Moon," *Proceedings of SPACE 90 Conference: Engineering, Construction and Operations in Space*, edited by S. W. Johnson, ASCE, New York, pp. 707–716.

Cohen, M. M. (1996), "First Mars Outpost Habitation Strategy," *Strategies for Mars: A Guide to Human Exploration*, Vol. 86: Science and Technology Series, American Astronautical Society, pp. 465–498.

Cohen, M. M. (1997), "Design Research Issues for an Interplanetary Habitat," Society of Automotive Engineers, Paper 972485.

Cohen, M. M., and Kennedy, K. J. (1997), "Habitats and Surface Construction Technology and Development Roadmap," *Government Sponsored Programs on Structures Technology*, edited by A. K. Noor and J. B. Malone, NASA CP-97–20624, NASA, Washington, D.C., July.

Collins, P. (2002), "Space Hotels: Civil Engineering's New Frontier," *Journal of Aerospace Engineering*, Vol. 15, No. 1, pp. 10–19.

Collins, P. (2006), "*The Economic Benefits of Space Tourism, JBIS*, Vol. 59, pp. 400–410.

Criswell, D. (1972), "Lunar Dust Motion," *Proceedings of 3rd Lunar Science Conference*, NASA, Washington, D.C., pp. 2671–2680.

Criswell, M. E., Sadeh, W. Z., and Abarbanel, J. E. (1996), "Design and Performance Criteria for Inflatable Structures in Space," *SPACE 96, Engineering, Construction, and Operations in Space, Proceedings of the ASCE*, edited by S. W. Johnson, New York, pp. 1045–1051.

Drake, R. M., and Richter, P. J. (1990), "Design and Construction of a Lunar Outpost Assembly Facility," *SPACE 90 Engineering, Construction, and Operations in Space, Proceedings of the ASCE*, edited by S. W. Johnson, New York, pp. 449–457.

Duke, M. B. (1994), "Consideration for the Utilization of In-Situ Resources in Life Support Systems on the Moon," Society of Automotive Engineers, Paper 941258, July.

Duke, M., and Benaroya, H. (1993), "Applied Mechanics of Lunar Exploration and Development," Special Section: Applied Mechanics of a Lunar Base, *Applied Mechanics Reviews*, Vol. 46, No. 6, pp. 272–277.

Ehrike, K. A. (1985), "Lunar Industrialization and Settlement—Birth of a Polyglobal Civilization," *Lunar Bases and Space Activities of the 21st Century*, edited by W. Mendell, Lunar and Planetary Inst., Houston, TX, pp. 827–855.

Eichold, A. (1996), "Conceptual Design of a Crater Lunar Base," Society of Automotive Engineers, Paper 961464, July.

Eichold, A. (2000), "Conceptual Design of a Crater Lunar Base," *Proceedings of the Return to the Moon II*, AIAA, Reston, VA, pp. 126–136.

Ettouney, M., and Benaroya, H. (1992), "Regolith Mechanics, Dynamics, and Foundations," *Journal of Aerospace Engineering*, Vol. 5, No. 2, pp. 214–229.

Ferrall, J. F., Ganapathi, G. B., Rohatgi, N. K., and Seshan, P. K. (1994), "Life Support Systems Analysis and Technical Trades for a Lunar Outpost," NASA TM 109927.

Fitch, J. M., and Bobenhausen, W. (1999), *American Building: The Environmental Forces That Shape It*, Oxford Univ. Press, New York.

Fujii, T., Midorikawa, Y., Shiba, M., and Nitta, K. (1990), "Human Requirements for Quality of Life in Lunar Base," Society of Automotive Engineers Paper 901207; *Journal of Aerospace*, pp. 365–375.

Galloway, R. G., and Lokaj, S. (eds.) (1994), *Engineering, Construction, and Operations in Space, Proceedings of the ASCE*, New York.

Galloway, R. G., and Lokaj, S. (eds.) (1998), *Engineering, Construction, and Operations in Space, Proceedings of the ASCE*, New York.

Goodings, D. J., Lin, C., Dick, R., Fourney, W. L., and Bernold, L. E. (1992), "Modeling the Effects of Chemical Explosives for Excavation on the Moon," *Journal of Aerospace Engineering*, Vol. 5, No. 1, Jan. pp. 44–58.

Gracia, V., and Casanova, I. (1998), "Sulfur Concrete: A Viable Alternative for Lunar Construction," *SPACE 98 Engineering, Construction, and Operations in Space, Proceedings of the ASCE*, New York, pp. 585–591.

Halajian, J. D. (1964), "Soil Behavior in a Low and Ultrahigh Vacuum," Contribution 64-WA/AV-14 to Winter Annual Meeting, American Society of Mechanical Engineers, New York.

Happel, J. A. (1992a), "The Design of Lunar Structures Using Indigenous Construction Materials," Master's Thesis, Civil Engineering Dept., Univ. of Colorado, Boulder, CO.

Happel, J. A. (1992b), "Prototype Lunar Base Construction Using Indigenous Materials," *SPACE 92 Engineering, Construction, and Operations in Space, Proceedings of the ASCE*, edited by W. Sadeh, New York, pp. 112–122.

Harper, L. D., and Connell, K. M. (1990), "The Lunar Outpost: Testbed for Colonization and Extraterrestrial Evolution," AIAA Paper 90–3850, Sept.

Harrison, R. A. (1992), "Cylindrical Fabric-Confined Soil Structures," *SPACE 92 Engineering, Construction, and Operations in Space, Proceedings of the ASCE*, New York, pp. 123–134.

Hempsell, M. (2006), "Space Tourism in the Context of a Diverse Market," *JBIS*, Vol. 59, pp. 411–416.

Hoffman, S. J., and Niehoff, J. C. (1985), "Preliminary Design of a Permanently Manned Lunar Surface Research Base, Lunar Bases and Space Activities of the 21st Century," *Proceedings of the Lunar and Planetary Institute*, pp. 69–76.

Hypes, W. D., and Wright, R. L. (1990), "A Survey of Surface Structures and Subsurface Developments for Lunar Bases," *SPACE 90 Engineering, Construction, and Operations in Space, Proceedings of the ASCE*, edited by S. W. Johnson, New York, pp. 468–479.

Joachim, C. E. (1988), "Extraterrestrial Excavation and Mining with Explosives," *Engineering, Construction and Operations in Space*, edited by S. W. Johnson and J. P. Wetzel, American Society of Civil Engineers, New York, pp. 333–343.

Johnson, S. W. (1964), "Criteria for the Design of Structures for a Permanent Lunar Base," Ph.D. Dissertation, Dept. of Civil Engineering, Univ. of Illinois, Urbana, IL.

Johnson, S. W. (1989), "Extraterrestrial Facilities Engineering," *1989 Yearbook*, Encyclopedia of Physical Science and Technology, Academic Press, New York.

Johnson, S. W. (ed.) (1996), *Engineering, Construction, and Operations in Space, Proceedings of the ASCE*, American Society of Civil Engineers, New York.

Johnson, S. W., and Chua, K. M. (1992), "Assessment of the Lunar Surface and in Situ Materials to Sustain Construction-Related Applications," Joint Workshop (DOE/LANL, NASA/JSC and LPI) on New Technologies for Lunar Resource Assessment, Santa Fe, New Mexico, April.

Johnson, S. W., and Chua, K. M. (1993), "Properties and Mechanics of the Lunar Regolith," *Applied Mechanics Review*, Vol. 46, No. 6, June.

Johnson, S. W., Chua, K. M., and Burns, J. O. (1995b), "Lunar Dust, Lunar Observatories, and Other Operations on the Moon," *Journal of British Interplanetary Society*, Vol. 48, pp. 87–92.

Johnson, S. W., Chua, K. M., and Carrier, III, W. D. (1995a), "Lunar Soil Mechanics," *Journal of British Interplanetary Society*, Vol. 48, pp. 43–48.

Johnson, S. W., and Leonard, R. S. (1985), "Evolution of Concepts for Lunar Bases," *Lunar Bases and Space Activities of the 21st Century, Proceedings of the Lunar and Planetary Institute*, pp. 47–56.

Johnson, S. W., Smith, J. A., Franklin, E. G., Moraski, L. K. and Teal, D. J. (1969), "Gravity and Atmosphere Pressure Effects on Crater Formation in Sand," *Journal of Geophysical Research*, Vol. 74, No. 20, Sept. 15, pp. 4838–4850.

Johnson, S. W., and Wetzel, J. P. (eds.) (1988), *Engineering, Construction, and Operations in Space, Proceedings of the ASCE*, American Society of Civil Engineers, New York.

Johnson, S. W., and Wetzel, J. P. (eds.) (1990a), *Engineering, Construction, and Operations in Space, Proceedings of the ASCE*, New York.

Johnson, S. W., and Wetzel, J. P. (1990b), "Lunar Astronomical Observatories: Design Studies," *Journal of Aerospace Engineering,* Vol. 3, No. 4, Oct.

Johnson, S. W., and Wetzel, J. P. (1990c), "Science and Engineering for Space: Technologies from SPACE 88," *Engineering, Construction, and Operations in Space, Proceedings of the ASCE*, American Society of Civil Engineers, New York.

Keller, J. R., and Ewert, M. K. (2000), "Lunar Dust Contamination Effects on Lunar Base Thermal Control Systems," Society of Automotive Engineers, Paper 2000–01–2405, July, p. 7.

Kelso, H. M., Hopkins, J., Morris, R., and Thomas, M. (1988), "Design of a Second Generation Lunar Base," *SPACE 88 Engineering, Construction, and Operations in Space, Proceedings of the ASCE*, edited by S. W. Johnson, American Society of Civil Engineers, New York, pp. 389–399.

Kennedy, K. J. (1992), "A Horizontal Inflatable Habitat for SEI, *"SPACE 92 Engineering, Construction, and Operations in Space, Proceedings of the ASCE*, American Society of Civil Engineers, New York, pp. 135–146.

Kennedy, K. J. (1999), "ISS TransHab: Architecture Description," Society of Automotive Engineers, Paper 1999–01–2143, July.

Khalili, E. N. (1989), "Lunar Structures Generated and Shielded with On-Site Materials," *Journal of Aerospace Engineering,* Vol. 2, No. 3, July, pp. 119–129.

King, C. B., Butterfield, A. J., Hyper, W. D., and Nealy, J. E. (1989), "A Concept for Using the External Tank from a NSTS for a Lunar Habitat," *Proceedings, 9th Biennial SSI/Princeton Conference on Space Manufacturing*, pp. 47–56.

Koscheyev, V. S., Greaves, I. A., Leon, G. R., Hubel, A., and Nelson, E. D., (1996), "Comfort and Heat Control During Extended Space Flights," Society of Automotive Engineers, Paper 961538, July.

Leonard, R. S., and Johnson, S. W. (1988), "Sulfur-Based Construction Materials for Lunar Construction," *SPACE 88 Engineering, Construction, and Operations in Space, Proceedings of the ASCE*, New York, pp. 1295–1307.

Lin, T. D. (1987), "Concrete for Lunar Base Construction," *Concrete International (ACI)*, Vol. 9, No. 7.

Lin, T. D., Senseney, J. A., Arp, L. D., and Lindbergh, C. (1989), "Concrete Lunar Base Investigation," *Journal of Aerospace Engineering*, Vol. 2, No. 1, Jan., pp. 10–19.

Lowman, P. D. (1985), "Lunar Bases: A Post-Apollo Evaluation," *Lunar Bases and Space Activities of the 21st Century, Proceedings of the Lunar and Planetary Institute*, Houston, TX, pp. 35–46.

Mangan, J. J. (1988), "The Expandable Platform as a Structure on the Moon," *SPACE 88 Engineering, Construction, and Operations in Space, Proceedings of the ASCE*, edited by S. W. Johnson, American Society of Civil Engineers, New York, pp. 375–388.

Mendell, W. W. (ed.) (1985), *Lunar Bases and Space Activities of the 21st Century*, Lunar and Planetary Inst., Houston.

Namba, H., Ishikawa, N., Kanamori, H., and Okada, T. (1988b), "Concrete Production Method for Construction of Lunar Bases," *SPACE 88 Engineering, Construction, and Operations in Space, Proceedings of the ASCE*, American Society of Civil Engineers, New York, pp. 169–177.

Namba, H., Yoshida, T., Matsumoto, S., Sugihara, K., and Kai, Y. (1988a), "Concrete Habitable Structure on the Moon," *SPACE 88 Engineering, Construction, and Operations in Space, Proceedings of the ASCE*, edited by S. W. Johnson, New York, pp. 178–189.

NASA (1989), "90-Day Study on Human Exploration of the Moon and Mars," NASA Rept., Nov.

NASA (2002), National Space Science Data Center, NASA Goddard Space Flight Center, http://nssdc.gsfc.nasa.gov/planetary/ice/ice _moon.html [retrieved 23 July, 2002].

Newman, D. J., Alexander, H. L., and Webbon, B. W. (1994), "Energetics and Mechanics for Partial Gravity Locomotion," *Aviat. Space and Environ. Med.*, Vol. 65, pp. 815–823.

Nowak, P. S., Criswell, M. E., and Sadeh, W. Z. (1990), "Inflatable Structures for a Lunar Base," *SPACE 90 Engineering, Construction, and Operations in Space, Proceedings of the ASCE*, American Society of Civil Engineers, New York, pp. 510–519.

Nowak, P. S., Sadeh, W. Z., and Criswell, M. E. (1992), "An Analysis of an Inflatable Module for Planetary Surfaces," *SPACE 92 Engineering, Construction, and Operations in Space, Proceedings of the ASCE*, edited by W. Sadeh, American Society of Civil Engineers, New York, pp. 78–87.

Office of the Chief of Engineers (1963), "Special Study of the Research and Development Effort Required to Provide a US Lunar Construction Capability," Dept. of the Army.

Okumura, M., Ohashi, Y., Ueno, T., Motoyui, S., and Murakawa, K. (1994), "Lunar Base Construction Using the Reinforced Earth Method with Geotextiles," *SPACE 94 Engineering, Construction, and Operations in Space, Proceedings of the ASCE*, American Society of Civil Engineers, New York, pp. 1106–1115.

Parkinson, B. (2006), "A Parametric Investigation of the Economics of Space Tourism," *JBIS*, Vol. 59, pp. 417–421.

Pieniazek, L. A., and Toups, L. (1990), "A Lunar Outpost Surface Systems Architecture," *SPACE 90 Engineering, Construction, and Operations in Space, Proceedings of the ASCE*, edited by S. W. Johnson, American Society of Civil Engineers, New York, pp. 480–489.

Podnieks, E. R. (1990), "Lunar Mining Outlook," AIAA Paper 90–3751, Sept.

Prael, R. E., Strottman, D. D., Strniste G. F., and Feldman W. C. (1990), "Radiation Exposure and Protection for Moon and Mars Missions," Los Alamos National Lab., Rept. LA-UR-90–1297, Los Alamos, NM, April.

Richelson, J. T. (2000), "Shooting the Moon," *Bulletin of the Atomic Scientists*, Sept./Oct. pp. 22–27.

Richter, P. J., and Drake, R. M. (1990), "A Preliminary Evaluation of Extraterrestrial Building Systems," *SPACE 90 Engineering, Construction, and Operations in Space, Proceedings of the ASCE*, edited by S. W. Johnson, American Society of Civil Engineers, New York, pp. 409–418.

Sadeh, W. Z., and Criswell, M. E. (1994), "A Generic Inflatable Structure for a Lunar/Martian Base," *SPACE 94 Engineering, Construction, and Operations in Space, Proceedings of the ASCE*, edited by S. W. Johnson, American Society of Civil Engineers, New York, pp. 1146–1156.

Sadeh, W. Z., Sture, S., and Miller, R. J. (eds.) (1992), *Engineering, Construction, and Operations in Space, Proceedings of the ASCE*, edited by S. W. Johnson, American Society of Civil Engineers, New York.

Sargent, R., and Hampson, K. (1996), "Challenges in the Construction of a Lunar Base," *SPACE 96, Engineering, Construction, and Operations in Space, Proceedings of the ASCE*, American Society of Civil Engineers, New York, pp. 881–888.

Schroeder, M. E., and Richter, P. J. (1994), "A Membrane Structure for a Lunar Assembly Building," *SPACE 94 Engineering, Construction, and Operations in Space, Proceedings of the ASCE*, edited by S. W. Johnson, American Society of Civil Engineers, New York, pp. 186–195.

Schroeder, M. E., Richter, P. J., and Day, J. (1994), "Design Techniques for Rectangular Lunar Modules," *SPACE 94 Engineering, Construction, and Operations in Space, Proceedings of the ASCE*, American Society of Civil Engineers, New York, pp. 176–185.

Seiheimer, H. E., and Johnson, S. W. (1969), "Adhesion of Comminuted Basalt Rock to Metal Alloys in Ultrahigh Vacuum," *Journal of Geophysical Research*, Vol. 74, No. 22, Oct. pp. 5321–5330.

Sherwood, B. (1990a), "Site Constraints for a Lunar Base," *SPACE 90 Engineering, Construction, and Operations in Space, Proceedings of the ASCE*, edited by S. W. Johnson, American Society of Civil Engineers, New York, pp. 984–993.

Sherwood, B. (1990b), "Robotic Lunar Surface Operations: Engineering Analysis for the Design, Emplacement, Checkout and Performance of Robotic Lunar Surface Systems," Boeing Co., Rept. D615–11901, NASA, Rept. NAS2–12108, Jan.

Silberberg, R., Tsao, C. H., Adams, Jr., J. H., and Letaw, J. R. (1985), "Radiation Transport of Cosmic Ray Nuclei in Lunar Material and Radiation Doses," *Lunar Bases and Space Activities of the 21st Century*, edited by W. W. Mendell, Lunar and Planetary Inst.

Simmerer, S. J. (1988), "Preparing to Bridge the Lunar Gap," *Journal of Aerospace Engineering*, Vol. 1, No. 2, April.

Simonson, L. C., Nealy, J. E., and Townsend, L. W. (1992), "Concepts and Strategies for Lunar Base Radiation Protection: Prefabricated Versus In-Situ Materials," Society of Automotive Engineers, Paper 921370; *Journal of Aerospace*, pp. 1348–1359.

Slane, F. A. (1994), "Engineering Implications of Levitating Lunar Dust," *SPACE 94 Engineering, Construction, and Operations in Space, Proceedings of the ASCE*, American Society of Civil Engineers, New York, pp. 1097–1105.

Smith, A. (1993), "Mechanics of Materials in Lunar Base Design," Applied Mechanics Reviews, edited by H. Benaroya, Vol. 46, No. 3, 1993, pp. 268–271.

Strenski, D., Yankee, S., Holasek, R., Pletka, B., and Hellawell, A. (1990), "Brick Design for the Lunar Surface," *SPACE 90 Engineering, Construction, and Operations in Space, Proceedings of the ASCE*, American Society of Civil Engineers, New York, pp. 458–467.

Sullivan, T. A. (1990), "Process Engineering Concerns in the Lunar Environment," AIAA Paper 90–3753, Sept.

Thangavelu, M. (1990), "MALEO: Modular Assembly in Low Earth Orbit, An Alternative Strategy for Lunar Base Development," 41st Congress of the International Astronautical Federation, Paper IAF90–443, Oct.

Toups, L. (1990), "A Survey of Lunar Construction Techniques," *SPACE 90, Engineering, Construction, and Operations in Space*, edited by S. W. Johnson, American Society of Civil Engineers, New York.

Vanderbilt, M. D., Criswell, M. E., and Sadeh, W. Z. (1988), "Structures for a Lunar Base," *SPACE 88 Engineering, Construction, and Operations in Space, Proceedings of the ASCE*, edited by S. W. Johnson, American Society of Civil Engineers, New York, pp. 352–361.

Van Susante, P. (2002), "Scenario Description of the Construction of a Lunar South Pole Infrared Telescope," *SPACE 02, Engineering, Construction, and Operations in Space*, edited by S. W. Johnson, American Society of Civil Engineers, New York.

Vanzani, V., Marzari, F., and Dotto, E. (1997), "Micrometeoroid Impacts on the Lunar Surface," *Lunar and Planetary Science Conference XXVIII*, Lunar and Planetary Inst., Houston, p. 2, http://www.lpi.usra.edu/meetings/lpsc97 [retrieved 23 July, 2002].

Walker, S. T., Alexander, R. A., and Tucker, S. P. (1995), "Thermal Control on the Lunar Surface," *JBIS*, Vol. 48, pp. 27–32.

Watson, P. M. (1988), "Explosives Research for Lunar Applications: A Review," *Engineering, Construction and Operations in Space*, edited by S. W. Johnson and J. P. Wetzel, American Society of Civil Engineers, New York, pp. 322–331.

Woodcock, G. R., et al., "Spa ce Transfer Concepts and Analysis for Exploration Missions, Final Technical Report, Phase 1," Boeing Co. D615–10030-2, 30 April 1991; NASA, NAS8–37857.

Yong, J. F., and Berger, R. L. (1988), "Cement-Based Materials for Planetary Materials," *SPACE 88, Engineering, Construction, and Operations in Space*, edited by S. W. Johnson, American Society of Civil Engineers, New York.

Bibliography

Abarbanel, J. E., and Criswell, M. E. (1997), "Design Development of an Inflatable Module for a Lunar/Martian Base," Society of Automotive Engineers, Paper 972487.

Aldrin, B., and McConnell, M. (1989), *Men from Earth*, Bantam Books, New York.

Alred, J. et al. (1989), "Lunar Outpost," NASA JSC-23613, Aug.

Balin, M. G., Likens, W. C., Finn, C. K., Bilardo, V. J., Jr., and Ng, Y. S. (1991), "Analysis of an Initial Lunar Outpost Life Support System Preliminary Design," Society of Automotive Engineers, Paper 911395; also *SAE 1991 Transactions, Journal of Aerospace*, pp. 999–1015.

Benaroya, H. (1998), "Economic and Technical Issues for Lunar Development," *Journal of Aerospace Engineering*, Vol. II, No. 4, Oct. pp. 111–118.

Benaroya, H., and Ettouney, M. (1992b), "Design and Construction Considerations for Lunar Outpost," *Journal of Aerospace Engineering*, Vol. 5, No. 3, pp. 261–273.

Brooks, C. G., Grimwood, J. M., and Swenson, Jr., L. S. (1979), "Chariots for Apollo: A History of Manned Lunar Spacecraft," NASA SP-4205.

Brooks, R. A. (1986), "A Robust Layered Control System for a Mobile Robot," *IEEE Journal of Robotics and Automation RA-2*, pp. 14–23.

Brooks, R. A. (1990), "Elephants Don't Play Chess," *Designing Autonomous Agents: Theory and Practice from Biology to Engineering and Back*, edited by P. Mae, Massachusetts Inst. of Technology Press, Cambridge, MA., pp. 3–15.

Budden, N. A. (ed.) (1994), "Catalog of Lunar and Mars Science Payloads," NASA RP 1345, Aug.

Carr, M. H., and Greeley, R. (1980), "Volcanic Features of Hawaii: A Basis for Comparison with Mars," NASA SP-403.

Casanova, I., and Aulesa, V. (2000), "Construction Materials from In-Situ Resources on the Moon and Mars," *Proceedings of Seventh International Conference and Exposition on Engineering, Construction, Operations and Business in Space*, American Society of Civil Engineers, pp. 638–644.

Celentano, J. T., and Amorelli, D. (1965), "Design Requirements for Manned Orbital Bases and Lunar Bases," *2nd Space Congress: New Dimensions in Space Technology, Canaveral Council of Technical Societies*, Cocoa Beach, FL, pp. 743–752.

Conway, L., Volz, R. A., and Walker, M. W. (1990), "Teleautonomous Systems: Projecting and Coordinating Intelligent Actions at a Distance," *IEEE Transactions on Robotics and Automation*, Vol., 6, No. 2, April, pp. 146–158.

Dick, R. D., Fourney, W. L., Goodings, D. J., Lin, Ch. P., and Bernold, L. E. (1992), "Use of Explosives on the Moon," *Journal of Aerospace Engineering.*, Vol. 5, No. 1, Jan., pp. 59–65.

Elrod, M. (1995), "Considerations of a Habitat Design," *JBIS*, Vol 48, pp. 38–42.

Gitelson, J. I., and Shepelev, Ye. Ya. (1996), "Creation of Life Support Systems for the Lunar Outpost and Planetary Bases: History, Present State of Research in Russia (Former Soviet Union) and Prospects," Society of Automotive Engineers, Paper SAE 961553.

Hart, P. A., Howe, S. D., Johnson, S. W., Leigh, G. G., and Leonard, R. S. (1990), "A Center for Extraterrestrial Engineering and Construction (CETEC)," *SPACE 90 Engineering, Construction, and Operations in Space, Proceedings of the ASCE*, New York, pp. 1198–1205.

Horiguchi, T., Scki, N., Yoneda, T., Hoshi, T., and Lin, T. D. (1998), "Behavior of Simulated Lunar Cement Mortar in Vacuum Environment," *Proceedings of Sixth International Conference and Exposition on Engineering, Construction, and Operations in Space*, American Society of Civil Engineers, pp. 571–576.

Howe, A. S. and Howe, J. W. (2000), "Applying Construction Automation Research to Extraterrestrial Building Projects," Society of Automotive Engineers, Paper 2000–01–2465.

Huang, X., and Bernold, L. E. (1993), "Towards an Adaptive Control Model for Robotic Backhoe Excavation," Transportation Research Record, No. 1406, National Research Council, National Academy Press, Washington, D.C. Oct. pp. 20–24.

Huang, X., and Bernold, L. E. (1994), "Control Model for Robotic Backhoe Excavation and Obstacle Handling," *Proceedings of the ASCE Specialty Conference on Robotics for Challenging Environments*, pp. 123–130.

Iwata, T. (1994), "Evolutionary Scenario of Lunar Manufacturing," *JBIS*, Vol. 47, pp. 539–542.

Kelly, T. J. (1990), "A Review of the Apollo Lunar Program and Its Lessons for Future Space Missions," AIAA Paper 90–3617, Sept.

Kemurdjian, A., and Khakhanov, U. A. (2000), "Development of Simulation Means for Gravity Forces," *Proceedings of Fourth International Conference and Exposition on Robotics for Challenging Environments*, pp. 220–225.

Lewis, J. S., and Lewis, R. A. (1987), *Space Resources: Breaking the Bonds of Earth*, Columbia Univ. Press, New York.

Lin, Ch. P., Goodings, D. J., Bernold, L. E., Dick, R. D., and Fourney, W. L. (1994), "Model Studies of Effects on Lunar Soil of Chemical Explosions," *Journal of Geotechnical Engineering*, Vol. 120, No. 10, Oct., pp. 1684–1703.

Mankins, J. C. (1995), "Strategies to Begin and Sustain a Lunar Exploration Program—Revisited," AIAA Paper 95–4060, Sept.

Moore, G. T., and Huebner-Moths, J. (1991), "Genesis II Advanced Lunar Outpost Human Factors Design Response," *IDEEA ONE*, pp. 517–531.

Moore, P. (1976), *Guide to the Moon*, Butterworth Press, London.

Nash, D. B., et al. (1989), "Science Exploration Opportunities for Manned Missions to the Moon, Mars, Phobos, and an Asteroid," Jet Propulsion Lab., 89–29, Pasadena, CA, June.

National Council on Radiation Protection (2000), "Radiation Protection Guidance for Activities in Low Earth Orbit," NCRP, Rept. 132, Dec.

Nelson, T. J., Olson, M. R., and Wood, H. C. (1998), "Long Delay Telecontrol of Lunar Equipment," *Proceedings of Sixth International Conference and Exposition on Engineering, Construction, and Operations in Space*, American Society of Civil Engineers, pp. 477–484.

Richter, T., Lorenc, S. J., and Bernold, L. E. (1998), "Cable Based Robotic Work Platform for Construction," *15th International Symposium on Automation and Robotics in Construction*, pp. 137–144.

Sadeh, W. Z., Criswell, M. E., and Abarbanel, J. E. (1996), "An Inflatable Module for Use on Mars," *The Case for Mars VI*, edited by K. R. McMillan, Science and Technology Series, Vol. 98, San Diego, Univelt, Inc., pp. 303–308.

Space Studies Board (2000), *Radiation and the International Space Station: Recommendations to Reduce Risk*, National Academy Press, Washington, D.C.

Wilson, J. W., et. al. (1991), "Transport Methods and Interactions for Space Radiations," NASA RP-1257.

Wilson, J. W., et. al. (1999), "Astronaut Exposures to Ionizing Radiation in a Lightly-Shielded Spacesuit," Society of Automotive Engineers, Paper 1999–01–2173.

Wilson, J. W., Miller, J., Konradi, A., and Cucinotta, F. A. (1997), "Shielding Strategies for Human Space Exploration," NASA CP 3360, Dec.

16

lunar habitat concepts

LARRY TOUPS
AND
KRISS J. KENNEDY

INTRODUCTION

THIS CHAPTER PRESENTS three integrated habitat concepts for first-generation lunar outposts. These concepts were developed by the NASA Constellation Lunar Architecture Team (CxAT-Lunar) over the period 2007–2008 in preparation for the Lunar Capability Concept Review program milestone. Thus they capture contemporary thinking by the most advanced space habitation team working today: a team of NASA space architects and engineers developing approaches both feasible and appropriate for initial lunar outpost buildup to implement the U.S. Vision for Space Exploration.

CxAT-Lunar habitation study objectives are to 1) identify promising habitation options that meet mission architecture objectives; 2) identify desirable habitation features; 3) understand operational constraints based on different habitation options; and 4) understand the cost and risks of different habitation options.

All three habitat concepts are Class I/II hybrids as defined in Chapter 15. They contain facilities and equipment for living support, science laboratories, extravehicular activity (EVA), logistics, and maintenance. They accommodate known pre-Phase A level requirements for mission exploration objectives; launch packaging and lander capability; delivery, unloading, and deployment; surface operations (crew, mission, EVA, science, and logistics); and sustainability, program risk, and technology readiness.

The chapter describes the concepts' configurations and functionality in several areas. Crew operations includes basic crew accommodations such as sleeping, eating, hygiene, and stowage. EVA operations include EVA capability in addition to the suitport airlock function, such as redundant airlock(s), suit maintenance, spares stowage, and suit stowage. Logistics operations include enhanced accommodations for 180-day duty tours, such as closed-loop life-support systems hardware, consumables stowage, spares stowage, interconnection to other habitable units, and a common interface mechanism for future growth and mating to pressurized rovers. Mission and science operations include outpost equipment and capabilities such as intra-vehicular activity glove box, life support, medical operations, and enhanced autonomy.

OVERVIEW

CxAT-Lunar surface mission campaign analysis focused on three primary scenarios, all of which have the chief purpose of establishing an outpost:

1) *Lunar scenario 1.0* (LS1) investigates sustaining a crew of four for six months with full outpost capability and the ability to perform long surface mission excursions using large mobility systems. Figures 1 and 2 depict notional outpost configurations for the LS1 scenario, showing the first two habitat types described next: horizontal cylinder and inflated torus.
2) *Lunar scenario 2.0* (LS2) investigates a mobile architecture approach focused on early exploration using two small pressurized rovers and a mobile logistics support capability. (Because of its emphasis on roving exploration rather than outpost buildup, LS2 habitat concepts are not discussed next.)
3) *Lunar scenario 3.0* (LS3) investigates a variation of LS1 that delivers a core habitation capability that is augmented later to achieve full outpost capability (Figure 3). This modular approach is the third habitat type that will be described.

This material is declared a work of the U.S. Government and is not subject to copyright protection in the United States.

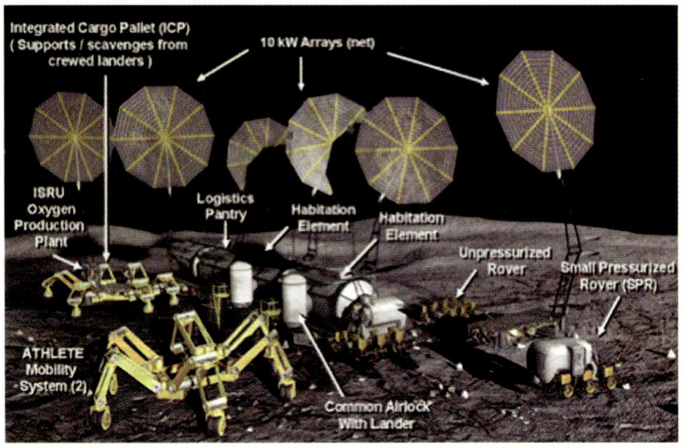

FIGURE 1 LS1 outpost concept, incorporating the horizontal-cylinder habitat type. Note various mobility and support systems: ATHLETE chassis, small pressurized rover, and power and supply unit.

FIGURE 2 LS1 outpost concept, incorporating the inflatable-torus habitat.

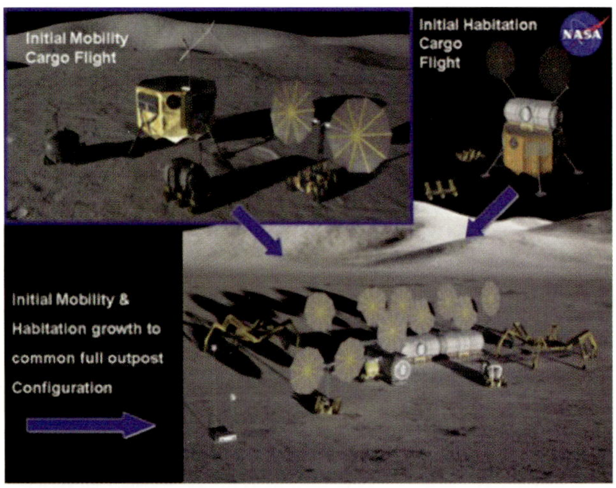

FIGURE 3 LS2 initial mobility concept and LS3 habitation emphasis concept both build up to a full outpost. LS3 concept (upper right) incorporates the core habitat type.

LS1 phasing is driven by developing a fully functional outpost as early as possible. Although sized for a polar location, its combination of robust energy storage and ability to relocate any outpost element enables deployment anywhere on the Moon. Mobility capabilities can be tailored to science objectives as needed. Two alternative habitat concepts support LS1: a hard-wall, aluminum horizontal cylinder and a larger inflatable torus.

LS1 is based on lunar surface system study findings from 2004–2006 and on capabilities of the *Altair* lander configuration under study in 2007. That lander configuration features a large, expendable descent module with flat top-deck used for either crewed ascent module or lunar surface systems payload cargo. With the lander deck more than 6 m above the lunar surface, access to and removal of payloads from the deck becomes a major systems architecture driver.

LS2 and LS3 are derivative concepts that attempt to balance early functionality with program affordability. These scenarios assume two missions per year on average, alternating between crewed and cargo missions. Buildup is structured so that it can be paused at any time to accommodate technical contingencies, funding stretch-out, or changed objectives (e.g., a sortie mode where the purpose is to visit widely dispersed sites before investing more permanent infrastructure at a given location). Modular buildup enhances the opportunity for international partners to contribute elements and systems, yielding an international lunar base. Outpost functionality is allocated so that a loss of any one element does not result in significant loss of critical outpost capability.

Both approaches build up to a capability similar to LS1 but with slightly less habitation volume and no ability to accommodate a crew during long eclipse periods. For both, a minimum-functionality outpost is deployed in a single cargo flight featuring either mobility emphasis (LS2) or core habitation emphasis (LS3). These configurations accommodate four-person crews for 14–28-day surface missions. At this point it would be feasible to introduce a "hold" in outpost buildup, deferring additional functionality until either international participation or available budget permitted. The LS2 and LS3 buildup manifests achieve similar end-states with redundant habitats, ability to sustain a crew for 180 days, and ability to support long-distance and -duration roving excursions.

All three scenarios are based on common contextual assumptions: 1) scenarios begin with a "boots-on-the-Moon" sortie mission in 2020; 2) support for an initial habitation mode and a later, outpost-complete mode with mobile exploration capability; 3) surface crew of four; 4) variable initial mission duration with eventual goal of 180-day surface stays; 5) top-loaded, flat-deck lander sized for a 10-m launch shroud, with 8.8-m-diameter payload deck 6.2 m above the surface; and 6) mobility/offloading capability based on ATHLETE (Chapter 18 in this volume discusses this mobility system). All three are also based on a design philosophy that enhances supportability. Lightweight materials and multifunction structures and packaging limit system mass. Reuse of elements, systems, components, and structural materials limits the total mass launched to the outpost. The *Altair* lander in particular could offer substantial potential for logistics avoidance if designed to be cannibalized for spares and structure to meet outpost resource needs.

HORIZONTAL-CYLINDER CONCEPT

Concept Description

The outpost habitat system comprises three common modules (one laboratory and two habitat units) plus pressurized logistics modules (PLM). Each aluminum-lithium (Al-Li) hard shell pressure vessel has 3.5-m internal diameter and 8.2-m internal length, yielding 78-m^3 pressurized volume (19.5 m^3/crew member). The floor area is about 21 m^2/element and, 64 m^2 total for all three modules together. For a four-person, 180-day surface mission the three-module outpost provides 0.33 m^3/crew member/day.

The modules are offloaded from the landers that bring them to the lunar surface for emplacement into an outpost configuration. This removes them from landing-zone hazards, positions them as desired for connection, and puts them close to the surface to facilitate EVA, pressurized rover, maintenance, and repair access.

Each module is mounted in an external structure frame that also integrates external subsystems [the power and supply unit (PSU)] to make it self-sufficient and is removed from the lander for surface emplacement as a single functional element (Figure 4). A common mating mechanism design is used to connect the modules and to dock small pressurized rovers (SPR) (Figure 5). Thermal radiators are integrated to the top of the structure frame; an articulating solar array and lunar communications terminal (LCT) antenna also deploy. The module hull is covered with multilayer insulation (MLI) for passive thermal protection and shielded by a composite fabric micrometeoroid and surface ejecta (MM/SE) protection barrier. A deployable, open-grated, aluminum isogrid dust porch and stairs provide access between habitats and surface. Because a key operations design goal is to optimize the height of the complex above the surface, the configuration shown in Figure 4 could actually be significantly lower than shown.

The Lab-1 unit is delivered first because it provides the airlock function and is needed for the initial science mission at the beginning of outpost emplacement. It contains a geological science lab, airlock, and EVA ops and maintenance provisions. The system has a 5.5 m^3 airlock with dust containment system and dust lock inside the habitat (see Chapter 17 for a detailed comparison of airlock configuration alternatives). With its PSU and LCT, Lab-1 is a self-sufficient, habitable mini-outpost (Figure 6).

Crew ops, mission ops, airlock, galley, wardroom, crew sleep area, and closed ECLS subsystems are in Hab-1. PLM-1 is retrofitted on the surface into Hab-2 to contain medical ops, biological science, exercise, and stowage functions. Each successive element is delivered with the logistics supplies needed to enable

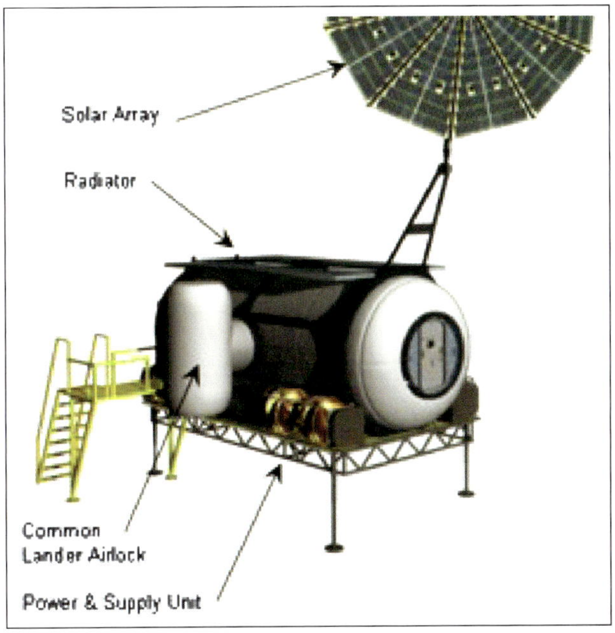

FIGURE 4 Common habitation element with integrated exterior subsystems is the basis for a modular outpost.

FIGURE 5 Small pressurized rover docks directly with Lab-1 module for IVA crew transfer.

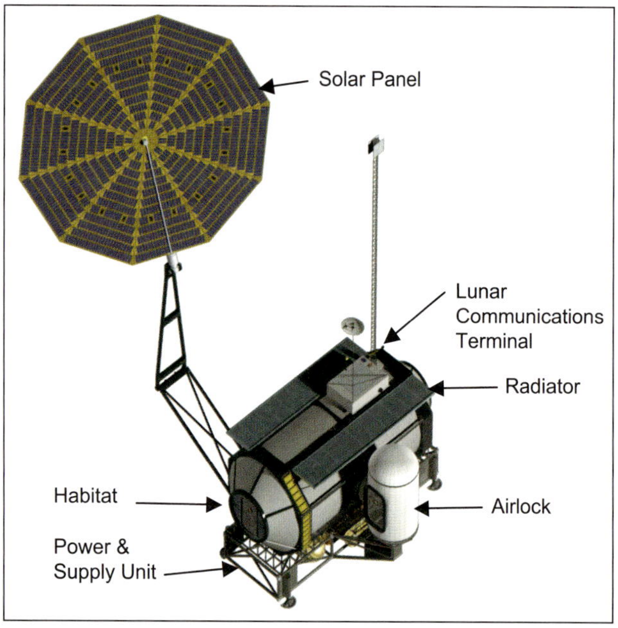

FIGURE 6 Self-contained Lab-1 element is first outpost unit.

14, 28, 45, and finally 180 crew-day missions. Stowed items are packaged in order of their use. Lab-1 carries up to 1979 kg [equivalent to 152 International Space Station (ISS) Crew Transfer Bags (CTBE)] of pressurized consumables; Hab-1 carries up to 2912 kg (224 CTBE). After initial delivery of the habitation elements, the outpost uses PLMs to deliver supplies.

Hatches are 1.0-m wide by 1.52-m tall submarine-style, pressure-assisted doors that open inward. An active/passive mating mechanism with ~1.8-m outside diameter encircles the hatch. Lab-1 and Hab-1 each have three pressurized hatches with mating ports, one on each end plus a lateral port. Hab-2 (the retrofitted PLM-1) could use the same configuration or have just two end-hatches with mating ports as would subsequent PLMs. In outpost mode (Figure 7), the modules are connected in-line (Lab-1 mates to Hab-1, which mates to Hab-2) and the element pass-through is normally left open during operations. Crew members use individual sleep bunks in Hab-1. Lab-1 and Hab-1 each devote their respective side ports to airlocks. The free end port in Lab-1 is used by the SPR, and the free end port in Hab-2 is used for PLMs after the first.

FIGURE 8 Mobile laboratory mode separates Lab-1 from outpost for remote excursions using ATHLETE and SPR.

The lab and hab can operate independently. In the mobile laboratory mode (Figure 8), the Lab-1 module/PSU element is attached to an ATHLETE mobility system, decoupled from the remaining elements of the outpost and moved to a new surface location where the crew can live while performing science operations. This mode is a flexible way to investigate remote sites.

The SPR provides speedier mobility for local excursions, backup mobility for returning to the outpost, and a way for the crew to divide into pairs while exploring. Having Lab-1 on hand provides IVA analytical

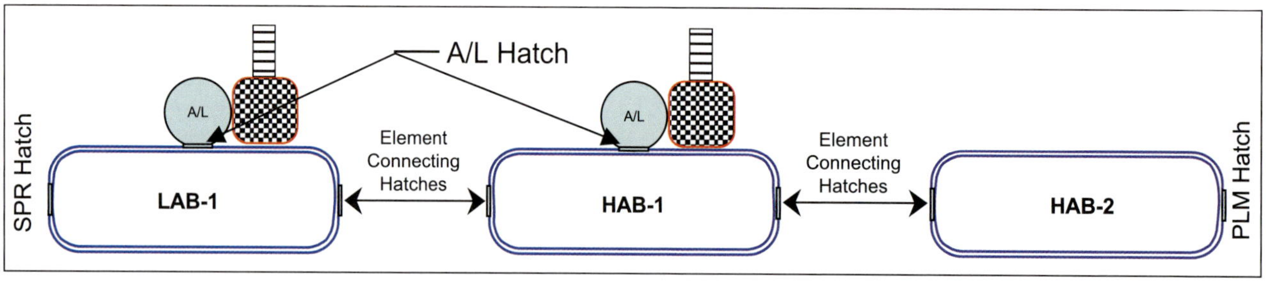

FIGURE 7 Outpost mode interconnect topology is straightforward.

laboratory capability, airlock, consumables, and volume for long excursions and fairly comfortable accommodations (Figure 9). Recumbent seat/workstations are used for crew rest and sleep periods.

Subsystems

The ECLS strategy for LS1 is to provide a distributed redundant system. Lab-1 provides a partially closed ECLS for the initial habitation when it is in mobile laboratory mode. The primary habitat (Hab-1) closes the air and water loops. Lab-1 and Hab-1 together provide redundancy. The ECLS system includes pressure control, air revitalization, water recovery and management, waste management, fire detection and suppression, and emergency subsystems. Air revitalization includes CO_2 removal and reduction, O_2 generation, trace contaminant control, ventilation and fans, airborne particulate control and monitoring, and atmosphere composition monitoring. Water recovery includes H_20 recovery from cabin condensate, gray-water collection, storage and distribution, and quality monitoring. Waste management includes urine collection and pretreatment, fecal collection, and trash collection. The waste and hygiene unit is in Hab-1 near the water recovery subsystem.

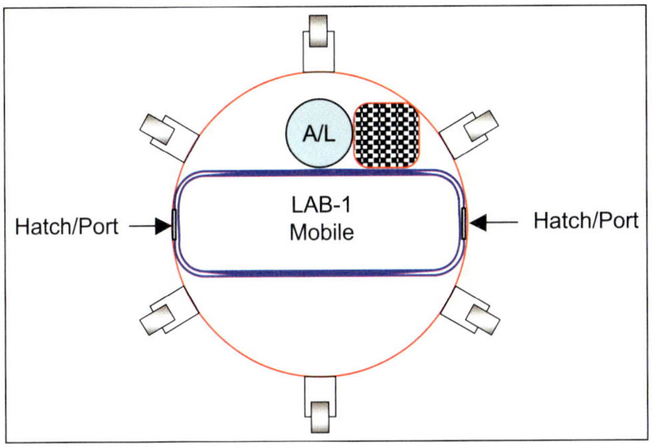

FIGURE 9 Mobile laboratory mode topology provides full-service lab capability on excursions.

Each habitat element has a power management and distribution subsystem (PMAD) that manages distribution of power to internal systems and interfaces with the external power system in the PSU. Power storage is in the PSU.

Each element has an active thermal control subsystem (ATCS) and a passive thermal control subsystem (PTCS). The dual-loop ATCS removes thermal loads from the element, gathering heat with coolant loops inside the module and transferring it to ~18 m^2 of body-mounted radiator panels atop the element. The ATCS is designed to support the thermal loads and for the polar location environment. The coolant is a 60/40 mixture of propylene glycol/water, and the lines are stainless steel. The design uses both ISS thermal control system heritage and technology developed for the Vision for Space Exploration. The PTCS reduces the thermal input from the environment through use of MLI.

The avionics system has primary and secondary computers and S-band software-defined radio subsystems (including antennas), distributed crew utility panels, EVA airlock interface panel, workstations, network server, mass memory units, networking bus and interfaces, redundancy management, and internal wireless. Avionics components are intended to be common with *Altair* lander avionics.

Radiation protection for solar proton events (SPE or solar flares) is a "water wall" surrounding the crew sleep area and filled with ~1000 kg of water (Figure 10). The crew is not protected from galactic cosmic radiation (GCR) other than incidentally by the structure, resupply tanks, subsystems, and water wall during rest/sleep intervals.

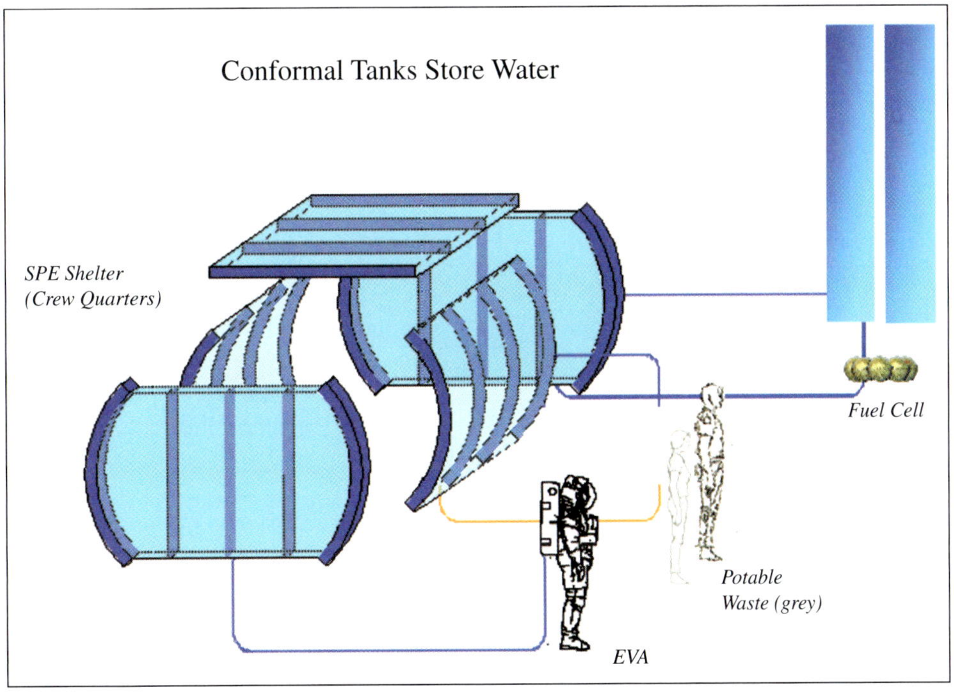

FIGURE 10 Sleep-area water wall provides solar-flare radiation shielding.

Internal Architecture

The horizontal-cylinder modules are smaller than ISS modules because their interior needs to be optimized for reach, clearance, and access in a gravity field. The internal layout is zoned by function in an effort to separate the working lab (noisy/dirty) from the living hab (quiet/clean) areas. However, EVA functions are integrated into both Lab-1 and Hab-1 units because of volume and mass constraints. Such commonality can have cost-avoidance benefits. Subsystems are distributed among the modules as volume and incremental functionality and redundancy permit. For example, avionics and power management are fully distributed, whereas the subsystems to close the life-support air and water loops arrive in Hab-1. Figure 11 shows the outpost internal layout, and Figure 12 shows a cutaway axonometric view of the outpost mode configuration.

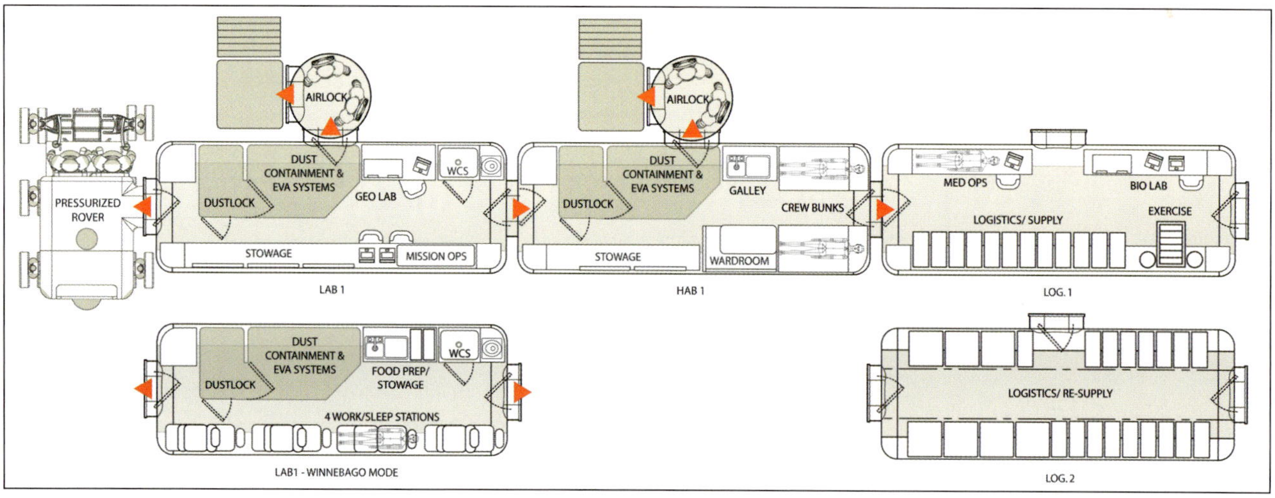

FIGURE 11 Internal plan of LS1 outpost using horizontal-cylinder modules. Main diagram shows outpost mode with SPR docked at left end. Diagrams below show Lab-1 in mobile laboratory mode and PLM logistics module before conversion to Hab-2.

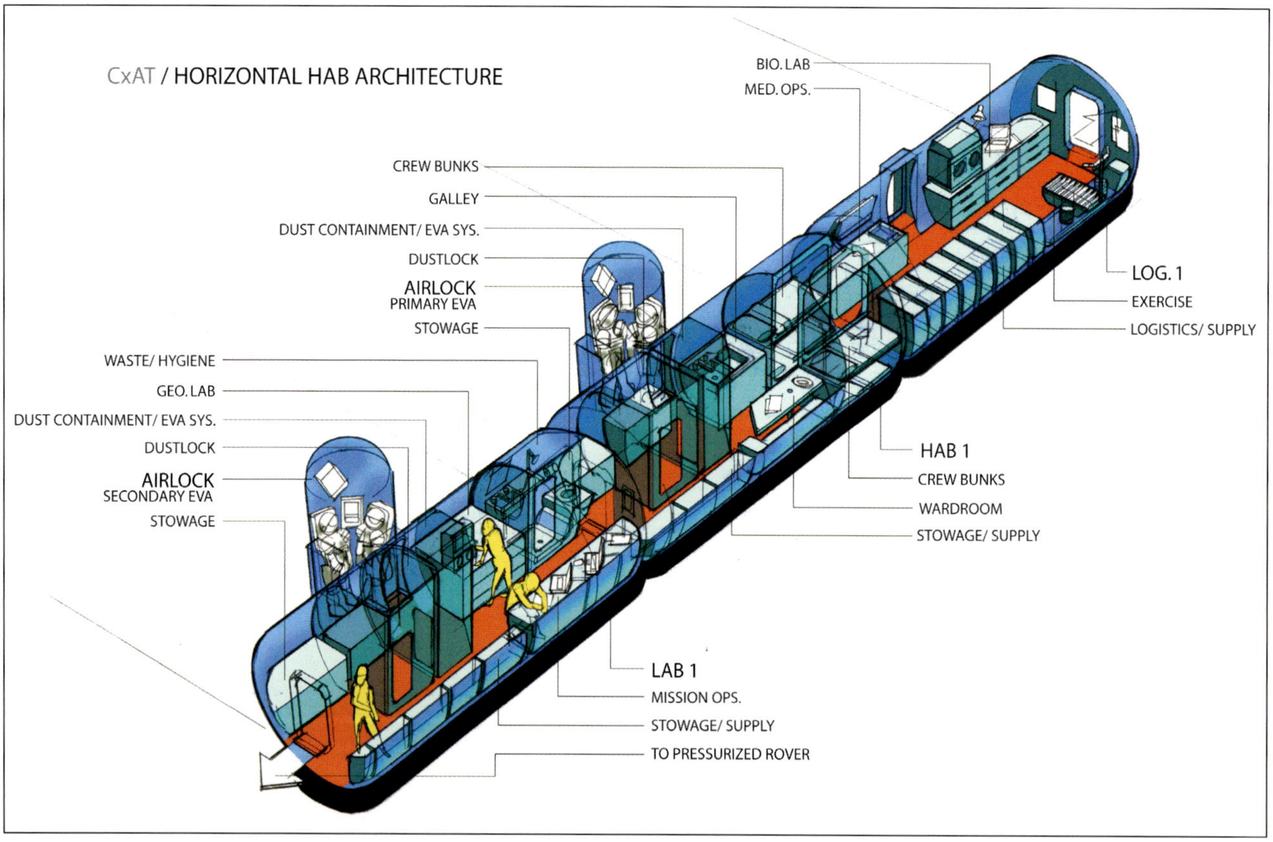

FIGURE 12 Cutaway axonometric view of LS1 outpost mode.

Lab-1 contains the geosciences lab, mission operations station, stowage, life-support and waste and hygiene equipment, and airlock with dustlock and dust containment system. The airlock accommodates two EVA suits and pass-through lock for science samples and equipment. The geoscience laboratory contains workbench space, glove box, and equipment stowage. The EVA ops area contains a maintenance area with EVA equipment stowage and spares, tools, and cleaning and repair equipment. Isolating the EVA function from the module by an intervening lunar dust containment area is believed to be critical to a healthy habitat environment.

Hab-1 contains four crew bunks, galley and wardroom, stowage, life-support equipment, and airlock (with dustlock and dust containment system as in Lab-1). Putting the waste management compartment in Lab-1 (necessary for initial outpost operation) has the benefit of separating it from the galley/wardroom and bunk locations to limit cross odors. Because of volume constraints, privacy curtains are utilized for the bunks.

Hab-2 (once retrofitted from PLM-1 as its supplies are distributed among the other modules and consumed) contains the biomedical/life sciences lab, crew health care and exercise equipment, logistics/supply stowage, and life-support equipment. Subsequent PLMs mate to the free end of this unit. In most cases the "visiting" PLM will be filled with waste for future disposal or recycling as its supplies are consumed.

Resource Summary

Unmargined power required for nominal outpost mode operations is 10.2 kW (Table 1). This drops to 2.0 kW during quiescent intervals while not occupied. The corresponding unmargined thermal load during nominal operations is 4.6 kW (aircooled) and 7.2 kW (cold plate). Mass properties are tabulated in Table 2. The outpost configuration total mass to the

TABLE 1 Power and thermal loads for the LS1 outpost horizontal-cylinder modules.

LOAD	HAB-1	LAB-1	HAB-2	TOTAL POWER OR THERMAL, W
Outpost power active, W_e	5115	4556	533	10,204
Outpost quiescent power, W_e	1066	625	323	2,014
Outpost air-cooled thermal, W_t	1025	3226	391	4,642
Outpost cold-plated thermal, W_t	4090	1330	142	7,231
Lab-1 in mobile laboratory mode, W	n/a	4556	n/a	4,556

TABLE 2 Horizontal-cylinder module mass properties, not including airlocks.

HABITAT SUBSYSTEM	HAB-1 MASS, kg	LAB-1 MASS, kg	HAB-2 MASS, kg	TOTAL OUTPOST MASS, kg
Structures	2,204	2,204	1,676	6,084
Shielding	331	233	233	797
Power	295	295	113	703
Thermal	243	222	45	510
Avionics	108	104	5	217
Life support	1,278	1,621	146	3,045
Airlock/suitport	600	600	N/A	1,200
Outfitting	136	1,260	424	1,820
Total dry mass	5,195	6,539	2,642	14,376
30% growth	1,559	1,962	793	4,313
Total mass with 30% growth	6,754	8,501	3,435	18,689

surface is 18.7 mt for the three outpost habitation units. The mobile lab mass is 8.5 mt of that total. Figure 13 shows the interface block diagram between the habitat and other outpost systems.

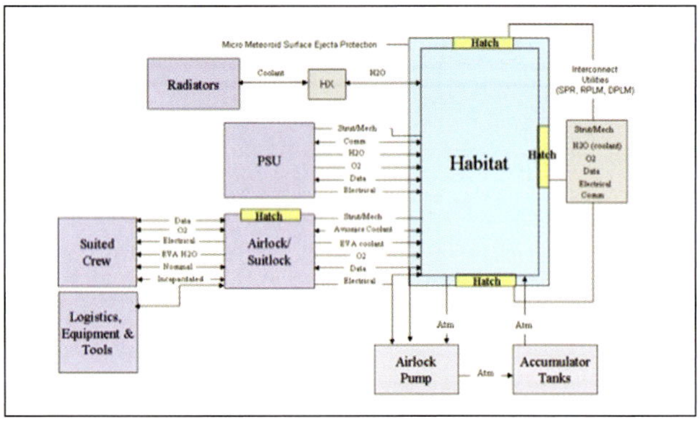

FIGURE 13 LS1 surface habitat external interfaces.

INFLATED-TORUS CONCEPT

Concept Description

The inflatable habitat concept provides more volume per unit mass than the horizontal-cylinder module type. However, because of the need for redundant pressure vessels at a permanent outpost, two inflatable modules are still required. The outpost configuration comprises two inflatable elements (one lab and one hab) and a PLM mated together (Figure 14). Each torus module provides 174 m^3 of pressurized volume, and so the total pressurized volume is 348 m^3 or 87 m^3/crew member. A PLM adds another 78 m^3 of stowage and supplies volume (the same PLM module as in the previous LS1 concept) for a total volume of 426 m^3. For a four-crew, 180-day mission, this concept provides 0.59 m^3/crewmember/day.

FIGURE 14 LS1 outpost configuration using inflatable-torus module type.

The inflated torus has an 8.5-m internal diameter with 3.6-m high structural core. The eight-longeron structural core is integrated via the support frame to the PSU. The packaged module is removed from the lander by ATHLETE. Each module has an inflatable, pre-integrated airlock. Lab and hab are designed to operate independently, consistent with the LS1 outpost scenario; the lab can be decoupled from the outpost and taken on excursions by ATHLETE. Each torus module has three docking ports: one mates with the other module, a second accommodates an inflatable airlock, and the third is available for SPR or PLM mating.

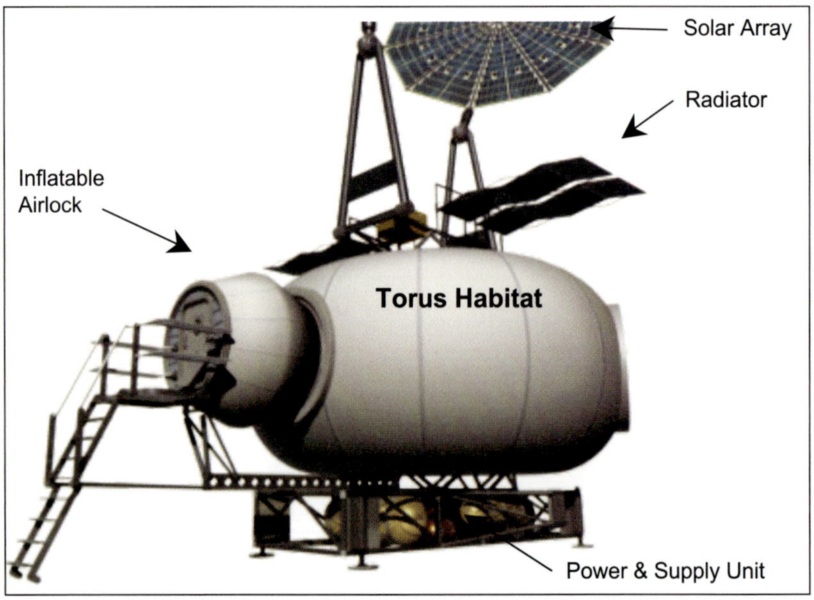

FIGURE 15 Inflated-torus module integrated with PSU.

Each habitat element is delivered fully outfitted, pre-integrated with a PSU via a 57-kg interface structure (Figure 15). Deployable radiator panels, solar array, and communications antennas are attached to the

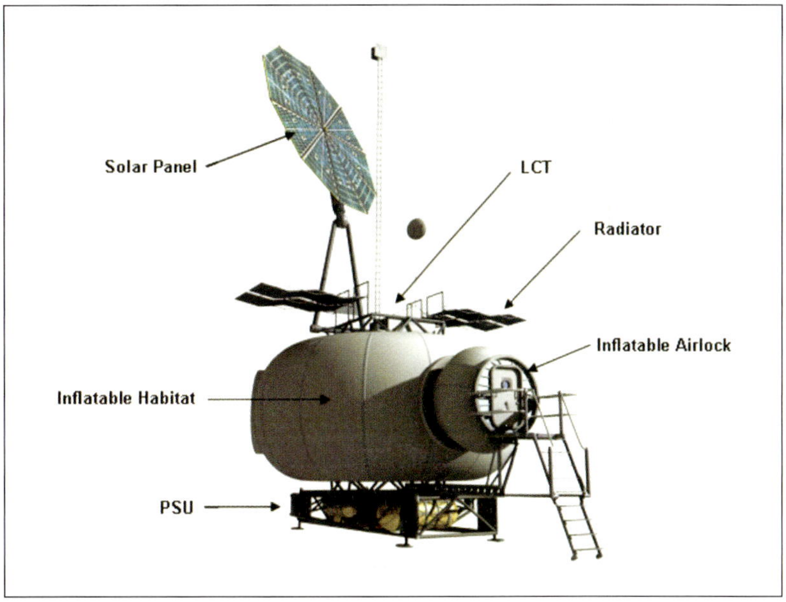

FIGURE 16 Inflated torus module with external systems deployed.

FIGURE 17 Mobile laboratory mode with ATHLETE carrying lab element for excursion science.

FIGURE 18 Lightweight rigid structural core is the module's backbone, resisting axial and circumferential tension loads.

core exterior structure (Figure 16). The Hab element is delivered to the lunar surface first.

In outpost mode, each module's PSU provides the structural connection to the lander payload platform and to ATHLETE, the electrical power storage subsystem, and jack stands for ground support and leveling. After emplacement, the modules are connected via pressurized mating interface to each other and to other habitat elements to configure the outpost. In mobile laboratory mode, ATHLETE attaches to the lab element PSU, and the module decouples from the outpost. ATHLETE roves to a location of interest for remote laboratory operations (Figure 17).

The inflatable torus concept is derived from the TransHab prototype developed by NASA in the 1990s (see Chapter 8). The toroidal shell expands outward from an internal rigid structural core (Figure 18) of shear panels and columns (longerons). The core structure is capped top and bottom by a gusseted, domed plate; an interface ring attached with clevis pins around the edge of the plate at the longeron joint captures and seals the open central edges of

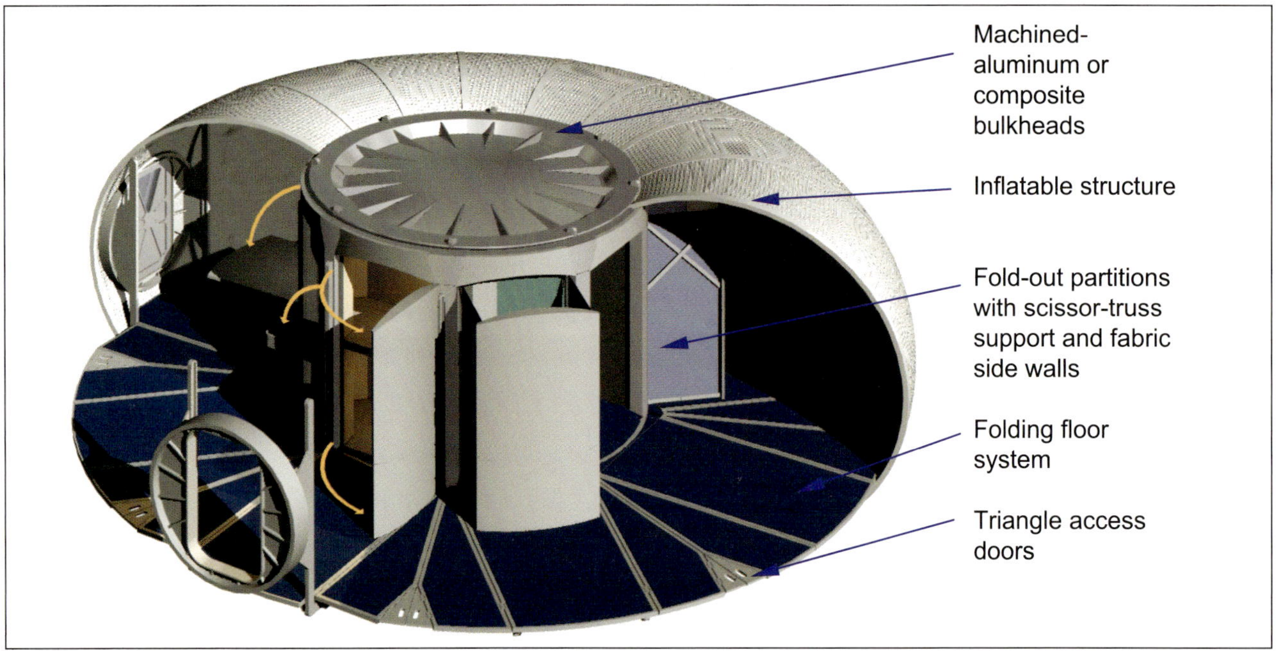

FIGURE 19 Core captures central open edges of inflatable toroidal shell at top and bottom and contains pre-integrated subsystems.

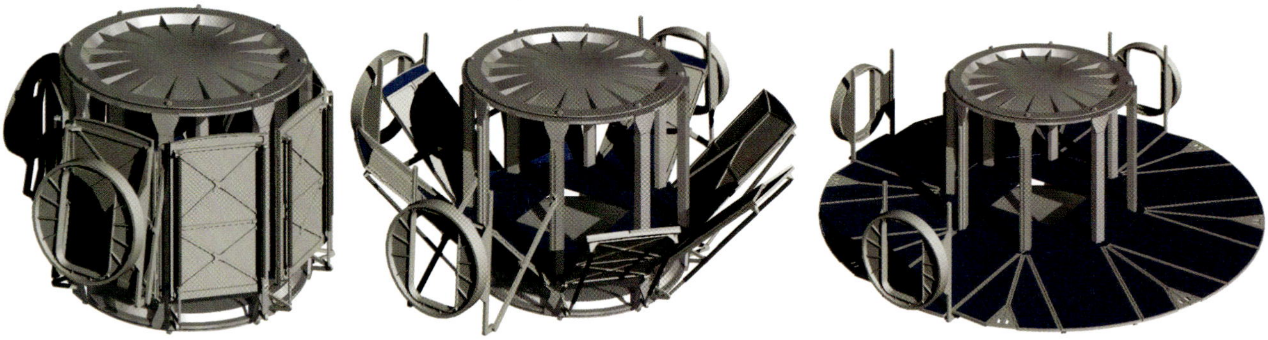

FIGURE 20 Floor system mechanism deploys as shell inflates (left to right), automatically positioning pre-integrated, pretested hull penetrations.

the inflatable bladder (Figure 19). The multilayered inflatable fabric system is optimized for lunar surface environmental conditions. Following the TransHab design, it comprises an external micrometeoroid/ejecta bumper layer, MLI thermal protection, Vectran or Kevlar® structural restraint layer, redundant pressure bladders, and internal scuff protection.

Habitat subsystems and deployable crew accommodations are pre-integrated in the rigid core, and the inflatable shell is packaged around it for transport. The flooring system uses panels, beams, and struts that deploy the floor as the unit inflates (Figure 20). Port and hatch assemblies are integrated as part of the floor system and hinged to the structural core. At these penetrations, the shell edges are captured and sealed as at the top and bottom of the core. Once integrated and acceptance tested, the uninflated shell is folded, packaged, and cinched down for transport. The module is integrated onto the PSU for final acceptance testing, launched, delivered to the surface by the *Altair* lander, positioned by ATHLETE, and mated to the outpost. After deployment, inflation, and verification, the core components are deployed (Figures 21 and 22). No pressurized consumables are carried in the inflatable's core.

Each habitat element is delivered with logistics supplies. Eight small logistics carriers (SLCs) can be carried in the core; additional SLCs are delivered on cargo flights. After habitat setup, PLMs are used for

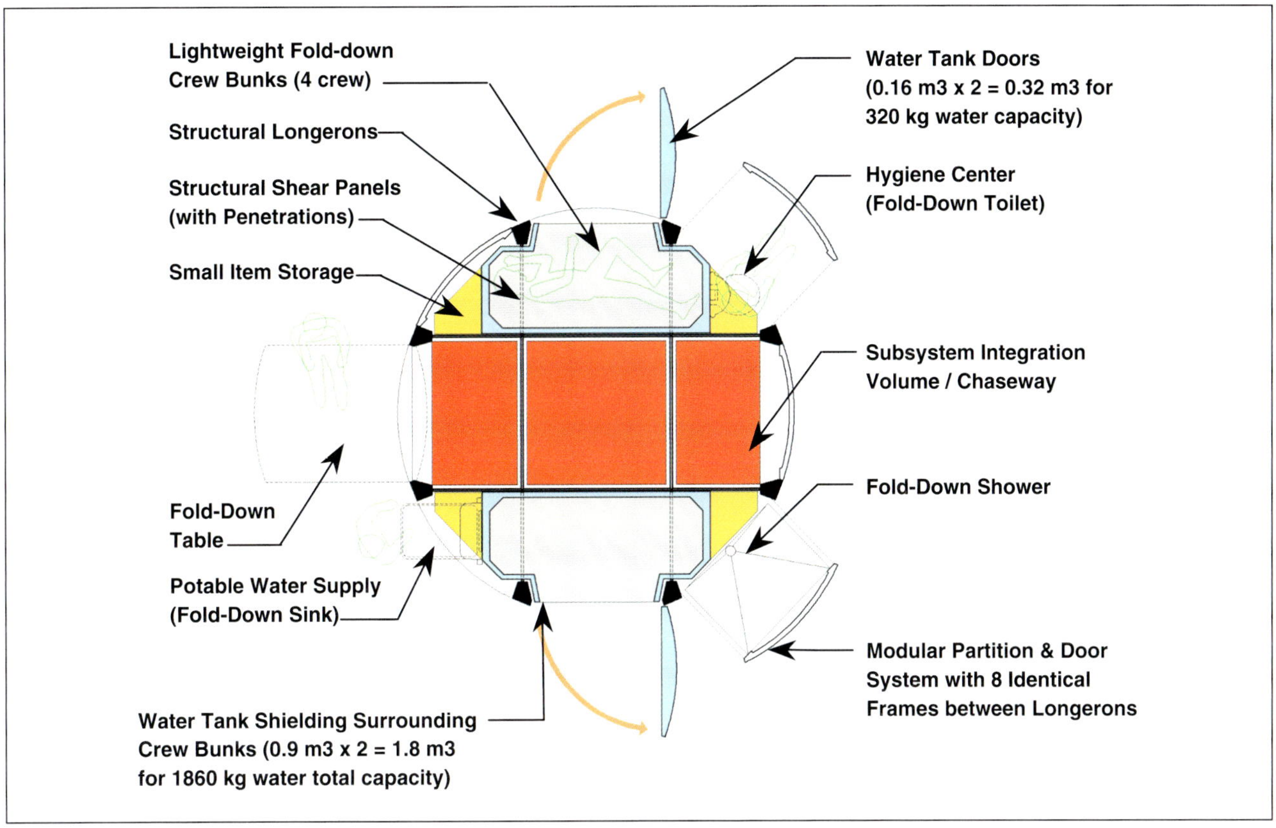

FIGURE 21 Plan view of major crew systems components pre-integrated into structure core.

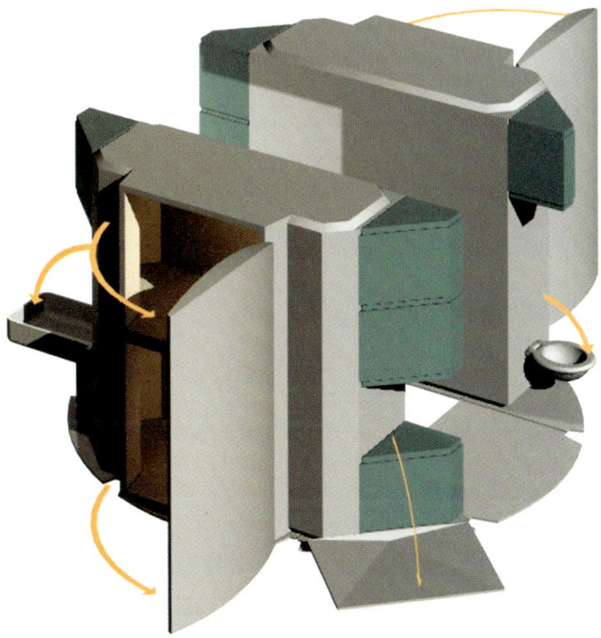

FIGURE 22 Core crew systems deployment.

resupply; they mate to either the free port of either hab or lab.

The subsystems strategy and functionality are the same as with the horizontal-cylinder LS1 module concept described earlier. Life-support system redundancy is distributed between both inflatable modules. When in mobile laboratory mode, the lab module provides partially closed ECLS. The hab ECLS subsystems close the air and water loops.

Internal Architecture

The overall outpost topology and floor plan are shown in Figure 23. Each element is delivered mostly outfitted. The lab module (Figure 24) contains mission ops, geoscience and bioscience labs; medical ops and exercise facilities; waste and hygiene systems; life-support equipment; stowage and maintenance; and inflatable airlock with dustlock and dust containment system. The hab module (Figure 25) contains crew ops: galley, wardroom, and pantry; waste and hygiene systems, life-support equipment, four crew bunks with integral SPE radiation protection; stowage; and inflatable airlock with dustlock and dust containment system. System details are the same as for the horizontal-cylinder LS1 concept described earlier.

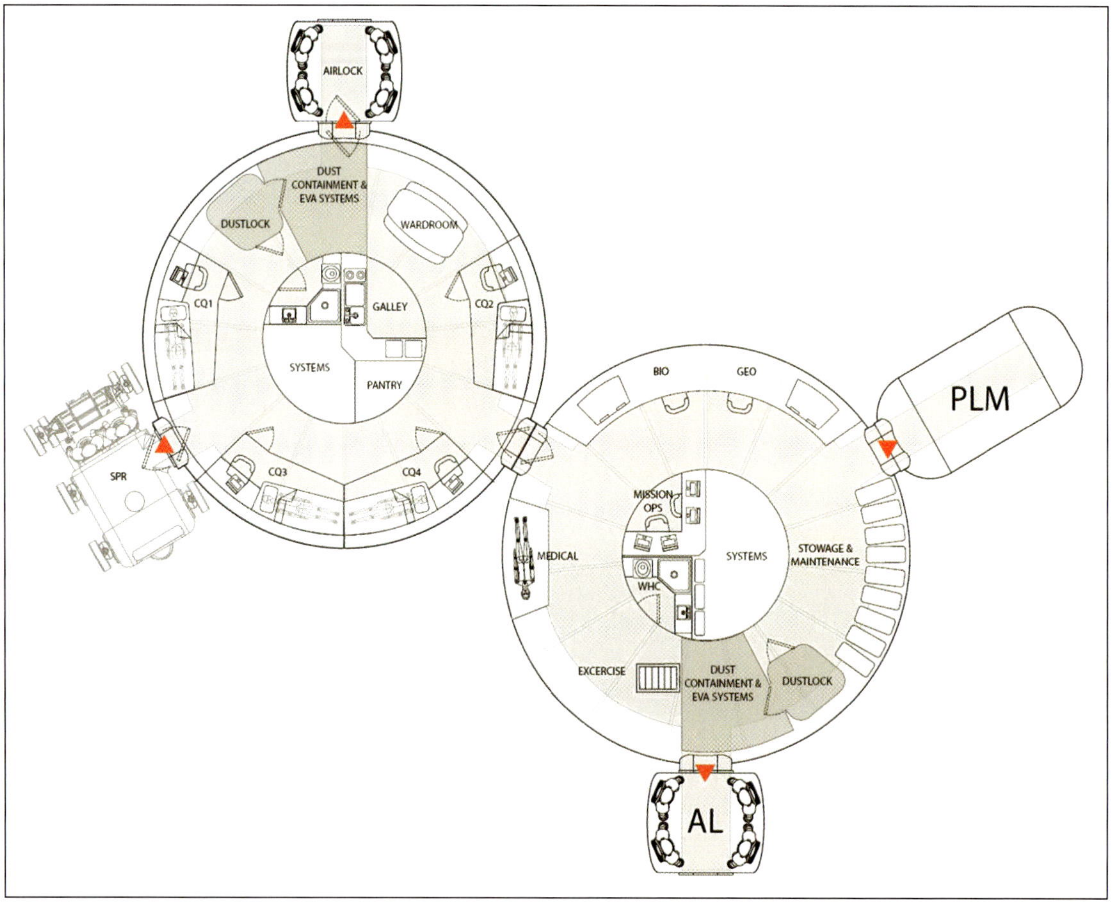

FIGURE 23 Topology and internal plan of LS1 outpost using inflated-torus modules, with inflatable airlocks, PLM, and SPR mated. Detailed plans are shown in Figures 24 and 25.

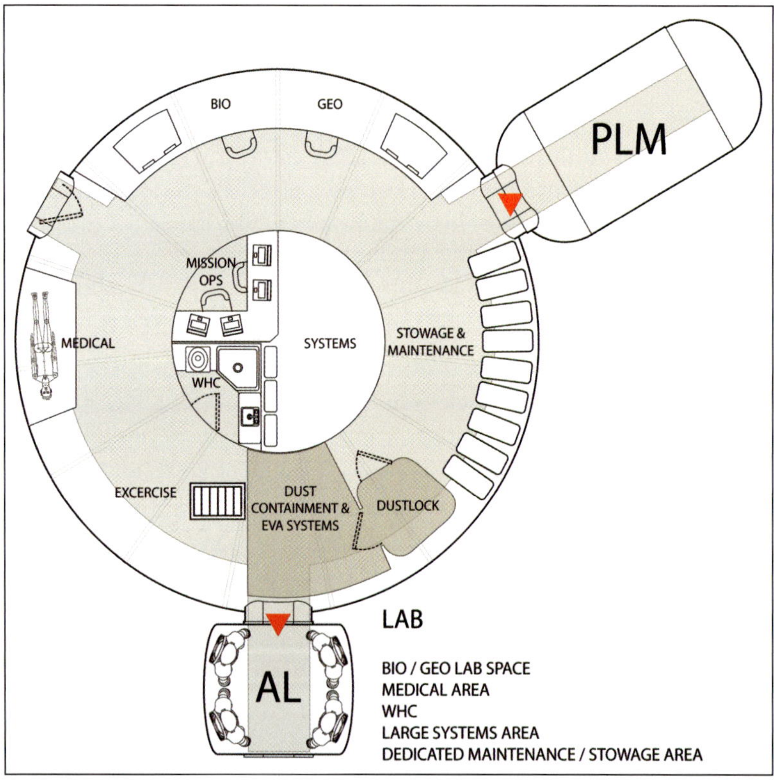

FIGURE 24 Lab-module internal plan.

Resource Summary

Unmargined power required for the nominal outpost operations is 9.5 kW (Table 3). Quiescent power required while unoccupied is 1.1 kW. Unmargined thermal loading is 4.2 kW (air cooled) and 5.2 kW (cold plate). Mass properties are tabulated in Table 4. Total mass delivered to the surface for the two inflatable habitation units is 18 mt. The mobile lab mass is 9.9 mt of that.

CORE-HABITAT CONCEPT

Concept Description

LS3 is a functional variation of the LS1 scenario, intended to attain early habitation within an affordable program funding profile. A crewed sortie to the outpost location provides the opportunity for human reconnaissance before primary cargo element deliveries begin.

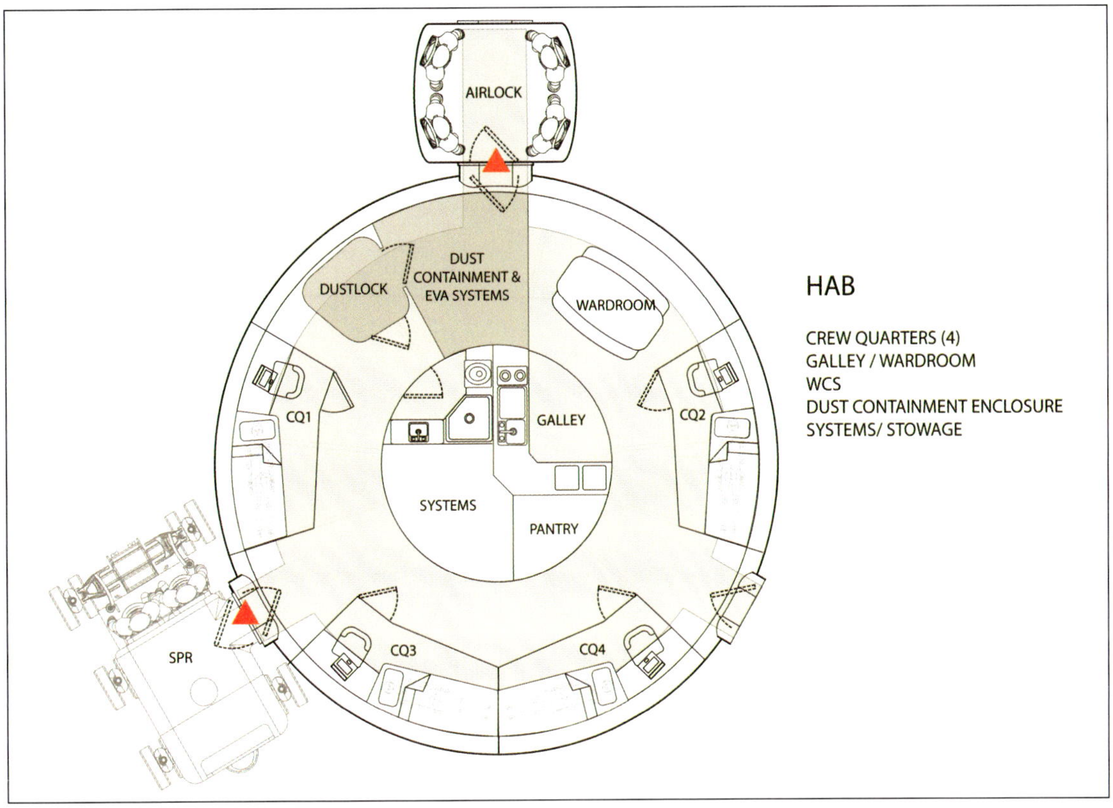

FIGURE 25 Hab-module internal plan.

A minimum functional outpost is then deployed in a single cargo mission bringing a core habitation element. The outpost remains integrated with the lander (Figure 26) for two years or more. (A small boom crane is used for offloading cargo from the payload deck, which is over 6 m above the surface.) At this point, options are preserved for accommodating four-crew, 14–28 day missions; for continued outpost buildup; or for deferral of buildup while awaiting international or commercial surface system contributions.

As with the habitat element concepts described earlier, the core-habitat element is supported by a PSU, powered by deployable solar arrays, and networked through a deployable LCT. Once a heavy-lift mobility system (e.g., ATHLETE) is delivered on a

FIGURE 26 Core-habitat concept initial outpost integrated on *Altair* lander.

TABLE 3 Power and thermal loads for the LS1 outpost inflated-torus modules.

LOAD	HAB	LAB	SUBSYSTEM TOTAL POWER OR THERMAL, W
Outpost power/active, We	5028	4471	9499
Outpost quiescent power, We	568	535	1103
Outpost air-cooled thermal, Wt	1028	3223	4251
Outpost cold-plated thermal, Wt	4000	1248	5248

| TABLE 4 Inflatable-torus module mass properties, not including airlocks.

HABITAT SYSTEM	HAB MASS, kg	LAB MASS, kg	OUTPOST TOTAL MASS, kg
Structures	2,621	2,621	5,242
Protection	566	459	1,025
Power	295	295	590
Thermal	363	331	694
Avionics	131	127	258
Life support	1,278	1,622	2,900
Initial inflation system	276	276	552
Airlock/suitport	600	600	1,200
Outfitting	136	1,270	1,406
Total dry mass	6,266	7,601	13,867
30% growth	1,880	2,280	4,160
Total with 30% growth	8,146	9,881	18,027

cargo flight, the core habitat is emplaced close to the surface (Figure 27) to facilitate routine operations.

The LS3 approach builds up to an end-state capability similar to LS1—redundant habitats, ability to sustain four crew for 180 days, support of long-distance and duration roving (Figure 28)—but with somewhat less habitation volume and no ability to accommodate crew during long eclipse periods. The scenario assumes two missions per year on average, alternating between crewed and cargo missions. Mobility capabilities can be tailored to emergent science objectives as needed. The core surface system technologies and outpost operations approach are applicable to Mars exploration.

The LS3 outpost habitation system comprises three main elements with hard-wall, horizontal-cylinder habitat modules: core habitat and two reusable pressurized logistics modules (RPLM-1 and RPLM-2) (Figure 29). The RPLMs are retrofitted on the surface into living and working areas to provide functions required for long-duration crewed missions. Each is delivered with both logistics supplies and the outfitting required to retrofit it. Disposable pressurized

FIGURE 27 Core-habitat concept emplaced on surface for initial outpost operations with SPR.

FIGURE 28 LS3 outpost end-state configuration. View shows three core modules plus expendable PLM.

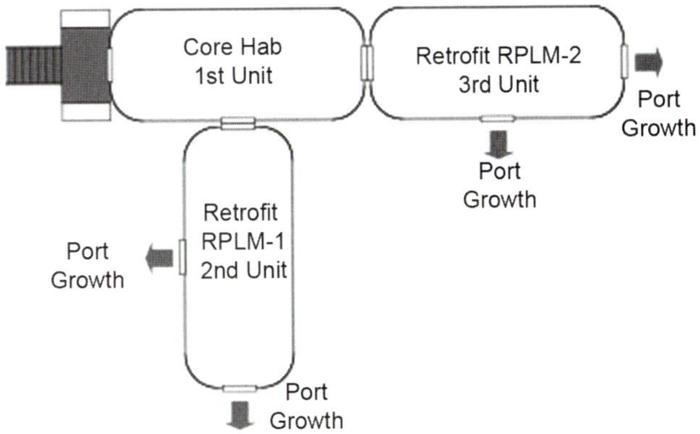

FIGURE 29 Core-habitat concept topology and floor plan.

logistics modules (DPLM) are periodically mated to the outpost habitation system to provide logistics resupply and trash disposal. Four-port, two-port, and three-port core-habitat configuration options were investigated, all based on leveraging commonality with the lander ascent module and airlock pressure vessels. The three-port option was baselined (Figure 30). Suitports are used for EVA egress (see Chapter 18 for a detailed discussion).

Each aluminum-lithium core-habitat module has 3.0-m internal diameter and 8.3-m internal length, yielding 55-m^3 pressured volume. Floor area is 2.3 × 7.7 m, 17.8-m^2/module. The total floor area of the outpost is 53.5 m^2, and the total volume is 165 m^3, which for a four-crew, 180-day mission is 0.23 m^3/crewmember/day.

PSU, radiators, deployable solar arrays, LCT, MLI insulation, fabric composite shield, and surface-access stairs are integrated with the module to comprise the complete element, as with the horizontal-cylinder concept described earlier. The three ports per module allow multidirectional outpost expansion while leaving two ports open for SPR docking.

The PSU frame structure has detachable wings and is intended to incorporate power system, modular logistics tanks, communications, and other systems. Two versions accommodate alternative outpost power storage technologies (Figure 31): the first accommodates batteries; the second accommodates evolution to regenerative fuel cells including water tanks.

The PSU provides the structural interfaces for all payload operations: 1) jack legs to support and level cargo on the surface; 2) 12 hard points for the *Altair* lander interface (16 hard points with PSU wings installed); and 3) four multifunction mating interfaces on each side for ATHLETE. Rather than driving over high payloads to hoist them from above, ATHLETE splits into two three-legged "Tri-ATHLETE" units that mate with the PSU along its edges and then operate together as a single ATHLETE chassis (Figure 32). In this mode the PSU frame acts structurally as part of ATHLETE.

Subsystems

Most subsystem descriptions for the core-element concept are common with the horizontal-cylinder and inflated-torus concepts described earlier. The core-element subsystems are sized for a full-up outpost

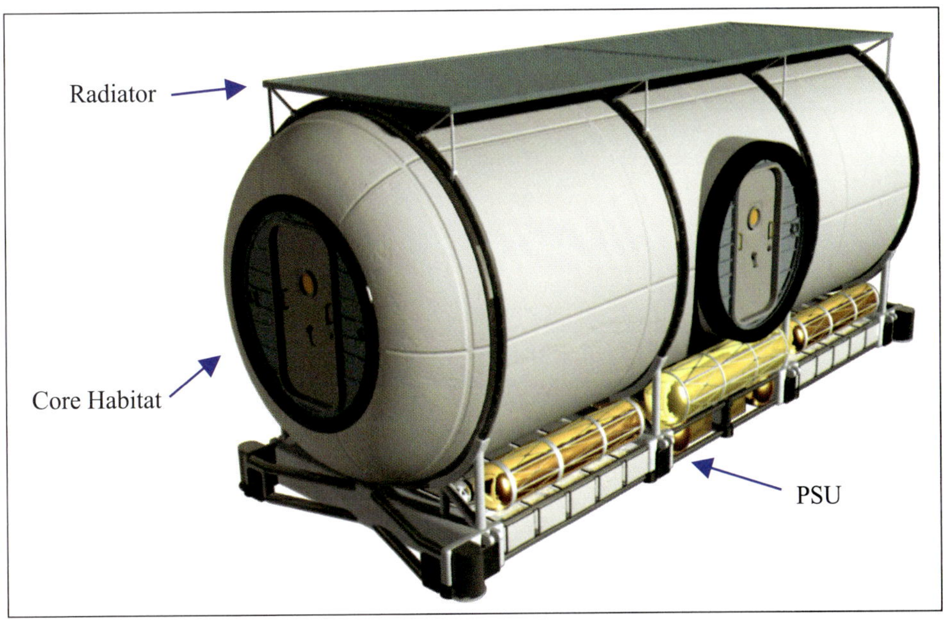

FIGURE 30 Core-habitat three-port option, integrated with PSU and external radiators.

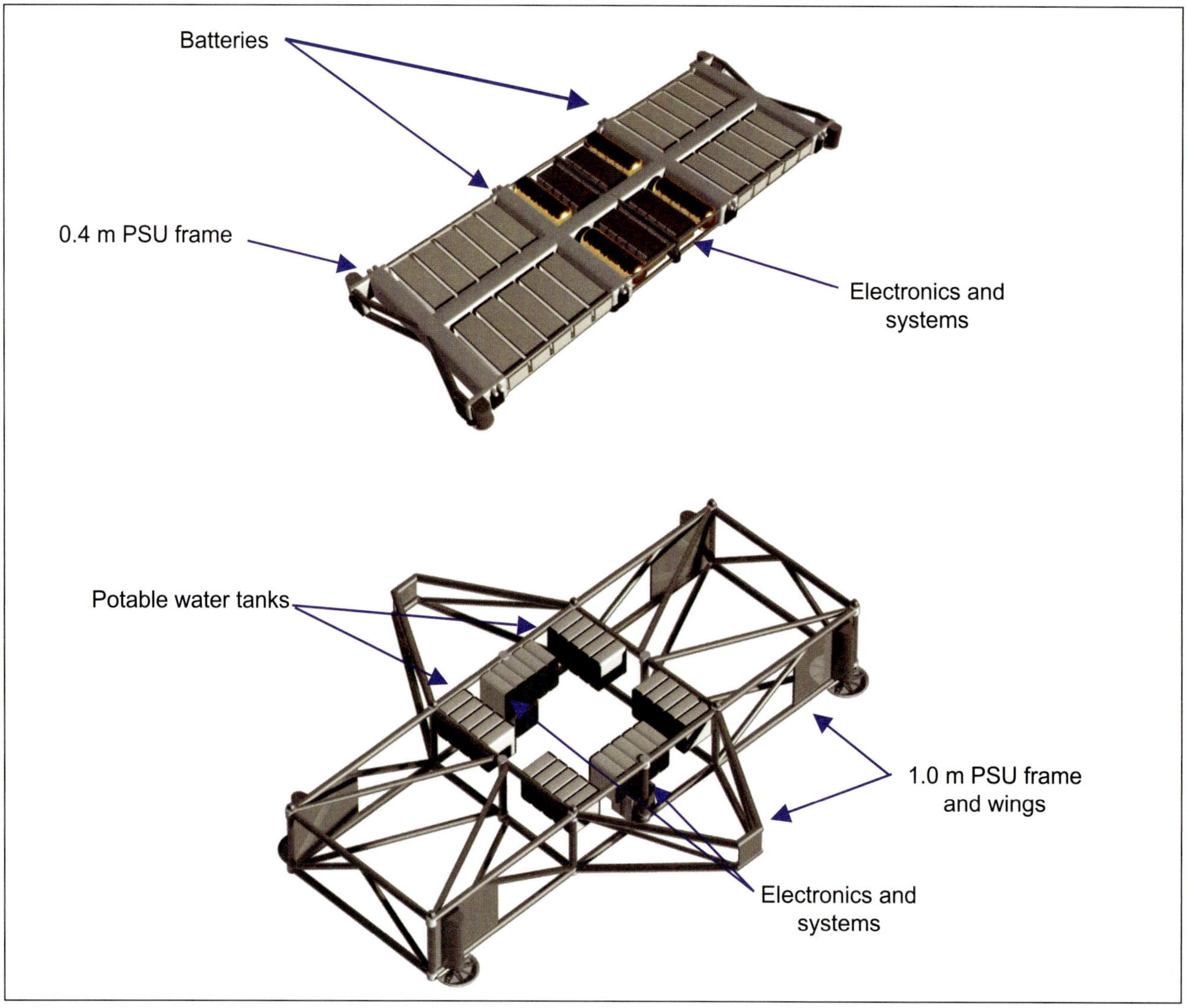

FIGURE 31 PSU configurations accommodate alternative power storage approaches: a) thin PSU holds batteries; and b) thick PSU shown with wings holds regenerative fuel cells and tanks.

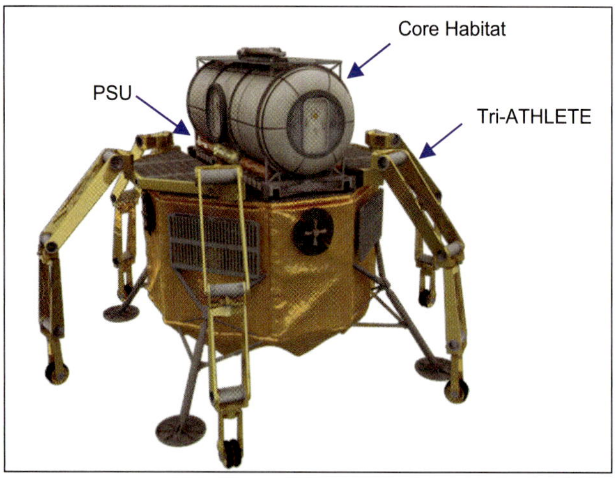

FIGURE 32 Dual Tri-ATHLETE units mated to PSU edges configure a complete offloading and surface transportation system.

capability incorporating two parasitic RPLMs. Because the core habitat hosts the primary operating systems, redundancy is provided within each subsystem rather than distributed among the modules as with the earlier two concepts. The spartan initial core habitat (Figure 33) is delivered with just enough logistics and spares to support a first short crewed outpost mission (Figure 34). Initially 12 M02 stowage bags are integrated where the geoscience lab will be set up, and 44 CTBE are integrated under the floor. A total of 92 CTBE of logistics items are delivered to the surface in the core habitat element.

One 6.5-m^3 suitlock accommodates two suits at a time. Additional egress capability is provided by one or more SPRs, which are designed with integral suitlocks. The suitlock assembly includes dust containment systems both internal to it and in the habitat interior.

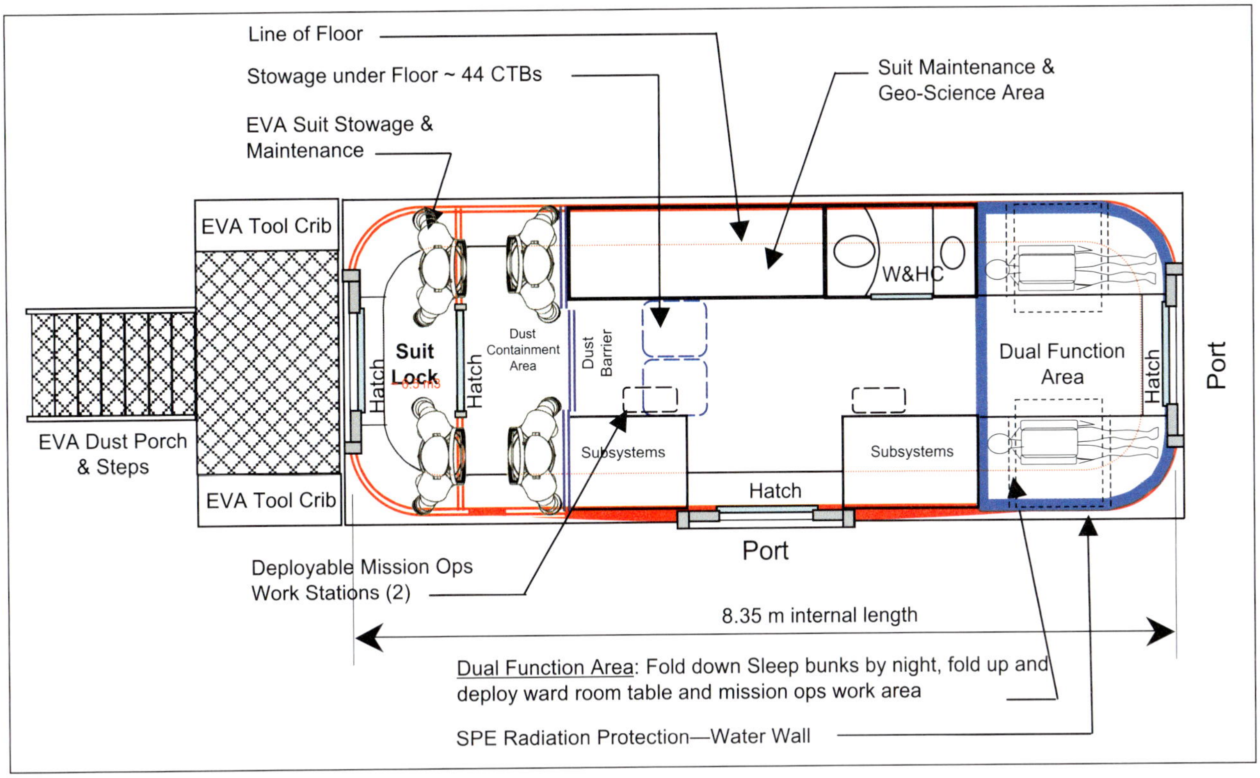

FIGURE 33 Initial core habitat has multi-use outfitting with no space to spare.

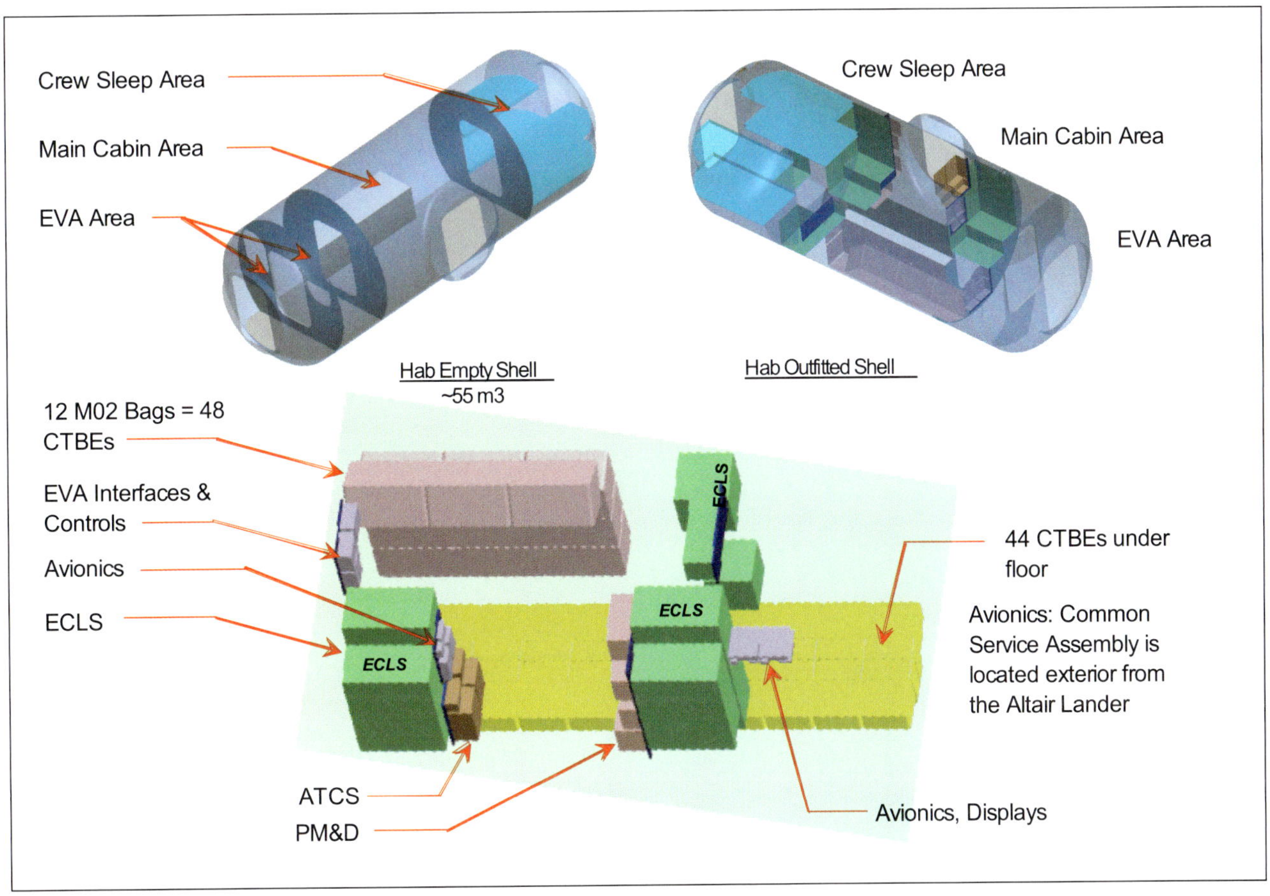

FIGURE 34 Core habitat is delivered with enough supplies for a first, short mission. Compact subsystem packaging is enabling.

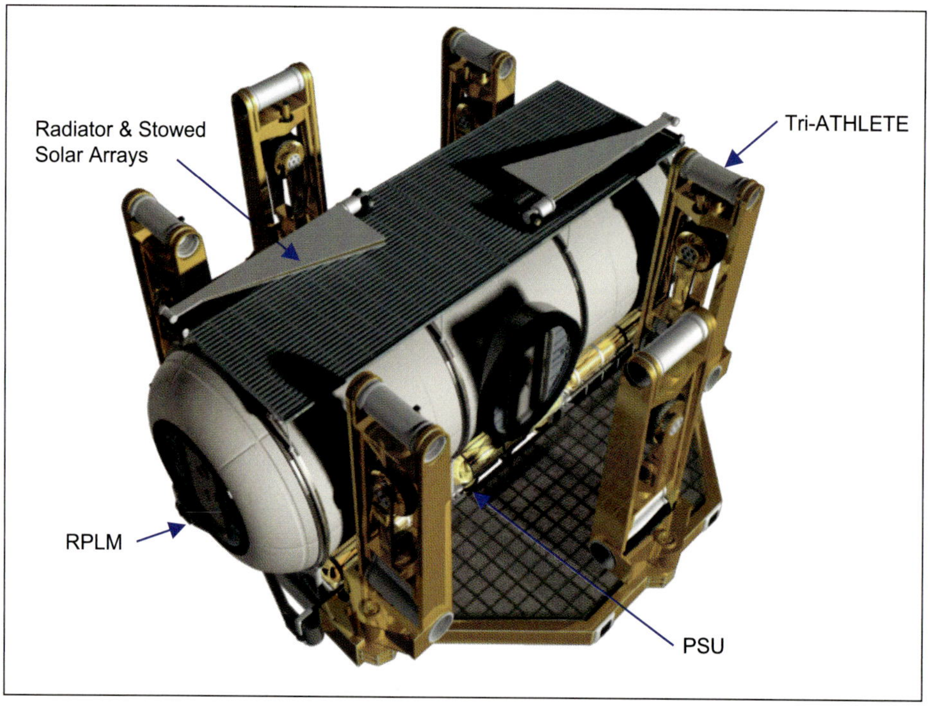

FIGURE 35 Tri-ATHLETE pre-integrated with RPLM PSU for delivery to the surface by *Altair*.

The core-habitat element provides primary PMAD for itself and the RPLM/DPLM elements. The PMAD includes two Dc-to-28 vdc converters, two power distribution units, two 50-switch power switching units, two portable equipment panels, and primary and secondary wiring. The power storage subsystem is in the PSU. Core-habitat ECLS is partially closed and hosts parasitic, minimal-capability ECLS in the RPLMs and DPLM. The waste and hygiene unit is in the core habitat near the water recovery subsystem.

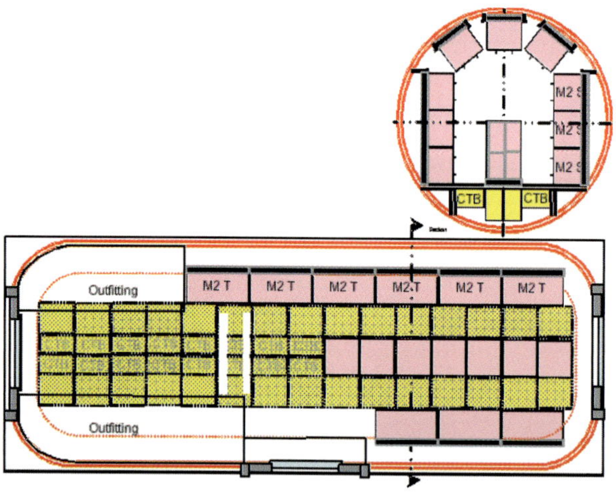

FIGURE 36 Each RPLM is packed with 248 CTBEs of logistics items for surface missions.

Outfitting is based on modular, deployable furnishings. Geoscience lab equipment is stowed and requires deployment, setup, and verification *in situ*. Crewmembers' individual sleep bunks are moved from the core habitat into the RPLM-1 crew ops unit once it arrives. Curtains provide visual privacy.

Reusable Pressurized Logistics Module

Reusable pressurized logistics module (RPLM) shells are copies of the core-habitat structure, including the three-port configuration; this minimizes nonrecurring development costs and enables an open-ended outpost architecture. A complete RPLM element includes a PSU. RPLM elements are delivered on cargo landers, pre-integrated with Tri-ATHLETEs (Figure 35).

The RPLM has passive external MLI insulation and MM/SE shielding like the core module. A water wall SPE shield is pre-integrated inside the RPLM-1 where the crew bunks get relocated upon commissioning. No long-term GCR protection is included.

Each RPLM has secondary PMAD with two converters, power distribution units, power switching units, and wiring. Each RPLM has its own ATCS. ECLS functionality is provided from the primary systems in the core module. Air distribution is provided into the RPLMs via intermodule ducting, intermodule fans, and intramodule fans. The RPLMs include O_2, CO_2, and trace contaminant sensors, and fire detection and suppression equipment. RPLM-1 avionics includes crew utility panels, workstation interface, and network bus interface.

Each RPLM is delivered filled with 40 M02 bags around the walls, seven MO2 bags in the temporary aisle, and 60 CTBEs below deck, for a total of 248 CTBEs of crew logistics items (Figure 36).

Internal Architecture

The initial core habitat contains four crew bunks with SPE water wall; waste and hygiene systems; habitat and life-support subsystems; stowage; and EVA suitlock with dust containment provisions. As the outpost grows, this module is retrofitted into an EVA operations and geosciences unit hosting the geoscience lab; waste and hygiene systems; suitlock with associated systems; and EVA suit maintenance.

RPLM-1 initially contains logistics supplies including spares, and some pre-integrated hardware such as the galley (Figure 37). Once retrofitted it becomes the crew operations unit hosting four crew bunks with SPE water wall; galley and wardroom; habitat and life-support subsystems; and stowage (Figure 38). During the retrofitting operation, the crew members first relocate their sleep stations from the core module; then set up the core-habitat EVA Maintenance area and geoscience lab; then relocate the food warmer to the crew operations unit; and then set up the wardroom table. The struts, cables, and fabric shelves that secured the logistics items are reused to create furniture, walls, tables, and other internal furnishings. The SPE water wall in the module is filled to shield the sleep area. In this interim configuration, the crew operations unit hosts the telemedicine workstation until RPLM-2 arrival.

Similarly, RPLM-2 initially contains logistics supplies and some pre-integrated hardware such as the exercise equipment (Figure 39). Once retrofitted, it becomes the science and medical operations unit hosting the biomedical/life sciences lab; crew health care and exercise equipment; logistics/supply; and habitat and life-support subsystems (Figure 40). During the retrofitting operation, the crew relocates the telemedicine workstation from the crew

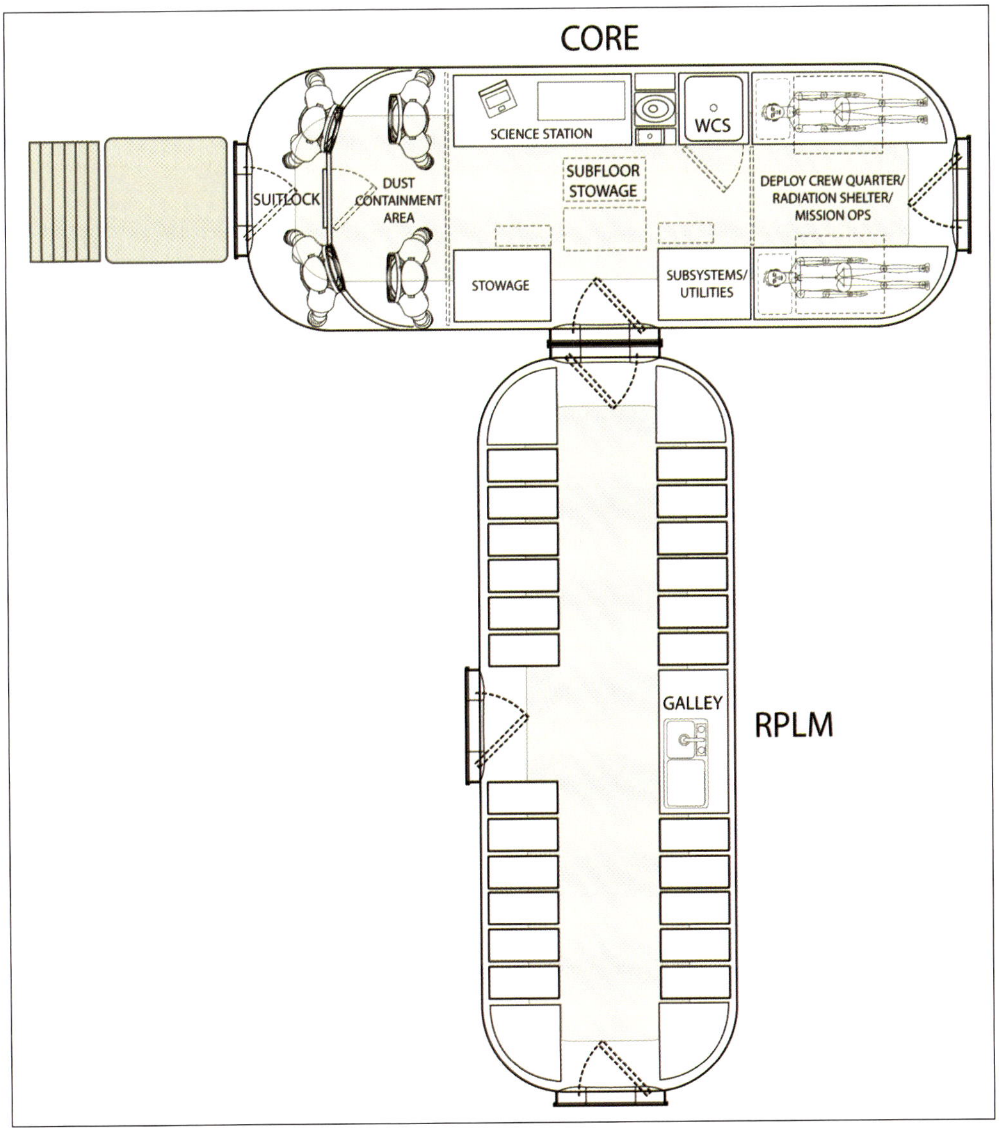

FIGURE 37 RPLM-1 arrives packed with logistics items and some pre-integrated crew ops subsystems like the galley.

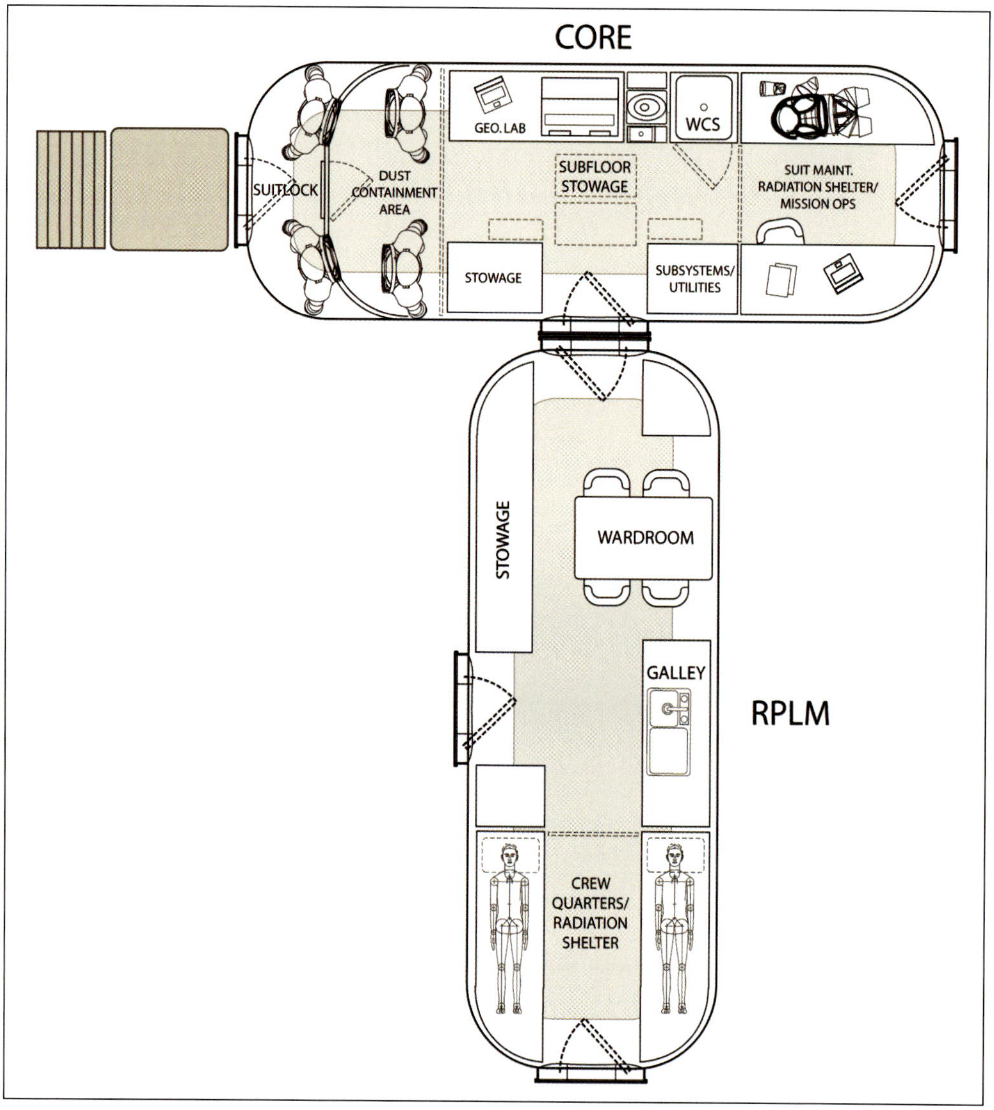

FIGURE 38 RPLM-1 is retrofitted into crew operations unit.

operations unit; sets up the bioscience and medical operations area; and sets up the exercise area. The medical ops subsystem comprises medical kits, environmental contingency kits, exercise/countermeasures equipment, and biomedical research equipment. The subsystem employs consultative telemedicine to monitor, maintain, and restore crew health, providing nominal and urgent care. Setup of the science and medical operations unit completes outpost outfitting.

Resource Summary

Unmargined power for nominal outpost operations is 9.6 kW (Table 5). The quiescent power required while unoccupied is 2.8 kW. Unmargined thermal load is 4.2 kW (air cooled) and 5.4 kW (cold plate). Mass properties are tabulated in Table 6. The total outpost mass delivered to the surface as payload is 17.3 mt. Figure 41 shows an interface block diagram for the core-habitat option, including suitlock.

CONCLUSION

The three early lunar habitation system concepts presented here were developed to examine implications of performance parameters under study for the *Ares V* launcher and *Altair* lander. They do not represent a NASA program baseline, nor do they imply specification of lunar scenario requirements. Studying them flushed out several surface habitat technology needs now integrated into the Exploration Technology Development Program.

Notable products of this study cycle include concepts for 1) the integrated inflatable-torus module; 2) palletized external subsystems to simplify module commissioning on the surface; 3) integration of lander and surface mobility concepts with an overall outpost architecture; and 4) staged retrofitting of logistics modules as primary outpost habitats. In particular, the way the PSU systems pallet and ATHLETE heavy-lifting mobile robot work together significantly mitigates the

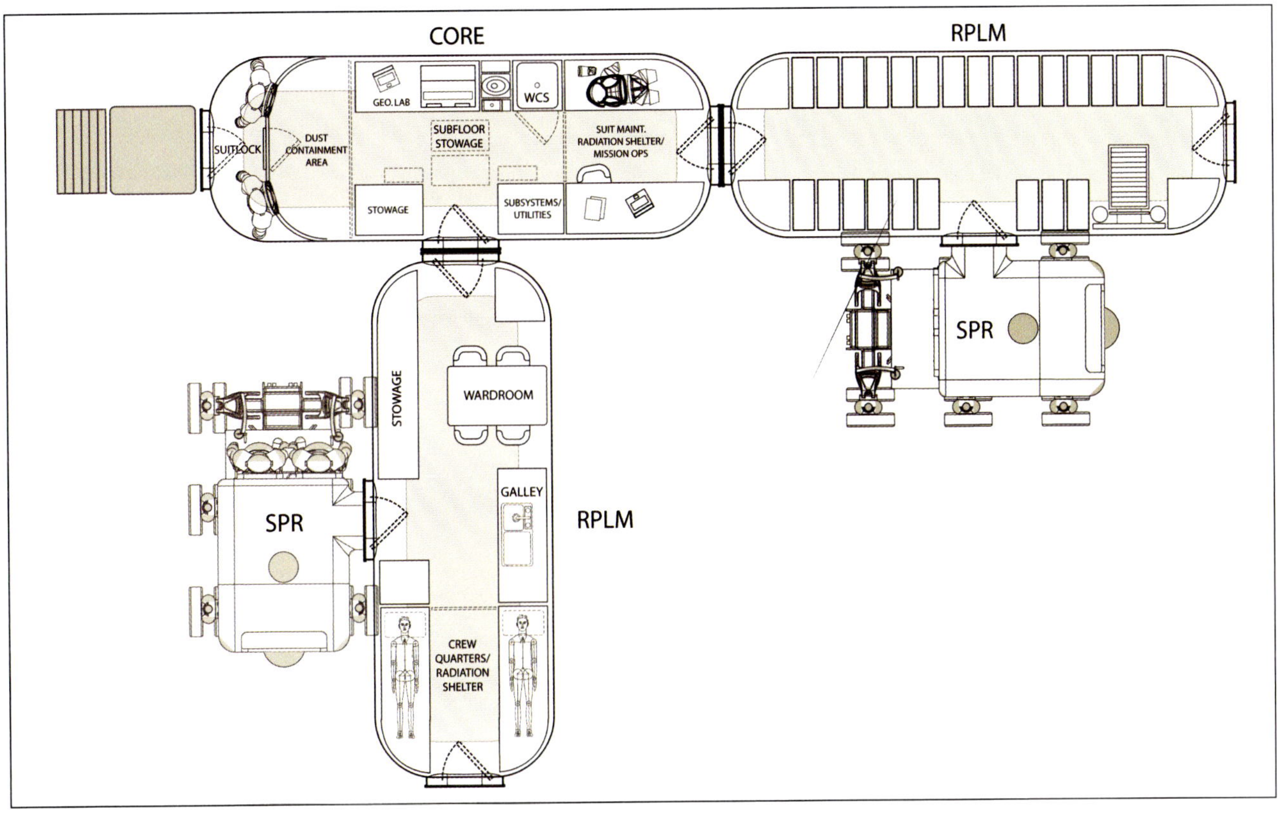

FIGURE 39 RPLM-2 arrives packed with logistics items and additional subsystems like crew exercise equipment. Early on, it can be used for SPR access.

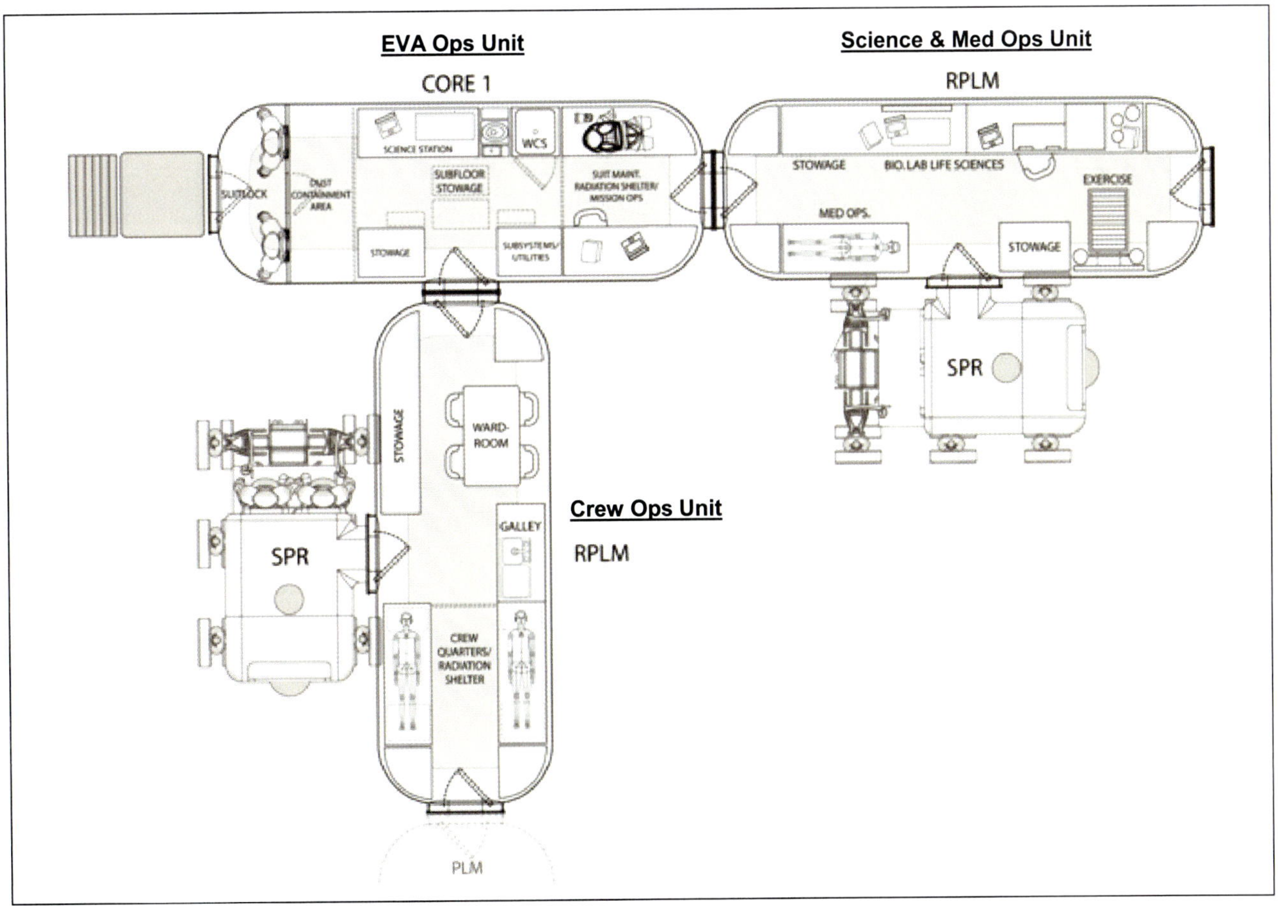

FIGURE 40 RPLM-2 retrofitting into science and medical operations unit completes outpost outfitting.

TABLE 5 LS3 outpost power and thermal loads.

	CORE HAB	RPLM-1	RPLM-2	TOTAL POWER OR THERMAL, W
Outpost power active, W_e	5748	1122	2772	9642
Outpost quiescent power, W_e	1133	836	836	2805
Outpost air-cooled thermal, W_t	1655	510	2047	4212
Outpost cold-plated thermal, W_t	3980	725	725	5430

Note: without the 30% growth

TABLE 6 Core module concept mass properties.

HABITAT SUBSYSTEM	CORE HAB MASS, kg	RPLM-1 MASS, kg	RPLM-2 MASS, kg	TOTAL OUTPOST MASS, kg
Structures	1,866	1,937	1,937	5,740
Protection	300	300	300	900
PM&D	277	268	268	813
ATCS	346	159	224	729
Avionics[a]	84	57	57	198
Life support	2,025	146	146	2,317
Airlock/suitport	462	0	0	462
Outfitting	217	1,020	936	2,173
Total dry mass	5,577	3,887	3,868	13,332
30% Growth	1,673	1,165	1,160	3,998
Total mass w/ 30%	7,250	5,052	5,027	17,330

[a] Avionics reuses the *Altair* lander Avionics Common Service Assembly; thus, the mass it not bookkept here.

complexity and risk of offloading, moving, emplacing, and commissioning habitat modules.

Areas requiring further study even for these concepts include 1) how to protect habitat systems from radiation using several meters of regolith; 2) subsystem accessibility for maintenance and repair; 3) equipment complement needed for various likely surface operations scenarios; and 4) dense packing of redundant subsystems sized to support multiple modules.

Constellation surface systems milestones like the Lunar Surface Concept Review and ongoing scenario analyses continue to drive out figures of merit, technical performance measures, desired features, functionality, operational concepts, risks, and cost estimates heading into that review.

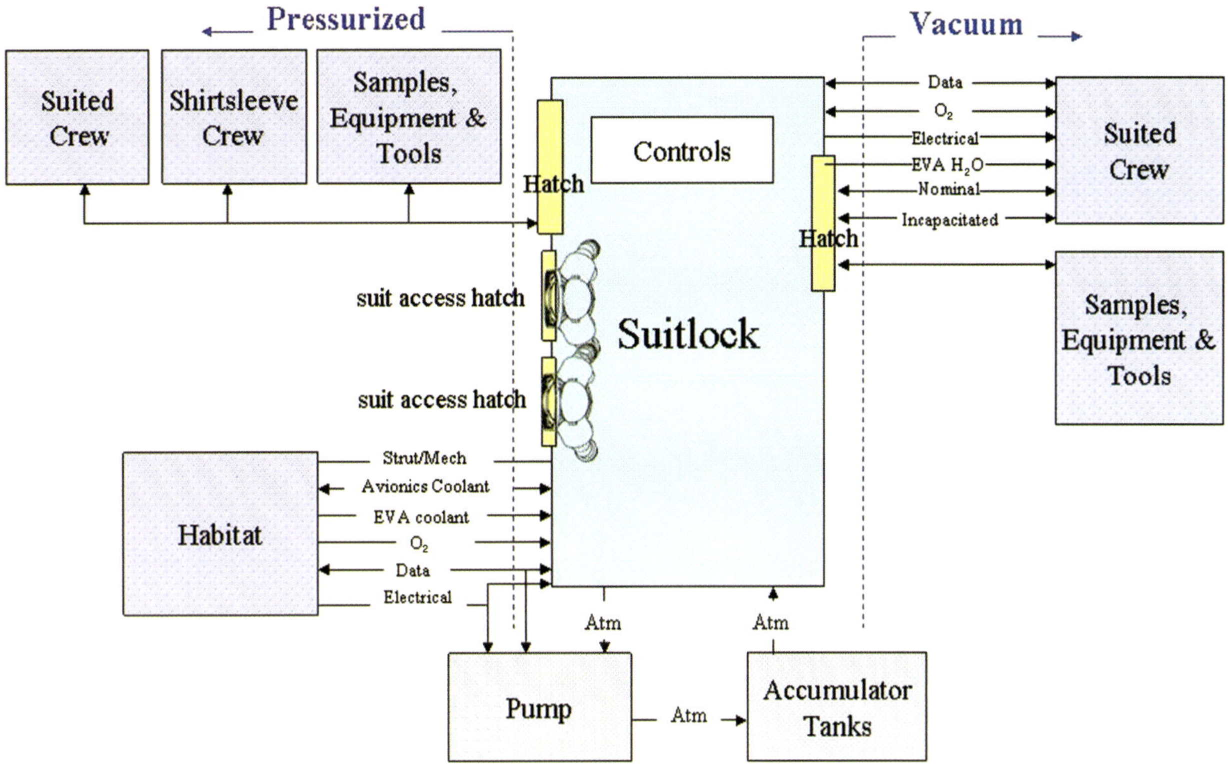

FIGURE 41 Interface block diagram for core-element supporting LS3 scenario, including EVA suitlock.

Acknowledgments

Several NASA centers participate in the CxAT Lunar Habitat Element Team: Johnson Space Center (JSC), Langley Research Center (LaRC), Marshall Space Flight Center (MSFC), Jet Propulsion Laboratory (JPL), and Glenn Research Center (GRC).

The Habitat Management Team comprises Larry Toups/JSC (lead); Kriss Kennedy/JSC; Brand Griffin/MSFC; John Dorsey/LaRC; Marianne Rudisill/LaRC; and Robert Howard/JSC.

The Habitat Team leads are A. Scott Howe, Design; Robert Howard, Human Systems Integration; Jennifer Green, Supportability Analysis; Kandyce Goodliff, Logistics Resupply; Jeff Jones, Medical Systems; Robert Trevino, Suitlock; Peggy Guirgis, EVA Systems; Amanda Carpenter, SE&I; John Dorsey and Tom Jones, Structures and Mechanisms; Ryan Stephan, Active Thermal Control; Bob Bagdigian, ECLSS; Oron Schmidt, Avionics; Scott Woodward, PM&D; and Chip Conlee and Evan Twyford, Internal Architecture.

Bibliography

NASA (2005), "Exploration Systems Architecture Study," NASA/TM-2005-214062, Nov.

NASA (2006a), "Constellation Architecture Requirements Document," CxP 70000 CARD.

NASA (2006b), "Human System Integration Requirements Document," Constellation Program 70024, Sept.

NASA (2007a), "Apollo Medical Operations Project: Recommendations to Improve Crew Health and Performance for Future Exploration Missions and Lunar Surface Operations," NASA/TM-2007-214755, Aug.

NASA (2007b), "NASA Space Flight Human System Standard Volume 1: Crew Health," NASA-STD-3001, March 5.

NASA (2008a), "Constellation Design Reference Missions and Operational Concepts," CxP 70007, Rev. B, Change 001, Jan.

NASA (2008b), "Constellation Program Systems Engineering Management Plan," CxP 70013, Rev. B, Dec.

17

lunar surface airlocks

BRAND NORMAN GRIFFIN

INTRODUCTION

TRADITIONAL CONCEPTS for lunar surface operations include use of an airlock to move crew or equipment between pressurized habitat and vacuum exterior without depressurizing the entire hab system. Early U.S. spacecraft accepted the penalty of depressurizing the whole spacecraft cabin for extravehicular activity (EVA) because they simply could not afford the additional weight, power, and volume of an airlock. Indeed, EVA on all six *Apollo* surface missions was done directly from the lunar module without an airlock. The first U.S. spacecraft airlock was on *Skylab* as a result of its long missions, repeated EVAs, and large cabin volume. The shuttle and International Space Station (ISS) include airlocks designed for a clean, weightless environment.

Airlocks for sustained, routine lunar EVA operations will be significantly different from these previously qualified designs. They will be used often (e.g., every other day) in a dusty, gravity-bound environment. Whereas weightless hatches can be round, hatches for a gravity environment must look more like submarine hatches to accommodate upright astronauts. *Apollo* exposed the hazards of bringing abrasive, potentially toxic lunar dust into the habitat: it affects crew respiratory and eye health, spacecraft cleanliness, and operation of pumps, filters, seals, hinges, and other mechanisms.

This chapter describes comparative analysis performed by the NASA CxAT Habitation Team (described in Chapter 16) to support lunar surface systems concept development for the Vision for Space Exploration. Three airlock types are compared: airlock (AL), defined as an independent pressure vessel with one hatch to the outside and another to the habitat; suitport (SP), which offers direct access from the habitat into an externally mounted, rear-entry suit; and suitlock (SL), which shares a pressure bulkhead with the habitat, allowing rear-entry suits to remain on the dusty side of a pressurized volume while the crew enters/exits the habitat. The SP concept trades poorly in general because it consigns suits to constant exterior exposure and complicates their maintenance. These aspects have not been studied extensively, although the SP is actively being considered for applications on small pressurized rovers. As concluded in the following sections, the AL is favored over the SL. Throughout the chapter, the term "airlock" represents the inclusive class of concepts, whereas the acronyms AL and SL refer to the specific concepts just defined.

For the comparative study, the key objectives are as follows:

1) Minimize loss of resources (make-up air).
2) Design for simple and safe operations.
3) Stress commonality (e.g., with ascent/descent vehicle and pressurized rover).
4) Minimize overall mass of the total architecture.
5) Minimize intrusion of lunar dust.
6) Accommodate transfer of crew, cargo, and an incapacitated crew member.

DEFINITION OF AL AND SL CONCEPTS

Features of the AL and SL concepts are compared in Table 1. ALs are separate pressure vessels, whereas SLs are structurally integral and share a common pressure bulkhead with the habitat. This difference leads to different entry/exit configurations for EVA crew. Using an AL, crew walk through a hatchway, and there are no special constraints on suit design. Using a SL, they "back up to" a bulkhead interface while wearing a specialized rear-entry suit and then exit directly into the habitat.

Copyright © 2008 by ASCE. Published by the American Institute of Aeronautics and Astronautics, Inc., with permission. This paper was originally presented as ASCE Paper "Lunar Habitat Airlock/Suitlock" in *Construction and Operations in Challenging Environments: Earth and Space 2008*.

TABLE 1 SL design is more complex than a traditional AL design.

FEATURE	AL	SL
Layout	Separate vertical cylinder, elliptical end domes, opposing hatches	Integrated vertical cylinder, elliptical end domes, entry hatch, two backpack hatches and an equipment hatch
Hatches and seals	One submarine hatch (other hatch is part of hab) 9.3 m (366 in.) of seal, 18.6 m (732 in.) for two airlocks	One submarine hatch, two suitlocks and one equipment hatch 13.7 m (541 in.) of seal, 27.4 m (1082 in.) for two suitports
Mass	510 kg	593 kg (25 kg for each suitport hatch accommodation, STS airlock hatch 32.7 kg)
Volume	4.25 m^3 (150 ft^3)	4.25 m^3 (150 ft^3)
Pumping	0.133 kWhr (ISS), 30 min	0.133 kWhr (ISS), 30 min
EVA aids	Handholds	Handholds, PLSS docking guides, height adjustment
Dust control	Suit brought into hab volume Equipment brought into hab volume	Suit remains in airlock except for repair and servicing Equipment brought into hab volume

LUNAR EVA ANTHROPOMETRICS

At this writing, there is not yet any official lunar EVA suit, but it is not necessary to have actual suit dimensions to compare airlock options. Spacesuits tend to conform to the size of the crew member, augmented by a back-mounted personal life-support system (PLSS). This means that astronaut population anthropometry can be "suit-factored" for approximate dimensions. Low lunar gravity will also influence suit sizing. Weightless crew members experience as much as 5-cm growth of the spinal column. This study assumes a less severe, 2.5-cm spine-lengthening reaction to ⅙-*g* lunar gravity. Figure 1 shows the lunar anthropometry assumed.

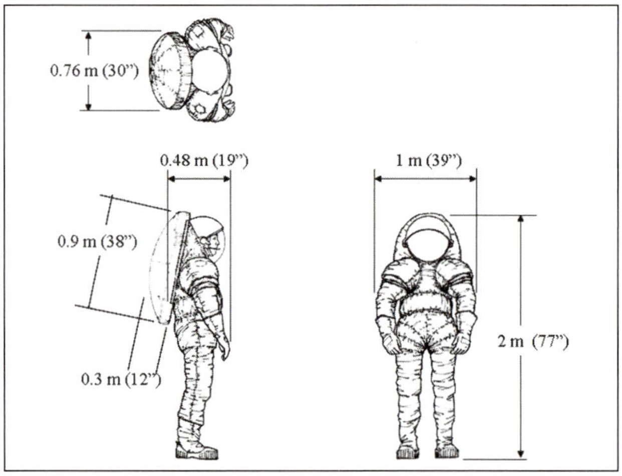

FIGURE 1 Reference anthropometry is based on the Mark III suit configuration.

A single suit configuration is needed for representative operations analysis and sizing. The *Apollo* extravehicular mobility unit (EMU), the only suit to walk on the Moon, is not considered because there have been significant advances in suit design since *Apollo* and because it was not designed for the rear-entry configuration required for an SL system. However, the Mark III suit design is based on rear entry and is used in simulated planet-surface field tests. The Mark III is scaled to fit the EVA population for this analysis.

Both airlock options must be designed to accommodate the size range of the crew population, but the SL system design must also accommodate a consequent 0.45-m (17.7-in.) height differential (Figure 2) for mating the PLSS hatch to the habitat bulkhead hatch. A simple step-or-stoop solution provides a common height interface, where shorter crew would use a step and taller crew would stoop slightly, to mate with the hatch. Because of pressure suit stiffness, stepping is favored over stooping. The step height shown in Figure 3 accommodates the range and favors the median population rather than extreme crew members.

Type and Number of Hatches

Airlock hatchways are structural openings through which EVA crew and cargo pass to move between

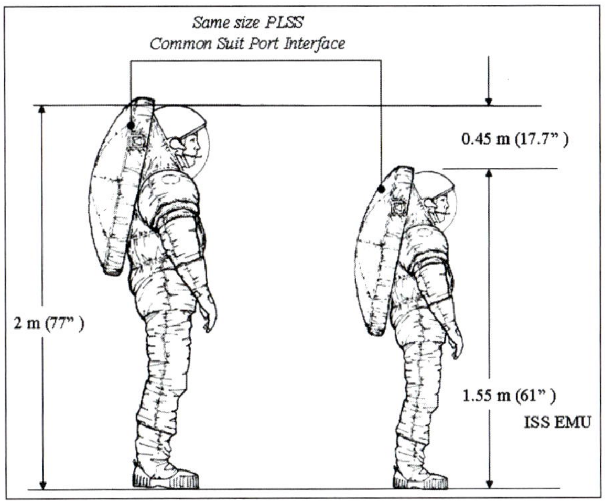

FIGURE 2 Crew population size variance leads to a wide range of suit sizes that must be accommodated by an airlock.

the lunar surface and the habitat interior. Hatches must also accommodate an incapacitated crew member in off-nominal operations. For hatches, the principal difference between AL and SL is that the SL has hatchways for rear-entry suits. This means that the AL can use one type of hatch for all three functions, whereas the SL requires three (Figure 4). And the AL system has one hatch total (the habitat provides the other), whereas the SL requires four hatches for a complete system.

Operable Seal Length

Airlocks are intended to preserve and retain air, and so leakage is an important performance metric. The seals in an operable hatch pose the greatest potential leak path, therefore seal length can be used as a proxy metric for leakage. Counterpressure hatches (closed against the internal pressure) have been used in spacecraft, but they are complex and heavy. Therefore, this analysis assumes pressure-assisted hatches instead. The exception is the SL suit hatch, which requires a special mechanism to seal suit against the bulkhead. This connection is especially challenging because EVA crews in "dirty" suits have to make a secure pressure seal inside a dirty SL. Even more significant is the likelihood of seal damage, surface scratching, and particulate interference of the seal from abrasive lunar dust. A high frequency of EVA exposes suit, airlock interior, and seals to increasing levels of contamination as the mission progresses.

There are two seals for the PLSS: one for the suit-to-bulkhead interface and the other for the interior

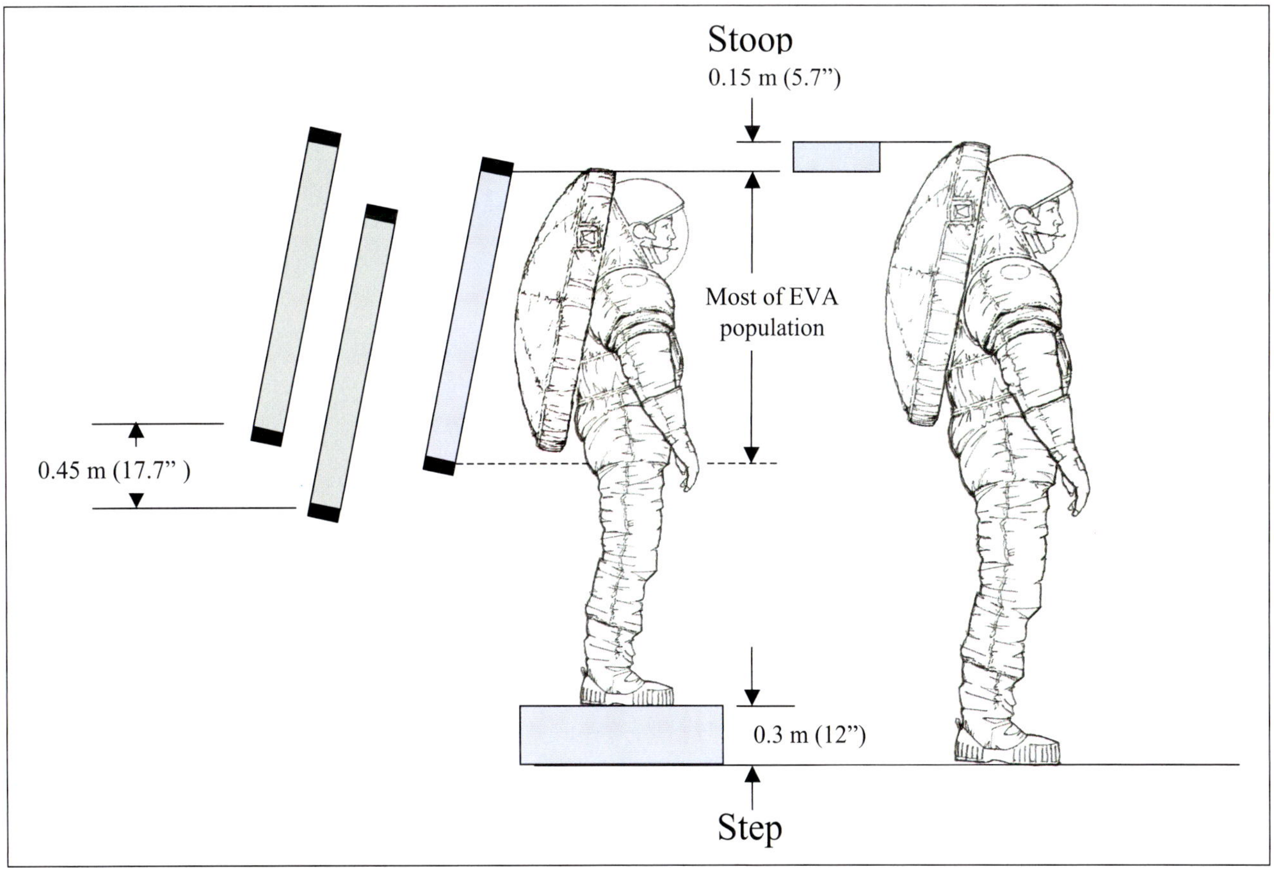

FIGURE 3 A simple step-or-stoop design accommodates the range of suit dimensions for the rear-entry SL.

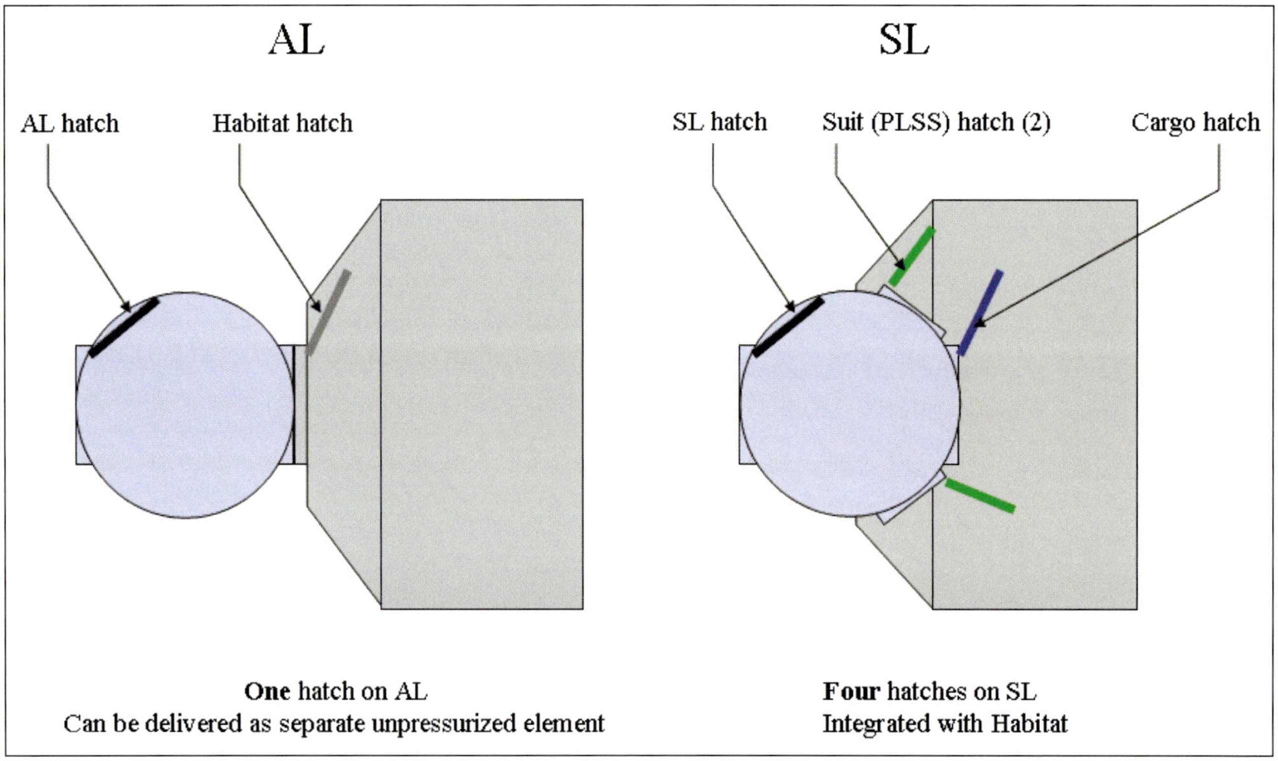

FIGURE 4 SL requires twice as many hatches and three different types of hatch.

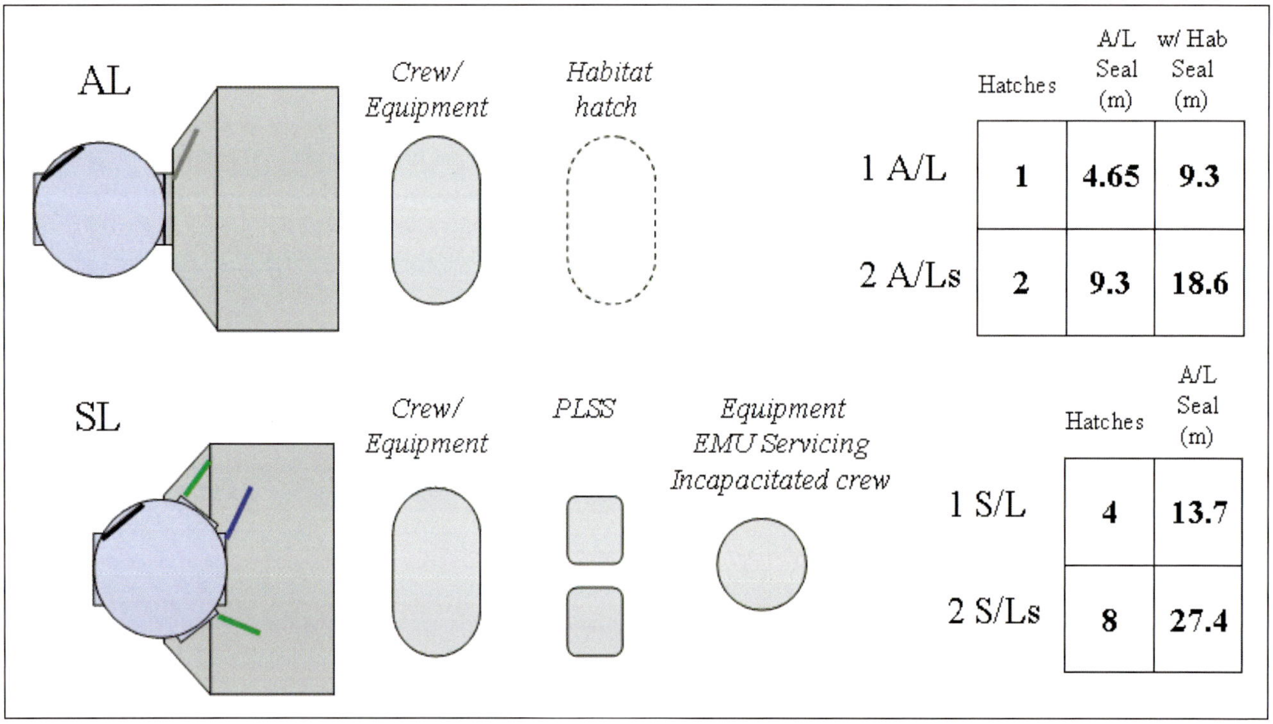

FIGURE 5 SL has far greater total length of seal than the AL and thus pays a higher price in leakage of cabin air.

hatch cover. Because the hatch cover is the primary seal, only one PLSS seal length is used in this analysis. Even so, the results show the SL has 4.4 m (47%) more seal length than an AL including its habitat hatch (Figure 5). Excluding the habitat hatchway makes the comparison more stark: the SL has 8.05 m (173%) more seal length. These percentages are derived from representative hatch sizes (Figure 6), but the trends remain consistent because the SL has four hatches.

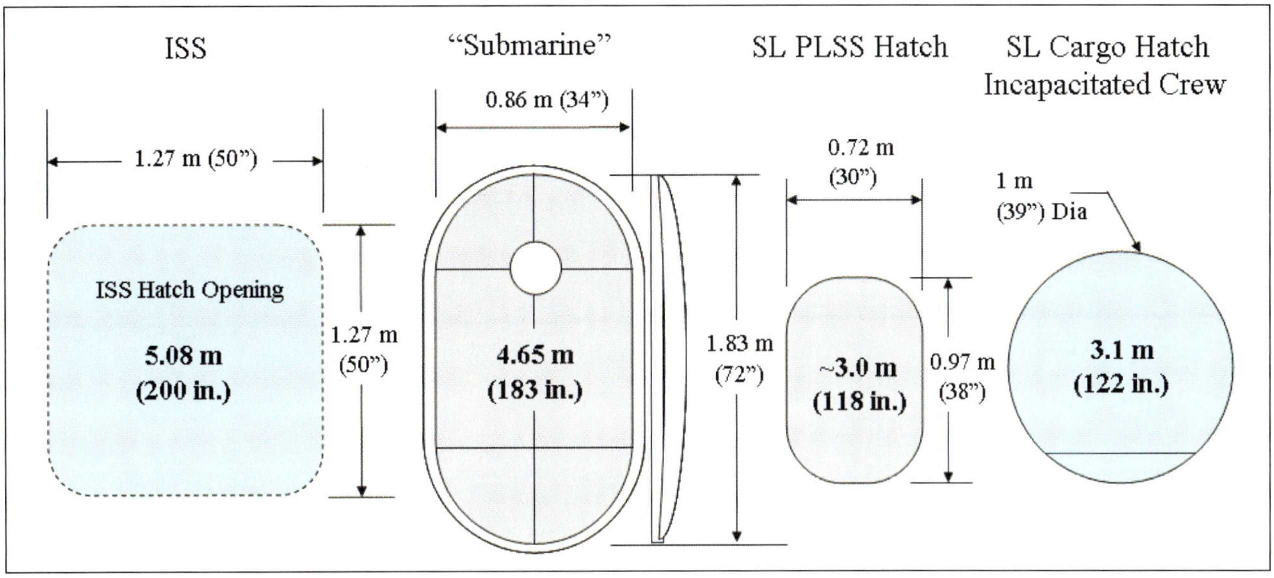

FIGURE 6 Representative hatch dimensions used in the comparative airlock analysis.

AIRLOCK SIZING

Because EVA uses the buddy system, airlocks should accommodate at least two crew members at once. In addition to surface habitats, landers and pressurized rovers are candidates for airlocks (Figure 7). For this analysis, habitat and lander are assumed, each designed for four crew, whereas the pressurized rover accommodates two. This means the choices are either an airlock for four (all at once) or two (cycled twice). Through increasing diameter, an AL can accommodate four; however, an SL is geometrically challenged because it has to provide four rear-entry hatches lined up shoulder to shoulder. The Mark III PLSS is 0.76 m (30 in.) wide. Allowing 0.03 m (3 in.) on either side for bulkhead structure and arm movement, it takes 3.7 m (144 in.) to accommodate four crew. With a hatch added to accommodate incapacitated crew, this is excessive.

Smaller airlocks are attractive for several reasons. They weigh less, package better, and pump less air, and thus cycle quicker and lose less (an airlock can lose 10% of its air in each cycle). Thus, the incentive is to make an airlock as small as possible while still allowing for inward hatch swing and some crew movement. Consequently, both AL and SL are sized for two crew members (Table 2).

	Lander w/Airlock	Lander w/o Airlock	Habitat w/ 1 Airlock	Habitat w/ 2 Airlocks	Pressurized Rover w/ Airlock	Pressurized Rover w/o Airlock
Two Crew Airlock	1 airlock per landing (cycle 2 crew in and out)	Reasonable option (like Apollo)	Cabin depress for emergency ingress/egress	Reasonable option	Reasonable option	Reasonable option (like Apollo)
Four Crew Airlock	2 or 4 crew at once Questionable option	Reasonable option (like Apollo)	Cabin depress for emergency ingress/egress	Excessive capability Questionable option	Two crew per rover (Not an option)	Two crew per rover (Not an option)
Two Crew Suitlock	1 suitlock per landing (cycle 2 crew in and out)	Reasonable option (like Apollo)	Cabin depress for emergency ingress/egress	Reasonable option	Reasonable option	Reasonable option (like Apollo)

Reasonable option Questionable option Not an option Reasonable option, but out of scope for study

FIGURE 7 Sizing airlocks for two simultaneous EVA crew supports multiple applications.

TABLE 2 Mass totals for the airlock ripple throughout an architecture comprising multiple landers and surface elements (A/L 510 kg[a] and S/L 594 kg[a]).

OPTION	MASS	TOTAL WITH PUMPS	NOTES
1 A/L for Hab	510	631.1	Depress Hab element of emergency ingress/egress
2 A/L for Hab	1020	1262.2	
1 S/L for Hab	594	715.1	Depress Hab element of emergency ingress/egress
2 S/L for Hab	1188	1430.2	
1 A/L for Hab + Lander A/L	3570	4417.7	Assumes 6 crew landers
2 A/L for Hab + Lander A/L	4080	5048.8	Assumes 6 crew landers
1 S/L for Hab + Lander S/L	4158	5005.7	Assumes 6 crew landers
2 S/L for Hab + Lander S/L	4752	5720.8	Assumes 6 crew landers
1 A/L for Hab + Lander A/L + P Rover A/L	4080	4927.7	Assumes 6 crew landers
2 A/L for Hab + Lander A/L + P Rover A/L	5100	6068.8	Assumes 6 crew landers
1 S/L for Hab + Lander A/L + P Rover S/L	4752	5599.7	Assumes 6 crew landers
2 S/L for Hab + Lander A/L + P Rover S/L	5346	6314.8	Assumes 6 crew landers
ISS Pump[b]	79.4		
Equalization valve[b]	6.9		
Controls[b]	14.8		
Tanks[c]	20		
Total	**121.1**		

[a]Extrapolation from ISS crew lock, JSC20466 Rev. B.

[b]Space Station *Freedom* Airlock Depress/Repress System.

Design and Performance, SAE 921378, D. James, July 13–16, 1992.

[c]Estimate

CREW AND CARGO TRANSLATION

Translation of crew and cargo through an AL and an SL is shown in Figure 8. This section expands the description for each option.

Airlock

For ingress, two EVA crew members step through an open hatch into the AL chamber. The hatch is closed and the chamber pressurized. Then, the hatch to the habitat is opened, and, depending on suit design, the crew begin doffing while in the airlock or in the habitat. For egress, suit inspection, preparation, and prebreathe can take place either in the habitat or AL. After checkout, the habitat hatch is closed, the airlock pumped down, pressure equalized, and the crew opens the outer hatch to the lunar surface.

Getting an incapacitated EVA crew member through the AL in lunar gravity is easier than it would be on Earth. Assume the incapacitated crew member is unconscious and the suit is still functioning at normal pressure. The pressure will tend to shape the suit similar to an outstretched person except for low-torque joints that are free to move. A modern lunar suit (including the crew member) could be about 185 kg (407 lb). On the Moon, the mass is the same, but feels like about 30 kg (66 lb) when being lifted. This load is manageable by a single person for short distances in an emergency. The incapacitated crew member is "positioned" inside the airlock, allowing the hatch to swing closed while the active crew member stands in the remaining volume. Once the chamber is pressurized and the inner hatch opened, a shirt-sleeve crew member assists in moving the incapacitated crew member inside. AL hatches do not need to be sized for this contingency

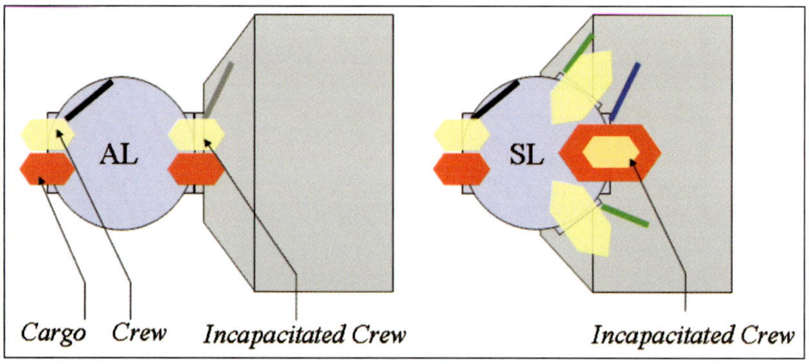

FIGURE 8 Crew and cargo translation operations are significantly different in the two concepts.

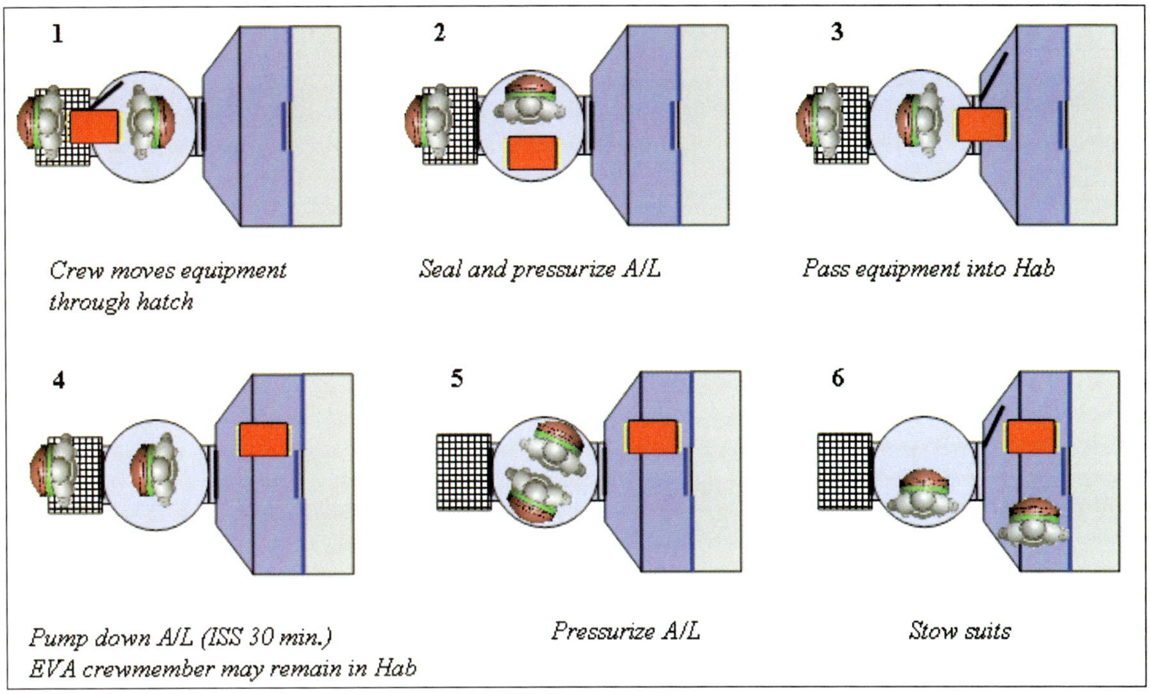

FIGURE 9 Six-step process ingresses hand-carried cargo through an AL.

because both hatches are already large enough to accommodate all operations.

There are two ways to move cargo through the AL. The first is for typical, moderate-size cargo on conventional EVAs (Figure 9). The second is for large cargo (e.g., rack-type equipment) that leaves no room for crew inside a small airlock. Such hardware is placed in the chamber by EVA crew, left alone to be pressurized, then transferred into the habitat by shirt-sleeve crew so that the AL can be resealed, pumped down, and cycled again by the waiting EVA crew. During the cargo cycle, the EVA crew use life-support umbilicals connected to the habitat exterior rather than consuming limited PLSS resources.

Suitlock

Ingressing the habitat using a SL begins like the AL scenario, but the procedure differs once inside the chamber. Each crew member backs into a rear-entry interface. Once the seal is secure, the pressure hatch enclosing the PLSS is opened, and then the PLSS itself is opened allowing the suited crew member to back out into the habitat. Egress with an SL first requires the crew to enter the "dustibule" to inspect the exterior of the suit, helmet, gloves, and boots. After this, the hatch to the habitat is sealed, the crew conducts a prebreathe and climbs into the suit, the PLSS hatch is sealed and verified, and then the outer PLSS cover hatch is closed, sealed, and verified. After checkout, pump-down, and pressure equalization, the EVA crew disconnects from the PLSS interface and egresses the SL.

As with the AL, an incapacitated crew member is positioned in the dustibule of the SL by the EVA buddy, who then closes the outer hatch. Because the incapacitated crew member can neither connect the PLSS interface nor exit the suit, a separate (cargo) hatch is required in the system, and this is a key difference from the AL. In this contingency scenario, the SL dustibule is pressurized, allowing the cargo hatch to be opened so that the incapacitated crew member can be brought inside. The active EVA crew member uses the PLSS interface to exit the suit as usual.

For cargo transfer (Figure 10), all items are transferred through the cargo hatch used for incapacitated crew. Another key difference from the AL design is that all cargo including small items must stay in the SL dustibule for retrieval after the chamber has been pressurized and the EVA crew has ingressed the habitat because only the suits can mate with the PLSS interface hatches.

SUIT INSPECTION, SERVICING, AND REPAIR

Because of the environment, type of work and frequency, suits require thorough inspection prior to EVA. Abrasive, statically clinging lunar dust easily abrades outer layers, scratches hard surfaces, degrades thermal properties, and interferes with integrity of seals. Therefore, the readiness of gloves, boots, visors, cameras, lights, displays, and thermal coatings needs to be verified for safe and productive excursions.

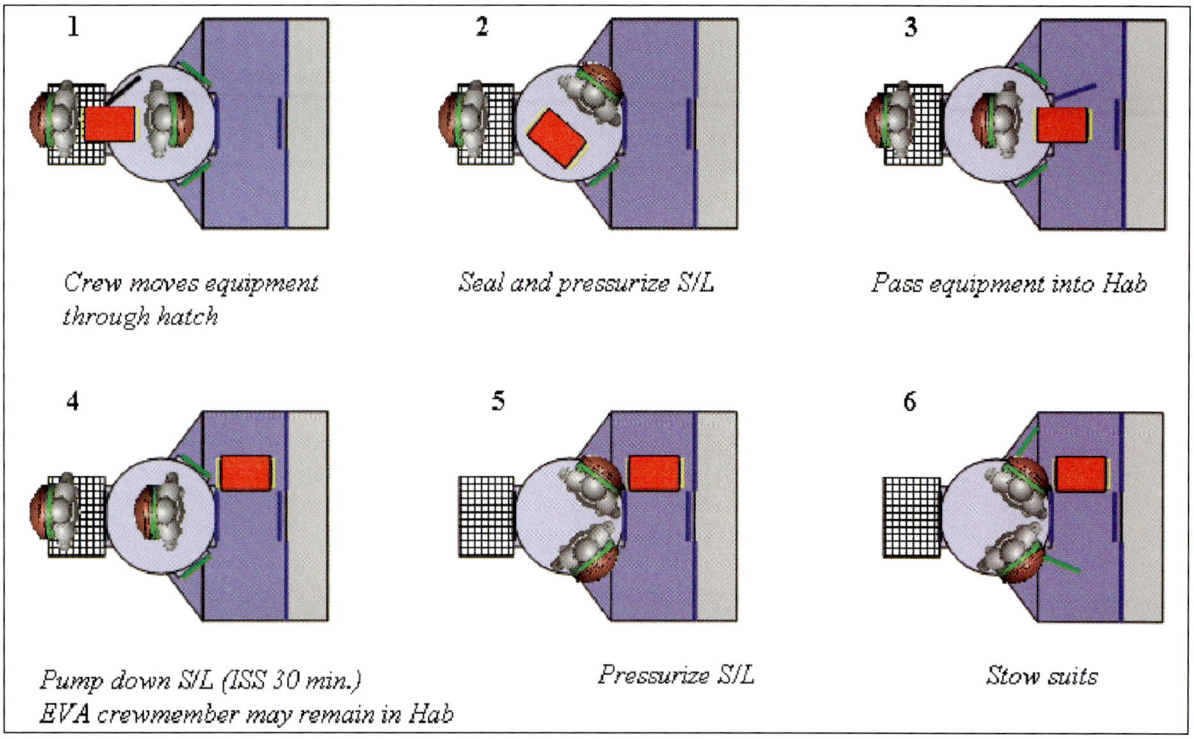

FIGURE 10 Six-step process ingresses cargo through an SL.

DUST CONTROL

Crew brush off loose dust outside before entering either AL or SL (Figure 11). A secondary level of dust control occurs in the dustibule. Crew in the SL exit the suit directly into the habitat while crew in the AL enter into a partitioned area inside the habitat. Although the suits are isolated from the rest of the habitat, EVA crew must enter this dusty area prior to each EVA to "preflight" the suit. Finally, for both AL and SL operations, dusty suits must be brought into the habitat periodically for scheduled or unscheduled maintenance.

SUIT STOWAGE, DON/DOFF

The AL can accommodate any suit type, whereas the SL requires a pressure-tight interface with a rear-entry suit. Suits can be stowed optionally in the AL (as is done with the shuttle airlock), whereas the SL is designed for suits to remain in the dustibule. With a crew of four and only one airlock (whether AL or SL), two suits are stowed in the habitat by default. Thus with either concept, there will always be two dusty suits inside the habitat (Figure 12).

Donning and doffing the suit is closely tied to how it is stowed. For the AL, an EVA anteroom is the most reasonable donning and doffing location. Volume estimated for this activity is shown in Figure 13. The SL is less forgiving. Rear entry requires both headroom and handrail above the helmet and accommodation of the angle needed to drive the legs down into the suit (Figure 14). These dimensions were discovered during lunar-gravity testing on NASA's KC-135.

SEPARATE VS INTEGRATED PRESSURE VESSEL

The SL shares a common structural bulkhead and is therefore integral with the habitat. The AL, however, is a separate "end-item" system. It can be delivered, connected, and removed separately from the habitat. Advantages of such a separate system are as follows: 1) flexibility in delivery manifesting on a lander; 2) ability to be relocated, e.g., removed from the ascent vehicle for installation on the surface habitat or moved from one surface location to another; 3) ability to be replaced because of wear, damage, or upgrade; 4) optimized geometry not tied to habitat dimensions; 5) potential commonality with other small crew cabs, e.g., on the ascent/descent vehicle; and 6) separate procurement, design, development, fabrication, test, and checkout.

Configurations other than those used in this comparative analysis are possible, but generally fall into two types: a full diameter or a vertical cylinder. NASA Ames Research Center has studied both SL configurations as shown in Figure 15. The full-diameter concept uses the end-dome of a cylindrically

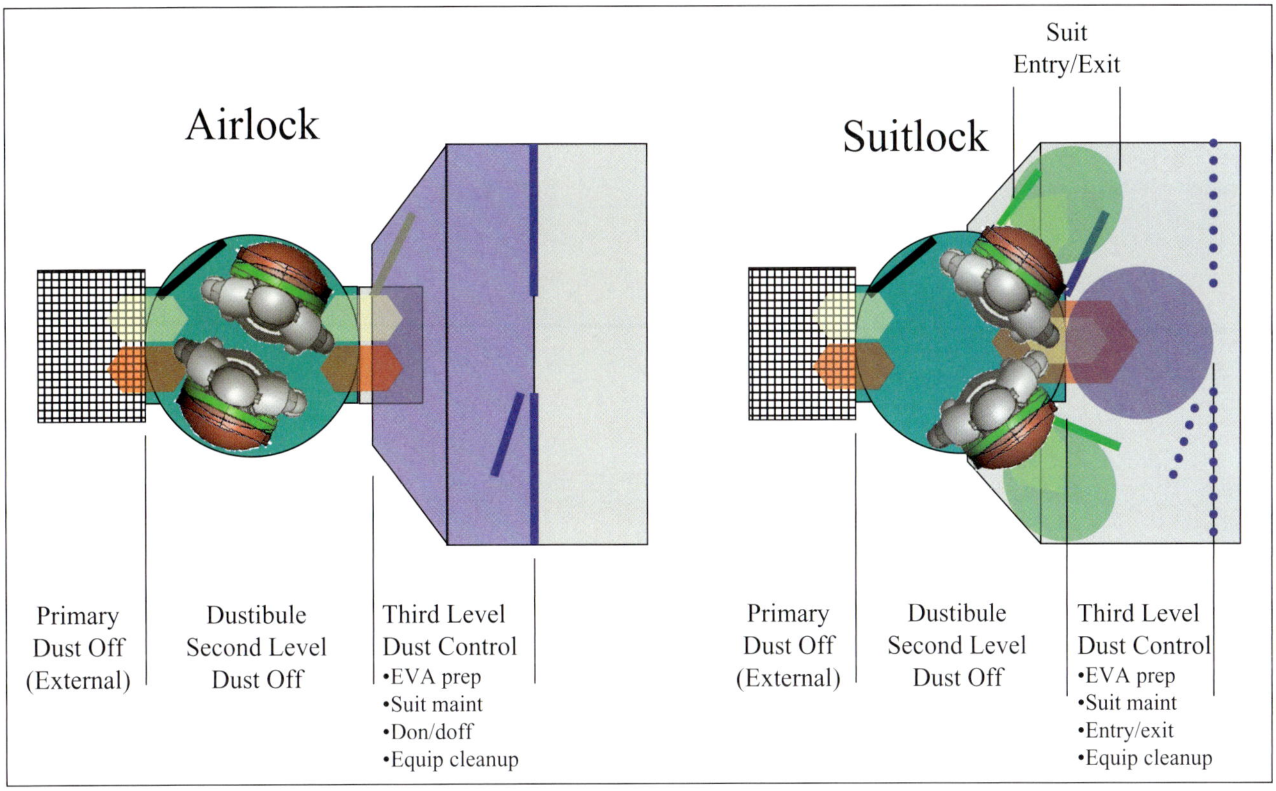

FIGURE 11 Dust control, like cargo transfer, complicates what might at first appear as simple airlock operations.

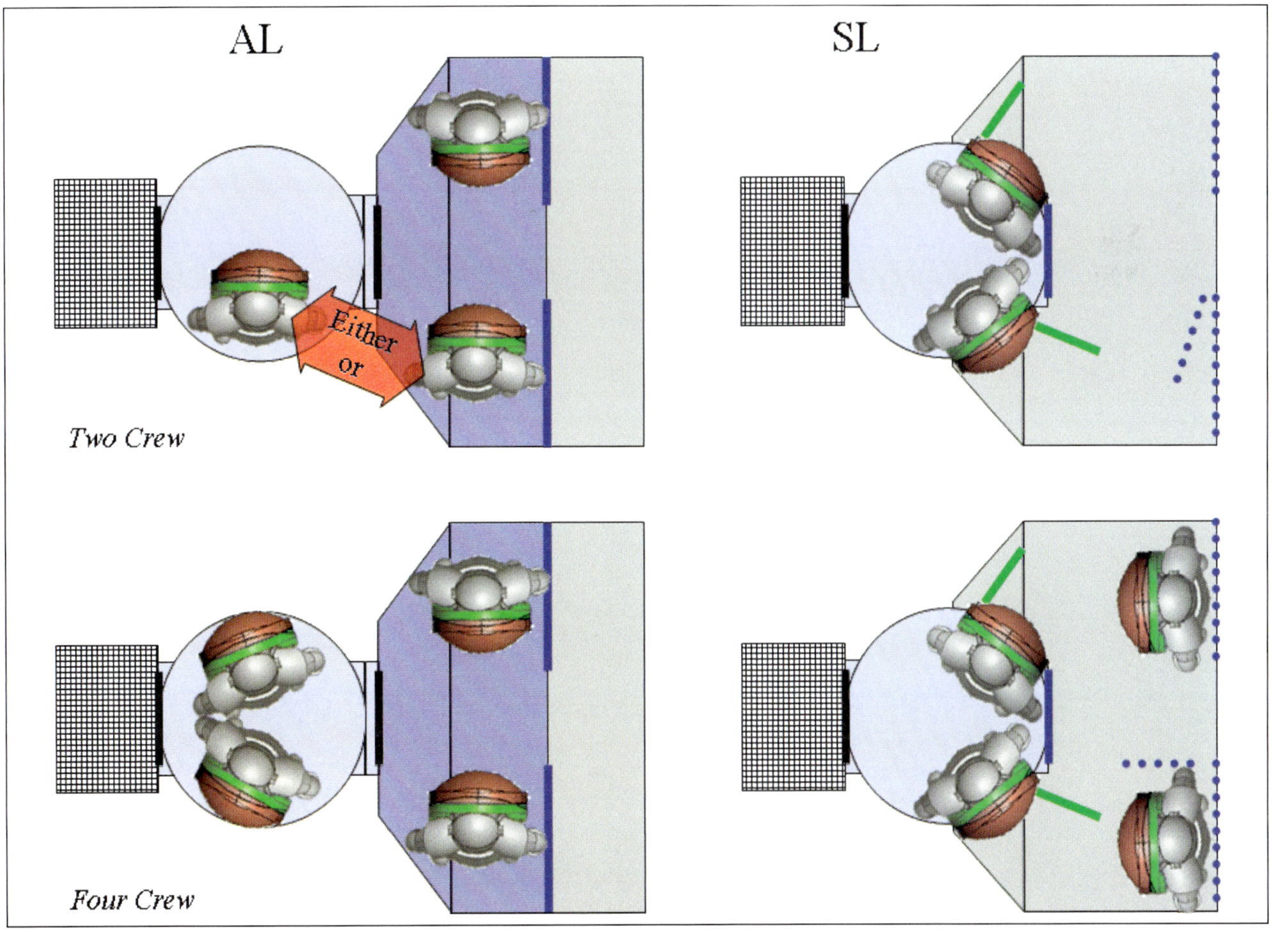

FIGURE 12 Systems using a single airlock for four crew stow dusty suits inside the habitat.

Suit Configuration	Don/Doff Technique	Donning Volume
Apollo A7LB (lunar EMU)	Dual-Plane Zipper	5.47 m³ (193 ft³)
Shuttle SSA	Single-Plane Horizontal Body Seal Closure	3.64 m³ (129 ft³)
Shuttle LES	Single-Plane Zipper	2.63 m³ (93 ft³)
MKIII*	Single-Plane Rear Entry	3.23 m³ (114 ft³) 4.78 m³ (ft³)**

* Most representative of lunar suit
** (Howard, 05)

FIGURE 13 Donning/doffing modern lunar suits consumes more volume than for shuttle suits in microgravity (Wilmington and Stanley 1992).

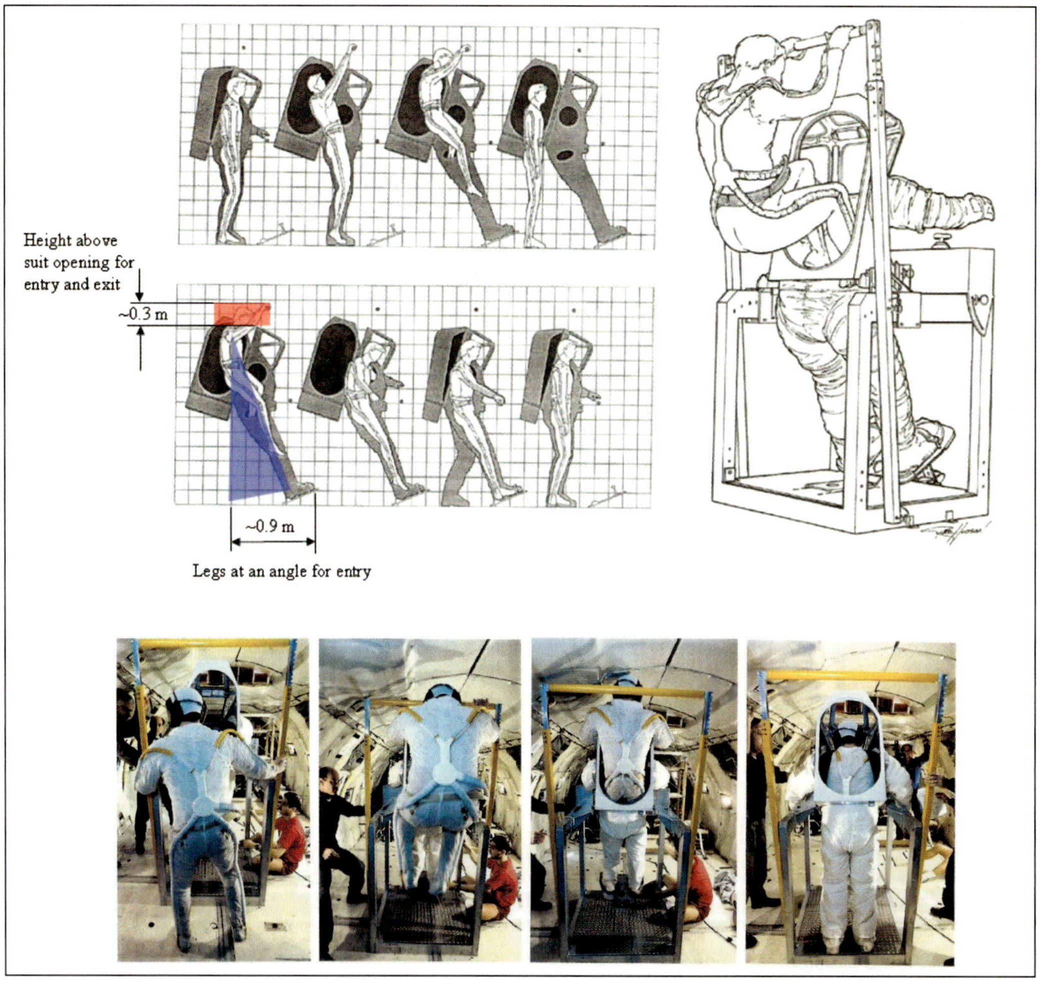

FIGURE 14 Donning/doffing requirements were developed during lunar-gravity testing.

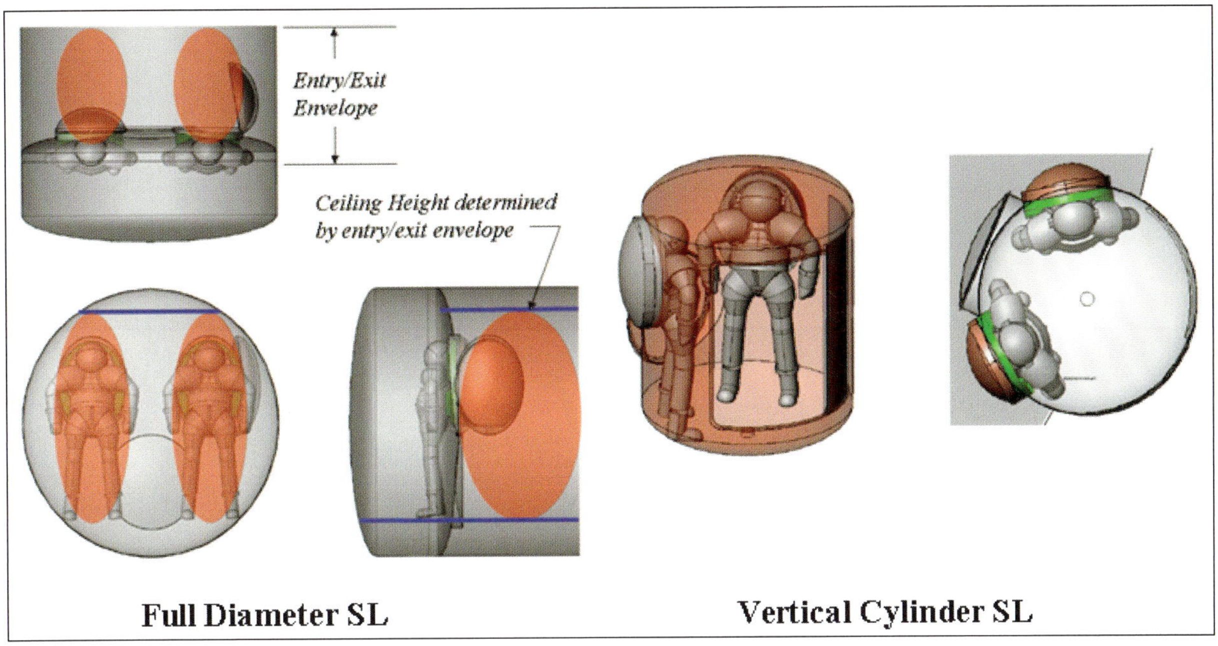

FIGURE 15 Bulkhead-partitioned habitat module or separate, vertical cylinder are the two most reasonable lunar surface airlock configurations.

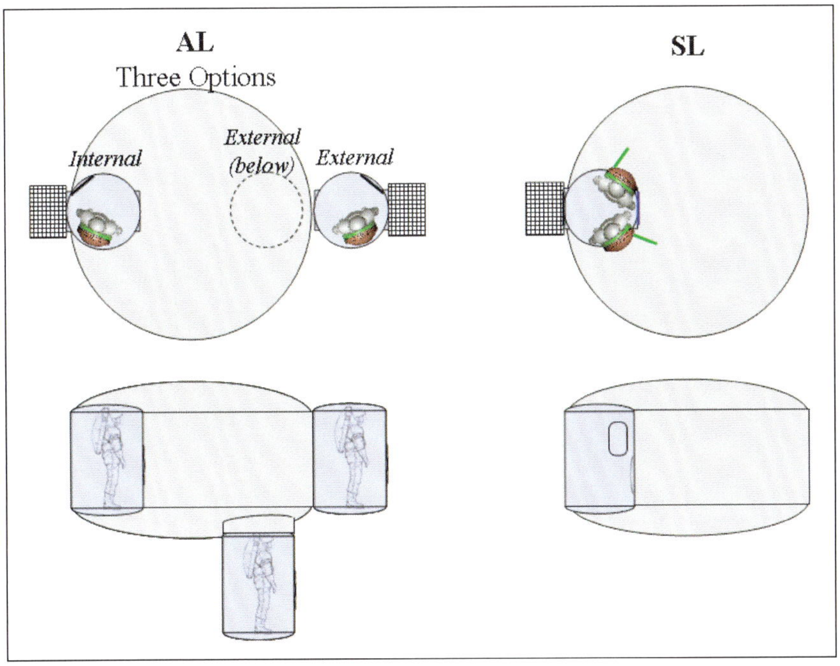

FIGURE 16 Separately fabricated AL offers more integration flexibility than the SL.

shaped, horizontally oriented habitat. With respect to volume, full end-dome concepts conforming to habitat dimensions typically result in oversized SLs, with increased pump time and air loss. The vertical-cylinder configuration is free for SL function optimization, but requires structurally shaping the habitat end-dome around the SL to provide adequate interior volume for entering and exiting the suit.

Another lunar outpost habitat configuration would be based on larger-diameter vertical habitats equal to the dynamic envelope of a large launch shroud. Figure 16 shows three options for accommodating a separable AL but only one for a separate-vessel SL (entirely within the larger pressure vessel) because it needs ample interior perimeter for suit entry and exit.

CONCLUSION

The SL concept is appealing because it is innovative and appears at first glance more efficient for cycle time and dust control than a conventional AL. However, the AL trades more favorably when the details are considered. It is lighter overall, has fewer leak paths, is far simpler to operate, uses the same, large hatch to transfer crew, cargo and incapacitated crew without delays, and has greater commonality with other small pressure vessels throughout an integrated system architecture (Table 3). The AL does not impose highly specialized design requirements on the suit or habitat structure, and it can be procured, fabricated, tested, checked out, and flight manifested independently from the rest of the habitat system.

TABLE 3 AL concept compares favorably to the SL concept overall, against the objectives listed in the introduction.

OBJECTIVE	FINDINGS	FAVORED
Minimize resources loss	Small volume best Pump required Pressure-assisted hatches minimize leak AL has minimum operable seal length	AL
Design for simple operations	AL 1 hatch vs SL 4 hatches AL eyes forward vs suitlock backing into PLSS dock AL does not require rear-entry suit Consolidated EVA support configuration AL does not require equipment for different crew heights	AL
Design for safety	Two means of ingress/egress AL (large hatch) better for incapacitated EVA crew member	AL
Seek commonality	AL with 1 hatch vs SL with 3 hatch types	AL
Minimize program mass	AL weighs less than SL AL only on habitat (none for lander and pressurized rover)	AL
Address lunar dust	AL and SL require IVA prep zone (dusty) SL offers better routine dust control	SL
Crew and cargo	Small AL and SL volume means cycling pumps for cargo transfer AL (large hatch) better for EVA handling	AL

The AL does not possess the intrinsic dust control configuration of the SL, but the operational effectiveness of SL dust control is questionable for practical scenarios. Because of extensive-surface EVAs, lunar suits are at continuous risk of wear, damage, and degraded performance. To verify readiness prior to each EVA, suits require close inspection on the "dusty" side of the SL. Such inspection routinely exposes shirt-sleeve crew and the "clean" side of the habitat to the dusty side, compromising the intended separation. Furthermore, EVAs for four crew are scheduled in alternating pairs caused by daily fatigue. But because contemporary suits are not personally interchangeable, alternating crews have to install their suits on the dusty side of the SL for the next EVA if only one SL is included in the design. Thus dusty suits are introduced into the habitable area after every EVA, just as with the AL. The intended advantages of the SL approach do not hold up well in operational scenarios, and so the simpler AL concept is better. |

References

Howard, R. (2005), "CEV Internal Volume Study Report," NASA, Oct. 21.

Wilmington, R. P., and Stanley, D. C. (1992), "Space Suit Donning Volumetric Evaluations," Lockheed, Technical Paper 30439 (JSC 26019), Houston, TX, Aug.

Bibliography

EVA Catalog: Tools and Equipment, NASA Johnson Space Center, 20466, Rev. B, Nov. 1993

Griffin, B. N. (1994), "A Space Suit for Productive and Safe Extravehicular Activity (EVA)," *Engineering, Construction, and Operations in Space*, Vol.2, edited by R. G. Galloway and S. Lokaj, American Society of Civil Engineers, New York, pp. 1372–1381.

Griffin, B. N., Spampinato, P., and Wilde, R. C. (1999), "Extravehicular (EVA) Systems," *Human Spaceflight: Mission Analysis and Design*, edited by Wiley J. Larson and Linda K. Pranke, McGraw–Hill, New York, pp. 707–737.

James, D. (1992), "Space Station Freedom Airlock Depress/Repress System Design and Performance," Society of Automotive Engineers, Warrendale, PA, Paper 921378, July.

Webbon, B., Luna, B., Brown, J., Gonzales, A., Jones, H., Koss, B., Smith, D. (2007), Society of Automotive Engineers, Warrendale, PA, ICES Paper 2007-01-3245.

18

habot concept

MARC M. COHEN
AND
ROSS A. TISDALE

INTRODUCTION

HABOT IS A CONTRACTION OF HABITAT AND ROBOT. It constitutes an approach that combines human and robotic exploration capabilities into a mobile lunar-basing architecture (Mankins 2000; Mankins 2001; Cohen 2003; Cohen 2004a). The Habot unit concept consists of a self-mobile habitat landed autonomously at a specific landing zone (LZ) on the Moon. It moves under its own power to a base site. More Habots follow over a period of one to two years, landing at the LZ and moving themselves to the base site. After verification of the Habot base, human crews arrive to conduct surface missions.

The Habot is not intended to serve as a crewed spacecraft in low Earth orbit, cislunar space, or lunar orbit. It is intended for crew use only on the lunar surface. The crew would travel to the Moon in a separate leg of the architecture [e.g., crew exploration vehicle (CEV) to lunar orbit and *Altair* lander down to the surface]. This system of vehicles could make common use of a nominal six-legged lander for both the crew lander descent stage and the Habot surface mobility system.

A modular, mobile base is an operationally attractive concept despite its high "systems overhead." This chapter explores several design implications of making the Habot concept real, in the process exposing and analyzing its benefits and limitations from the configuration and accommodations standpoints.

MISSION ARCHITECTURE

The design parti (point of departure) for the Habot begins from the analysis of advantages of a mobile base vs a stationary base, as presented in Table 1. This analysis leads to the question: why not make the entire base mobile? Then all of the assets, resources, reliability, and redundancy of the lunar mission move with the exploration crew. This approach means that the laboratory facility would travel with the explorers, affording them the capability to conduct complex and sophisticated scientific assays and analyses on site, without a need to "return to the base." With Habot, wherever the crew may roam, they are at home at the base.

The nominal mission timeline is 100 days, allocated to a primary mission lasting through two lunar diurnal cycles (56 Earth days). There are eight Earth days' margin for liftoff from the Moon and 36 Earth days reserve. The initial crew size is four. The baseline number of crew missions is 10, for a total campaign crew time of 560 Earth days, and total capability for 1000 crew days on the Moon during those 10 missions.

After landing, each lander jettisons its terminal descent propulsion system and roves a safe distance of 5–10 km away from the LZ to a preselected base site. The mobile modules can operate in an autonomous or teleoperated mode to navigate the lunar surface. At the base site, the modules can connect autonomously or telerobotically into a base cluster, making pressure port connections among themselves. Once enough units arrive to form a multimodule pressurized lunar base, they assemble at a site of scientific or technical interest and make the vital interconnections for pressurized access, communications, data, life support, etc. After the base is verified as safe for crew, the first lunar expedition crew launches from Earth. The crew lands and transfers to the base, where they set up housekeeping and begin their work. When the

This material is declared a work of the U.S. Government and is not subject to copyright protection in the United States.

TABLE 1 Comparison of stationary and mobile lunar-base approaches.

STATIONARY LUNAR BASE	MOBILE LUNAR BASE
ADVANTAGES	
Accumulation of assets in one location; economy of agglomeration Economies of scale in one location Potential for larger, permanently situated habitats (e.g., inflatables) Potential to use regolith as shielding Ability to situate power supply permanently (e.g., nuclear reactor) Less complexity—no need to make everything mobile	Greater mobility of assets affords superior exploration opportunities and operations Makes best advantage of commonality and economy of scale through repetitive production Ability to modularize habitat modules to match launch vehicle capacity Greater system redundancy Ability to bring the science lab to multiple sites of interest—excursions are not "just picking up rocks" Single type of EVA access module for both excursion rovers and base Can establish LZ at each new location, and moving units from the LZ is not a burden.
DISADVANTAGES	
Program risk of putting fixed base in "wrong" location and needing to support distant science field operations Necessity for dissimilar heavy equipment movers and pressurized rovers Cost and burden of moving all modules and equipment from LZ Risk of stranding a rover excursion crew far from the base Inflexibility/impracticality of accessing widely separated science sites/objectives	System overhead of making everything mobile System overhead of dividing major mission functions into small units Risk of roving base not returning to the ascent vehicle Risk of failure to reconnect units at new base location Increased unit complexity Must carry its own radiation shielding (Cohen 2004c)

crew members complete their mission, they launch from the lunar surface in their lander's ascent stage, rendezvous with the CEV in lunar orbit, and return to Earth. In the subsequent weeks or months, the Habots can simply await the next crew or separate and move across the lunar surface to a new location of scientific interest before the crew arrives. A new Habot logistics module could land near the second base site to resupply consumables and bring new equipment.

During the lunar day, the crew conducts the exploration portion of the mission, working from the base. In addition to the base cluster, the concept includes two or three excursion Habots. The crew uses the excursion Habots as pressurized rovers, meeting and docking as necessary for various crew operations and procedures. As the lunar day approaches its end, the excursion Habots return to the base cluster, where they remain except for contingency or emergency operations. During the lunar night, the crew members stay primarily in this multichamber lunar base and pursue work they can perform with minimal need for EVA or rover excursions, such as sample analysis, equipment repair, and preparation of reports. This arrangement constitutes the baseline configuration for the cluster.

It is also possible for the crew members to drive or travel with the Habots when the base migrates, although their presence is not required for this operation. The crew can also use individual units as pressurized rovers to explore the lunar environment nearby. Figure 1 shows articulated legs carrying manipulator devices that can pick up rocks. A hexagonal "benzene ring" cluster configuration appears in the middle ground at the right.

The speed at which the Habots move is secondary to safe navigation and movement over varied and rough terrain. Acceptable top speed might be only about 1 km/hr, although in difficult terrain 0.1 km/hr might be the limit. Without crew present to consume life support and other consumables, or waste time waiting to arrive at a scientific destination, the slow end of the speed range is available to ensure predictable, reliable, and safe operations. Moving at low speed not only reduces risk but also the need for power for acceleration,

FIGURE 1 Habot mobile lunar base concept (artist: Pat Rawlings).

with its attendant mass penalties. The crew, meanwhile, waits and prepares safely on Earth until the Habots arrive at the new base location.

MOBILITY SYSTEM

A mobile lunar base allows explorers to bring the base close to sites of scientific interest to make the most complete investigation without the severe constraints and limitations imposed by rover traverses and EVA sortie time. A further advantage is that it is possible to land new mobile modules with new equipment, supplies, or logistical support in the path of the moving ensemble. These new units could then join the caravan to continue on the journey or provide a cache of supplies for the crew to pick up along the way. The mobility system is enabling for the Habot concept. It is a challenging problem because it requires a durable and reliable transportation system to carry 10 mt across rough terrain without danger of overturning or losing control.

A mobility system prototype that could meet Habot requirements is the all-terrain hex-limbed extraterrestrial explorer (ATHLETE) rover developed by Brian Wilcox and Jaret Matthews at the NASA Jet Propulsion Laboratory (JPL). ATHLETE is a "wheels-on-legs" mobility system that combines the best features of wheels and legs: first as landing gear for a lander, then using wheels over traversable substrates, and finally using the legs to walk out of terrain too soft or difficult for rolling. ATHLETE carrying capacity, ground clearance, and dynamic stability are impressive. Figure 2 shows Brian Wilcox with one of the one-third-scale ATHLETE prototypes.

FIGURE 2 NASA-JPL multipurpose mobile platform prototype ATHLETE.

Contemporary work by NASA to determine the Constellation lunar surface systems architecture recognizes the operational need for large-scale element mobility (e.g., lander offloading, lander relocation, and separation of LZ from outpost assets) and the potential benefit of combining mobility with habitat modularity as pioneered by the original Habot concept. Mobility has emerged as a key capability for the initial outpost phase of lunar exploration. Technology development has continued on the ATHLETE chassis, and several concepts using versions of ATHLETE as landing gear, habitat platform, offloader, and truck have been evaluated.

Figures 3 and 4 show ATHLETE field tests conducted in 2008 that demonstrated level driving on uneven terrain, mutual acquisition and navigation by multiple chassis, and precise mutual positioning of payloads for interface mating. All three capabilities underlie the Habot mobile base concept and provide confidence in the viability of the basic architecture.

The mobility system interacts with the Habot base configuration in profound ways to achieve the full intent of the concept, wherein each unit is mobile and the base is an integrated "system of systems." The integration challenge is to match the base elements to a mobility system that can align and place them correctly within the overall configuration plan. The design challenge includes both the architectural issue of how to combine the modules into a practical base and the technical issue of how to place them precisely into this configuration. Smaller units are more amenable to mobility both for short, one-time traverses and for repeated long hauls.

HABOT REPETITIVE PRODUCTION

A fundamental driver behind the Habot concept is that—aside from conventional methods like reducing launch mass and mission scope—repetitive production of common elements is a well-known way to control costs. Despite the various tools to chip away at cost such as value engineering, earned value management, and life-cycle cost analysis, no miracle awaits for reducing launch costs, development costs, or fabrication costs of advanced space hardware by half or more. The "non-NASA" way to reduce fabrication and operating costs over the long term is to make a vehicle that is simple, reliable, and produced in numbers significant enough to achieve economy of scale.

FIGURE 3 One-third-scale ATHLETEs carry habitat mock-ups in field tests in 2008.

FIGURE 4 Multiple ATHLETEs demonstrate the ability to "find" each other and precisely position habitat payloads for berthing, exemplifying a key technology for Habot-derived architectures.

To understand the importance of repetitive production and the economies that it affords, it is useful to compare space module production to other industries. Table 2 compares three industrial production models: the automobile industry, the commercial aircraft industry, and the International Space Station (ISS) viewed as an industry. There are two comparisons to consider: commonality and the number and size of modules, whether they are similar or dissimilar, common or noncommon. The ISS ensemble, for example, consists of virtually one-of-a-kind modules. The two competing philosophies are integration and modularization.

Integration Philosophy

This argument states it is most cost efficient in terms of systems engineering and integration (SE&I) to combine all functions into large modules that can be built, integrated, and then transported. In this way, there is just one type of module or unit, and SE&I costs are not duplicated or "wasted" on a second module type. Some estimates of SE&I costs as a

TABLE 2 Comparison of automobile and aircraft industries to ISS considered as an industry.

AUTOMOBILE PRODUCTION	AIRCRAFT PRODUCTION	ISS MODULE PRODUCTION
$1–3 Billion in R&D, tooling, and facilities per vehicle type (e.g., cars and light trucks) 1 Production facility Production rate: 50 cars per hour; ~160,000 per year Standardized operation and maintenance; small service stations can maintain and repair any vehicle	$6–8 Billion in R&D, tooling, and facilities per aircraft type (e.g., 787) 1 Production facility Production rate: 100 to 200 planes per year Standardized operation and maintenance; regular airport service depots can maintain and repair the aircraft	$15–20 Billion in R&D, tooling, and facilities, for 5 module types (each is unique) 5 Production facilities Production rate: 5 modules + 3 nodes in ~10 years Operations and monitoring for each module is unique Differences in modules create vast complexities in integration and maintenance that require huge ground support

fraction of the total for ISS modules run as high as 75%. In the end, there will be as many as five different ISS modules not counting nodes and airlock.

Modularization Philosophy

This argument states it is more economical, reliable, and safe to spread the risk of a human lunar mission among multiple modules with a very high degree of commonality. In this view, the ability to mass produce common modules can reduce the cost and simplify the process of SE&I, despite some variation in the functions assigned to each module. Given a truly common module that can be adapted to house a variety of functions, a lunar exploration program would become much more manageable. A small common module in the range of 5–10 mt would be a candidate for launch on the Ares V for a "direct-to-the-Moon" mission without assembly or rendezvous in Earth orbit. The Habot architecture is consistent with this modularization approach.

HABOT-BASE ARCHITECTURE

The core of a Habot base consists of basically identical mobile modules customized just enough to support the crew's various living and working activities. The configuration emerges from the set of module types and conversely informs the design of configuration-specific arrangements.

Habot Module Types

The original Mankins treatment (Mankins 2000; Mankins 2001) proposes six Habot modules forming a modular, integrated lunar outpost cluster. The Habot base would consist of more module types deriving from the same basic pressure vessel, platform, and chassis. Together, these modules comprise the complete living and working environment: Figure 5 shows the general layout of a Habot basic module that could be adapted to many functions:

- EVA access facility, including no-volume suitport for egress, NASA robonaut anthropomorphic telerobot, and EVA suit stowage and maintenance
- Ward room (backup command and communications center)
- Crew cabin (sleeping quarters) in two or more modules
- Life sciences laboratory
- Physical sciences laboratory (including cupola and observatory)
- Bioregenerative life-support laboratory
- Physical/chemical life-support laboratory
- Excursion Habot (rover) for local exploration
- Fuel and logistics depot
- Powerbot mobile energy unit w/reactor or other source

Base-Plan Configurations

Figure 6 presents Mankins' early concepts for the base cluster. In both diagrams, a detached Habot rover unit (excursion Habot) appears in proximity to the airlock/docking module. Mankins chose the hexagonal plan geometry as a deterministic way to evolve a hexagonal grid, three-axis plan. However, when imposed upon pressure vessel design, it poses serious technical problems. Flat surfaces are not optimal for a pressure vessel, so the six faces require stiffening, which adds considerable structural mass.

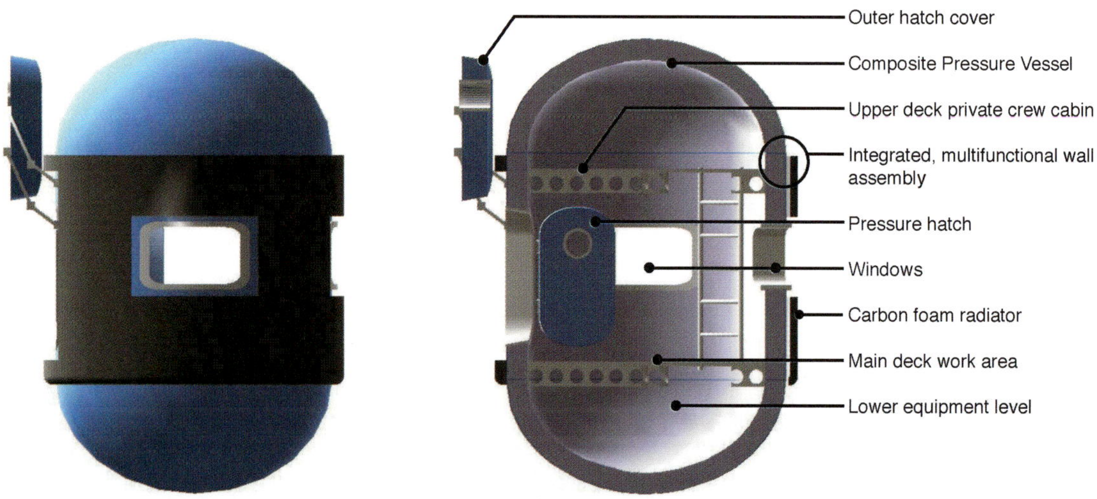

FIGURE 5 General concept for a Habot basic module (artist: Ross A. Tisdale).

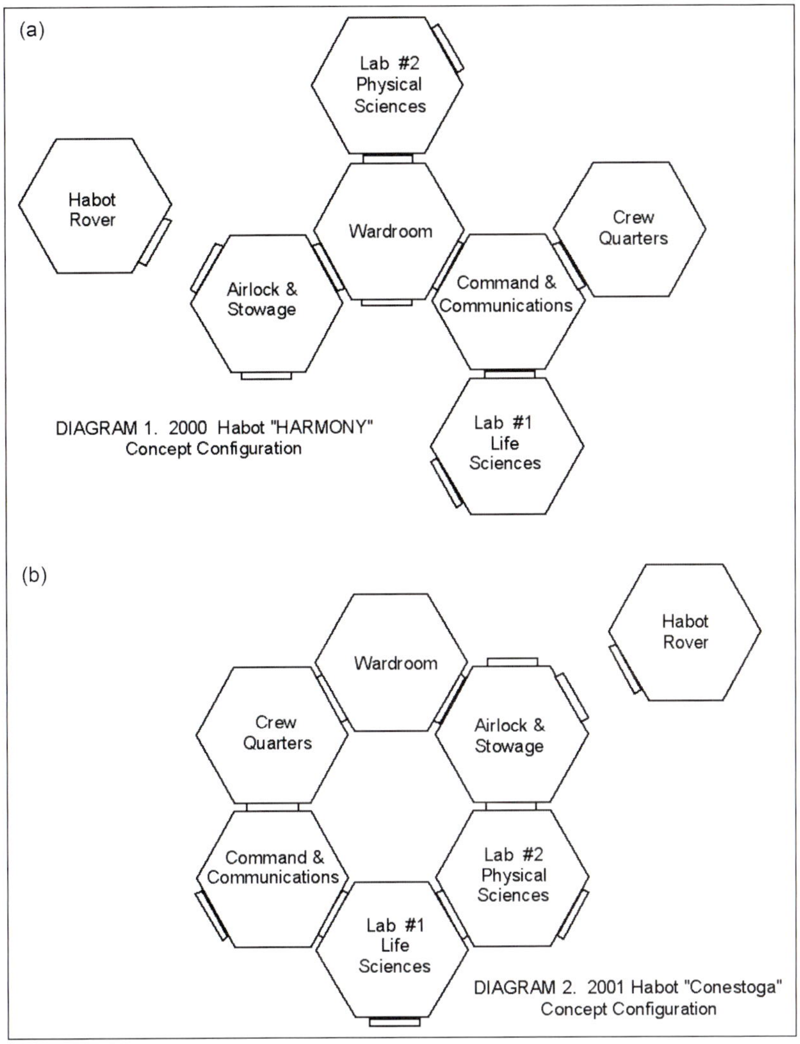

FIGURE 6 Mankins' original configuration concepts for the Habot base cluster: a) 2000 Habot Harmony concept configuration and b) 2001 Habot Conestoga concept configuration.

In subsequent module plan conceptual designs, the floor plans are circular to represent an upright cylinder, consistent with Figure 5.

OPEN-ENDED PLAN Figure 6a shows the relatively open-ended Harmony configuration Mankins presented in the original 2000 Habot paper. Module labels correspond to designations in the original concept. This plan has the potential advantage of being able to grow organically on axis within the constraints of 60 deg/120 deg/180 deg geometry. It would allow replacement Habot units to be attached at almost any open port as the plan expands, contracts, or changes.

CLOSED-LOOP PLAN Figure 6b presents the closed-loop Conestoga (benzene ring) configuration that appears in Figure 1. The labels show the authors' interpretation of those modules into the ring configuration. The benzene ring affords the surest means of achieving dual access and dual remote egress from all of the core modules. However, the benzene ring concept poses the additional challenge of the "last module docked" between two pressure ports angled 120 deg apart. Although U.S. Patent 4,728,060 [triangular-tetrahedral space station ("Space Station Architecture" 1998)] claims lateral docking of a module between two nodes, no approach to achieve it has been developed. In any case, the case shown in the figure requires alignment of both pressure ports at an obtuse angle rather than along one axis.

CONFIGURATION ANALYSIS

A first-order architectural analysis of Habot-base configurations addresses both the clustering of modules to form a base and the internal configuration of typical modules. Module and base geometry must make best use of limited area and volume. Yet individual Habot modules are so small that interaction of module-scale and base-scale geometries becomes critical to the system-of-systems design.

Configuration Criteria

The analysis uses nine base configuration criteria:

1) Provide circulation access to all parts of the base. Dual access is preferable.
2) Provide egress from all parts of the base; ideally, provide dual remote means of egress for fire safety and escape in emergencies.
3) Provide efficient, utilitarian amount of equipment functional area and volume; the shapes of these areas and volumes must be compatible with the circulation.
4) Provide the ability to create suitable workspaces and social areas within the base.
5) Provide ease of mobility for assembly and disassembly of the base.
6) Provide efficient thermal view angles for body-mounted radiators.
7) Provide modules for docking excursion Habots and accommodating EVA access.
8) Minimize number of pressure port angle variations.
9) Economize docking ports.

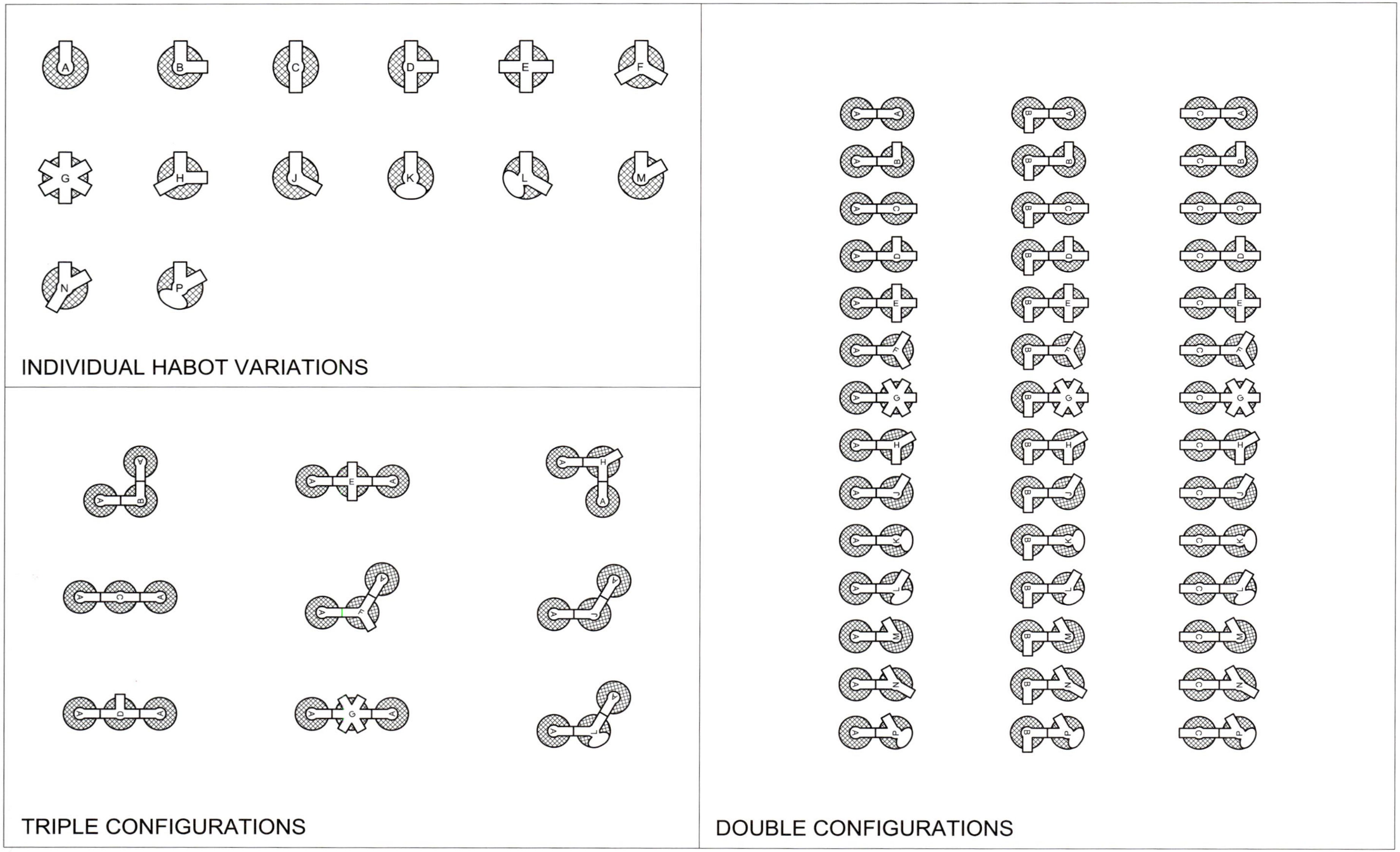

FIGURE 7 Individual module variations (upper-left section) and pairing patterns (remainder of figure). Letters refer to plan types discussed in text and compared in Table 3.

Floor Plans

Exterior wall thickness would be approximately 20–25 cm to accommodate radiation shielding, thermal insulation, body-mounted thermal radiators, and micrometeoroid protection. This wall thickness would enclose a floor plan in the range of 3.5–4.5-m interior diameter. This dimension would allow the Habot module to fit within the 5-m-diam shroud of a Delta IV Heavy or Atlas V launch vehicle or a larger Ares V rocket.

FLOOR AREA ANALYSIS The upper-left quadrant of Figure 7 shows a figure/ground pattern analysis of the circulation/equipment allocations of nine alternative Habot plans. For this exercise, interior diameter is set at 3.75 m based on an earlier study for a Habot module of 50 m^3 total pressurized volume (Cohen 2004b). This diameter yields 11-m^2 interior floor area (exclusive of module wall thickness). This is even less than the 4-m version's 12.5 m^2, so that the architectural layout is more challenging but nonetheless indicates it might be possible to work within these volume and mass constraints. Essential dimensions for the 3.75-m-diam plan types are in Figure 8.

CIRCULATION AREA The subtractive quality of circulation area becomes significant as the number of docking/berthing/circulation pressure ports increases. Even with the minimum number of two ports shown in plans A, B, C, and J, the circulation area severely diminishes the equipment solid packaging area. For plan E, with four pressure ports at 90 deg, the area left for equipment is merely residual between the pressure ports. This strongly suggests that an interior diameter of 4 m or less is impractical for feasible Habot design.

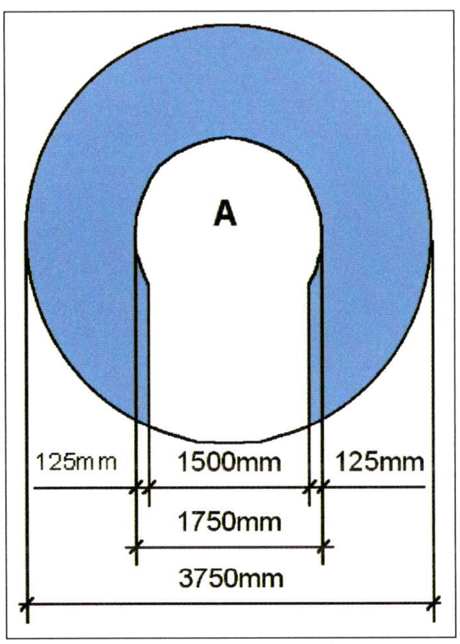

FIGURE 8 Circulation area (white) and usable equipment area (blue) within a single-port module.

Circulation is not a total liability, however. It provides a variety of useful functions besides allowing people and hardware to move around in a module or the base. Inside a confined spacecraft habitat, circulation area and volume provide working room in front of equipment racks, de facto social space, and the important perception of spaciousness, which researchers can model quantitatively (Wise 1988).

Plan composition rules for comparative analysis are as follows:

- Each plan type preserves a center circulation area 1750 mm in diameter.
- This center circle connects to at least one passage, 1500 mm wide, leading to a pressure port (standard ISS 1250 mm wide); total circulation area equals the center circle plus the area of all passages.
- Usable equipment area equals the total floor area minus the circulation area.
- Each additional pressure port in the outer wall requires a dedicated passageway.

Habot Plan Types and Areas

The most obvious observation is the inverse relationship between circulation area and solid equipment area: the more circulation area, the less area and volume available for equipment or other functions.

TYPE A Here is a minimal circulation arrangement with one pressure port entry and one passage to the center circle. Practical application as a living and working environment would be limited to an excursion Habot or rover in which the second means of egress is the EVA airlock. As in Figure 6, Plan A can also be applicable for a logistics module that would dock to one spare pressure port on the base cluster.

TYPE B L-shape divides the equipment area into a major "three-quarters" area and a minor "one-quarter" area. However, three one-quarter areas do not equal one three-quarters area because of the strong impact of circulation. The ratio of circulation to equipment area is essentially 1:1. Plan B represents the minimal arrangement for a base assembly pattern based on 90-deg bends in the circulation pattern. It would be used in an orthogonal scheme.

TYPE C Straight passage widens in the middle. Plan C represents the simplest connecting unit that affords dual access and dual remote egress. This plan divides the equipment area into two bilaterally symmetric sides. The ratio of circulation to equipment area is essentially 1:1.

TYPE D The T-shape attaches a module to the side of a linear arrangement of Habot units. This plan creates two smaller one-quarter areas and one larger half area.

TYPE E The orthogonal cross axis divides the floor into four equal areas. Plan E allows two linear arrangements of Habots to cross at the center. This cruciform plan would afford four end units at which to connect excursion Habots or airlock units, compared to the two available in a single linear arrangement.

TYPE F This conveys a radially symmetric arrangement of three pressure ports 120 deg apart. The three passages divide the equipment area into three equal-third portions. The ratio of usable equipment area to circulation area is approximately 1:2. This plan reflects the nodes in Mankins' Harmony base cluster. They would also be used in a "benzene ring" plan to allow attachment of excursion Habots to the third port.

TYPE G This type displays the maximum circulation area possible, with six pressure ports and six passages 60 deg apart around the circumference. At less than 5%, floor area available for equipment is nearly nonexistent. However, this plan might be appropriate for a central circulation hub, exercise or recreation area, or as a social area with stowable furnishings and equipment.

TYPE H This presents a hybrid option combining a 90-deg bend with an obtuse angle. This plan would be used as a node in a scheme that combined two or more of the prominent angles (60, 90, or 120 deg). The circulation pattern results in a one-quarter area and two bilaterally symmetric 135-deg oblique areas, each slightly larger than 120-deg equal-third areas.

TABLE 3 Properties of nine Habot floor plan variations for 3.75-M-diameter module version.

PLAN TYPE	NAME	CIRCULATION AREA	"USABLE" EQUIPMENT SOLID AREA	PERCENT CIRCULATION AREA	PERCENT "USABLE" EQUIPMENT SOLID AREA	RATIO OF "USABLE" AREA TO CIRCULATION AREA	REMARKS
A	Single	4.04	7.00	36.6	63.4	1.73	Occurs only as Habot rover w/ airlock
B	Double 90 deg "L"	5.60	5.44	50.7	49.3	0.97	Equivalent to plans C and J
C	Double 180 deg "Fat I"	5.64	5.40	51.1	48.9	0.96	Equivalent to plans B and J
D	Triple 90 deg "T"	7.16	3.88	64.9	35.1	0.54	—
E	Quad 90 deg "+"	8.71	2.33	78.9	21.1	0.27	Node for cruciform plan
F	Triple 120 deg	7.27	3.77	65.9	34.1	0.52	Essential node for Harmony
G	Hex 60 deg	10.64	0.40	96.4	3.6	0.04	Not feasible, except for a nonequipment area
H	Triple "Y" 90 deg/ 135 deg	7.19	3.85	65.1	34.9	0.54	Occurs only in hybrid plan
J	Double 120 deg	5.63	5.41	50.7	49.3	0.97	Key unit for benzene ring, equivalent to plans B and C

TYPE J This is the benzene-ring equivalent of a straight-through passageway. The 120-deg angle between pressure ports is essentially equivalent to the straight-passage plan C; the ratio of circulation to equipment area is essentially 1:1. The two floor areas at 120 and 240 deg suggest uses such as galley and dining area, with galley and food storage in the larger part and wardroom table against the wall in the smaller part.

Table 3 tabulates area properties of each of the nine variations. It shows the inverse relationship between the number of pressure ports (with their associated circulation aisles) and usable equipment area. With two ports each, plans B, C, and J afford the highest ratio of usable equipment area to circulation area.

Figure 9 charts the five evaluation metrics from Table 3 for each configuration. Each data point has meaning only in comparison to the other points on the same line. Figure 9 suggests plan types that score consistently highest will perform best. These plan types include most notably Plan A Single, Plan B Double 90 deg L, Plan C Double 180 deg T, and Plan J Double 120 deg. The plan that scores significantly worst is Plan G Hex 60 deg.

Habot Pairings

The next step is to look at possible plan combinations. Pairing examines at a molecular level how building blocks of a base cluster can join. It is not practical to analyze all possible permutations nor to attempt a deductive approach. Instead, this analysis takes the inductive approach of postulating a range of plan types and ways in which pairs can combine. The remaining sections of Figure 7 illustrate such selected pairings of Habot plan variations. These pairings could be incorporated into larger base configurations or comprise an "outpost" habitat. The analysis demonstrates several architectural design precepts:

- The net architectural building area (American Institute of Architects 1994 or later edition) can be defined as the gross building area minus circulation, service, and utility areas.
- Paired modules would function as adjacent rooms, providing a more diversified geometry.
- Joining unlike modules can create a visually stimulating arrangement, introducing variety into a confined and isolated lunar-base interior.
- Pairings show that a complex pattern of solid and open areas can develop in a geometrically diverse configuration
- Ninety-degree modules suggest an orthogonal pattern of development, and 120-deg modules suggest a hexagonal pattern.

Suitport EVA Access Facility

The EVA spacesuit, airlock, and support system enable the Habot crew to fulfill scientific and engineering mission objectives by egressing their habitat. The

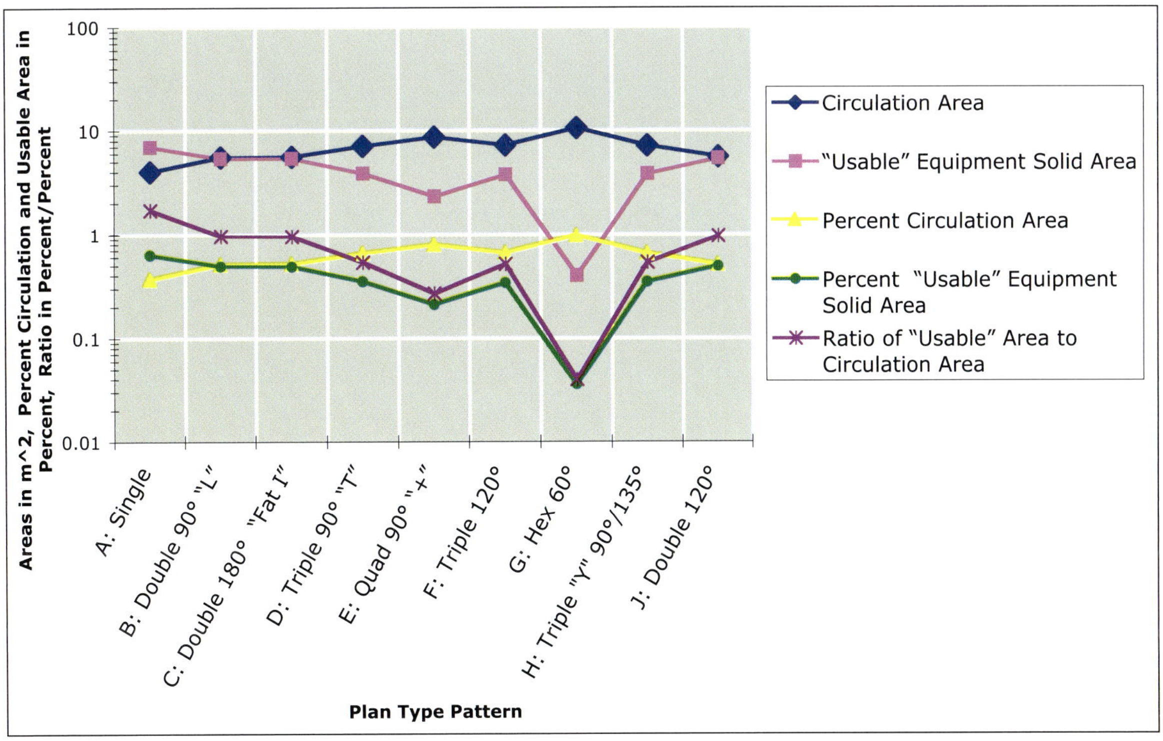

FIGURE 9 Comparative evaluation summary for the nine Habot plan types.

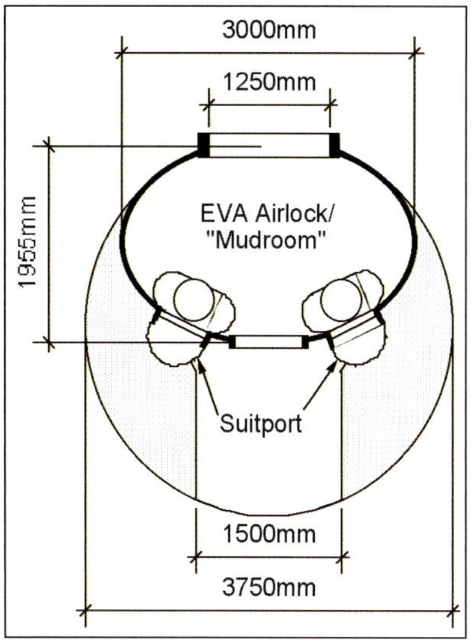

FIGURE 10 Plan view of Habot EVA access module, showing airlock with two suitports.

NASA Exploration Systems Architecture Study (NASA 2005) included the suitport as an alternative to the conventional pumpdown airlock for a lunar base or habitat (see Chapter 18 for a more extensive discussion of this option). The suitport approach could provide an efficient functional concept for spacesuit egress and ingress ("Suitport" 1989) by avoiding a pumpdown chamber. Figure 10 shows a schematic plan view of an EVA access module. The excursion Habot would be similar insofar as it would include airlock and suitports.

Thermal View Angle

In addition to the mobility system and architectural plan analyses, thermal view angle is an important consideration. Heat generated internally by the occupants (sensible heat), lights, avionics, and motors and other equipment must be actively removed. Because the module is like a thermos bottle in the vacuum of space, the only practical rejection technique is radiation. Cooling loops transport the heat to externally mounted radiator panels for rejection to the black sky. Wrapped around the unit shown in Figure 5 is a body-mounted radiator. The pink hatched segments of the external radiators indicate that they have active coolant flowing through them to reject heat.

INCIDENT SUN ANGLE As the Moon rotates, the Sun appears to move through the sky much as on Earth (albeit far more slowly), rising in the east and setting in the west for equatorial and midlatitude sites. By having wrap-around radiators, neither individual Habot modules nor a base assemblage are orientation-constrained. External radiator plumbing is zoned circumferentially into 45-deg sections, thereby operating radiator panels selectively to not face the Sun.

During the lunar morning, when the Sun illuminates the east-facing radiators, they would absorb more heat than they could reject, so they are not used. The west and north radiators provide the cooling instead. Then, as the Sun moves across the sky throughout the lunar day, the system progressively switches to

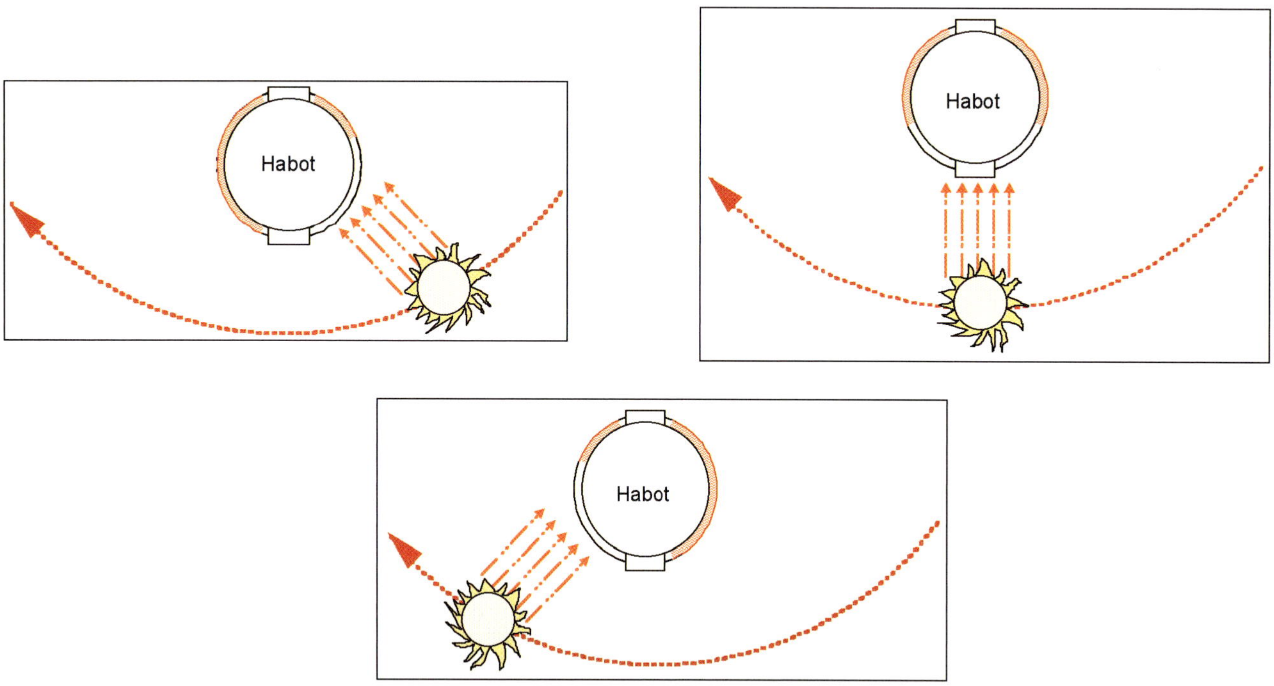

FIGURE 11 Wrap-around radiation panels make the Habot orientation insensitive for heat rejection with a system penalty of approximately two times the radiator area otherwise required.

the radiators least-exposed to solar incidence. As the Sun rises higher in the sky, and insolation on the east radiators diminishes, the Habot begins rejecting heat toward the east, while continuing on the west. In the lunar afternoon, the west radiators are shunted, and the east radiators take over.

HEAT-REJECTION VIEW ANGLE A second consideration is the radiator view angle. Heat can only be rejected to targets cooler than the radiator, so that Habot radiators that "see" each other will be inefficient. As Figure 11 shows, without the view-factor consideration, about two-thirds of the body-mounted radiator area could be active at any time, but cluster-base configurations will reduce this further.

Habot Module Connection Patterns

Figure 12 presents a systematic exploration of how Habot plan types can be combined into base configurations. It draws upon previous work by Sherwood and Capps (1990) on module cluster topology analysis (see Chapter 11). It identifies four fundamental connection angles: straight (180 deg), 60 deg, 90 deg, and 120 deg. These four angles are used to build four configuration types: open, closed loop,

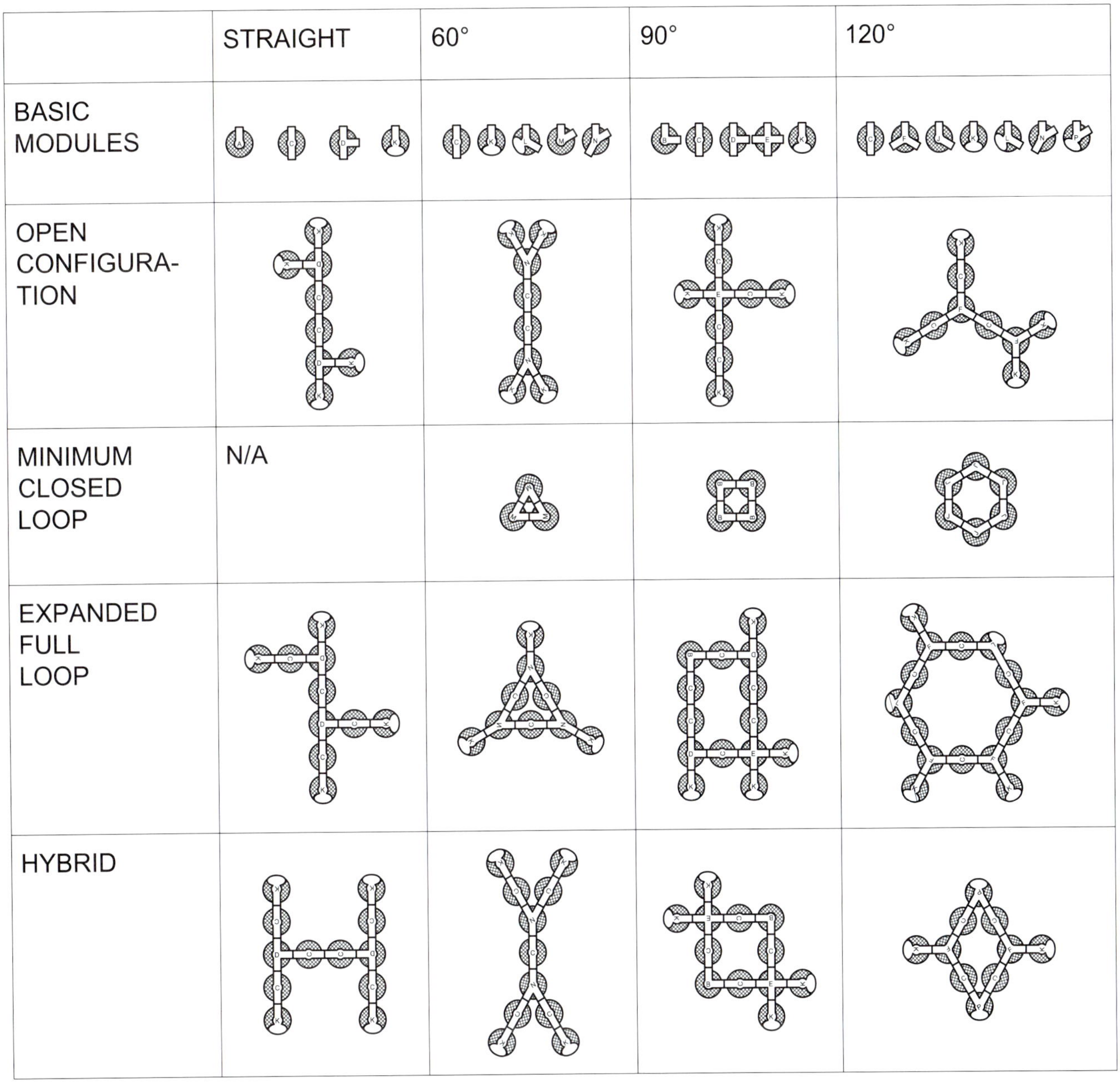

FIGURE 12 Habot base configuration analysis matrix. White ovals in end-units designate suitports (EVA access or excursion Habots units). (drawing by Ross A. Tisdale).

expanded loop, and hybrid. The analysis reveals several relationships:

- Open configurations for straight and 90 deg are similar.
- Open-loop configurations for 60 and 120 deg are similar.
- Minimum closed-loop configurations for 60, 90, and 120 deg are all simple polygons: respectively, triangle, square, and hexagon.
- Expanded full loop configurations for all four angles develop from adding 180 deg Plan C units.
- Hybrid configurations combine loops and multiple angles using additional Plan C units.

BASE CONFIGURATION EVALUATION

Table 4 tabulates an evaluation of the four fundamental patterns against the nine configuration criteria detailed next. This evaluation is intended to be as fundamental as possible, by addressing the generic pattern rather than any specific configuration within that pattern family. Six units were assumed as the minimum, exclusive of excursion Habots.

Although docking modules were included as criterion 7, they turned out to be not a discriminator. Similarly, dual remote egress was almost equally feasible for all module patterns, if an excursion Habot is always available at each distal docking port. However, for the straight/linear pattern, it is not possible to guarantee an excursion Habot will always be present there, so that pattern scores slightly lower to reflect this uncertainty.

Nine Evaluation Criteria

1) *Circulation access*—Goal: circulation access to all parts of the base. Dual access is preferable, especially in case of loss of safe access through one hatchway. This criterion measures how many of the six core modules have dual-pressurized access at all times from other permanently attached core modules. For example, in the open-plan straight configuration, the two end modules do not have permanent pressurized access. Because four of six modules meet the criterion, the rating is 0.66. Loop plans are better than open plans. A docked excursion Habot does not afford this kind of permanent pressurized access, although it might do so in an emergency. (The crew would enter the excursion Habot at a different pressure port and drive it to the module they are trying to reach.)

2) *Dual remote egress*—Goal: egress from all parts of the base. Ideally, the configuration provides dual remote means of egress. This criterion directly measures safety in case of evacuation of any module because of fire, decompression, contamination, or other such hazards (Raasch et al. 1985; Cohen 2001). Excursion Habot units can afford a means of egress; however, the evaluation credits only half the total number of excursion Habot egresses, penalizing the straight/linear configuration in particular.

3) *Efficient equipment areas*—Goal: maximum amount and shape utility of equipment functional areas and volumes, compatible with circulation. This criterion measures how useful the overall floor plan is at providing equipment area in relation to circulation area. The number comes directly from Table 2 as the ratio of equipment area to circulation area.

4) *Work spaces/social areas*—Goal: suitable workspaces and social areas within the base. This criterion assumes the best module for work and/or social areas is not a circulation hub. Although a circulation hub with three, four, or more hatches might seem like a logical meeting place, the problem is that excessive "through circulation" can be disruptive, especially to group work or team activities. In addition, for social settings such as wardroom and galley, the small equipment area in a circulation hub might not provide sufficient space for wardroom table and food preparation. Habot modules are, however, large enough to set the wardroom table to one side of a type B, C, or J plan.

5) *Ease of assembly*—Goal: ease of mobile access for assembling and disassembling the base cluster. This criterion examines the sequence of assembling units into the base cluster, evaluating the "puzzle-making" aspect of each configuration: how easily is the last unit aligned and docked? The ideal configuration avoids having to align a unit between two other modules, especially with simultaneous docking ports at differing angles.

6) *Qualitative thermal view angles*—Goal: efficient view angles for body-mounted radiators. The best base configuration is the straight or linear configuration, where the body-mounted radiators face orthogonal to the circulation axis and have a clear view away from adjacent modules. Other configurations are all compromised in this respect; although all include some

TABLE 4 Evaluation of the four major configuration patterns against nine configuration criteria.

CONFIGURATION CRITERIA	STRAIGHT LINEAR PATTERN	60-DEG PATTERN TRIANGLE	90-DEG PATTERN (SQUARE BOX)	120-DEG PATTERN (HEX)	REMARKS
1) Circulation access/dual access	0.67	1.00	1.00	1.00	Gives the percentage of core modules with permanent, pressurized dual access
2) Dual remote egress	0.80	1.00	1.00	1.00	Gives the percentage of core modules with pressurized dual remote access, allowing 50% of end modules to use excursion habots
3) Efficient equipment functional areas	0.82	0.75	0.77	0.82	Gives the ratio of equipment area/ circulation area * the number of plan types from Table 2
4) Work spaces/ social areas	0.67	0.50	0.60	0.50	Gives the ratio of core modules with ONLY 2 ports/total number of modules
5) Ease of assembly	1.00	0.33	0.80	0.66	Gives the proportion of modules that can make a 0-deg linear connection to other modules
6) Qualitative thermal view angles	1.00	0.30	0.50	0.70	Gives the ratio of core modules that have clear, unobstructed thermal view angles for body-mounted radiators
7) Docking modules	1.00	1.00	1.00	1.00	Gives the ability to provide docking for 4 excursion Habots or EVA access units
8) Minimal pressure port variations	0.50	0.33	0.50	0.50	Gives the variation in number of different angles of pressure ports
9) Docking port efficiency	0.86	0.80	0.83	0.86	Gives the ratio of number modules*2 ideal ports/number of actual ports
Sum	7.32	6.01	7.03	7.01	$= a+b \cdots +n$
Product	0.13	0.01	0.08	0.08	$= a*b \cdots *n$
Arithmetic mean	0.81	0.65	0.77	0.77	$= (a+b \cdots +n)/n$
Harmonic mean	**0.16**	**0.02**	**0.10**	**0.10**	= Product/arithmetic mean

radiators facing away from the cluster, effective heat rejection throughout the lunar day requires pumping coolant around the entire base.

7) *Docking modules*—Goal: ability to dock excursion Habots or EVA access units. Maximum score corresponds to four available docking ports.

8) *Minimal pressure port variations*—Goal: minimum variation of pressure port arrangements. This criterion measures the number of different port angles times the number of modules at each angle.

9) *Docking port efficiency*—Goal: efficient use of docking ports. This criterion assesses the number of docking ports necessary to build the base configuration, compared to an ideal average allotment of two ports per module. It is calculated by dividing the actual number of ports into the number of modules times two.

Evaluation Summary

To integrate a comprehensive evaluation, the individual scores listed in Table 4 can be combined in different ways. Each combination reveals different findings, and the evaluation also yields qualitative lessons about the treatment of differing module types. For this initial analysis, the scores are unweighted because without a programmatic context there is no basis for claiming, for example, that a safety-related criterion like "dual remote egress" is three times more important than some other criterion.

The first combination simply sums the individual scores for each pattern. The scores are very close, and so this appears not particularly revealing. The second combination simply multiplies the individual scores, which accentuates their differences, particularly for patterns suffering from at least one very low individual score. This is the approach used in reliability calculations. The resulting products range from 0.01 to 0.13, a ratio of 13:1. At 0.08, the hexagonal and rectangular patterns score close to the 0.075 average; the straight pattern performs much better than those two, and the triangular pattern performs poorly.

The arithmetic mean results in the same ranking. The final combination is the harmonic mean, which uses arithmetic mean to normalize product; this calculation reduces the effect of criteria with very similar scores. Again, the ranking is the same: the straight/linear pattern scores highest (harmonic mean 0.16); the 90-deg rectangular pattern and the 120-deg hexagonal pattern tie for second place (harmonic mean 0.10); and the 60-deg triangular pattern scores lowest (harmonic mean 0.02). Compared to the middle-scoring patterns, the straight/linear pattern scores 60% higher, and the triangular pattern scores 80% lower. Figure 13 plots the results tabulated in Table 4, graphically showing the variances and alignments of the four pattern types. The evaluation indicates the 60-deg triangular pattern should be eliminated.

Viable Base Configurations

Figure 14 presents 10 representative Habot base geometries based upon the three surviving patterns, selected from many that are possible. This sample uses various module floor plans to achieve diverse layouts. Each docking module has three pressure ports spaced either at 90 or 120 deg. Rather than being used primarily for internal circulation, these end hubs might best serve as external circulation nodes for docking excursion/rovers and EVA access facilities. The triple 120-deg Plan F type emerged as the favorite excursion/rover docking hub.

Figures 14a–14d show experiments with dense-pack clusters of modules based on two assumptions: presumed advantage in maximizing access ports and use of gimbaled, tracking radiators to sidestep the constraining issue of body-mounted view factor. Figures 14a, 14b, and 14d compare three short straight/linear pattern configurations, with docking modules offset to the sides to form cruciform plans. Figures 14a and 14b explore 90- and 120-deg hubs respectively for end-module connection. Figure 14d takes the short cruciform pattern in the hybrid direction of combining it with a rectangular (square) loop. This pattern creates a square circulation grid throughout the cluster. Figure 14c shows a 120-deg hexagonal configuration with a "hollow benzene ring."

Figures 14f–14i show four experiments in the straight/linear pattern. Figure 14f is the most pure—a straight series with 120-deg hubs at each end providing four mobile unit docking ports. Figure 14g hybridizes this configuration by adding two units to one side to form a small square. Such a cluster at the center may offer operational advantages over a long string of modules, and Figure 14f could add extra modules mid-series without having to remove and relocate an end hub. Figure 14h does the same thing but with the central cluster offsetting the two halves of the linear pattern, increasing the richness of the interior environment using the same kit of parts.

Figure 14j uses three pairs of 120-deg hubs to achieve a hollow-triangle base configuration, and Figure 14e offers an inverted version of this plan. Despite the poor performance of 60-deg triangular patterns per se, these examples show how better-performing cluster arrangements can nonetheless achieve triangular geometries, by using 120-deg connectors to form either ring or branch topologies.

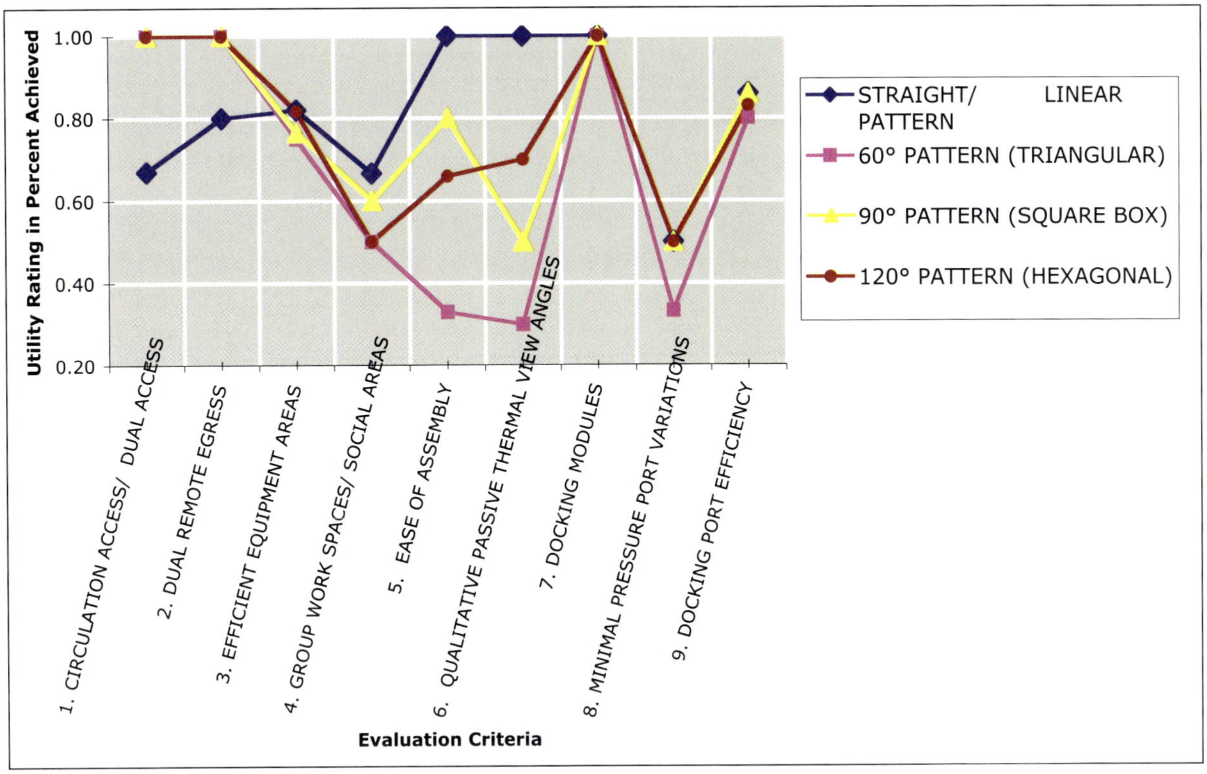

FIGURE 13 Plot of evaluation results from Figure 11.

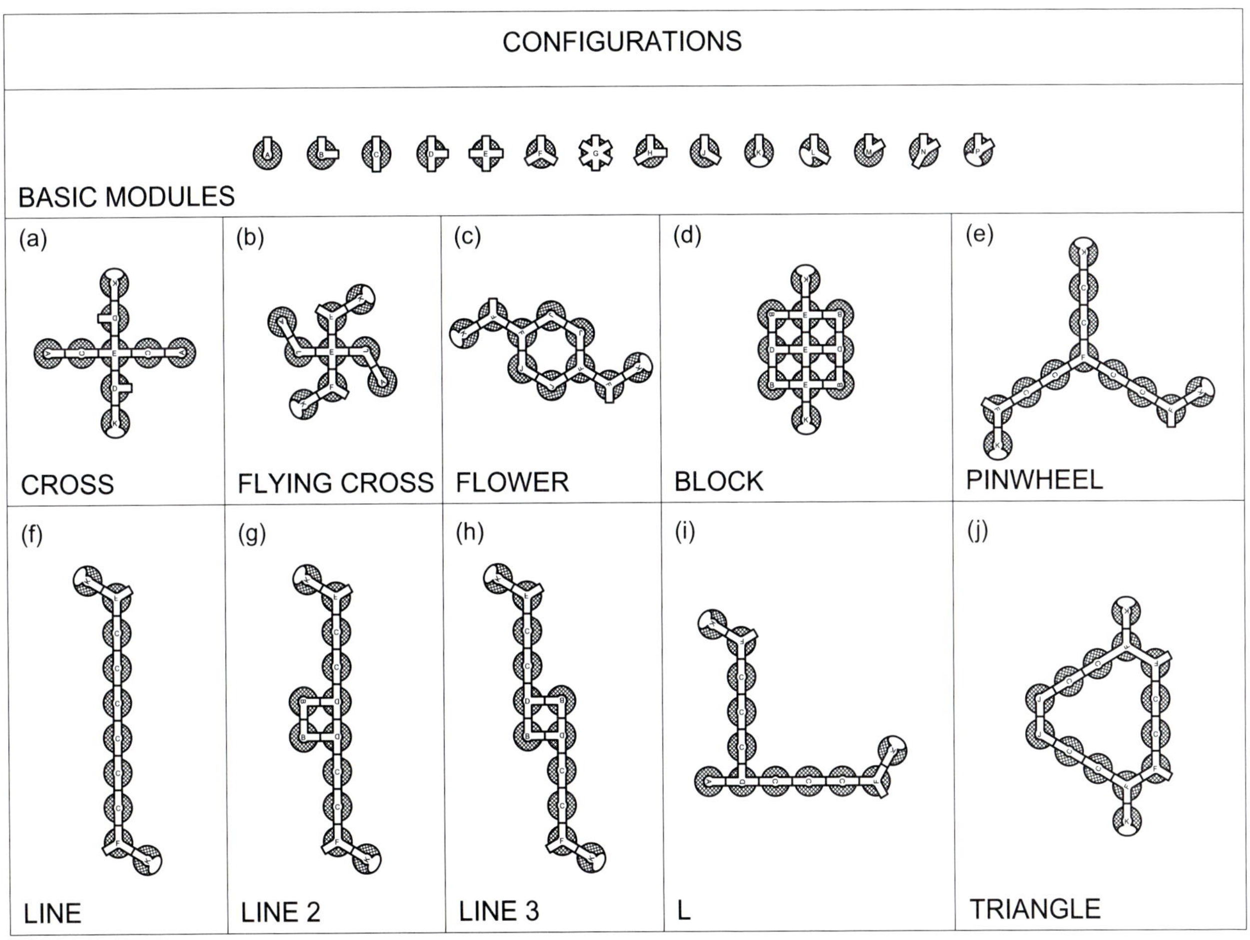

FIGURE 14 Representative viable Habot-base configuration patterns. Suitport locations are designated by small white ovals.

The conclusion from both levels of analysis—module plan and base configuration—is that simpler solutions are best. The straight "Fat I" passage plan C module type and its logical extension to the straight/linear pattern for overall base configuration scored best. Two particular variations add value: docking modules with 120-deg pressure ports and simple offsets in the line of modules to create local clusters.

CONCLUSION

The Habot case demonstrates how detailed preliminary analysis can be used to bound the viable design space for novel concepts, helping focus subsequent trade studies and concept development. The cost of developing and operating lunar infrastructure is quite high, and so many architectural solutions will be proposed and evaluated before system designs are committed. "Scrubbing" the Habot concept as demonstrated here and developing enabling technologies (e.g., ATHLETE) enable the community to move on to the next higher level of systems engineering challenges: hardware penalty of dividing a habitable system into small modules with multiple ports; variable thermal performance of mobile modules; integrity of repeated large-seal connections in the lunar dust environment; and internal packaging efficiency (e.g., ratio of equipment to circulation space). |

References

American Inst. of Architects (1994 or latest ed.), "Architectural Areas of Buildings," D101 (1980 or latest ed.), *Handbook of Professional Practice*, Vol. 4, American Inst. of Architects, Washington, D.C.

Cohen, M. M. (2000), "Pressurized Rover Airlocks," Society of Automotive Engineers, Paper 2000–01–2389, July.

Cohen, M. M. (2001), "Space Laboratories," Society of Automotive Engineers, Paper 2001–01–2142, July.

Cohen, M. M. (2003), "Mobile Lunar and Planetary Bases," AIAA Paper 2003–6280, Sept.

Cohen, M. M. (2004a), "Mobile Lunar Base Concepts," 2004 Space Technology and Applications International Forum, Paper 2004-STAIF-291, Feb: AIP CP-699, American Inst. of Physics. College Park, MD; pp. 845–853.

Cohen, M. M. (2004b), "Habot Multivariate Design Model Pilot Study," Society of Automotive Engineers, Paper 2004–01–2366.

Cohen, M. M. (2004c), "Carbon Radiation Shielding for the Habot Mobile Lunar Base," Society of Automotive Engineers, Paper 2004–01–2323.

Mankins, J. C. (2000), "Modular Architecture Options for Lunar Exploration and Development," International Astronautical Congress, Paper IAA-00-IAA.13.2.05, Oct.

Mankins, J. C. (2001), "Modular Architecture Options for Lunar Exploration and Development," *Space Technology*, Vol. 21, pp. 53–64.

NASA (2005), "Exploration Systems Architecture Study," NASA, Final Rep., Dec., http://www.nasa.gov/mission_pages/exploration/news/ESAS_report.html [retrieved 11 June 2006].

Raasch, R. F., Peercy, R. L., Jr., and Rockoff, L. A. (1985), "Space Station Crew Safety Alternative Study," Final Report; *Vol. II: Threat Development*, NASA CR-3855, NASA Scientific and Technical Information Branch, Washington, D.C., June.

Sherwood, B., and Capps, S. (1990), "Long-Duration Habitat Trade Study," NASA Study Contract NAS8–37857, March 23.

"Space Station Architecture, Module, Berthing Hub, Shell Assembly, Berthing Mechanism, and Utility Connection Channel" (1988), U.S. Patent 4,728,060, March 1.

"Suitport EVA Access Facility" (1989), U.S. Patent 4,842,224, June 27.

Wise, J. (1988), "The Quantitative Modeling of Human Spatial Habitability," NASA CR-177501, Aug.

19

roving laboratory

ANDREAS VOGLER
ARTURO VITTORI
STEPHEN RANSOM
AND
LORIS GRANZIERA

INTRODUCTION

PROPOSALS BY NASA AND ESA for human and scientific exploration of the Moon and Mars have led to various concepts for pressurized rovers capable of traversing the rugged surfaces of these planetary bodies. Such a vehicle is a self-contained mobile environment, supporting its crew in reasonable comfort so that they can work efficiently and safely. Because roving astronauts are likely to be several days, or possibly weeks, away from a main base of operations, they will be prepared to dispense with some "home comforts" much the same way *Apollo* astronauts accepted their very basic environment.

This chapter describes a pressurized rover concept called MarsCruiserOne (MCO) that blends crew habitability and space engineering requirements with those imposed by planetary protection related to extravehicular activity (EVA) and recovery of surface samples. Planetary protection poses especially interesting and challenging issues. The design team took an approach based on combining known and proven technologies with challenge technologies, allowing development of new ideas for mobile surface habitats.

Although MCO was initially proposed as a European contribution to a human Mars surface exploration mission, the vehicle concept is applicable also to long-duration lunar exploration. Pressurized rover development could support early lunar-return human missions. Technologies to benefit the MCO would need to be developed to a level of readiness that could be implemented at least four to five years before launch. Development of traction systems, high-performance fuel cells, EVA suits, and contamination protection techniques are primary enabling issues.

Other items that must be developed concern techniques to repair and maintain the vehicle and methods to replenish gases and liquids needed for life support, propulsion, and power. Allocating the main interior functions via an interior design in the early concept stage helps identify geometrical conflicts and potential manufacturing problems, as well as provide insight into options for prototype testing. Additionally, MarsCruiserOne is conceived as a European technology and innovation platform for the aerospace and automotive industries as well as other, nonaerospace fields of technology.

MISSION SCENARIOS

The MCO is dimensioned to be launched in a 5-m-diameter fairing on an *Ariane* rocket (Figure 1). Employed for scientific exploration purposes, the MCO carries a crew of at least two astronauts. It is capable of performing *in situ* analytical and robotic sampling tasks as well as photographic tasks while stationary and in motion. A sample storage system and glove box support scientific experiments enroute. Visual, infrared (IR), and other sensor data generated by the measurements and analyses are processed, stored, and transmitted to the main base or to Earth via Mars relay satellite. The MCO performs excursions away from the main base of up to 20 sols (Martian days), of which 10 sols might be spent at an exploration site. The crew performs EVA surface tasks at any time during the excursion.

The MCO can deploy and retrieve small payloads such as stationary, *in situ* scientific packages, rovers, or aerobots placed at places of interest. These subvehicles explore regions along the MCO path to reveal potential hazards or obstacles or acquire data from regions too hazardous for or inaccessible to the MCO or EVA crew. The subvehicles are remotely guided by crew from within the MCO or on EVA or operate autonomously. They might rendezvous with the MCO

Copyright © 2008 by the authors. Published by the American Institute of Aeronautics and Astronautics, Inc., with permission.

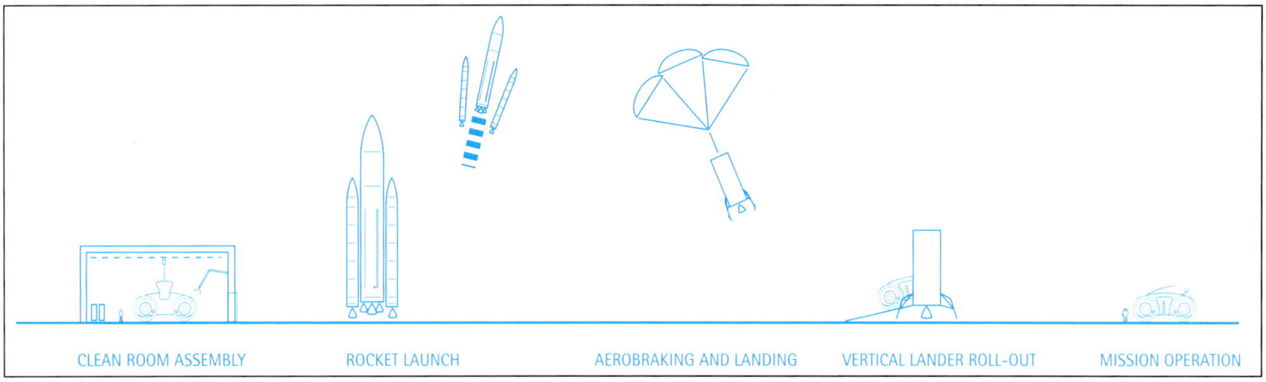

FIGURE 1 Assembly and transportation of the MarsCruiserOne rover (courtesy of Architecture and Vision).

further enroute upon completion of their mini-missions.

The MCO is designed to rescue six astronauts at the main base in the event of catastrophic failure, transfering them to a Mars ascent vehicle for return to orbit. In this contingency, the MCO could support six astronauts for seven days. The astronauts would have EVA suits appropriate for launching to their transfer ship and would wear them throughout such an evacuation procedure. Mechanical counterpressure suits would package better than conventional pressure suits. Astronauts could also be accommodated for up to seven days while repairs are made to main base elements. The MCO layout and surface-EVA suit provisions allow two astronauts to perform surface EVA tasks at any one time.

FIGURE 2 Perspective view of the MCO showing the rim-driven, hubless, omnidirectional wheels, docking/berthing airlocks within the wheels and the large front window (courtesy of Architecture and Vision).

VEHICLE DESCRIPTION

The MCO consists of a pressurized compartment for habitation, exploration planning, and scientific investigation, and a mobility system based on large, adaptable wheels (Figure 2). The pressurized volume is optimized for the crew living and working environment, whereas most vehicle subsystems such as power supply, environmental conditioning system, etc. are located outside the pressure vessel. Overall layout and vehicle size are defined by launcher payload envelope requirements, crew habitation requirements for both normal and contingency missions, and power requirements.

The MCO is designed to traverse the Martian surface at an average speed of 5–10 km/hr, controlled autonomously or by the astronauts, and navigated using a Mars global positioning system. Autonomous cruising enables overnight driving to reach remote locations efficiently.

Dimensions

The MCO is sized to fit within a launcher payload envelope 5 m in diameter and 10 m in length. It is transported in a container fitted to the launch-vehicle upper stage; the primary structural interface is an upper-stage adapter at the rear of the MCO and trunnions integrated within the body of the vehicle. To maximize crew habitable volume within this envelope and provide sufficient height for crew working inside, MCO is based on a large-diameter, hubless, rim-driven wheel design. This allows use of the space inside the wheel envelope and side locations for docking/berthing ports and airlocks. The pressure hull consists of two cones located between the wheels, connected by an oval tube.

Most previously proposed pressurized rovers envisage a simple cylindrical crew cabin mounted above the locomotion system. These concepts result in the center of mass and astronaut access points being located high above the Martian surface. The MCO configuration allows both the center of mass and astronaut surface EVA access points to be located closer to the surface (Figure 3). The "symmetry" of the MCO about its longitudinal axis also reduces any offset of its center of mass with respect to the longitudinal axis of a potential launch vehicle (Figure 4).

Wheels and Suspension

The MCO is propelled by four large, independent, electrically driven and suspended wheels. Analysis of the locomotion subsystem reveals that larger wheel diameters reduce the vehicle's overall power requirements and consequently the size of the fuel-cell system used to provide electrical energy. These factors help limit overall vehicle mass. The design proposes hubless wheels. The inner rim is fixed, and the outer rim is driven by linear motors. Omnidirectional ("Meccanum") wheels with 23 rollers arranged at a 45-deg angle around the rim allow the rover to be driven sideways and diagonally. Rover steering is controlled by the speed and the direction of rotation of the independent wheels, controlled electronically. The ability to drive laterally allows sideways docking with a habitat or another rover. It also allows the rover to approach geologically interesting sites like the edge of a cliff or crater. The rollers are made of carbon-fiber filament for minimum weight.

To reduce the packaged volume, the wheels might be moved laterally. The wheels would be extended

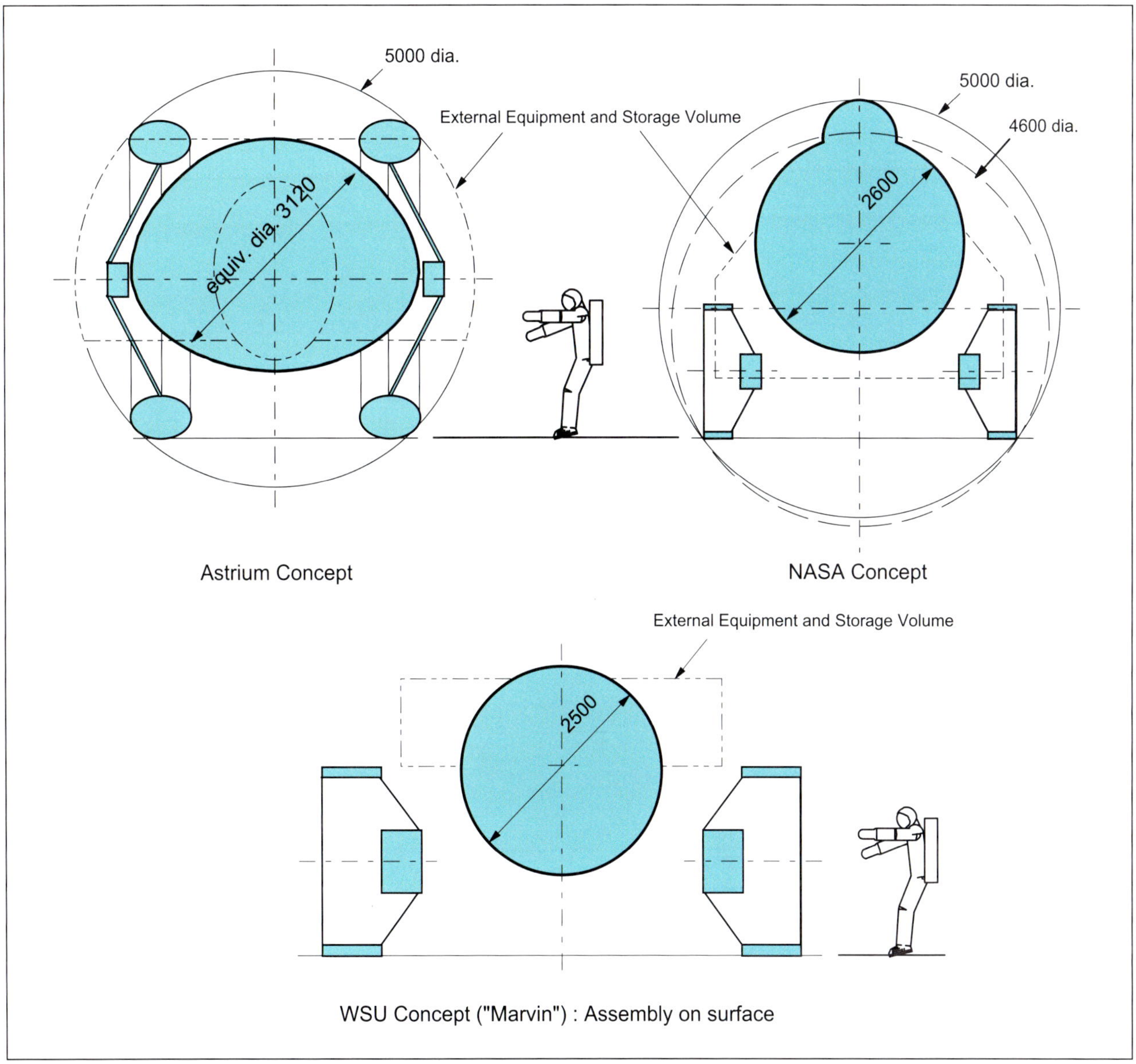

FIGURE 3 Comparison of rover concept cross sections (courtesy of Astrium et al. 2002).

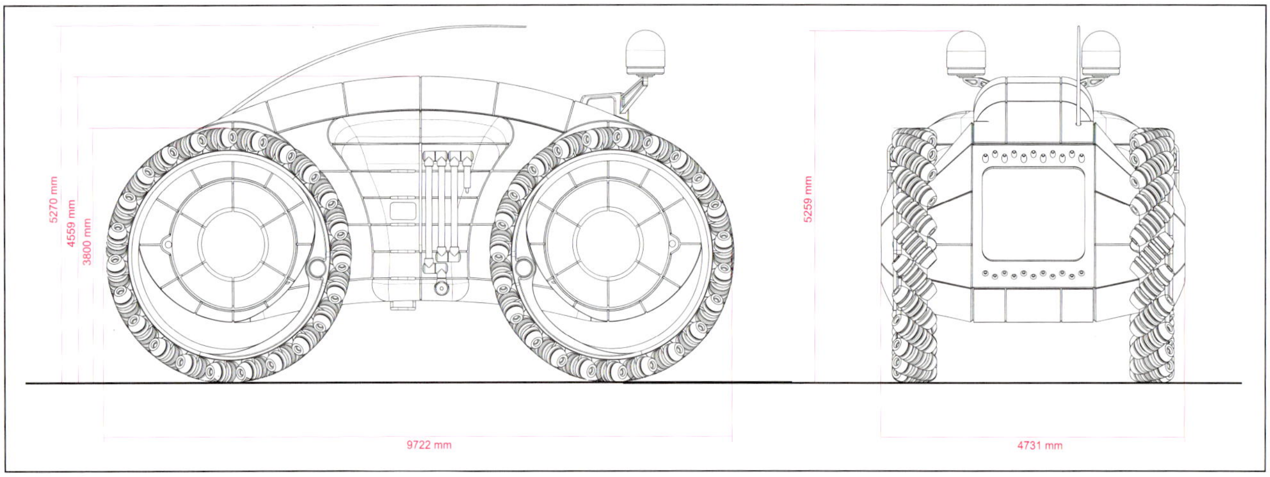

FIGURE 4 Overall dimensions of MCO in operating configuration. The rover is arranged symmetrically to control launch mass distribution.

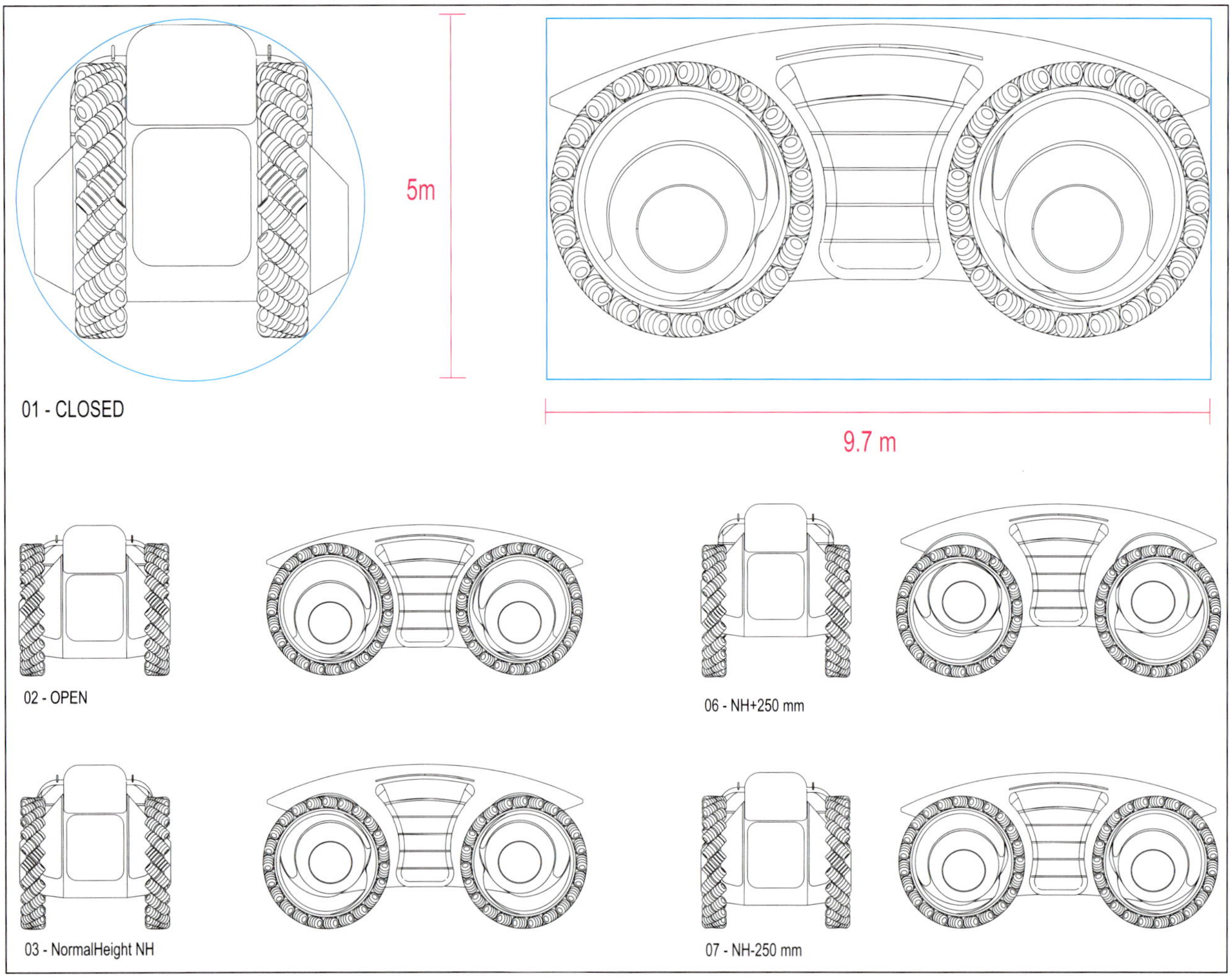

FIGURE 5 MCO with wheels in transport position (top) and maximum positions in relation to normal driving height. Each wheel is controlled individually.

sideways after landing. Rotating the axis, which is fixed to the main body and the nonturning rim, allows the wheel to move in a vertical direction, thus allowing an adjustment of +/– 250 mm in each direction (Figure 5).

The vehicle is designed to roll over obstacles between the wheels of up to 500 mm in height, and each individual wheel can roll over obstacles up to 300 mm in height. Individual vertical adjustment of the wheels keeps all four wheels in touch with the

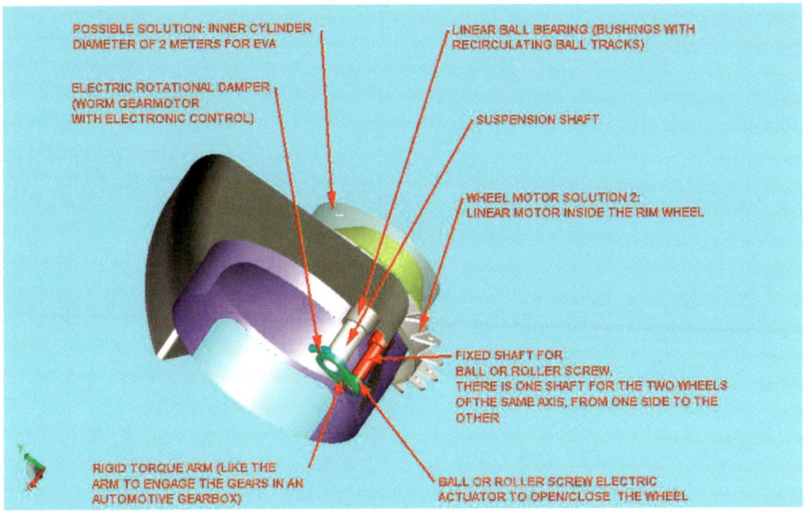

FIGURE 6 Suspension, dampening, and wheel adjustment mechanism uses electric actuators (courtesy of G-Engineering).

ground and also allows the rover to level itself horizontally while traveling along slopes of up to 7-deg inclination. A further advantage of the big wheel design is that crevasses up to 1 m in width might be traversed if safe. The suspension and dampening system uses technology similar to that used for many years in Formula One racing cars. A rotational damper offset from the wheels' axes constantly checks wheel movement and dampens the shocks (Figure 6).

Manipulator Subsystem

The primary manipulator subsystem is installed alongside the externally mounted science facility rack (Figure 7). The arm has a 3-m reach and can be fitted with various end effectors. It is equipped with an all-purpose gripper and vision system comprising several cameras. The manipulator can be relocated to other positions on the MCO to assist, for example, external inspection tasks. Samples taken from the Martian surface can be directly deposited into an examination glovebox within the sealed science facility, which an astronaut can then access from inside the vehicle.

The arm and camera can be deployed vertically, and the camera head rotated through 180 deg, to provide panoramic views of the surrounding landscape. This feature could satisfy both science and public interest, as evidenced by the large viewership of Jet Propulsion Laboratory (JPL) Mars mission Web sites. Provision for video transmission of laboratory excursions would be a key objective of the smaller accompanying rovers, which would photograph the MCO as it rolls across the Martian plains.

Rover Access

Astronaut access from rover to other elements is via airlock hatches integrated into the front right side and rear left side of the MCO. The hatch dimensions are based on the NASA standard airlock designed for the International Space Station (ISS). For docking the dust covers are opened, and the rover moves sideways to the mating assembly. The mating is controlled by opto-electronics.

FIGURE 7 MCO near Husband Hill. The six-degrees-of-freedom, seven-joint robot arm is visible on the side of the central science facility (courtesy of Architecture and Vision, astronauts by Max Grueter, background image by NASA/Jet Propulsion Laboratory-Caltech/Cornell).

Accommodating two docking berths without compromising interior volume is a principal benefit of the MCO concept. It provides redundancy of course, but also the ability to connect several rovers into a contiguous volume. This opens the possibility of a "rover-only" mission, as discussed in some mission scenarios.

Consideration of surface EVA preparations, and stringent requirements to avoid Earth–Mars–Earth planetary contamination, led to the need to provide a large volume for crew to prep both for egress and for ingress. This drove selection of the suitport concept as described by Steinsiek et al. (2003). The suits are attached to the outside of the rover inside the wheel hubs and protected by dust covers. Thus, the outside of the suit always stays outside the habitable volume, avoiding dust and cross-contamination inside the crew cab. The suits are entered through the backpack (Figure 8). A fold-out staircase allows access to the ground (see also Figure 7).

Vehicle Subsystems

Vehicle subsystems are primarily housed in compartments arranged between the pressurized volume and the outer chassis of the vehicle. The environmental control and life-support subsystem assumptions are based on further development of ISS techniques. The thermal control subsystem uses both passive and active cooling systems. Surface heat-pipe radiators are accommodated on the external surface of the MCO.

The vehicle is controlled by the crew from two seated positions located in the forward part of the cabin. Astronaut helmet-mounted displays could also enable visual and voice command inputs, based on technology developed for military aircraft applications. Many of the vehicle control components needed for the MCO are currently being developed within the automobile industry for terrestrial applications.

Radiation protection poses a serious challenge for pressurized rover designs. One option is to use onboard water for radiation protection, which involves complications from distributing water through the walls and verifying that the water remains potable. In the MCO design presented here, water tanks are located above and under the cylindrical volumes. In addition, most equipment is arranged around the central tubular volume, which helps protect the crew in a "safe-haven" mode. Adding polyethylene inside the honeycomb sandwich material used for the cabin structure would provide even more protection, especially important for lunar missions where there is no planetary atmosphere.

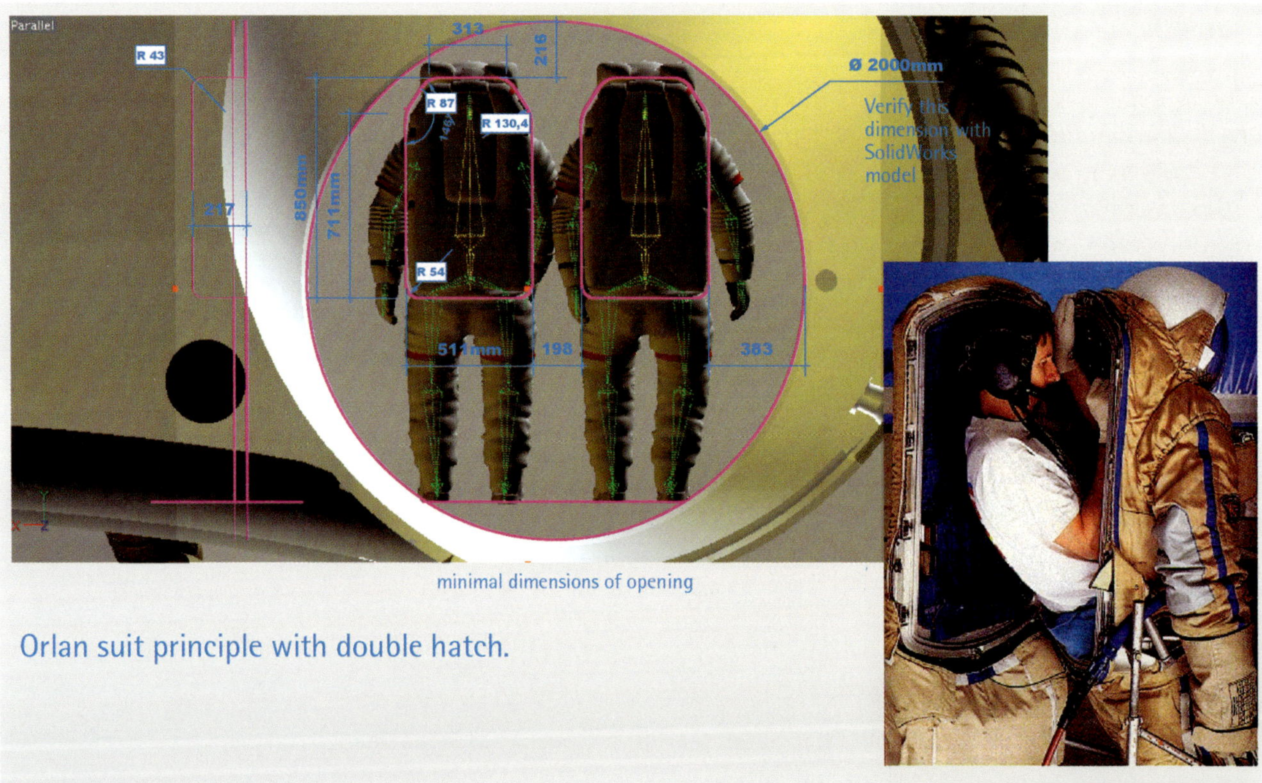

FIGURE 8 Airlock is located in the center of the hubless wheels and employs the suitport concept. The rear-entry Russian Orlan suit is shown for reference.

INTERIOR LAYOUT

Designing the interior is an iterative process. Recognizing that a specific rover design must be done together with the rest of the surface architecture, the layout presented here is based on principal ergonomic dimensions.

Volume

MCO pressurized volume above floor level is 41.2 m^3. Crew accommodations requirements are based on Stilwell et al. (1999). Assuming a 40 crew-day, shuttle-like mission with minimal hygiene provisions, a volume of approximately 6 m^3 is needed for stowage. Oxygen and nitrogen tanks are located outside the pressurized volume, above the laboratory section. Water tanks are above and below the cylindrical sections at both ends. This leaves a habitable volume of approximately 36 m^3, which meets the NASA-STD-3000 recommendation for a crew of two, even beyond six months (NASA 1995). Arno (1999) calculated a pressurized volume of 23 m^3 for a seven-day rover mission with three crew members. For 40 crew-days with two crew members, 20.7 m^3 would result using the same assumptions. The proposed design thus has ample space also to accommodate additional functions and science equipment.

Interior Design

When planning a human habitat, it is important to establish design guidelines and strategies early. This is especially true for human space systems because of their high degree of design complexity. The design question *What should the rover interior look like?* forced the team to develop an integrated concept early, rather than assembling the interior design from separate "design pieces." It proved helpful to work with design analogs from the earliest stage. Aircraft, yachts, and campers are applicable design analogs for the highly confined rover environment and help visualize concepts to identify potential problems and solutions. Such analogs offer many useful ideas for rover design. Soft interior padding avoids unsafe, hard edges, and can zone surfaces with color. Rims and edges kept within reach allow the handholds needed in mobile environments without appearing overtly technical or added-on, as protruding bars and handles usually do. Fold-up tables and screens are often used. Seats in aircraft and cars are well developed, offering good ergonomics and comfort. Door handles are integrated into the wall but big enough to be comfortably handled while wearing gloves. In most transportation design today a "non-bolt" and "non-nuts" approach, where visible screws are avoided, is used for closeouts. This is an important development in civil-vehicle design over the last several decades. Snap-on panels and outfitting make a nice and safe space, but also provide easy access to the structure when removed.

In addition to the distribution of functions in plan described next, the MCO design is vertically zoned. This is important because confined spaces fill up quickly with stowed and in-use items. Clear vertical zoning defines the space, improves habitability, and increases safety. The zones correspond to human ergonomics as well as the view from eye level while standing, sitting, reclining, and lying down, and establish functional and aesthetic continuity throughout the interior. Emphasizing horizontal order optically widens the space, and avoiding storage-container protrusions prevents accidents. The interior sketch shown in Figure 9 illustrates some of these points.

FIGURE 9 Sketch of MCO interior. Corners are made soft, and vertical and horizontal zoning organizes the space (courtesy of Architecture and Vision, Alessandro Natalini).

Functional Layout

The MCO cabin basically has three sections: conical cylinders on both ends with a tubular interconnection between them (Figures 10 and 11). The front cylinder is used for the cockpit and also for the main connection airlock, whereas the rear cylinder is used for EVA on one side and additional docking berth. Both fore and aft cylinders are identical geometrically, and so different layouts of docking/berthing and airlocks could be accommodated. The cockpit section contains driving seats and movable storage that can be located in front of the hatches when the rover is not docked. The rear cylinder also has a large window like the cockpit. It works together with a central utility unit to accommodate hygiene and cooking functions. The cylinders can be visually divided from the central tube by textile closures. All three odor-generating functions (airlock, toilet, and kitchen) are separated from working and sleeping areas. They are also temporally used functions, so that the kitchen and the airlock especially can share volume on a noninterference basis. Storage provisions and one glovebox are located in the tubular connection between the cylinders. The tube has a conical indentation at its center that allows space for the glovebox. The glovebox accesses samples stored outside the hull.

Functional Units

Functional units like cockpit, airlocks, and glovebox must fulfill technical and safety requirements, but also serve as elements of a human habitat. Therefore their functions have to be well understood by habitat designers trying to integrate them into a confined environment. Patterns of movement through the interior zones, ingress and egress through the hatches, convertibility of the cockpit seats, and how the suitports and berthing ports are used are among the subjects primed for detailed dimensional studies working with full-scale mock-ups.

COCKPIT The cockpit contains the driver's seat, steering unit, electronic equipment, and storage and working surfaces to one side. It provides a clear visual environment by enhancing horizontal lines and offering clean, nonreflecting surfaces with practical storage (Figure 12). The forward view outside is unobstructed, interior light is controlled to avoid glare, and good visibility of display and control units is supported by heads-up displays and monitors that provide views all around the vicinity of the MCO. A good view is available from the seats, but also from a standing position. Remote-sensing technologies allow driving in darkness or poor visibility caused by dust storms. Moving in and out of the seats is fast, whether in shirt sleeves or pressure suit.

DRIVING/SLEEPING SEAT UNIT The seat units need special design consideration because they must be highly convertible, that is, provide comfortable seating in daily clothes, leisure clothes, and pressure suit, and be used for driving, working, and sleeping. Design analogues include first-class seats of various airlines, but new concepts for more flexible seatscapes are possible.

STOWAGE The complex shape of the MCO hull requires customized incorporation of stowage provisions. Textile-based stowage systems are prime candidates for accommodating the 6-m^3 design allocation discussed earlier. About 50% of the crew accommodation volume is needed for food and waste collection, and an exchangeable storage scheme beneath the floor panels would allow replacing food with waste volume as the mission progressed.

GLOVEBOX AND LABORATORY AREA The central portion of the cabin contains the glovebox working area. Internal access is via gloves or robotic manipulators; external access is via the manipulator arm that delivers samples to the storage container. An internal mechanism delivers samples to the analytical compartment within the glovebox. The glovebox subsystem is equipped with analytical geological instruments, for example, microscope, electron microscope, spectrograph, saws and polishers for preparing samples, and ultrasonic equipment. This allows sophisticated, *in situ* laboratory analysis to be done without bringing samples inside the habitable volume, preserving the interplanetary contamination boundary between Earth and Mars material.

TOILET AND HYGIENE All waste and waste water are processed and sterilized at the main base; no waste is dumped on the Martian surface. This design feature accommodates both biological planetary protection and general wilderness preservation considerations. Thus, the MCO's toilet system is highly dependent on the main base's systems, and only waste stabilization (no waste treatment) is planned onboard the rover.

The toilet and hygiene facilities have minimal dimensions and are integrated into the central utility block in the rear cylindrical volume. The toilet separates solid from liquid waste by a division in the bowl. Solid waste is collected in biodegradable bags to be carried to the main base, and waste water is suctioned into a hygienic container system. In 40 crew-days about 5 kg of feces and 65 liters of urine are collected. Drying feces in a composting toilet is a subject of further study. Interior cabin air is pulled through the toilet system and then filtered to remove odor. Minimal surface gaps make it easy to clean.

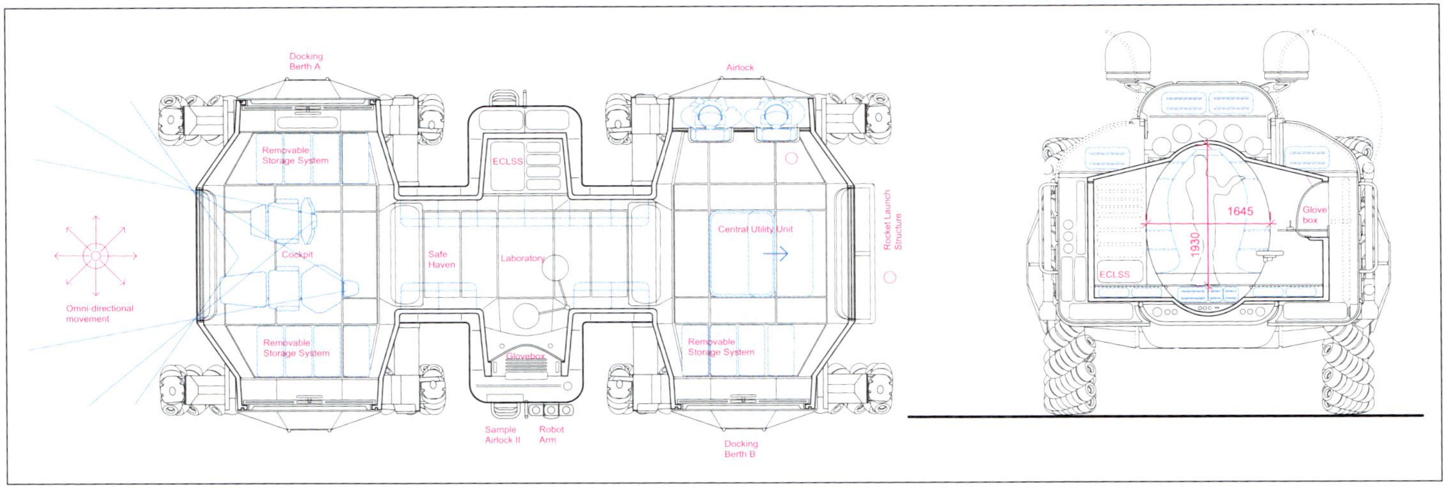

FIGURE 10 MCO plan and cross section: cockpit with driving seats (left); laboratory with glovebox (center); central utility unit containing cooking and hygiene facilities like shower and foldaway toilet (right).

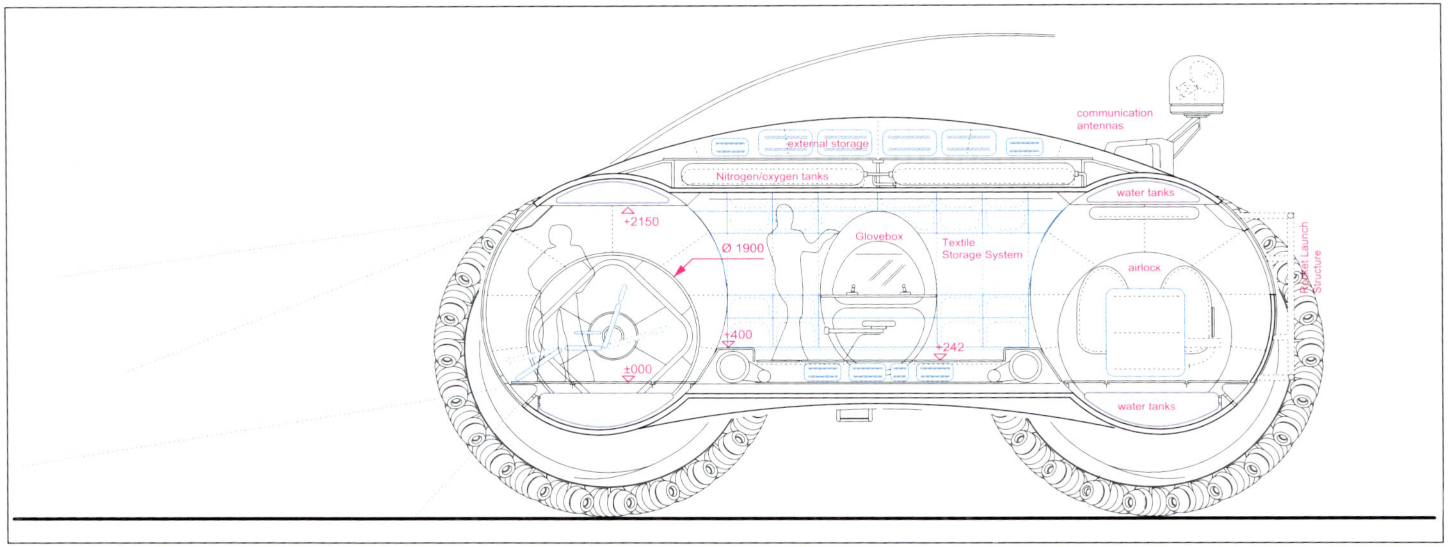

FIGURE 11 MCO longitudinal section, showing water tanks above and below the cylindrical volumes and gas bottles above the tubular connection.

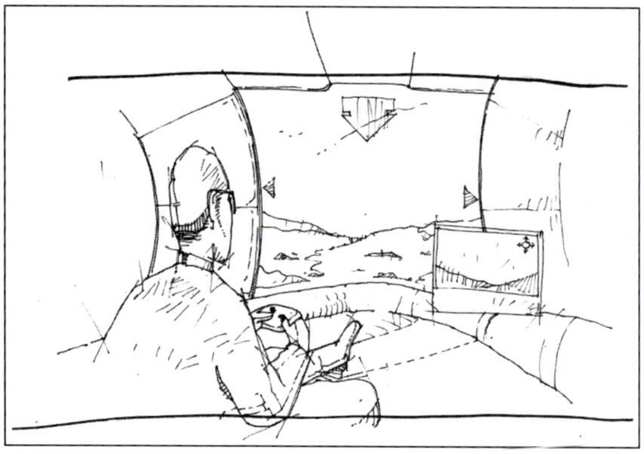

FIGURE 12 Sketch of cockpit with wireless drive control (courtesy of Architecture and Vision, Alessandro Natalini).

The hygiene compartment has a sink for hand washing and storage for utilities. Body wash is via wet towels that can be washed at the main base or a suction-based shower like the personal hygiene assistant developed by the Technical University of Munich (Vogler 2000). The main base would have larger hygiene facilities and a full-body cleansing system (Hoffmann 2001).

GALLEY AND FOOD PREPARATION Food preparation for a rover mission is done in advance at the main base, where the food is then vacuum-packed in portions to stay fresh for as long as 20 days. The requirement for an onboard refrigerator needs to be determined. Food is warmed in a microwave oven. The crew members take a personal set of dishes from the main base and clean them after use. The galley thus can be kept minimal, like an airline galley, but with special attention to waste disposal.

EVA AIRLOCK ACCESS The surface EVA subsystem is designed around a rear-entry suit. A 69-kN/m^2 cabin atmosphere allows 26-kN/m^2 spacesuit operations without prebreathing. Access and maneuvering space on both sides for the suitport awaits further exploration using full-scale mock-ups. In the proposed design this access space is minimized, assuming the use of a mechanical counterpressure suit. The standard assumption is that EVA is conducted only using the buddy system in which both crew members go EVA simultaneously. Scenarios for emergencies and moving an injured crew member back inside also require mock-up-based testing. During driving and non-EVA operations, the exterior portion of the suitports, including the suits themselves, is protected by a dust cover. EVA prep includes opening the dust cover and deploying a folding staircase to reach the ground. Tool stowage is provided in the rear of the rover.

CONCLUSION

The MCO functionality is an important component of human Mars surface exploration. The MCO and derivative concepts will be further developed symbiotically with a main base configuration and its habitat subsystems. An MCO-like, pressurized excursion rover is dependent on the base habitat's waste recovery and other life-support subsystems, but at the same time can be an important extension of the main habitat. It offers a crew of two a temporary, independent habitat, thus introducing spatial variety and relief from the confined living situation of the base. Despite its risks, mobile exploration will likely bring welcome relief from sensory deprivation for the explorers, but also provide a mission purpose with structured time for crew members who remain at the base during the sortie. Going exploring with the rover is likely to become a favorite activity within the overall mission. In addition to its mobile purpose, the MCO is a "lifeboat," it can provide emergency shelter for the whole crew for up to a week if life support is disrupted in the base habitat. Close interdependency with base habitat subsystems necessitates integrated planning.

Further development of the large, hubless wheel subsystem opens possibilities for effective use of the rover footprint for cabin volume. Studying existing designs and patents for hubless, rim-driven vehicles provides inspiration and a technology basis. Use of the volume within the wheel envelope is especially promising for space applications because it allows a more compact design, and large wheels have the advantage for unknown terrain of high grip with low pressure.

An ever-present design challenge for a small, mobile cabin is providing well-designed, comfortable, and safe activity zones, while accommodating sufficient stowage provisions and still preserving an open, spacious feeling. The stowage requirement is mission-dependent and requires detailed operations analysis to be fully understood. For example, in the present design, underfloor volume is used for rover subsystems. Models, mock-ups, and prototypes can be used to refine conceptual designs like the MCO. In addition, other industries that design compact vehicle interiors, e.g., for automobiles and aircraft, offer inspiration, expertise, images to help communicate design options, and even suppliers for relevant mock-up equipment.

References

Arno, R. (1999), "Planetary Surface Vehicles," *Human Spaceflight: Mission Analysis and Design*, edited by W. J. Larson and L. K. Pranke, McGraw–Hill, New York, pp. 447–476.

Astrium, Alenia Aerospazio, EADS Launch Vehicles and DLR (2002), "European Mars Missions Architecture Study," Executive Summary and Final Report to ESA, 14566/00/NL/WK, July.

Hoffman, S. J. (ed.) (2001), *The Mars Surface Reference Mission: A Description of Human and Robotic Surface Activities*, NASA/TP-2001-209371.

NASA (1995), "Man-Systems Integration Standards," NASA-STD-3000, Rev. B, July.

Steinsiek, F., Ransom, S., and Boettcher, J. (2003), "Mars EVA Suit Airlock (MESA)," SAE, AIAA, AIChE, ASME, 33rd International Conference on Environmental Systems, Paper 2003-01-2448, July.

Stilwell, D., Boutros, R., and Connolly, J. (1999), "Crew Accommodations," *Human Spaceflight: Mission Analysis and Design*, edited by W. J. Larson and L. K. Pranke, McGraw–Hill, New York, pp. 575–606.

Vogler, A. (2000), "Micro-G-Architecture—A Transdisciplinary Education, Research and Product Development Project for Engineers and Architect," Society of Automotive Engineers, Paper 2000-01-2328, July.

Bibliography

Allen, C. S., et al. (2003), "Guidelines and Capabilities for Designing Human Missions," NASA/TM-2003-210785, Jan.

Vogler, A., Ransom, S., Vittori, A., and Foth, W. P. (2005), "Mobile Pressurized Laboratory Design Study," Society of Automotive Engineers, Paper 2005-01-3051, July.

20

flat-floor inflatable structures

JAMES D. LOWE

INTRODUCTION

INFLATABLE STRUCTURES FIGURE PROMINENTLY IN CLASS II concepts for planet-surface habitats. In *The Case for Mars,* Zubrin and Wagner (1996) present an interesting discussion about pressurized buildings for use on Mars. They discuss at length how the difference between interior and exterior pressure is so great that some sort of massive anchoring and foundation system would be required to hold such a building against the ground to provide inhabitants with a flat floor. The basic problem is that the external atmospheric pressure is almost zero—indeed it is zero on the Moon—but the interior carries Earth-normal atmospheric pressure. The roof and walls literally want to explode away unless they are suitably restrained; said another way, the tremendous net force caused by the differential pressure makes the habitat "want to" be a sphere.

This is by no means a newly discovered problem. For example, in *Survival on the Moon*, a popular work from the 1960s, Masik (1966) suggests that instead of anchoring the walls of a hemispherical or semicylindrical building to lunar rock and having the lunar surface form the floor's backbone, one could just attach a flat floor with "very stout beams" to the walls. This creates an integral container to hold the atmosphere, and the degree of stoutness of the floor beams defines the flatness of the floor by limiting its deformation.

Speranza (2004) presents arguably the most iconic version of contemporary thinking regarding inflatable habitats. He presents a 6-m-diameter inflatable hemisphere held snugly to the ground by a network of 19 external straps radiating from the top of the hemisphere and anchored to the ground. There does not appear to be a "stout floor" to prevent the flat-floor from bending; the straps take care of this problem by pulling the envelope down against the ground.

An assumption implicit in typical designs appears to be that spheres, cylinders, and possibly toroids will be the geometries used to form the basis of inflatable planet surface buildings. If sectioned to produce a flat-floored building, then anchoring would be needed to compensate for the loss of envelope symmetry that stabilizes the structure against the internal pressure.

This chapter presents an alternative inflatable envelope geometry based on regular prismatic polyhedra, which could be used to build inflatable habitats that yield relatively flat floors without external anchoring. The basic pressure vessel structure concept consists of an asymmetrical inflatable envelope superstructure bonded to rigid floor panels. The chapter also examines some of the resulting loads in the envelope and floor; shows that simple materials could resist those stresses and strains; discusses some options for packing deflated forms for transport; and looks at human scaling concerns for organizing the interior space. Much of this approach is inspired by contemporary knowledge of tensegral structures (Beukers and van Hinte 1999) and airship technology (Khoury and Gillett, 1999).

Copyright © 2008 by James D. Lowe. Published by the American Institute of Aeronautics and Astronautics, Inc., with permission.

GEOMETRICAL PRINCIPLES

Take a circular sheet, fold it in half, and seal the curved edge. This forms a flat polyhedron with two edges, two vertices, and two surfaces, and satisfies Euler's polyhedron equation. Figure 1 shows a paper mock-up of this semicircular polyhedron.

When inflated, the flat semicircular shape deforms into a highly three-dimensional crescent. Figure 2 shows an inflated model made from thin aluminized polyester with clear packing tape used to seal the seams. The circle used to make the model has a 4-ft (1.2-m) diameter. By inspection, the diameter upon which the circle was folded transforms from a straight edge to a concave curve. The wrinkling results from the initially flat sheets deforming into curved surfaces.

Consider again the folded-circle polyhedron we started with, and imagine pulling one of its surfaces perpendicularly away from the other. Figure 3 shows a cardboard mock-up of the new semicylindrical polyhedron that results from the transformation. In this model, the semicircular surfaces are separated by a distance equal to one-half the radius of the originating circle. This polyhedron has six edges, four vertices, and four surfaces, and also satisfies Euler's polyhedron equation. Figures 4 and 5 show what the semicylindrical polyhedron looks like when inflated.

The model of Figure 5 is made from thin aluminized polyester. The diameter is 4 ft (1.2 m), and the

FIGURE 1 Transformation of a circular plane to a semicircular polyhedron.

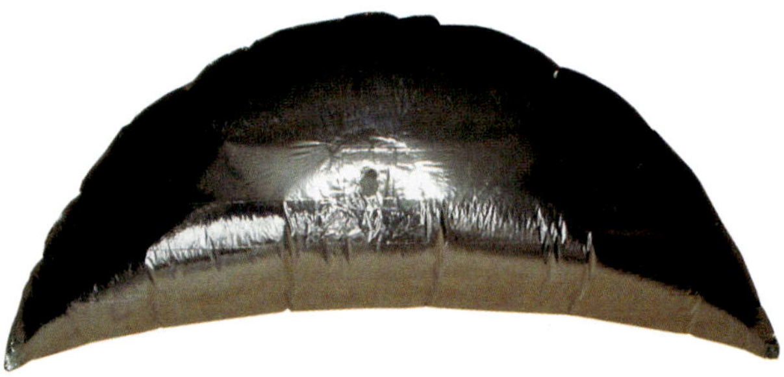

FIGURE 2 Inflated semicircular polyhedron.

FIGURE 3 Semicylindrical polyhedron.

FIGURE 4 Inflated semicylindrical polyhedron.

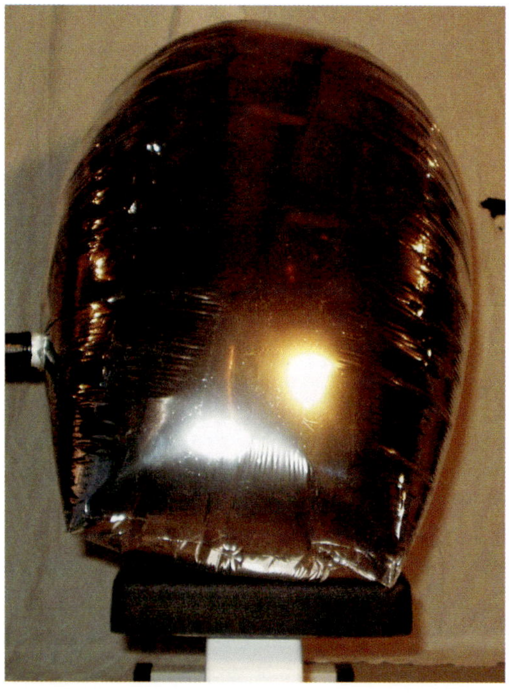

FIGURE 5 End view of inflated semicylindrical polyhedron.

separation between the semicircular surfaces is 1 ft (0.3 m). By inspection, the once-concave lower edge of the inflated crescent has transformed into a surface with almost imperceptible concave arc, albeit with local upward curvature along its length caused by the ballooning out of the semicircular side walls.

An interesting question arises: at what separation between the two semicircular planes does the initial crescent edge go flat, beyond which the lower surface bulges outward, and the semicylindrical polyhedron begins its transition to an overall cylinder when inflated?

Consider a semicylinder polyhedron with radius r' and distance l separating its semicircular surfaces. The height h at any location in the semicylinder polyhedron is given by $h = r' \cos\theta$, where θ is 0 deg at the center—and highest point—of the semicylinder. By inspection of the model, the cross section of the inflated semicylindrical polyhedron approximates a circle at its midsection. If the radius of this circle is r at some θ (differential pressure across the envelope membrane), then the circumference is $C = 2\pi r$. It is rather straightforward to determine the relationship between r and r' because C, to first approximation, is also equal to $2h + 2l$:

$$r = (r' \cos\theta + l)/\pi$$

This equation can be used to plot the ratio of the diameter $d = 2r$ of the circular cross section of a pressurized model to the unpressurized height h of the original semicylindrical polyhedron at any position θ:

$$d/h = 2(l + r' \cos\theta)/(\pi r' \cos\theta)$$

Figure 6 plots d/h for $r' = 1$ and a variety of values of l/r'. The figure shows that for $l/r' = 1/2$, d and h are very close to being equal in the range of 0 deg $\leq |\theta| \leq 40$ deg. In fact, at $\theta = 0$, $d/h = 3/\pi$. Figure 6 also shows that as the ratio of l/r' becomes much larger than one-half, the previously observed crescent shape disappears, and the floor membrane generally begins to bulge out, thereby pushing the inflated shape toward an overall "pudgy" sphere or cylinder depending on the value of l.

This mathematical model is very simple and only a first approximation. Elasticity is not explicitly accounted for, and the equations also should not be considered highly valid near the edges and corners of the envelope.

Figure 6 shows that at some value of l/r' slightly larger than one-half, d/h goes to 1 at $\theta = 0$. This value can be found using the d/h equation.

$$l = \pi/2 - 1 = 0.571 \text{ for } d/h = 1, \theta = 0, r' = 1$$

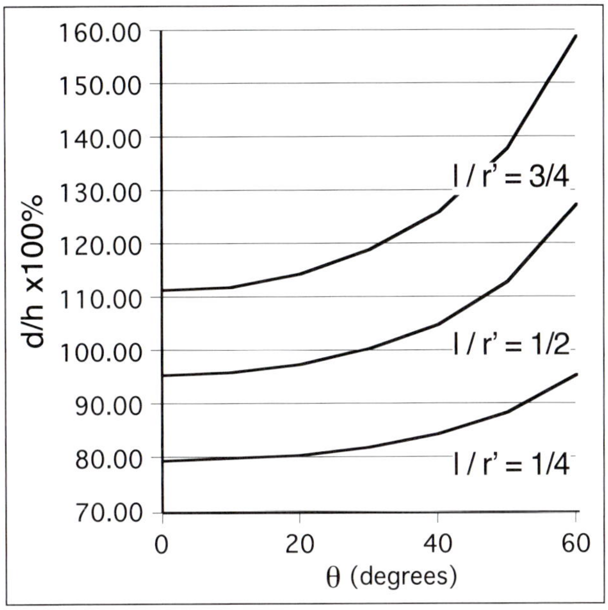

FIGURE 6 Parametric relationships for inflated polyhedra (see text for explanation).

The model of the semicylindrical polyhedron uses an l/r' of one-half, which is close to producing a d/h of 1 at $\theta = 0$. Given the approximate nature of the mathematics, this choice of l/r' both flattens out the crescent to maximum effect and allows floor plates to be squares of the same size, a useful property discussed in the following sections.

INTERNAL PRESSURE

Internal pressure is the dominant parameter for sizing the envelope and floor plates. A very low difference between internal and external pressures is used to maintain the shape of Earth-bound, pneumatic inflatable structures like blimp envelopes or inflatables covering sports fields. On the Moon or Mars, where there is no or essentially no native atmosphere, the pressure differential is high, and the structure must withstand this. The lower the internal pressure of such a habitation structure, the lower the strength (and therefore mass) of its component parts.

How low can the pressure be and still support human life without adverse side effects? There are a number of possibilities for internal atmospheric pressure, but no commonly accepted standard. Several choices are tabulated in Table 1, summarized from Reed and Coulter (1999). For reference, Kennedy et al. (2001) state the TransHab inflatable module prototype was designed to provide inhabitants with full standard sea-level atmospheric pressure.

TABLE 1 Possible choices for internal atmospheric pressure.

PRESSURE LEVEL	VALUE, kPa
Minimum to support human life	25 to 26
American spacesuit pressure	29.6 {4.3 psi}
Russian spacesuit pressure	39.2 {5.7 psi}
Suggested by Reed and Coulter (1999)	70 to 80 {10.2 to 11.6 psi}
Standard sea-level atmospheric pressure	101.3

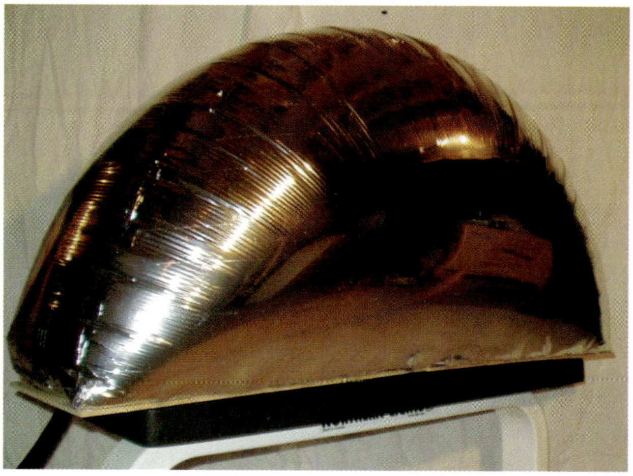

FIGURE 7 Inflated semicylindrical polyhedron with rigid floor.

FLOORING

A rigid floor can be substituted for the bottom of the inflated semicylindrical polyhedron to provide a flat living surface for inhabitants. Because the bottom of the envelope does not have significant curvature, it lends itself to complete flattening through the structural substitution of rigid plates. Figure 7 shows an inflated model with a floor. The floor comprises four separate square panels. They are not connected to each other, only to the envelope. In the case of the model, the floor panels are made of quarter-inch-thick (6.4-mm) medium-density fiberboard and attached to the polyhedron's side walls with strips of packing tape.

On a full-size structure the floor plates would slip into pockets attached to the envelope's bottom panel (Figure 8). The pockets would hold the floor panels against the envelope and be tailored to distribute the load up and through the sidewalls. The plates would resist the internal pressure in bending, while the envelope would provide pneumatic integrity. The panels themselves could be made in a variety of ways: homogeneous plates as implied in Figure 7, honeycomb sandwich plates, or possibly collapsible x-brace truss frames. As discussed in the floor-strength section next, the choice of construction likely depends on the size of the envelope. Small habitats could be built with reasonably sized honeycomb floor plates, but large habitats could require an x-brace truss frame because pure honeycomb structures have scale-up limitations.

The total force exerted by the internal air pressure on each floor panel can easily be calculated by multiplying the internal pressure by the area of a panel. Table 2 tabulates the results for a selection of internal pressures extracted from Table 1.

Table 2 clearly shows why internal pressures below standard atmospheric are beneficial: the total load on the floor becomes quite substantial as the size of the structure increases. For comparison purposes, note a 2005, four-door Honda Civic sedan weighs in at around 11.4 kN (2560 lb). On the smallest floor panel shown, with $l = 4$ ft (1.2 m) and $r' = 8$ ft (2.4 m), each floor plate is subjected to roughly the total load of four Honda Civics. The next section, on floor strength, discusses bending considerations induced by these loads.

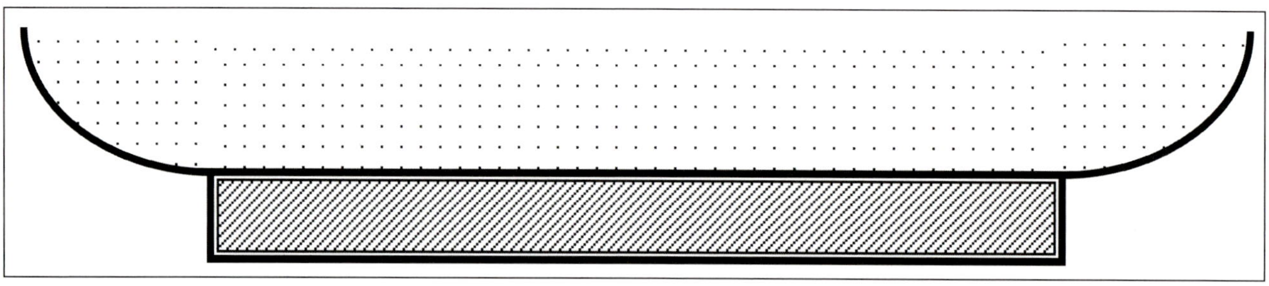

FIGURE 8 Cross-section sketch through an inflated semicylinder with floor panels held in place inside pockets.

TABLE 2 Total force exerted on a single floor panel of a floored, inflated semicylindrical polyhedron.

FLOOR WIDTH $l = r'/2$	PRESSURE DIFFERENTIAL		
	29.6 kPa {4.3 psi}	75 kPa {10.9 psi}	101.3 kPa {14.7 psi}
1.22 m {4 ft}	44.1 kN {9.91 klb}	111.6 kN {25.1 klb}	150.8 kN {33.9 klb}
1.83 m {6 ft}	99.1 kN {22.3 klb}	251.2 kN {56.5 klb}	339.2 kN {76.2 klb}
2.44 m {8 ft}	176.2 kN {39.6 klb}	446.5 kN {100.5 klb}	603.1 kN {135.5 klb}

[a]q = pressure differential across the membrane.

Figures 9 and 10 illustrate some ways a deflated structure can be stowed for transport. Figure 9 simply shows that with small, individual square plates, the floor can be readily packed into a compact volume. In this situation, the floor plates would be slipped into the envelope's pockets prior to inflation. Figure 10 shows that the floor plates could be pre-inserted into the pockets and the entire deflated structure rolled into a cube. Inflation would cause the structure to unroll into deployed position. Clearly, these packing strategies are idealizations that would likely be modified to take into account actual materials.

FIGURE 9 Deflated envelope with floor plates stacked on top.

FIGURE 10 Deflated envelope folded into a cube formed by the floor panels.

FLOOR STRENGTH

As referenced earlier, Masik (1966) claimed "stout beams" would be necessary across the floor of this kind of structure to prevent the floor from bending under the load of the internal pressure. In the case of the semicylindrical polyhedron, floor plates fulfill this role.

Accurately calculating the strength, mass, and thickness of the floor plates—as well as the envelope discussed in the next section—requires finite element analysis. However, for a first approximation that helps develop an understanding of the overall nature of the problem, we can use classical, closed-form, analytical approximations. In this analysis the floor panels are assumed to be simply supported plates, supported along the two opposing edges that slide into the pockets shown in Figure 7. There is no support along the other two edges of the panels; they are free. Equations for the bending of such a panel (subjected to a uniformly distributed pressure), along with the methodology outlined in HexWebTM (n.d.), are used for preliminary sizing.

The floor panels are assumed to be an aluminum honeycomb sandwich with top and bottom plies of 0.5-mm aluminum alloy 5252 H24 and honeycomb core of 3003 aluminum with 6-mm cell size and 83-kg/m^3 density. Total sandwich thickness was determined with the assumption that neither the maximum bending stress of the plies nor the maximum shear experienced by the honeycomb is allowed to exceed 75% of their yield values.

Tables 3 and 4 show the minimum required thickness and mass respectively of the floor panels for several inflated semicylindrical habitats. In all cases in Table 3, the floor panel thickness is deep enough to make the maximum floor deformation negligible.

For lunar-base transportation logistics, it is important to minimize panel thickness for packaging, but also to minimize panel mass. The figures clearly indicate that if a spacesuit-like atmospheric pressure is acceptable for such a habitat, then the floor plates can be fairly thin and of low mass, even for large-scale versions; however, requiring full sea-level atmospheric pressure causes the mass of the plates to increase dramatically. Four plates for a structure with $q = 101.3$ kPa and $r' = 4.88$ m would have a mass of nearly 500 kg. This analysis assumes the plate sandwich technology is scalable up to the largest-size structures under

TABLE 3 Minimum floor-panel thickness.

FLOOR WIDTH $l = r'/2$	PRESSURE DIFFERENTIAL		
	29.6 kPa {4.3 psi}	75 kPa {10.9 psi}	101.3 kPa {14.7 psi}
1.22 m {4 ft}	3.3 cm {1.3 in.}	8.2 cm {3.2 in.}	11.1 cm {4.4 in.}
1.83 m {6 ft}	4.9 cm {1.9 in.}	12.3 cm {4.8 in.}	16.6 cm {6.5 in.}
2.44 m {8 ft}	6.5 cm {2.6 in.}	16.4 cm {6.4 in.}	22.1 cm {8.7 in.}

TABLE 4 Mass of a single floor panel.

FLOOR WIDTH $l = r'/2$	PRESSURE DIFFERENTIAL		
	29.6 kPa {4.3 psi}	75 kPa {10.9 psi}	101.3 kPa {14.7 psi}
1.22 m {4 ft}	8.0 kg	14.1 kg	17.6 kg
1.83 m {6 ft}	22.4 kg	43.0 kg	54.9 kg
2.44 m {8 ft}	47.8 kg	96.5 kg	124.7 kg

consideration. However, unlimited scalability might not be possible as implied here; stiffer advanced-technology panels, or even modular panels, might be necessary. No attempt has been made to find the optimum plate size and material makeup for any of the inflatable sizes, as this is application specific.

ENVELOPE HOOP STRESS

The hoop stress in the wall of an inflated cylinder of radius R is given by the equation

$$S_h = q\frac{R}{t}$$

where q is the differential pressure and t is the thickness of the cylinder wall. The longitudinal stress in the same inflated cylinder is given by the equation

$$S_l = q\frac{R}{2t}$$

For an inflated sphere of radius R, the hoop stress and the longitudinal stress are identical and are also given by the equation

$$S_h = S_l = q\frac{R}{2t}$$

the same as the longitudinal stress of an inflated cylinder.

Although the inflated semicylindrical polyhedron is neither a cylinder nor a sphere, in the center it approximates a cylinder of radius r. Because this is the location of maximum radius, and from the equations we can see that the hoop stress is the larger of the two, the cylinder hoop stress equation can be used to calculate the necessary thickness of the envelope material. Table 5 tabulates values calculated for the minimum envelope thickness t, for a variety of inflated semicylindrical polyhedrons assuming the envelope is made from polyester sheeting.

TABLE 5 Approximate minimum envelope thickness of an unfloored, inflated semicylindrical polyhedron assuming a polyester envelope.

FLOOR WIDTH $l = r'/2$	PRESSURE DIFFERENTIAL		
	29.6 kPa {4.3 psi}	75 kPa {10.9 psi}	101.3 kPa {14.7 psi}
1.22 m {4 ft}	0.8 mm {0.031 in.}	2.0 mm {0.079 in.}	2.7 mm {0.106 in.}
1.83 m {6 ft}	1.2 mm {0.047 in.}	3.0 mm {0.118 in.}	4.1 mm {0.161 in.}
2.44 m {8 ft}	1.6 mm {0.063 in.}	4.1 mm {0.161 in.}	5.5 mm {0.217 in.}

An actual planet-surface inflatable structure would more likely use a composite of several materials to fully address space-rating requirements (Kennedy et al. 2001). Polyester typically has an ultimate tensile strength in the range 6,000–32,500 psi (41–224 MPa), yield strength in the range 5,100–7,900 psi (35–54 Mpa), and comes in thicknesses up to about 4 mm. The results of the calculations shown in Table 5 assume a material with ultimate tensile strength of 25,000 psi (172 Mpa) and a factor of safety of four, as recommended in Kennedy et al. (2001). The data presented in the table merely give an idea of the minimum thickness of material required for a commonplace material; hence, the table entries indicate that no new kinds of materials would be required to contain the internal atmosphere. Further design studies would consider the impact of environmental temperature, exposure to various types and intensities of radiation, abrasion caused by dust and debris, and insults caused by handling and deployment. All of these factors impact the ultimate design and construction of the envelope, but preliminary calculations shown in Table 5, and the existence of precedents such as Transhab (Kennedy et al. 2001) and Bigelow Aerospace modules (Belfiore 2005) suggest the inflated semicylindrical polyhedron envelope is feasible.

HUMAN SCALING

One of the problems with buildings with curved walls (e.g., cylinders and spheres) is that the curvature makes it difficult to accommodate human activities and equipment, furnishings, and storage. For habitats located in even reduced gravity, it is most convenient to have vertical walls. Semicylindrical habitats are not as challenging in this regard as spheres as long as they are of sufficient size.

Figure 11 illustrates a section through the smallest of the inflated semicylindrical polyhedra. The figure shows that the "six foot limit width" of the habitat is quite broad. The lower horizontal line in the section sketch is the 30-inch (0.76-m) line—about desktop height (Ching 1987)—indicating that most of the region where a person cannot stand upright could at least be used to accommodate a desk or table. Figure 12 shows an approximate superposition of all three sizes of inflated semicylindrical habitat discussed in this chapter.

Figure 12 illustrates that interior vertical space increases significantly once r' goes to 16 ft (4.9 m). Even at $r' = 12$ ft (3.7 m), designers should begin considering how to best make use of the vertical space; a loft is one possibility. An interesting consideration not yet fully understood is the effect of gravitational

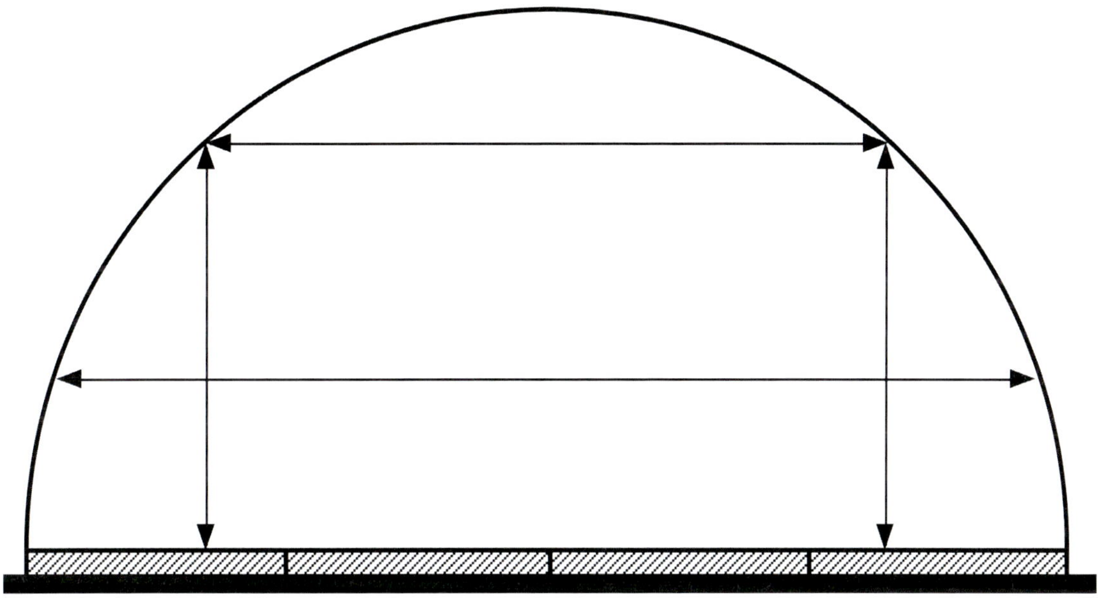

FIGURE 11 Section diagram through an $r' = 8$ ft habitat, showing head clearance and typical table height (horizontal lines).

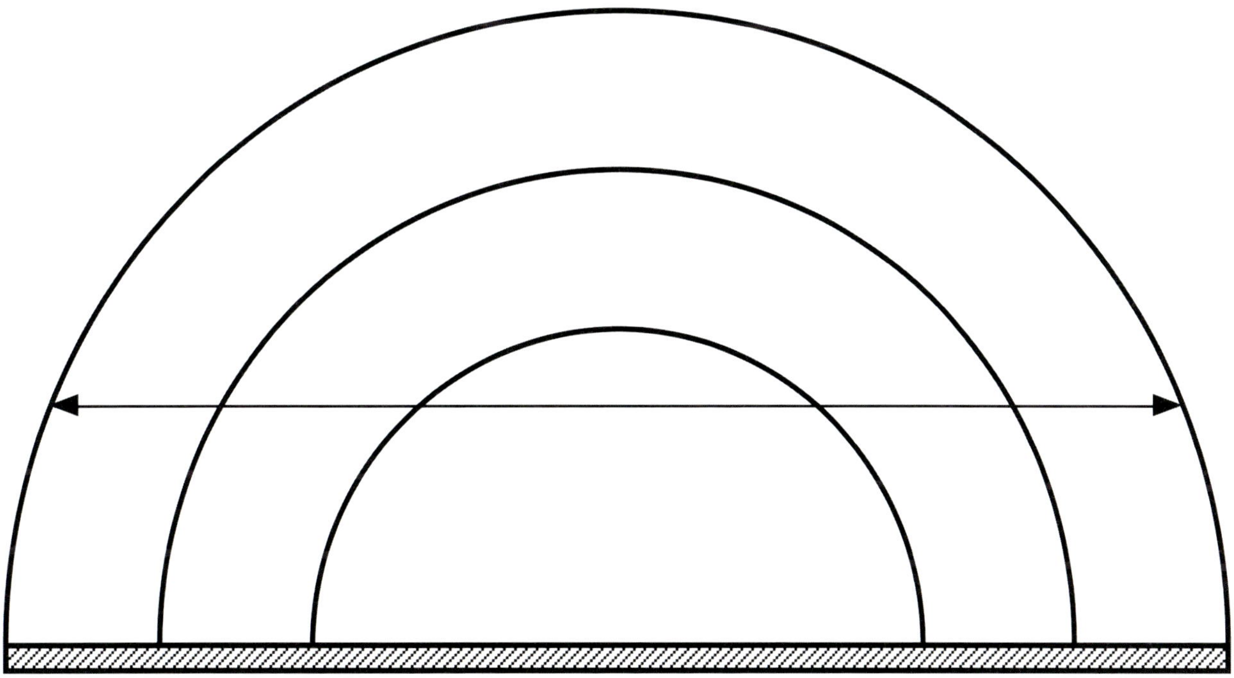

FIGURE 12 Head clearance for multiple, superimposed sections, corresponding to examples in Tables 2–5.

strength on ceiling height. As Kennedy (1992) discusses, dynamic ceiling height standards might need to be increased on the Moon because $\frac{1}{6}$ g probably leads to more vertical rise during routine movement. Because Mars' gravitational field is 0.38 g, any effect would not be as pronounced there. Further analysis is necessary to quantify desirable internal ceiling height.

Nonetheless, at r' = 16 ft (4.9 m), a full-fledged second story appears reasonable because there would be a large amount of unutilized overhead space without one. Questions regarding how to best utilize such overhead space are among those to be considered when mapping an activity program into such a habitat.

APPLICATIONS

There are a number of ways an inflated semicylindrical polyhedron could be used. A larger structure could be made by connecting together a number of these inflatables, one after another, to form a "longhouse" configuration. Where two inflated polyhedrons come together, the intersection would be replaced by tension straps, webbing, or a curtain of cables so that inhabitants could pass freely between bays, but without sacrificing the geometric properties of the single bay that make a flat floor possible. This load-bearing webbing is much like the internal load curtain in a blimp that hangs the gondola from the inside top surface of the envelope. Such a structure could be extended beyond four bays as necessary because length does not affect the overall structural characteristics. This sort of structure could be used as a stand-alone form or as a habitable extension for a lander equipped with access tunnels in a manner similar to that shown in Frassanito's Mars lander illustrations (Zukowsky 1999). Any such structure would require an airlock or pressure ports to allow EVA.

It would also be possible to use this structure type as a deployable shelter, fulfilling a role similar to the Martian pressurized ball tent proposed by Cockell (2004). Similarities between the requirements for that function and those of Archigram's Cushicle (Cook 1999) are striking.

CONCLUSION

The inflated, semicylindrical polyhedron form could be useful for lunar or Mars surface applications where flat-floor configurations are desirable for outfitting. Commonplace aerospace materials have sufficient physical properties to make such a geometry viable for structures with uninflated radius r' in the range of 8–12 ft (2.4–3.7 m), even when containing sea-level atmospheric pressure against vacuum. Multiple-bay structures are extensible to arbitrary lengths as long as suitably spaced, internal load-bearing tension webs or curtains are incorporated. Structures with larger radii appear possible, but further study of alternative floor-panel material systems is required. Finite element methods can detail stresses and strains in the various geometry elements of this structure system. Knowing they are not limited to spheres and cylinders, space architects can consider how flat-floor structures can support human space activity programs. |

References

Belfiore, M. (2005), "The Five-Billion Star Hotel," *Popular Science*, Vol. 266, March 2005, pp. 50–57.

Beukers, A., and van Hinte, E. (1999), "Tensegrity Hero," *Lightness: The Inevitable Renaissance of Minimum Energy Structures*, 010 publishers, Rotterdam, the Netherlands, pp. 32–37.

Ching, F. D. K. (1987), *Interior Design Illustrated*, Van Nostrand Reinhold, New York, pp. 62–69.

Cockell, C. S. (2004), "The Desert Crossings of Mars," *Martian Expedition Planning*, American Astronautical Society, San Diego, CA, pp. 163–173.

Cook, P. (1999), *Archigram*, Princeton Architectural Press, New York, pp. 64, 65.

HexWebTM (n.d.), "Honeycomb Sandwich Design Technology," 7586_HexWeb_Sand_Design.pdf, http://www.hexcel.com/NR/rdonlyres/80127A98-7DF2-4D06-A7B3-7EFF685966D2/0/7586_HexWeb_Sand_Design.pdf [accessed June 2006].

Kennedy, K. J. (1992), "A Horizontal Inflatable Habitat for SEI," *Engineering, Construction and Operations in Space III, Proceedings of the Third International Conference on Engineering, Construction and Operations in Space*, edited by Willy Z. Sadeh, Stein Stuve, and Russell J. Millers, Vol. 1, 1992, pp. 135–145.

Kennedy, K. J., Raboin, J., Spexarth, G., and Valle, G. (2001), "Inflatable Habitats," *Gossamer Spacecraft: Membrane and Inflatable Structures Technology for Space Applications*, edited by Christopher H. M. Jenkins, Vol. 191, AIAA, Reston, VA, pp. 527–552.

Khoury and Gillett (1999), *Airship Technology*, Cambridge Aerospace Series #10, Cambridge Univ. Press, Cambridge, England, U.K.

Maisak, L. (1966), *Survival on the Moon*, Macmillan, New York, pp. 54–57.

Reed, R. D., and Coulter, G. R. (1999), "Physiology of Spaceflight," *Human Spaceflight: Mission Analysis and Design*, edited by Wiley J. Larsen and Linda K. Pranke, McGraw-Hill, New York, p. 107.

Speranza, C. (2004), "Soil Anchorages for Mars and Moon Future Bases Inflatable Habitats" (Poster #9), 2nd European Workshop on Inflatable Space Structures, European Space Agency, June.

Zubrin, R, and Wagner, R. (1996), *The Case for Mars*, Simon & Schuster, New York, pp. 178–181.

Zukowsky, J. (1999), *Space Architecture: The Works of John Frassanito & Associates for NASA*, Menges, London, p. 77.

21

TRIGON modular robotic construction system

A. SCOTT HOWE
AND
IAN GIBSON

INTRODUCTION

PLANETARY SURFACE STRUCTURES can be categorized (Kennedy 2002). With Class I (pre-integrated) construction, ready-to-use habitats are delivered to the surface and occupied without any additional preparation. Class I structures are limited by the payload size constraints of the transportation systems used to deliver them and are most appropriate for short missions. With Class II (kit-of-parts) structures, prefabricated materials are delivered to the surface and inflated, assembled, or deployed into larger habitats. Class II structures are not as limited by size, but require effort expended on site for deployment and outfitting for habitation and will probably be used for more permanent outposts with longer occupancy. Class III (derived *in situ*) structures use native material to establish larger and yet more permanent settlements. Cohen and Kennedy (1997) explain that each type is preparatory for the next, but overlap is likely. In addition, because human "blue collar" labor is not practical for assembly as a result of the complexity of maintaining a safe environment for labor in harsh conditions, much of the construction will need to be automated.

One form of Class I structure that can also bridge over into Class II is the Habot concept (Cohen 2004) described in Chapter 18. Habots are pre-integrated habitats with robotic mobility for migrating across a planetary surface. Habot units delivered to the surface could find each other and connect to form a larger (Class II) base, and units could function as rovers, obviating the need for specialized surface transport vehicles. The original Habot concept used a legged walking mobility system too slow for a human crew. Howe and Howe (2005) derived a wheeled version called Mobitat (Mobile + Habitat), where mobility would be optimized and higher speeds achieved. However, the original Mobitat concept compromised mobility system functions, making it large and ungainly.

This chapter describes a vision of an integrated planet-surface architecture based on a flexible, modular structure system concept called TRIGON (Transformable Robotic Infrastructure-Generating Object Network). TRIGON is conceived as a core construction system for primary (Lai and Howe 2003) and secondary (Yip and Howe 2003) planetary structures, and for planet-surface mobility, that bridges from Class II to Class III. It consists of modular panels that self-assemble from a partially connected, compact stack densely packaged for delivery. Secondary functionality is incorporated into the panels through plug-in payloads carried by the panels during self-assembly (Howe 2005).

Among many potential applications described in the chapter, TRIGON overcomes some inadequacies of both the Habot concept and Mobitat concept (Howe, 2002) as the basis for a mobile infrastructure family. The conceptual designs described here integrate approaches for payload packaging, delivery, deployment, mobile planetary-base construction, rover functions, and reconfiguration.

Copyright © 2008 by the authors. Published by the American Institute of Aeronautics and Astronautics, with permission.

TRIGON CONCEPT

This design investigation of a modular mobile habitat centers on the fundamental TRIGON concept: modular panels that self-assemble by tumbling across each other, carrying specialized payloads into final usage locations as they move. This section describes 1) TRIGON panel self-mobility; 2) specialty payloads such as inflatable shielding and wheel mobility/suspension subsystems; and 3) core elements, special items such as pressure hatches, inflatable pressure barriers, and legged mobility systems that cannot be confined to payload panels but are designed to interface with TRIGON panel edges.

Stackable Modular Panels

The TRIGON system consists of polygonal panels with uniform edges. Built into these edges are actuators for linking to other panels. The actuators are hinged with an offset axis in such a way that any two panels can be oriented from 0 to 360 deg with respect to each other. This allows the panels to be stacked for compact transportation. Figure 1 shows a simple volume constructed of TRIGON triangular panels and illustrates how the panels can be folded and stacked.

The edge manipulator actuators—called connector arms—control the panels' revolute motion with respect to neighboring panels. Two panels mate by nesting their connector arms; one panel takes hierarchical dominance by moving its arms to the outside, and the other panel moves its arms to the inside so that both sets of arms mutually clasp. The revolute actuator in each arm set can then cause the panels to reconfigure their mutual angle.

Using this conceit, TRIGON panels can be added to a structure at any location, using motion of the connector arms to latch onto already completed portions of the structure and travel or "tumble" end over end, avoiding other traveling panels to find their own specified locations (Figure 2). Triangles can travel over square structure (and visa versa) in pairs (Howe

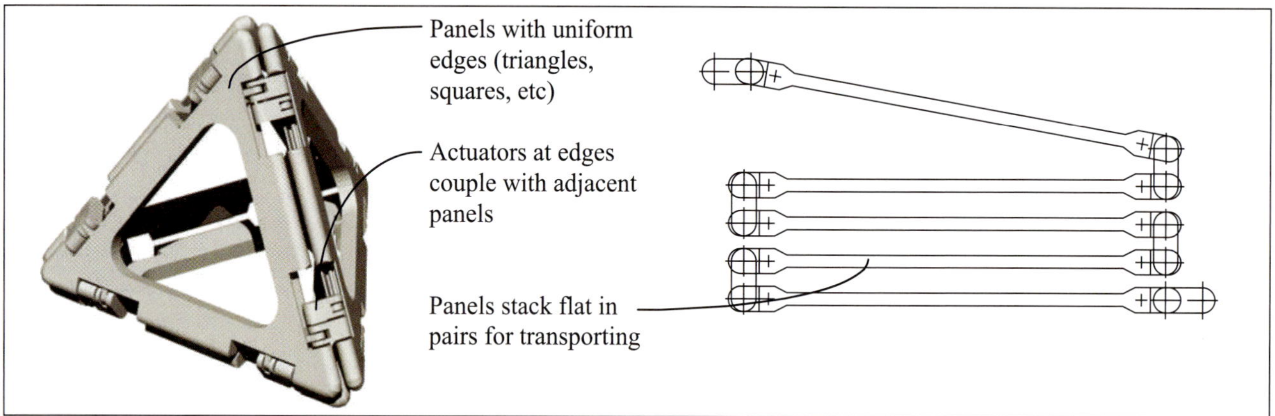

FIGURE 1 TRIGON system: stackable panels with common, robotic edge interfaces.

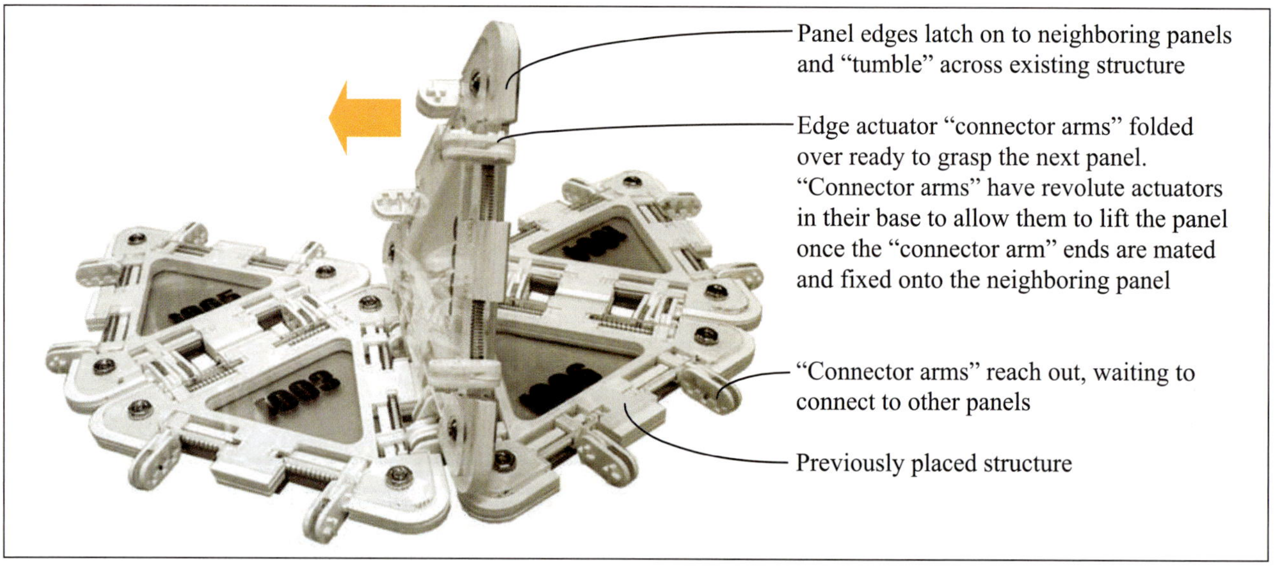

FIGURE 2 Functional mock-up showing how panels tumble over each other to self-assemble larger structures.

TABLE 1 Key parameters of various TRIGON panel designs showing maximum possible torques.

PARAMETERS	30-CM SQUARE	30-CM TRIANGLE	1-M SQUARE (THIN)	1-M TRIANGLE (THIN)	1-M SQUARE (THICK)	1-M TRIANGLE (THICK)
Nominal panel edge, m	0.30	0.30	1.00	1.00	1.00	1.00
Payload edge, m	0.20	0.20	0.70	0.70	0.70	0.70
Panel thickness, m	0.03	0.03	0.05	0.05	0.10	0.10
Panel mass, kg	4.05	2.49	17.9	11.7	281	183
Payload mass, kg	3.24	0.66	17.2	3.49	270	54.9
Torque, N-m	10.7	10.7	172	172	2700	2700
Mechanism count (motors + CSF harmonic drives in parallel)	Motor × 2 *CSF-8* × 2	Motor × 2 *CSF-8* × 2	Motor × 4 *CSF-11* × 4	Motor × 4 *CSF-11* × 4	Motor × 4 *CSF-25* × 4	Motor × 4 *CSF-25* × 4
Gear ratio	1:100	1:100	1:100	1:100	1:80	1:80
Maximum panel density, kg/m^3	2700	2700	700	700	5500	5500

and Gibson 2006a). Using only square and equilateral triangular panels, self-assembling domes, cylinders, trusses, and many other stable construction geometries can be assembled autonomously or via teleoperation.

A parametric model allows designers to work bottoms up from the mechanism, or top down from required panel size, to determine an optimum system (Howe and Gibson 2006b). Some examples of TRIGON panel designs are shown in Table 1, illustrating a large variety of sizes and load-bearing capacities possible depending on edge length, panel thickness, and number of mechanisms. The torque is calculated for 1-*g* conditions, assuming a worst-case scenario of lifting one square panel, with payload, vertically upward. Maximum panel density is calculated based on the envelope of a square panel nominal edge multiplied by the panel thickness, using the maximum panel and payload mass. (The density does not refer to maximum material density.)

Ideally, each panel would have a minimal number of mechanisms. The mechanism counts shown in Table 1 represent the largest quantity possible for each panel edge, to illustrate the amount of torque achievable with respect to the panel thickness.

Modular Plug-in Payloads

All of the mechanisms, actuators, sensors, and controllers required for the TRIGON panel primary motion function are located in a narrow zone parallel to the panel edge. A power and communications network runs through the connector arms of mated panels. This leaves a clear zone in the middle of each panel for specialty "payload" panels that plug in to power-communication ports to draw power as needed from the host panel for deployable manipulators, motors, instruments, mobility systems, etc. (Figure 3).

The possibilities for these payloads are virtually unlimited. Here we describe, in diagrammatic terms, several payload functions to give a TRIGON assemblage mobility, insulation, and protection from cosmic radiation. The list could also include antennas, instruments, and deployable manufacturing cells that could occupy the walls of a "cassette factory" (Howe 2005).

***IN SITU* SHIELDING PAYLOAD** One example of a specialty payload is thermal insulation panels or dense radiation shielding. Because TRIGON panels have gaps between them to accommodate grasping points for panel tumbling, the shielding application requires large, "floppy" coverings that overlap the gaps. Figure 4 shows a concept for an inflatable shield that folds up compactly into the body of the payload panel while "tumbling," but can be filled with potable or waste water, or regolith, once deployed. The shielding cells can overlap in multiple layers to create uniform thickness. Another shielding concept is "hoppers" that use compressed gas to expand inflatable ribs out from each TRIGON panel, creating a pocket to hold regolith while constricting a dump hole in the bottom. Moving belts made of TRIGON panels can then be used to fill the pockets with regolith found on the site. In this concept, if the structure needs to be relocated, its mass can be significantly reduced by deflating the hopper ribs, allowing the dump holes to relax and drain the regolith held in the pockets.

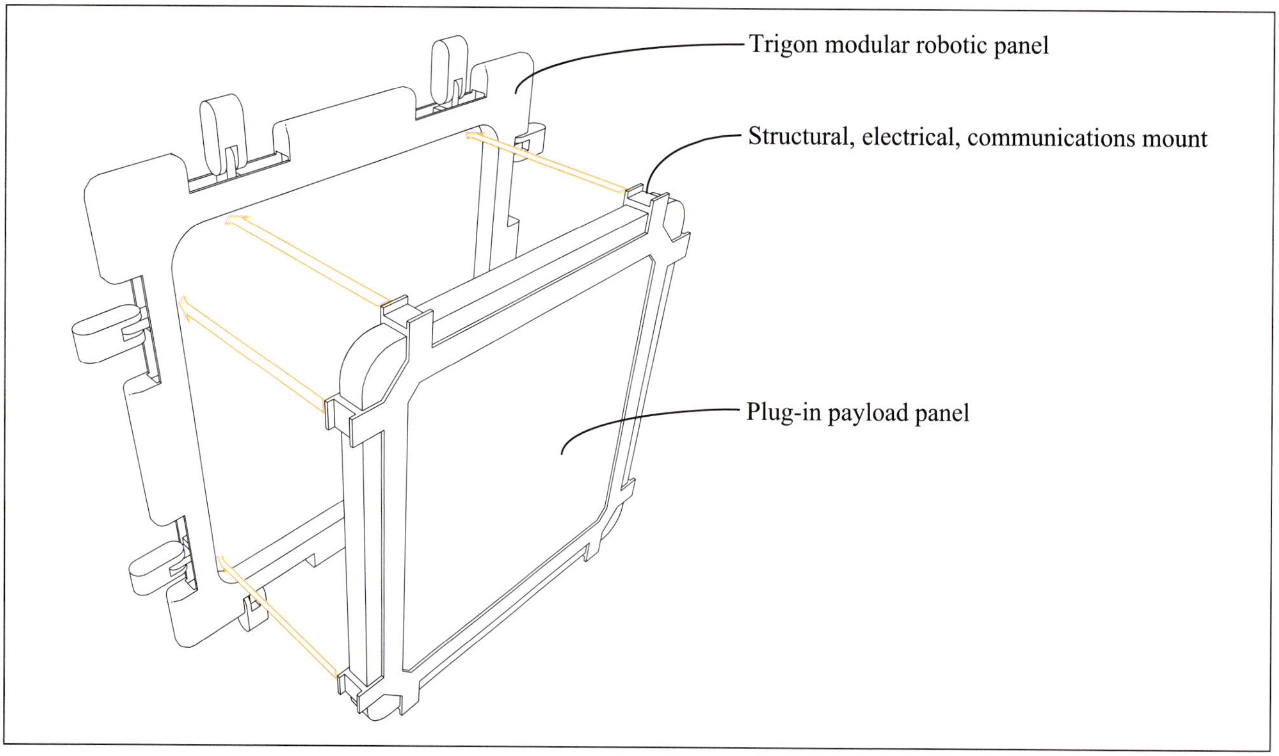

FIGURE 3 TRIGON payload panel is parasitic on host panel.

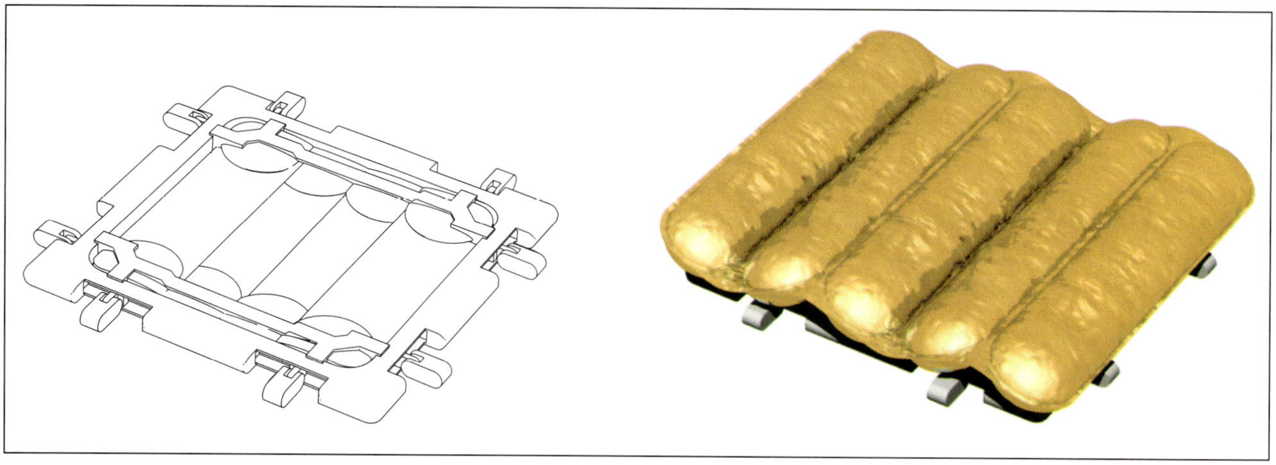

FIGURE 4 Shielding payload panels: left, folded and right, filled or inflated.

The advantage of inflatable or postfilled shielding bags is that the filling material need not be carried or imported from Earth, but can be found *in situ*. Some of the images shown in this chapter are rendered with thin inflatable shields, but shielding can be as bulky as needed, depending on the design of the bag and how it attaches to the payload panel structure to transfer inertial force loads. Postfilled shielding bags would need either a local mechanism for permanent filling and sealing or central or distributed supply-return infrastructure (e.g., where shielding bladders are used as wastewater tanks).

WHEELED MOBILITY PAYLOAD A key panel payload would be a mobility subsystem. Such a subsystem could be specially designed as a core element that interfaces with the TRIGON system but is too bulky to fit into a payload panel (see what follows). Or, a deployable wheel and suspension subsystem could conceivably be designed to fold flat into the body of two payload panels, so that it could be moved across a TRIGON structure in the normal tumble mode for final placement. The loaded TRIGON panels can be stacked back to back and moved via robotic pair tumbling to any required location on the structure. [For simplicity, TRIGON panels are typically shown tumbling singly across structure, whereas they would actually function in robotic pairs (Howe and Gibson 2006a).] This design has a motor for forward and reverse mobility and actuators for turning, deployment, and tire inflation.

A series of schematic diagrams illustrates the deployment concept (Figure 5). The sequence begins with the pair of panels opening, with double-wishbone suspension and deflated wheel packed into storage position in the two panels. The two TRIGON panels hinge at the top, allowing the double-wishbone suspension to expand outward for deployment. The TRIGON panel connector arms and suspension act in

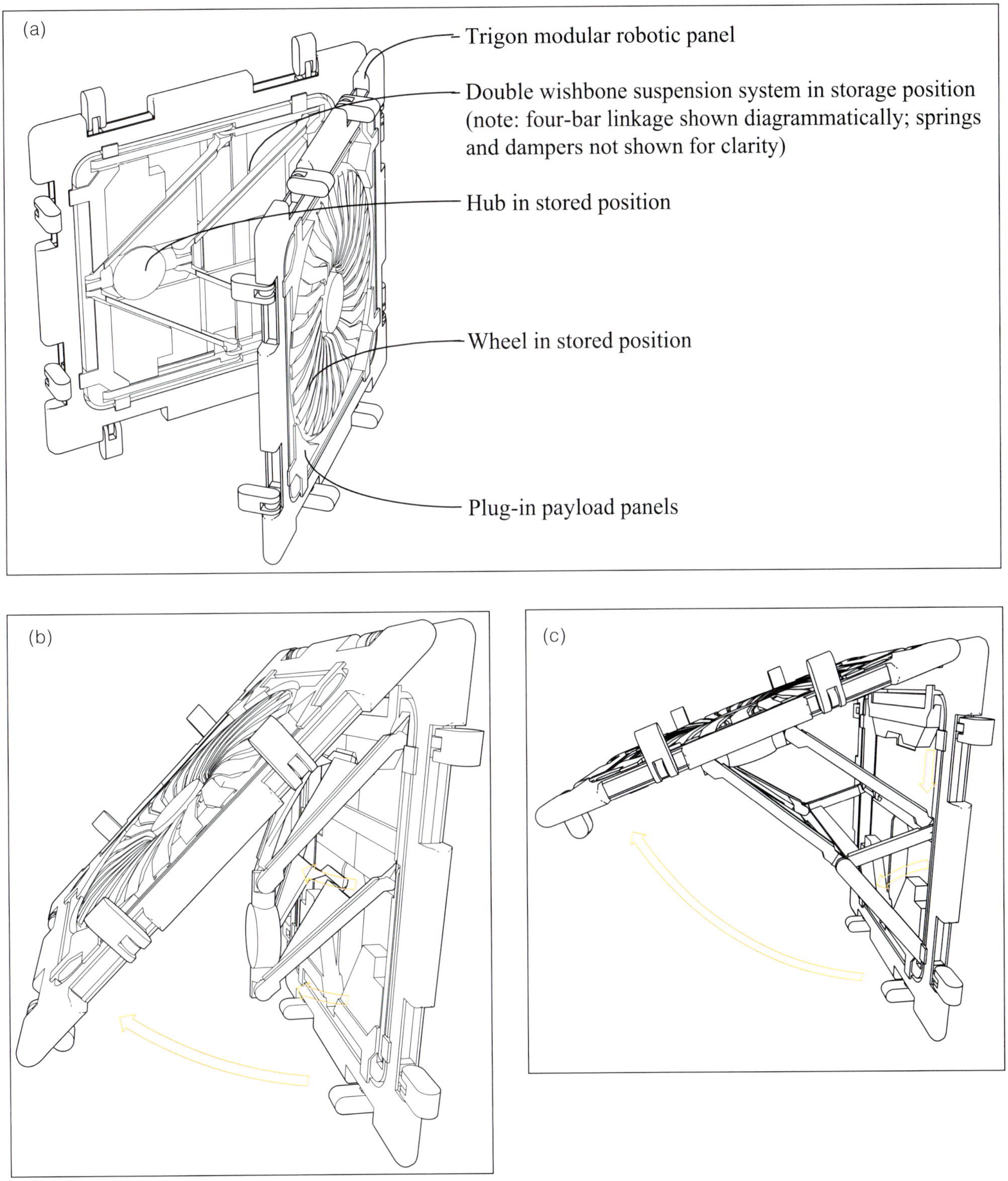

FIGURE 5 Mobility subsystem as a TRIGON plug-in payload: a) wheel payload (shown open) includes suspension and wheel assemblies; b) suspension deployment; c) mobility subsystem deployed, wheel still undeployed.

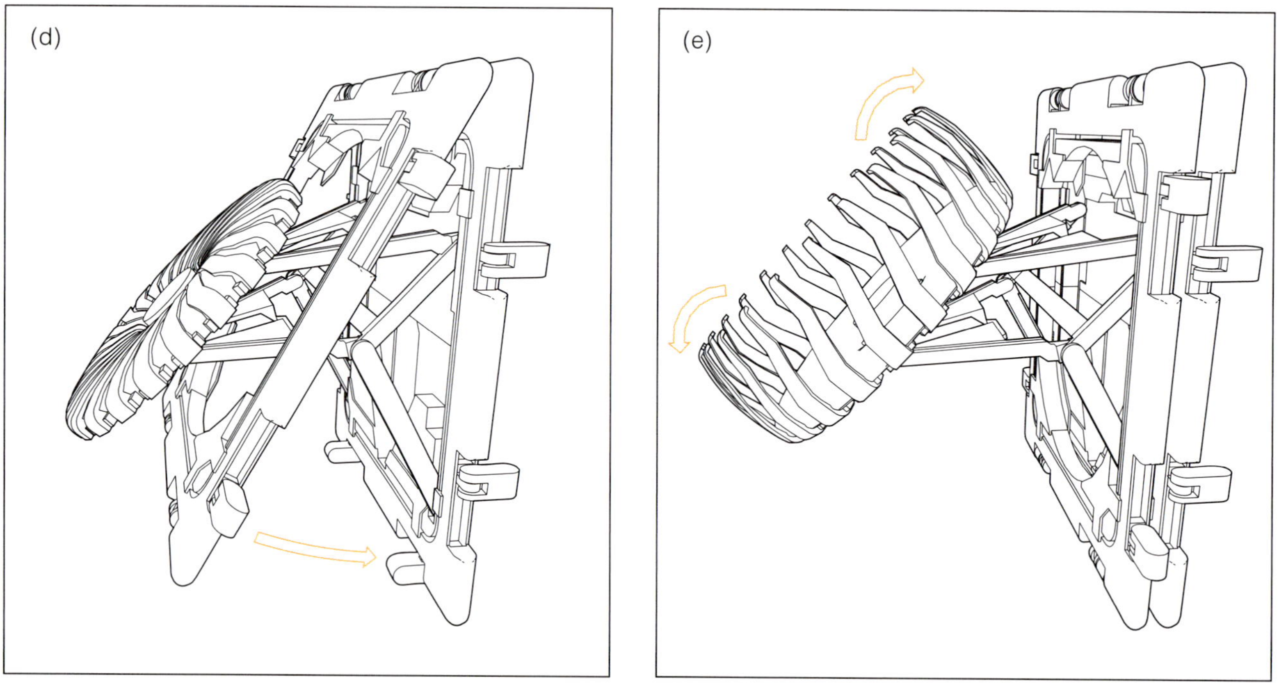

FIGURE 5 (continued) Mobility subsystem as a TRIGON plug-in payload: d) wheel deployment, wheel panel returned to stack; and e) treads deployed.

concert to bring the hub to meet and dock with the wheel. Once the suspension hub has docked with the wheel, the wheel payload releases it and folds back into place back to back with the panel containing the suspension. Once the wheel-suspension subsystem is in place, the tire can inflate. The tire consists of a nonstretch gas- or liquid-inflated bladder, with fold-out metal treads that open up like a flower. The tread tips have a spring-loaded band that keeps them wrapped tightly around the tire.

The suspension system mounts can be set at any position between 0 and 90 deg to accommodate various hull angles. Figure 6 shows the wheel system set at 90, 45, and 0 deg. These three settings would be the most common, but other settings in between are possible.

Construction algorithms devised for assembling rovers and wheeled habitats need to take into account the asymmetry of the plug-in panels because they have a dedicated mounting orientation. When the robotic pair tumbles into place, it might need to do additional positioning maneuvers to orient correctly.

Core Elements

Core elements are large or bulky items and equipment—including powerplants, inflatable pressure bladders, pressure hatches and connector tunnels, windows and viewports, and subsystems like unfolding legged mechanisms—that cannot fit into a payload panel, yet are designed to interface structurally and functionally with TRIGON panels.

INFLATABLE PRESSURE BARRIER LINING Using the TRIGON system, the pressure barrier critical for human survival need not be rigid or thick, as in traditional designs, to ensure pressure containment. The TRIGON system allocates the strength function to payload panels (deployable, inflatable, or post-filled), so that the pressure barrier need only be a thin membrane that can support the tremendous pressure differential across the very small dimensions of gaps and discontinuities in and between the structure panels.

In an inflatable pressure vessel application, the TRIGON system would provide a hard surface for secondary structure and mobility subsystem attachment. TRIGON panels can also provide secondary structure support for standard, hard-shell pressure vessels or inflatables designed with appropriate interfaces. Such large pressure vessels would be treated as core elements in the system, using foldout TRIGON mobility systems and other equipment as needed.

A thin-walled pressure barrier liner would comprise multiple, redundant sealed barriers with a thin restraint layer for support; provide continuous, reinforced seams at pressure ports and viewports; and incorporate hard

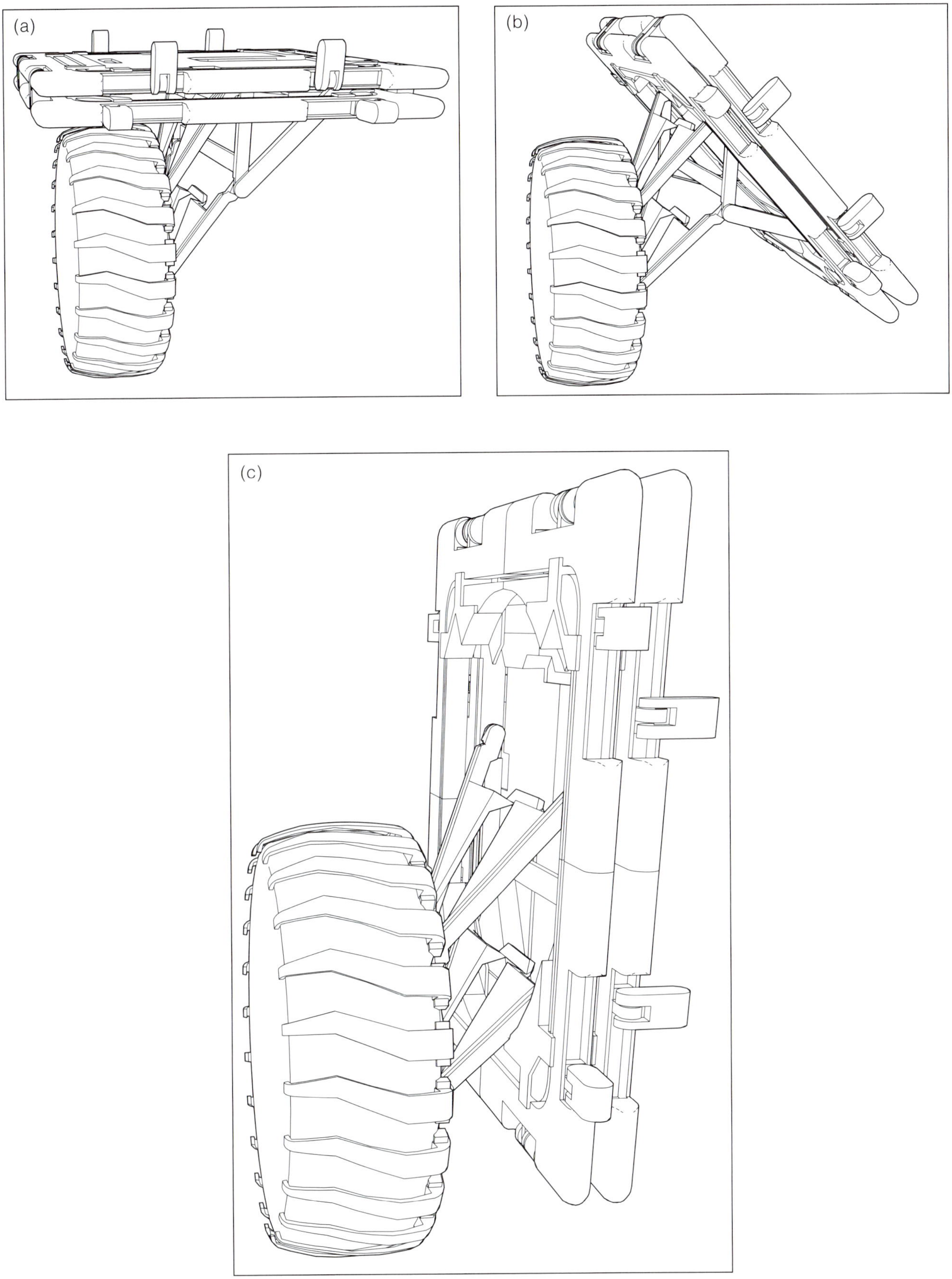

FIGURE 6 Mobility subsystem can accommodate any hull-mounting angle: a) 90-deg, b) 45-deg, and c) 0-deg wheel settings shown.

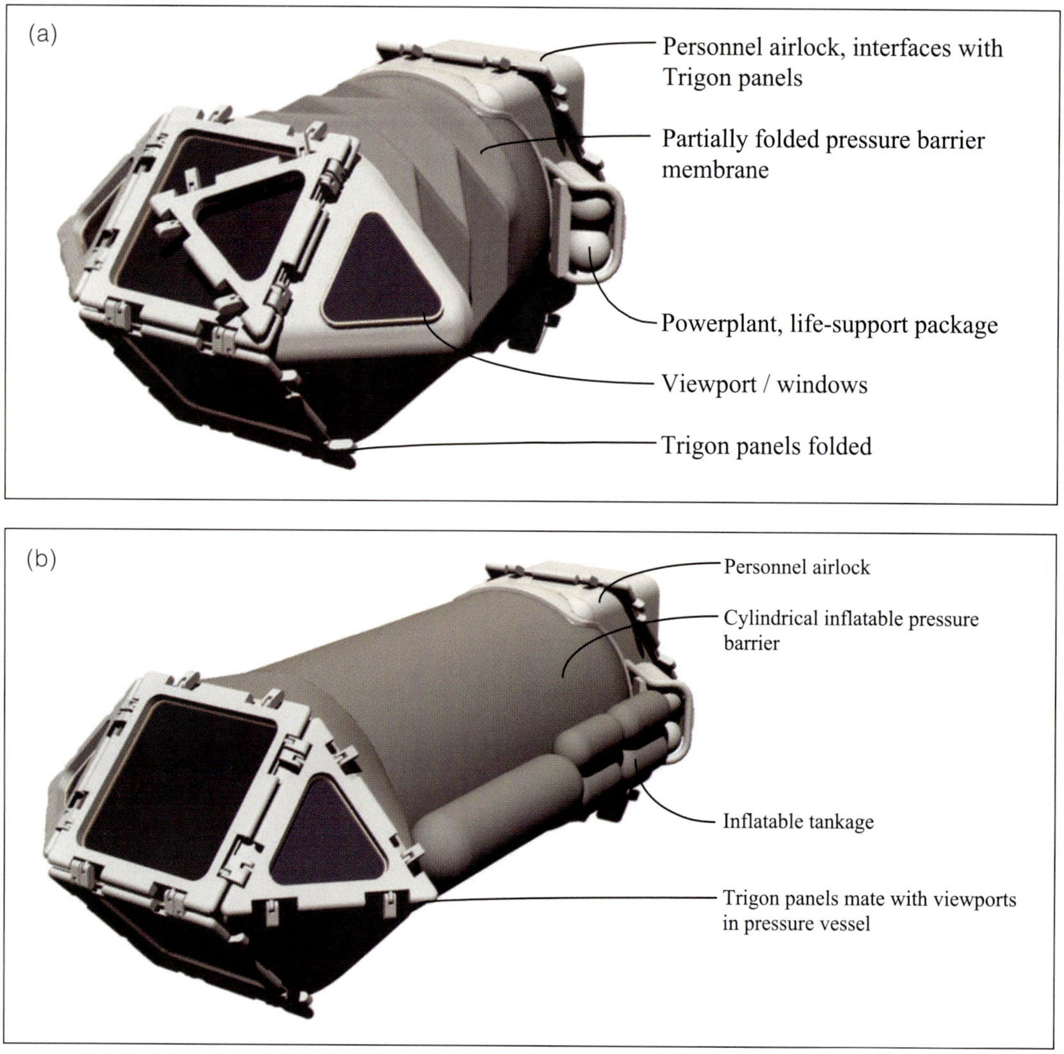

FIGURE 7 Inflatable pressure barrier for a rover: a) partially folded and b) fully deployed.

points for mounting internal outfitting. A significant challenge for inflatable membranes is how to fold them so that they unfold cleanly and without producing stress points. Figure 7 shows a scheme for a folding inflatable membrane pressure liner for a rover (see what follows). Inflatable membrane pressure liners can have built-in supplementary inflatable bags for shielding, tankage, or extra protection at TRIGON panel gaps. Note that pressure ports and viewports are permanently attached to the membrane.

PRESSURE HATCHES AND VIEWPORTS Pressure hatches are sized according to ergonomic and equipment dimensional constraints and therefore might be subject to standards different from those that determine optimum TRIGON panel size (Howe and Gibson 2006b). In addition, the pressure hatch assembly would be a hard structure that the inflatable pressure barrier liner would be permanently attached to. However, because the hatch penetrates the TRIGON shell, it needs edge interfaces that engage the panel edges, including compatible connector arms. Figure 8 shows a hatch assembly, including interfacing panel edges and extensible pressure tunnel connector. Figure 9 shows a personnel airlock for a rover, designed into a package that includes powerplant, life-support subsystem with oxygen tanks and other equipment. Note these elements are conceptual explorations of potential TRIGON applications. The hatch assembly personnel airlock (Cohen 2001) would be manufactured as an integrated unit permanently mounted to the inflatable pressure barrier lining and tested under controlled factory conditions on Earth, and then transported to the planetary surface as a package bundled with the necessary TRIGON panels and specialty payloads to fit out the entire rover end item. Windows and viewports permanently mounted to an inflatable pressure barrier lining would similarly be fit out as core elements. Standardized core elements like pressure hatch assemblies might be salvaged for new-use reconfigurations, assuming development of *in situ* techniques for bonding to pressure liners.

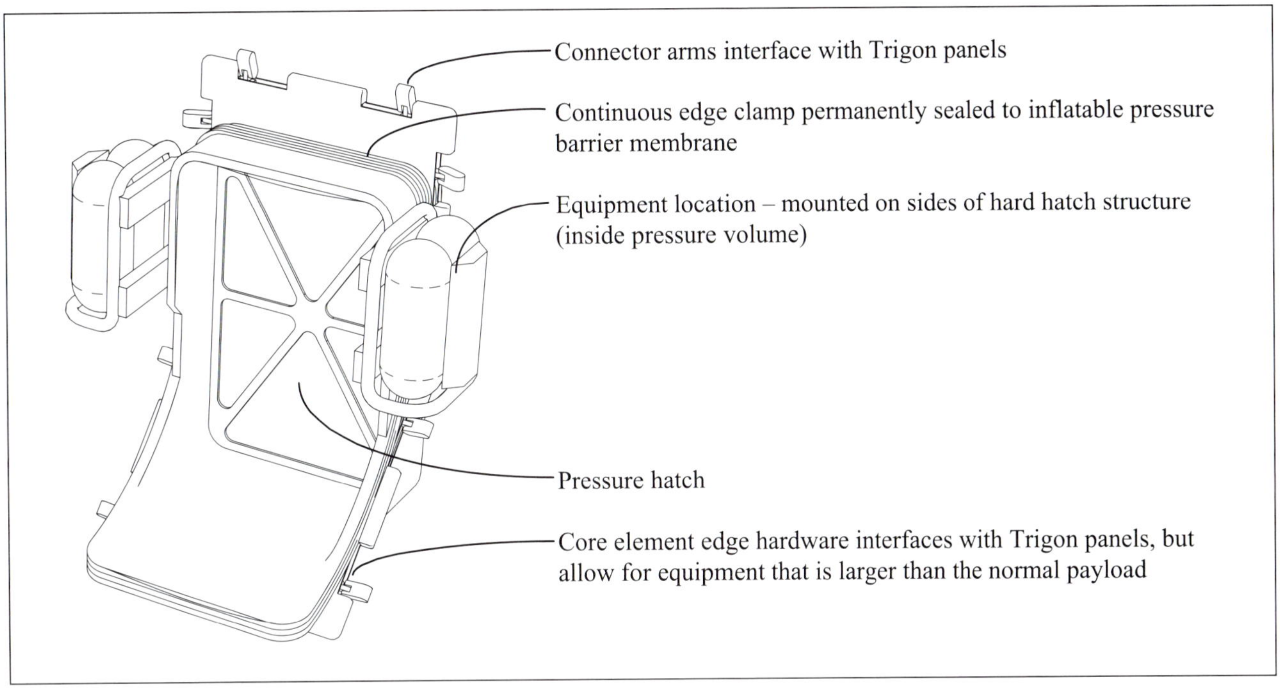

FIGURE 8 Pressure hatch assembly (interior view).

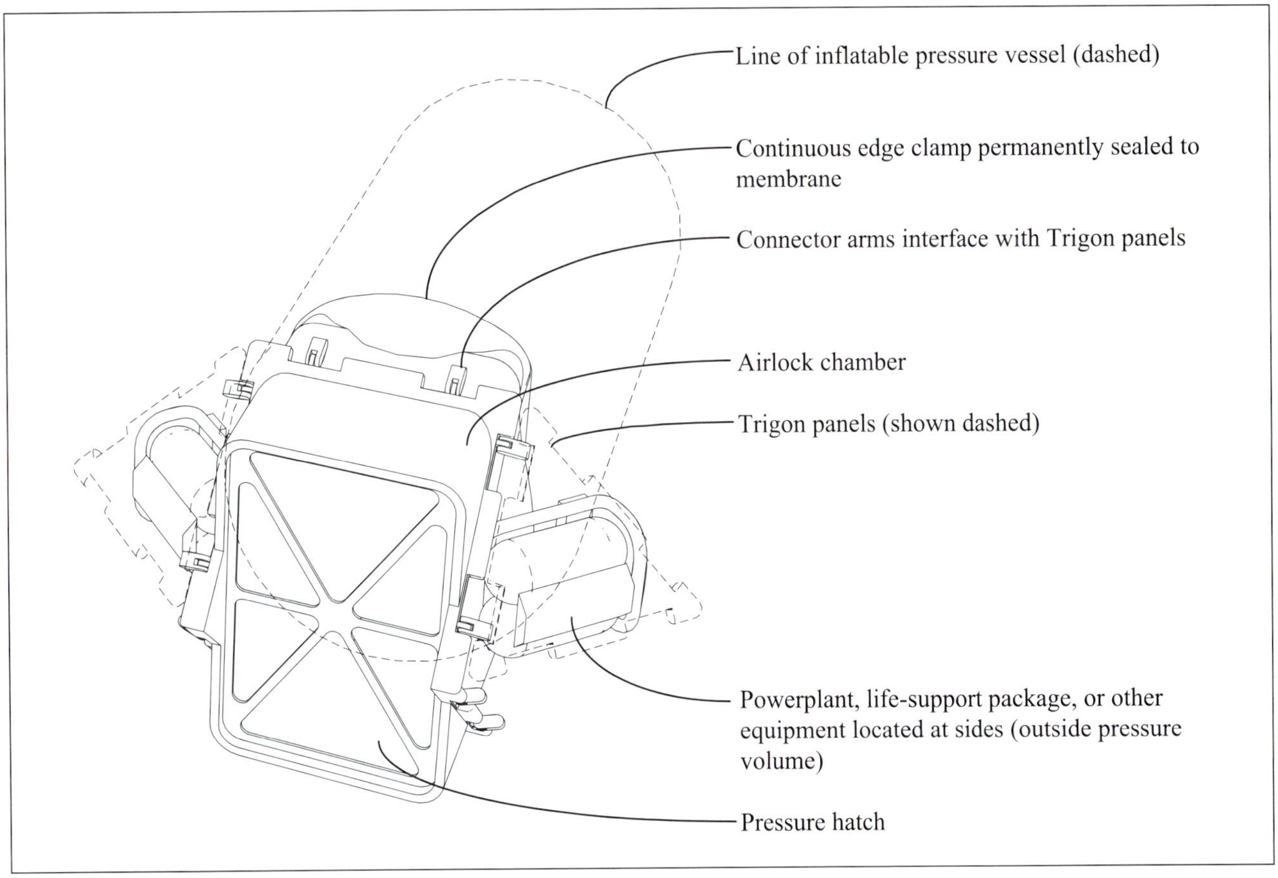

FIGURE 9 Personnel airlock assembly.

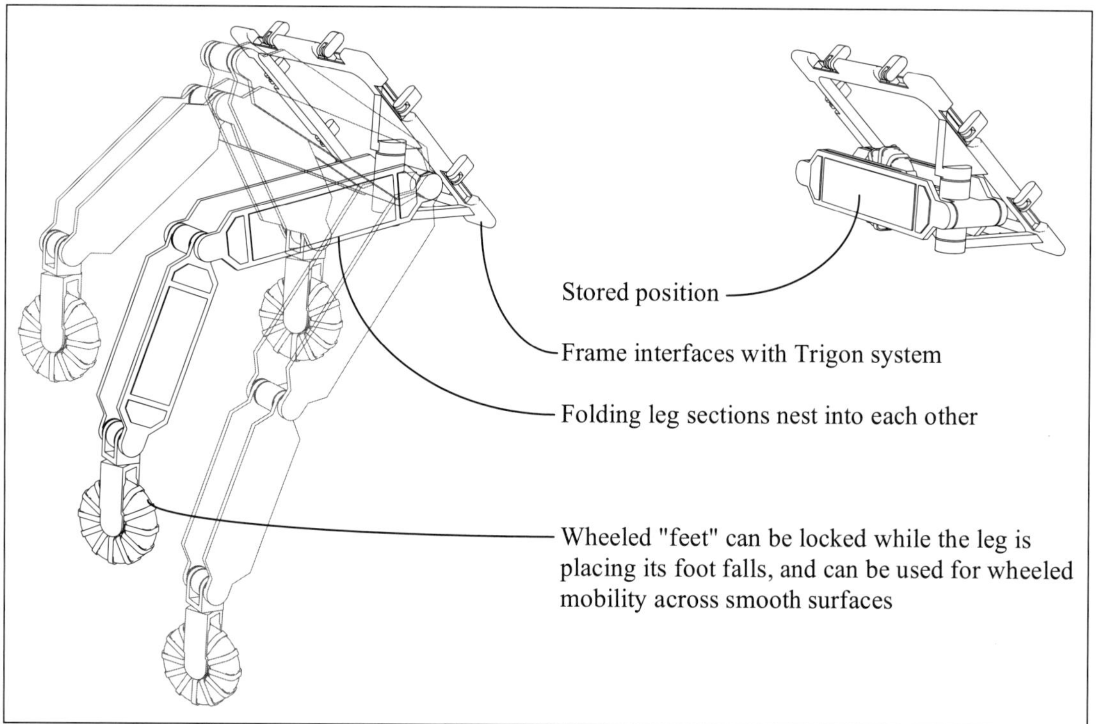

FIGURE 10 Leg in extended (left) and compact folded form (right).

LEGGED MOBILITY SYSTEM NASA's Jet Propulsion Laboratory has developed a multilegged mobility system prototype called the All-Terrain Hex-Limbed Extra-Terrestrial Explorer (ATHLETE) comprising a hexagonal platform with wheeled legs that fold compactly for transport. Though ATHLETE is a specific configuration, the folding-leg technology could be used compatibly, in a modular way, with the TRIGON modular construction system.

Figure 10 shows such a wheeled leg, in both extended and folded positions, mounted to a TRIGON-compatible chassis. The leg consists of three sections that fold into each other, operated by three revolute actuators at the "knees," and an additional vertically oriented actuator that points the leg. Fine maneuvering control is affected by a fifth revolute actuator that redirects the wheel. The legs can be used to walk over adverse terrain, using the wheels as feet, or to roll on more favorable ground. The legs allow leveling compensation for uneven terrain including slopes.

TRIGON-BASED WHEELED ROVERS

Using the modular elements just described, vehicle and habitat conceptual configurations can be explored. Figures 11 shows a small pressurized rover. For comparison purposes, Table 2 tabulates an estimate of the number of TRIGON panels and core elements needed. Figure 12 shows how the hatch core element is integrated into the overall configuration of TRIGON panels. Figure 13 illustrates a scenario with a pair of such rovers at work on the surface of Mars.

TABLE 2 Estimated number of TRIGON panels and core elements in a small, pressurized rover.

ELEMENT DESCRIPTION	QUANTITY
TRIGON triangle panels	10
TRIGON square panels	32
Triangular-shielding payload panels	6
Square-shielding payload panels	18
Suspension payload panels (including spares)	6
Wheel payload panels (including spares)	6
Antenna payload panel	1
Personnel airlock/powerplant/life-support core element	1
Inflatable pressure barrier lining core element (including two triangular viewports and two square viewports)	1

A concept for a six-wheeled rover-type habitat is shown in Figure 14. This rover habitat carries four

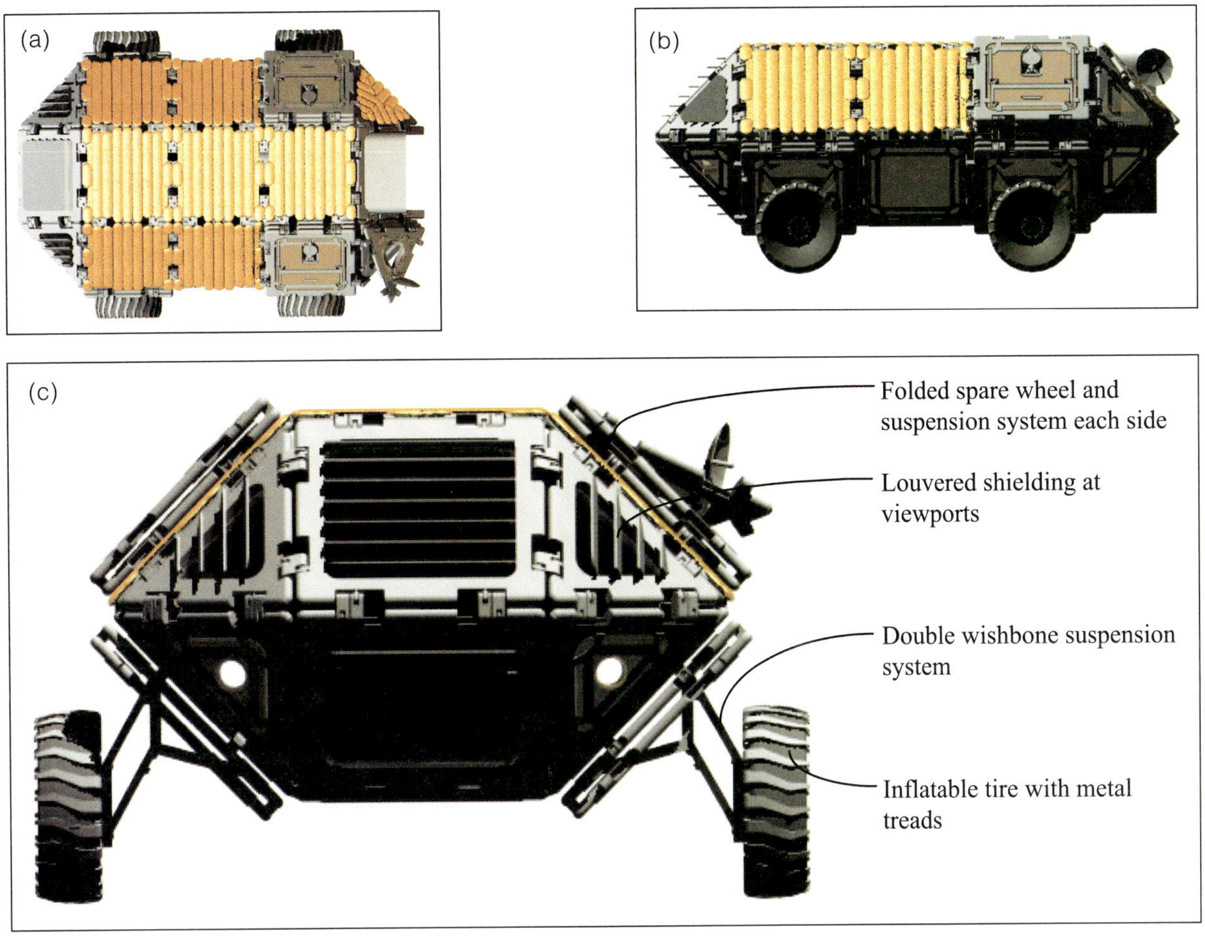

FIGURE 11 TRIGON-based pressurized rover concept: a) top view, b) side view, and c) front view. Some inflatable shielding omitted for clarity.

FIGURE 12 Pressurized rover, rear view showing pressure hatch (inflatable shielding omitted for clarity).

FIGURE 13 Pressurized rovers on Mars, showing utility trailer (courtesy of Chris Howe Design).

extra suspension-wheel subsystems, two hatches, and an airlock core unit combined with powerplant and life-support system. An estimate of the number of TRIGON panels and core elements for this concept is tabulated in Table 3. The large number of panels begins to reveal the potential hardware economy the TRIGON modularity provides. Multiple rover habitats could dock in various configurations: side by side, or side hatch to rear airlock hatch, using an extendable pressurized tunnel (Figure 15). This mode is essential for practical traverse scenarios, because rovers will almost certainly travel in pairs for safety; providing shirt-sleeve access between the rovers simplifies intercrew operations while stopped and yields a larger overall habitable volume for both crews.

A third wheeled infrastructure concept is an unpiloted, modular, workhorse tractor that can lift and carry cargo and provide generic mobility for scientific, construction, mining, and exploration applications. Figure 16 shows the concept: a) an unengaged tractor with partially opened "jaws," and b) the tractor closing in on a cargo crate also constructed of TRIGON panels. The tractor can lock onto the crate and lift it to twice its own height. The jaws can be widened for double- and triple-wide loads, or to insert dozer blades, drilling units, or other modular implements. An estimate of the number of panels and core elements is tabulated in Table 4.

FIGURE 14 TRIGON-based six-wheeled rover-habitat concept (shielding omitted for clarity).

TRIGON-BASED LEGGED ROVERS

TRIGON-based mobility system concepts for habitats and vehicles are based on ATHLETE-type folding legs that fall into the category of core elements, that is, too large to fit within the dimensions of a payload panel. Figure 17 shows one concept: a four-legged pressurized rover with eccentric pressure ports. This

TABLE 3 Estimated number of TRIGON panels and core elements in a six-wheeled rover habitat.

ELEMENT DESCRIPTION	QUANTITY
TRIGON triangle panels	10
TRIGON square panels	56
Triangular-shielding payload panels	8
Square-shielding payload panels	32
Suspension payload panels (including spares)	10
Wheel payload panels (including spares)	10
Antenna payload panel	1
Pressure port core element	2
Personnel airlock/powerplant/ life-support core element	1
Inflatable pressure barrier lining core element (including four square viewports)	1

FIGURE 15 Two rover-habitats docked, for contiguous habitat on a joint traverse.

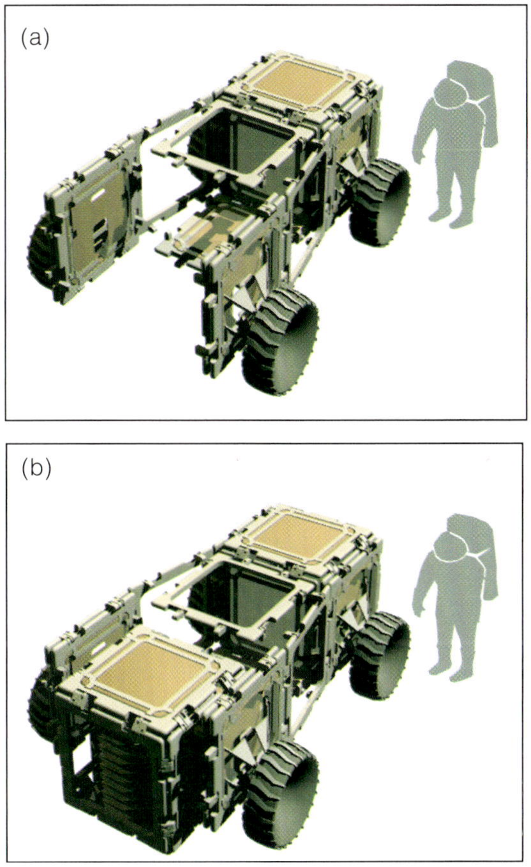

FIGURE 16 TRIGON-based modular tractor a) without implements and b) preparing to lift cargo crate.

TABLE 4 Estimated number of TRIGON panels and core elements in a modular tractor.

ELEMENT DESCRIPTION	QUANTITY
TRIGON square panels	16
Square payload panels	4
Suspension payload panels	4
Wheel payload panels	4
Powerplant core element	1

rover would function like the original Habot concept, where one vehicle would be dedicated to the airlock function, and other vehicles in the group would each also have dedicated functions.

Such a four-legged mobility system would have little trouble with rolling mobility, but would be less stable with only four legs for walking in rough terrain. In a large obstacle step-over scenario, three of the legs would need to position themselves such that the center of gravity of the rover is inside the subtended triangle while the fourth leg lifts to take the next step. This means the free leg must be carefully placed to ensure a new stable-triangle stance can be maintained while the next leg repositions. An estimate of the number of TRIGON panels and core elements for the four-legged pressurized rover is tabulated in Table 5.

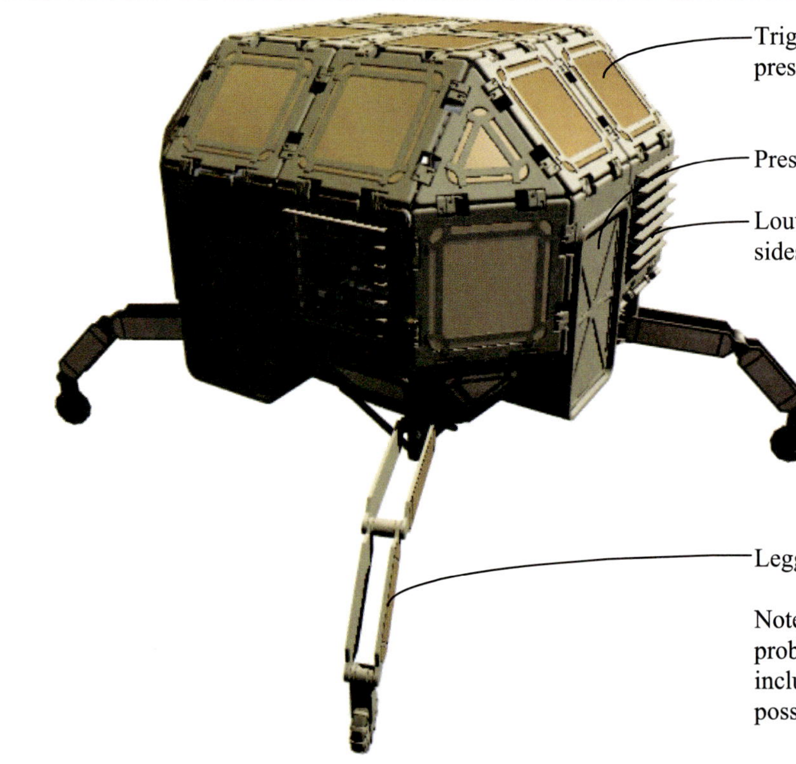

FIGURE 17 Four-legged pressurized rover.

TABLE 5 Estimated number of TRIGON panels and core elements in a four-legged, Habot-type pressurized rover.

ELEMENT DESCRIPTION	QUANTITY
TRIGON triangle panels	10
TRIGON square panels	28
Triangular shielding payload panels	8
Square shielding payload panels	28
Antenna payload panel	1
Pressure port core element (including powerplant/life support)	4
Leg core element	4
Inflatable pressure barrier lining core element (including 4 square viewports)	1

Mobitat2 is an eight-legged rover concept. In the precursor, Mobitat concept, rigid, pre-integrated, pressurized modules were hung from maximally flexible mobility platforms. These modules could be detached from the mobility platforms, placed at fixed locations, and connected to other modules via pressure ports, thus freeing the mobility platforms for other uses like drilling, construction, excavation, and lifting. Mobitat2 uses a different approach: compact, fold-out secondary structures carried as plug-in TRIGON payloads and deployed where needed during self-assembly. Instead of a hard-wall pressure vessel, Mobitat2 uses an inflatable bladder inner lining, prefitted to pressure ports and other openings, and inflated within a shell of TRIGON panels. Pressure ports are centered on each of the four sides. Figures 18 and 19 show Mobitat2 in both mobility mode and parked position with legs folded underneath. The maneuverability of the legs in both modes allows very fine adjustment in six degrees of freedom

FIGURE 18 Mobitat2 in walking/rolling mode.

FIGURE 19 Mobitat2 in parked position, with legs tucked underneath (shielding omitted for clarity).

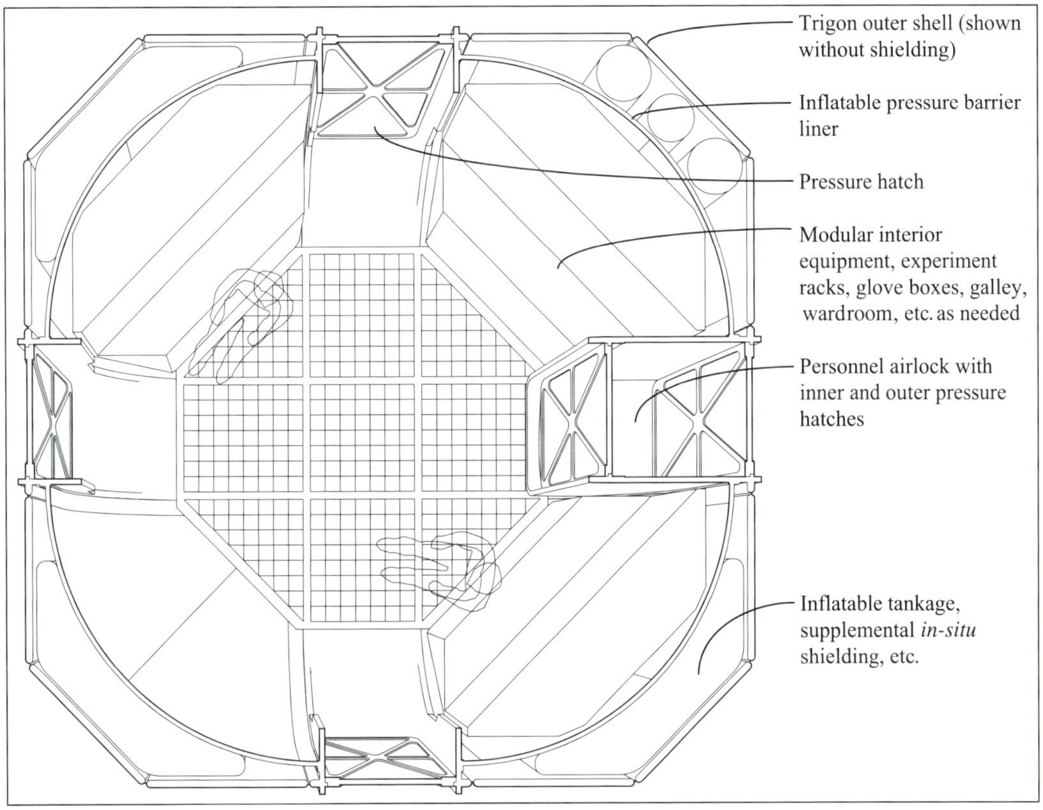

FIGURE 20 Mobitat2 schematic plan (shielding omitted for clarity).

(horizontal plane, vertical, yaw, pitch, and roll) to ensure leveling and lining up pressure ports properly for mating. A legged habitat also is able to adapt to extreme slopes or terrain variations.

Figure 20 shows a plan view of Mobitat2, with a near-cylindrical inflatable-membrane pressure-barrier lining. Because the TRIGON panels create a square with mitered corners, this pressure barrier shape leaves gaps that can be used for inflatable tankage, extra shielding, or other equipment that can tolerate vacuum. Figure 21 shows a sectional

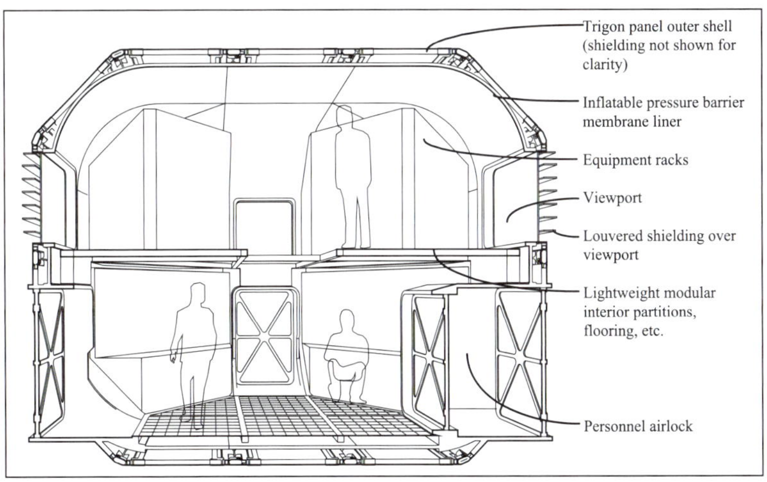

FIGURE 21 Mobitat2 schematic section (shielding omitted for clarity).

FIGURE 22 Moon-base scenario using Mobitat2 and other TRIGON-based elements (Courtesy of Chris Howe Design).

view of Mobitat2, including a lightweight floor deck that partitions the volume into two floors, or one floor plus loft.

An estimate of the number of TRIGON panels and core elements in Mobitat2 is shown in Table 6. Figure 22 shows a Moon-base concept using Mobitat2 units, both in place and on the move, along with a small pressurized rover and utility trailer. Shielding is deployed in parked habitats to protect the crew, but dumped or shed when an unoccupied unit is being moved. This scenario indicates the integrated architecture vision enabled by the TRIGON modular infrastructure concept.

TABLE 6 Estimated number of TRIGON panels and core elements in a Mobitat2 eight-legged habitat system.

ELEMENT DESCRIPTION	QUANTITY
TRIGON triangle panels	10
TRIGON square panels	66
Triangular-shielding payload panels	8
Square-shielding payload panels	66
Antenna payload panel	1
Pressure port core element	2
Leg core element	8
Personal airlock/powerplant/life-support core element	2
Inflatable pressure barrier lining core element (including four square viewports)	1

TRIGON-BASED LANDERS

TRIGON core elements could conceivably also include modular propulsion systems, in which rocket motors, propellant tanks, maneuvering thrusters, guidance and navigation subsystems, antennas, and other lander systems are configured to interface with TRIGON-based systems. Figure 23 shows representative preliminary configurations for several such propulsion core elements. (These are conceptual layouts only, whose propulsive performance parameters have not been analyzed or iterated.)

TRIGON support frames allow landers to be configured from modular collections of propulsion core elements, mobility panels, and payloads including stacked panels and core elements being delivered to a surface site (Figure 24).

When the lander reaches the surface, the payload packages unfurl or detach and self-assemble into target infrastructure elements like those just described. The lander motor and core can be cannibalized for its TRIGON frame elements, or maybe even adapted directly for surface uses requiring plumbing, like an *in situ* resource utilization (ISRU) processing plant. Once assembled, such a plant could be fitted with wheel panels to follow other mobile infrastructure elements to new locations.

COMPARISON AND CONCLUSIONS

In Table 1, several viable panel dimensions and masses were presented, showing a range of capacities available to designers. For the concepts presented in this

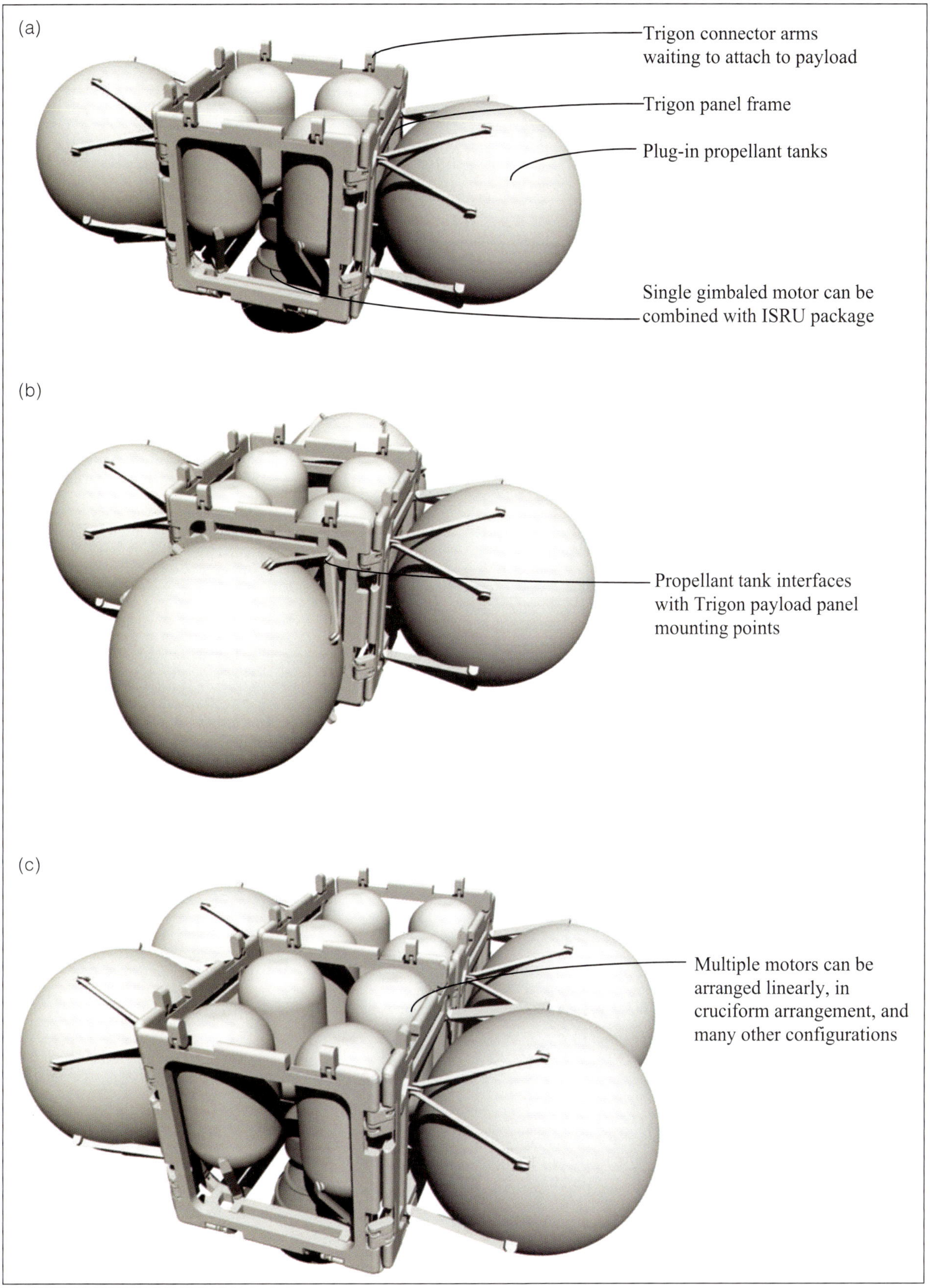

FIGURE 23 TRIGON-based propulsion core elements: a) basic, modular propulsion element using strap-on propellant tanks; b) additional propellant tanks added symmetrically; and c) ganged system with multiple motors and additional tanks.

FIGURE 24 Mobitat2 folded into compact form for delivery by TRIGON-based lander module(s).

chapter, the 1-m panel (thin) was used. Further investigation would reveal the best actual panel sizes and thicknesses; mass estimates for core elements, inflatable liners, and other equipment; and power requirements. Analysis tools are in place for panel optimization (Howe and Gibson 2006b) and robotic functionality analysis (Howe and Gibson 2006a). Other issues requiring further study include power generation; dust protection for mechanical connections; shielding the gaps; outfitting, self-assembly, and verification of functional components; simplification and minimization of mechanisms; placement of unmechanized panels where reconfiguration is unlikely; integration of inflation/deflation and liquid supply/return systems; and fully engineered design of mobility subsystems.

TABLE 7 Comparison of elements needed for TRIGON-based infrastructure concepts.

INFRASTRUCTURE ELEMENT	NUMBER OF PANELS	NUMBERS OF CORE ELEMENTS	PRESSURIZED VOLUME, ESTIMATED, M^3
Small pressurized rover	42	2	8.7
Six-wheeled pressurized rover/habitat	66	4	23
Unpressurized rover/tractor	16	1	—
Four-legged pressurized rover	38	9	21
Mobitat2 eight-legged habitat	76	13	56

The various configurations presented can be compared in terms of the number of panels and core elements used (Table 7). Note how the legged mobility systems have more core elements. Because core elements cannot be packaged within TRIGON panels, they are harder to stow and cannot be manipulated by TRIGON self-construction; panels would have to manipulate around them. Wheeled mobility subsystems can be packaged into TRIGON payload panels, but their volume must necessarily be constrained to do so.

The TRIGON modular robotic construction architecture appears applicable to multiple surface infrastructure needs. The concept's flexibility and capacity to reconfigure itself into various typologies to adapt to new needs both illustrate the merit of continued development of such modular architectures to support advanced equipment transformation scenarios. For example, is it conceivable that a vehicle with compact core elements and panel stacks might be configured first for landing, then for surface operations and crew occupancy, and then again later for ascent to be relocated? Would this make an effective surface exploration architecture? How much configuration transformation and re-outfitting can occur autonomously, and how would it be functionally verified prior to each next use? Detailed subsystem outfitting simulations and sparing scenarios need to be developed, and the practicalities of autonomous reconfiguration must be studied (For a point of comparison regarding outfitting complexity, see Chapter 4.) In the extreme, the TRIGON architecture might be the conceptual foundation for self-constructing, self-assembling, self-manufacturing, and self-replicating building systems delivered to planetary surfaces as seed factories that autonomously establish an outpost before the arrival later of human crews (Howe 2005).

The full potential of the TRIGON architecture—reconfigurable, identical structural components—is not yet understood. |

References

Cohen, M. M. (2001), "Airlocks for Pressurized Rovers," NASA TSP-ARC-14557.

Cohen, M. M. (2004), "Mobile Lunar Base Concepts," *Space Technology and Applications International Forum - STAIF 2004: Conference on Thermophysics in Microgravity; Conference on Commercial/Civil Next Generation Space Transportation; 21st Symposium on Space Nuclear Power and Propulsion; Conference on Human Space Exploration; 2nd Symposium on Space Colonization; 1st Symposium on New Frontiers and Future Concepts*, AIP CP-699, edited by M. S. El-Genk, American Inst. of Physics, College Park, MD, pp. 845–853.

Cohen, M. M., and Kennedy, K. J. (1997), "Habitats and Surface Construction Technology and Development Roadmap," *Government Sponsored Programs on Structures Technology*, edited by A. Noor and J. Malone, NASA, Washington, D.C., pp. 75–96.

Howe, A. S. (2002), "The Ultimate Construction Toy: Applying Kit-of-Parts Theory to Habitat and Vehicle Design," AIAA Paper 2002-6116, Oct.

Howe, A. S. (2005), "Cassette Factories and Robotic Bricks: a Roadmap for Establishing Deep Space Infrastructures," Society of Automotive Engineers, Paper 2005-01-2911, July; also *Proceedings of the 35th International Conference on Environmental Systems (ICES2005)*, Society of Automotive Engineers, Warrendale, PA.

Howe, A. S., and Gibson, I. (2006a), "TRIGON Robotic Pairs," AIAA Paper 2006-7407, Sept.

Howe, A. S., and Gibson, I. (2006b), "TRIGON Panel Size Optimization Studies," AIAA Paper 2006-7328, Sept.

Howe, A. S., and Howe, J. W. (2005), "Plug-in Hardware Concepts for Mobile Modular Surface Habitats," AIAA Paper 2005-2673, Jan.–Feb.

Kennedy, K. J. (2002), "The Vernacular of Space Architecture," AIAA Paper 2002-6102, Oct.

Lai, Y., and Howe, A. S. (2003), "A Kit-of-Parts Approach to Pressure Vessels for Planetary Surface Construction," AIAA Paper 2003-6281, Sept.

Yip, W., and Howe, A. S. (2003), "Deployable Secondary Support Structures for Planetary Construction," AIAA Paper 2003-6282, Sept.

22

lunar-base site design

BRENT SHERWOOD

INTRODUCTION

A SITE PLAN is a fundamental tool for organizing architecture. By addressing the largest dimensions and relationships of a project—the largest scale—it relates the project directly to the larger world. The site plan reflects and guides daily and developmental activities within the site and governs its maturation over time. A successful site plan reconciles conflicting requirements so as to avoid inefficiency or unsatisfactory accommodation or haphazard growth that would lead to them. By establishing a framework for integrating specific architectural elements, the site plan ensures that those elements work together to serve larger goals. The site plan is a dynamic document, evolving in response to real conditions when first conceived and later as built lessons are learned. A good site plan is resilient enough to accommodate changing requirements over long timescales while still expressing its originally conceived integrity and order. Some urban plans on Earth have served, recognizably intact, for millennia.

Many artists' concepts of lunar bases have been published and used to illustrate papers, proposals, presentations, and program plans. By showing explicit relationships among base elements, each such image implicitly records a site plan. Most of these implicit site plans intentionally feature a particular aspect of the base, evincing consideration of only a few of the integration constraints we can expect. Because the site plan plays such a fundamental role, if superficial or unresolved, it limits a scenario's credibility.

This chapter describes a method to develop an integrated lunar-base site plan. Developing a site plan is neither a mysterious nor accidental activity. A straightforward, repeatable method practiced commonly by the architectural profession is outlined and demonstrated here. This particular plan was needed to support a detailed study of automation and robotics for lunar surface operations (Woodcock et al. 1990). The overall study goal was to evaluate the feasibility of emplacing, operating, and maintaining an early, oxygen-producing lunar base before extensive or continuous human presence. The example described here limns several issues relevant to lunar bases and provides an example of how to produce a well-founded planetary-base site plan.

METHOD

The essential problem of the site plan consists of accommodating simultaneously the conflicting needs of all of the site elements. If the multivariate relationships among all of the elements could be identified and quantified and penalty functions accurately assessed, then linear programming could be used confidently to arrive at an optimal site arrangement. However, two factors confound this approach. The first is that for sites involving human activities, the relationships are so extensive and complex that they defy numerical modeling; an experienced architect provides more useful results than any algorithm. Admittedly, for the case of laying out an early lunar base, this complicating "human factor" is less prominent than other enabling and quantifiable engineering considerations.

The second confounding factor is that, during the very earliest study phases of such an unprecedented project as a lunar base, available details about the elements are quite insufficient to enable useful

Copyright © 2008 by Brent Sherwood. Published by the American Institute of Aeronautics and Astronautics, Inc., with permission.

quantification of their interrelationships. Again, the synthesizing approach of an experienced architect is the only practical solution to working with incomplete data.

The four-step method given next comprises a tool to record, organize, and finally visualize the most critical relationships of the site.

1) The first step is to define the site's "architectural program": the activities the site will support, the goals it must achieve, and finally the aspirations it should embody.
2) The second step is to define as completely as possible the site elements, those discrete pieces of the site architecture that impose unique, intrinsic requirements. These can be hardware systems (like buildings), siteworks (like roads), or more abstract elements (like a required activity space among the buildings). We include geographical, topographical, and geological information specific to the chosen site as part of the element definition.
3) The third step is to identity connections among the elements, constraining how they might, or must, be arranged within the site. An effective way to record these is to list them as part of the element description of each element they affect.
4) The fourth step is to generate a quantitative proximity diagram that graphically represents the relative sizes of the elements as well as the relative importance (and interference) of element connections. Iteratively optimizing the proximity diagram automatically generates the skeleton of a workable site plan.

PROGRAM AND ELEMENTS

The early, solar-powered lunar base analyzed by the source study fulfills a four-fold program. Justification for this choice of program is reported in Woodcock et al. (1990):

- Produce oxygen (for space transportation propellant) from lunar resources [lunar-derived liquid oxygen (LLOX)]
- Facilitate *in situ* human investigation of lunar science and engineering
- Be able to grow in capacity and capability over time
- Perform early heavy, dangerous, and routine work robotically

The primary reason for the incomplete nature of typical lunar-base site plans is lack of sufficiently complete, consistent element detail. A useful site plan cannot be generated until everything large enough or important enough to affect it is defined to comparable levels of detail. That includes all buildings, activity spaces, vehicles, industrial equipment, and utilities. The Robotic Lunar Surface Operations Study (Sherwood 1990b) was charged with developing one reasonable scenario down through several layers of engineering detail. The disadvantage of a point-design approach is that the underlying concept might not be universally relevant. But a tremendous compensating advantage is that a point-design can achieve a level of detail that requires all of the elements to be well understood, iterated, and consistent. Such detail exposes to analysis many significant engineering issues.

Engineering concept designs for the base elements, and their major connection constraints, are detailed in Sherwood (1990a). The elements can be divided into four classes: fixed, mobile, utility, and sitework. Table 1 lists salient characteristics of the lunar-base elements in these four categories and quantities for those that matter for conceptual site planning. Figure 1 shows the major fixed elements, and Figure 2 shows the mobile robots. The function, requirements, dimensions, capacities, modularity, and maintenance scenarios for all of the elements were codesigned iteratively, and so they are all tightly coupled with each other and with the base design described next. Two examples of critical interrelationships among major elements are described in Figure 3.

To the list of elements, we can append strategic provisions, for example, topological expansion constraints for long-term base growth: habitation, industry, mining, spaceport, and wilderness functions all introduce unique considerations treated in what follows. This surprisingly long list of elements and provisions comprises an approximately minimum set for a lunar base capable of supporting a handful of people (intermittently) and producing enough LLOX for the lander to make four round trips per year between the surface and lunar orbit (100 mt per year). It takes 13 landings, each delivering 30 mt, over almost four years, to get all of the necessary equipment to the site from Earth (Buddington 1990). One clear conclusion is that even an early, modestly productive lunar base is a substantial undertaking.

A site at the southern edge of the Mare Tranquilitatis, near the *Apollo 11* Tranquility Base, was attractive for our base program for several reasons. Its geology is empirically known through *in situ* human observation and returned samples. Its mature mare regolith is comparatively rich in ilmenite and is level and well comminuted by meteorite gardening. Thus a usable soil feedstock, low in rock content, can be scraped up with minimal mining infrastructure. The location at the

TABLE 1 Catalog of lunar-base site elements, quantities, and characteristics is the first data product required to develop an integrated site plan (courtesy of Boeing Co.).

CATEGORY	ELEMENT	QUANTITY	DESCRIPTION
Fixed elements	Lander	Up to 3	Cryogenic reusable, 30-mt payload capacity down, manned or unmanned operation
	Habitat system	1	Hab module, two airlocks, connecting node with cupola above, pressurized workshop module
	Radiation shelter	1	Regolith-filled, modular vaulted, aluminum rib-and-corrugated-shell structure envelops hab system, provides unpressurized truck access to workshop module hatch
	Solar (PV) array	24	20-kWe output, deployable, sun-tracking (east-west) rigid panel, anchored into regolith
	Regenerable fuel-cell (RFC) module	2	Self-contained unit, 20-kWe output, 50% overall efficiency, for night-time power
	Oxygen reactor	3	Batch-mode, fluidized-bed, uses hydrogen reduction of ilmenite
	LLOX depot	1 per pad	Station with liquefaction, refrigeration, redundant storage, pumping
Mobile robots	Straddler	2	Robotic mobile gantry: creeping speed for lifting, moving, positioning, grading, mining
	Truck	2	Slow speed, with outrigger-stabilized high-reach boom and utility suite, front-loader, rear tow with utility trailer suite, robotic with on-board operator station
	Rover	2	Light-duty, moderate speed, robotic or manned operation, for site survey and crew transport
Utilities	Radiator module	3+5 per pad	Module for ganged use; three for habitat system and five for each LLOX depot, with fixed deployable sunshade oriented east-west
	Debris barrier	12 per pad	Deployable rigid panels anchored into regolith downrange of lander approach path to protect base elements by intercepting ejecta from lander exhaust
	Hopper	22	With dump chute, holds 27-mt sorted regolith
	Storage shed	Optional	For shaded equipment parking, same structures as radiation shelter
	Guidance beacons	TBD	For lander approach targeting and local site navigation by mobile robots
	Communication transceivers	TBD	Interlink all mobile elements and link base to low lunar orbit and Earth
	LLOX terminal	1 per pad	Buried valve box and conditioning equipment for servicing lander with umbilical connectors for electrical, signal, and LLOX
	Gas lines	N/A	Conduct gaseous oxygen from oxygen reactor to LLOX depot
	Liquid/vapor lines	N/A	Connect LLOX depot to LLOX terminal
	Power substation	1	Regulates industrial load and crossover distribution to Habitat power system
	Power, data and grounding cables	N/A	Throughout site, linking all fixed elements
	Sensor heads	N/A	Monitor critical views
	Local lights	N/A	Augment Earth light during lunar night for critical areas
	End effectors and tools	N/A	For robot manipulators and crew
Siteworks	Spaceport	Up to 3	Landing pad paved with sieved stones
	Foundations	N/A	Undisturbed, naturally consolidated regolith exposed by scraping off 1-m overburden
	Open workyard	1	Paved area for equipment staging, disassembly, and reconfiguring
	Connecting roads	N/A	Leveled, paved with compacted gravel for dust control
	Deposition sites	N/A	Berms incrementally built up from depositing rocks, gangue, spent oxygen-reactor solids

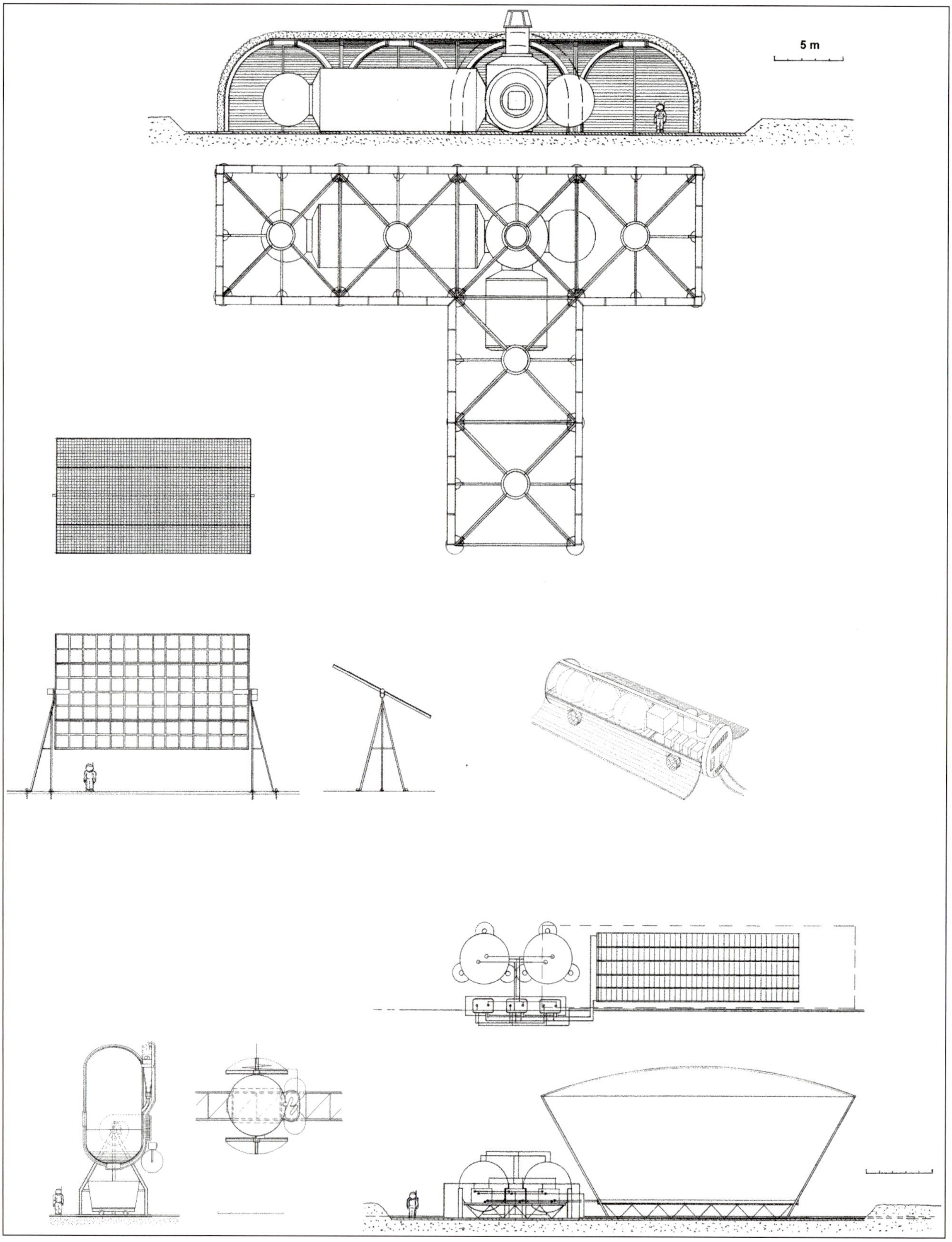

FIGURE 1 Major fixed lunar-base elements: Top—habitat system enclosed by regolith-containing radiation shelter, section and plan. Middle—power systems including tracking solar panels, plan and elevations; regenerable fuel-cell power storage unit, perspective. Bottom—ilmenite-reduction oxygen reactor, plan and section; LLOX depot, plan and elevation. All drawings to same scale (courtesy of NASA and Boeing Co.).

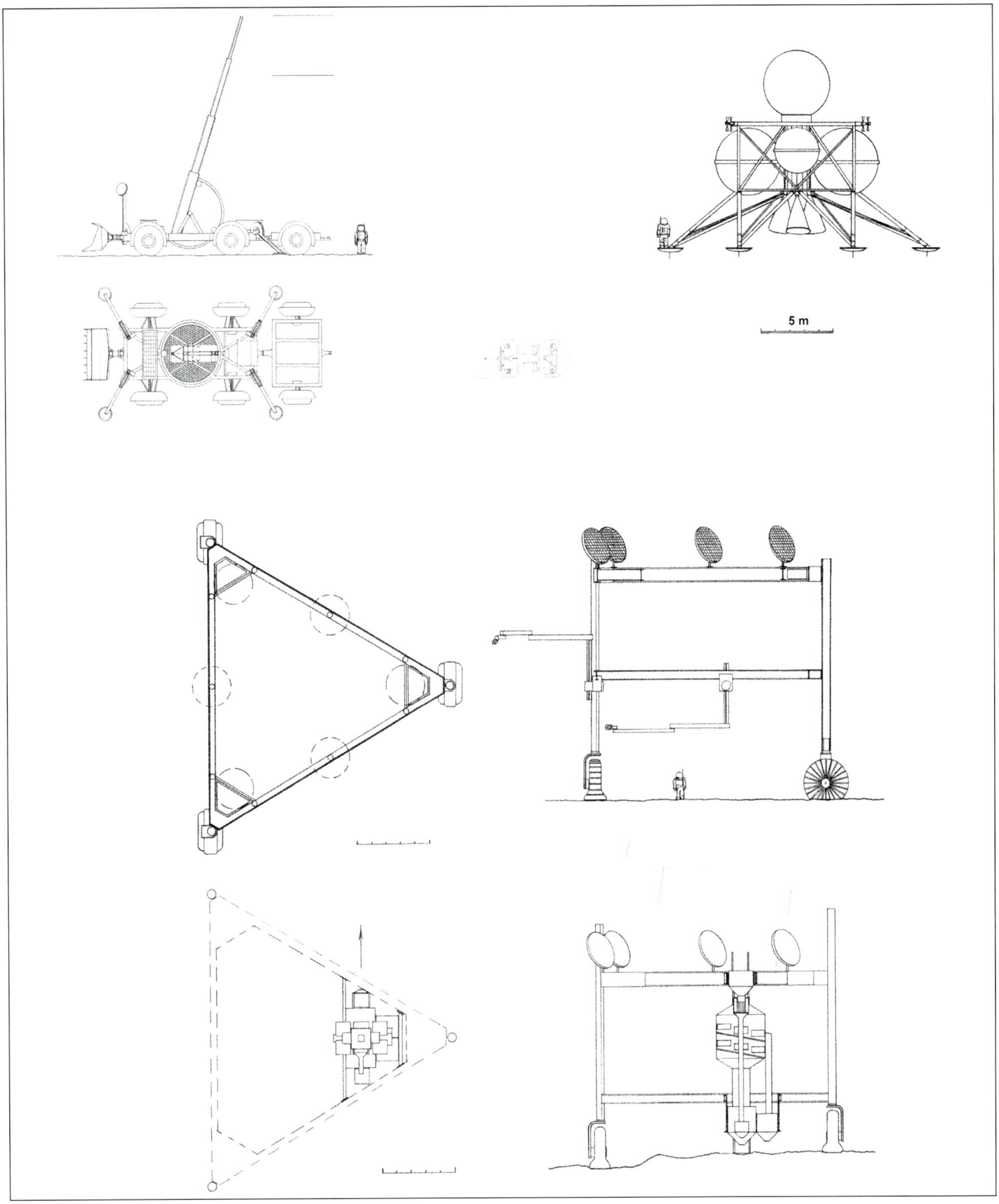

FIGURE 2 Mobile lunar-base robots: Top left—high-reach truck, elevation and plan. Top middle—unpressurized rover, plan. Top right—reusable lunar lander, front elevation. Middle—straddler, plan and elevation. Bottom—straddler with mining attachment, plan and section. All drawings to same scale (same scale as Figure 1) (Images by the author, reprinted courtesy of NASA and Boeing Co.).

edge of the mare allows mining growth to the north and exploration of highland regions to the south. The near side, near-equatorial site allows simple, continuous communication with Earth and minimizes propulsive requirements for transportation operations to and from orbit, which also preadapts the base for mass-driver launches to libration point targets in case a large-scale industrial future develops.

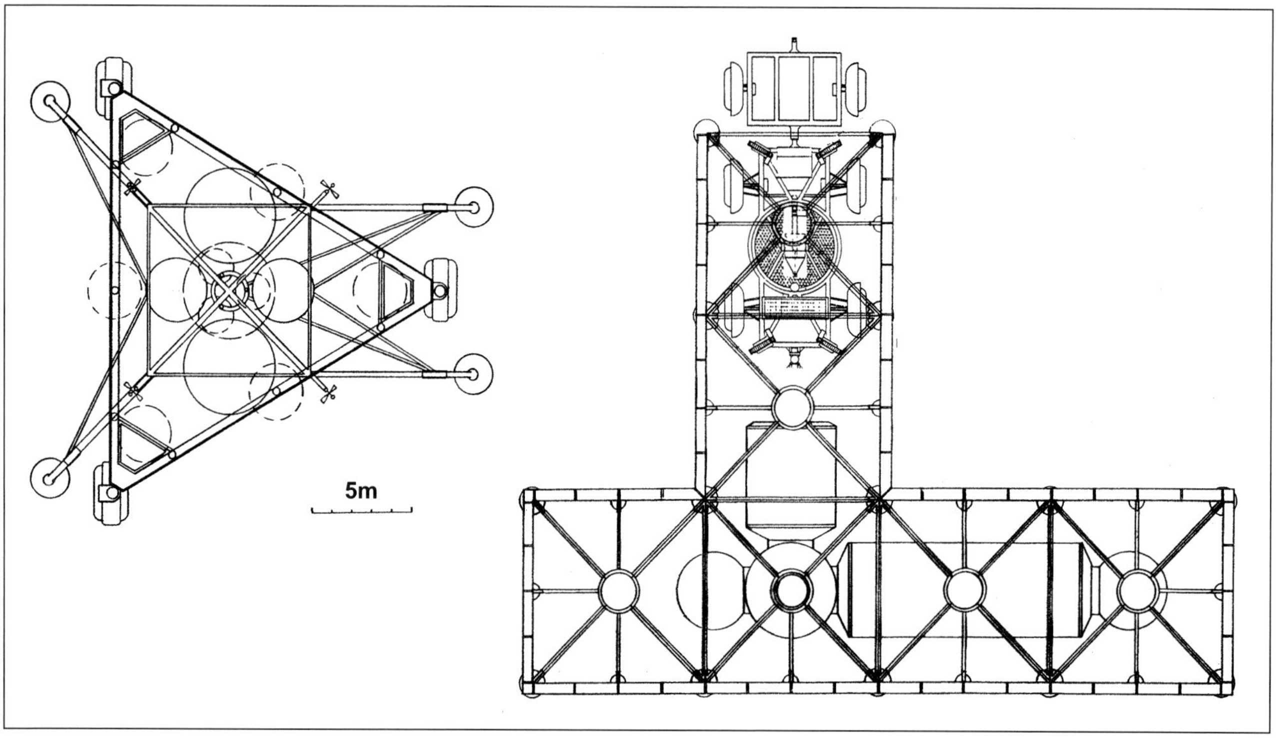

FIGURE 3 All base elements are designed iteratively to work together: Left—plan view of straddler over lander and radial asymmetry of lander allows straddler to drive off after landing, remove payloads from subsequent landings, and relocate a crippled lander. Right—plan view of truck in unpressurized garage for access to workshop module hatch; failed mechanisms throughout the base are exchanged robotically and introduced into the workshop for shirtsleeve repair. Drawings to same scale (Images by the author, reprinted courtesy of NASA and Boeing Co.).

PROXIMITY DIAGRAM

For surface sites, the proximity diagram should be two dimensional. (For space systems in microgravity, the proximity diagram must be three dimensional.) Figure 4 graphically catalogs the plan area of all major base elements collected together, as circles at the same scale. This diagram allows them to be directly compared, reveals the significant difference between the size of an individual element and the space multiple units will take up, and begins to show how some elements depend on each other (e.g., how much space is required for the paved unloading area around the lander).

Figure 5 arranges the element circles into a proximity diagram. The lines represent functional connections between pairs of elements; line thickness represents the relative importance of the connection. Ranking the connections this way arises from engineering judgments, based on relaxing proximity strengths one at a time and estimating the resulting complications. For example, grounding cables might be required to connect all base elements (on the Moon, only common "chassis" grounding is practical because of the anhydrous regolith), to prevent potentially hazardous differential charging. The total cable length should be minimized. But such cables are far cheaper than cryogenic fluid transfer lines, which are heavier, require more maintenance attention, and introduce the hidden cost of additional refrigeration power per unit length. So keeping fluid lines short and accessible is a good trade even if it means having longer grounding cables.

The rules of a proximity diagram are simple: connection lines should be kept as short as possible, and none should cross. Satisfying both of these rules strictly is hardly ever possible, but the distinction between a good solution and a poor one is usually obvious. Often, iteration between the proximity diagram and the element set produces better results. For example, without a "workyard" element, the lunar-base diagram would show a tangled web of activity lines through its center, representing physical vehicular access among all of the other elements. Explicitly requiring a workyard replaces that tangle with a definable element, which can be designed proactively and within which mutual access interferences can be resolved by operations scheduling rather than by permanent facility design.

When generating the proximity diagram, it is critical to accommodate peculiar requirements that might influence the ultimate physical element arrangement, if such constraints are so fundamental

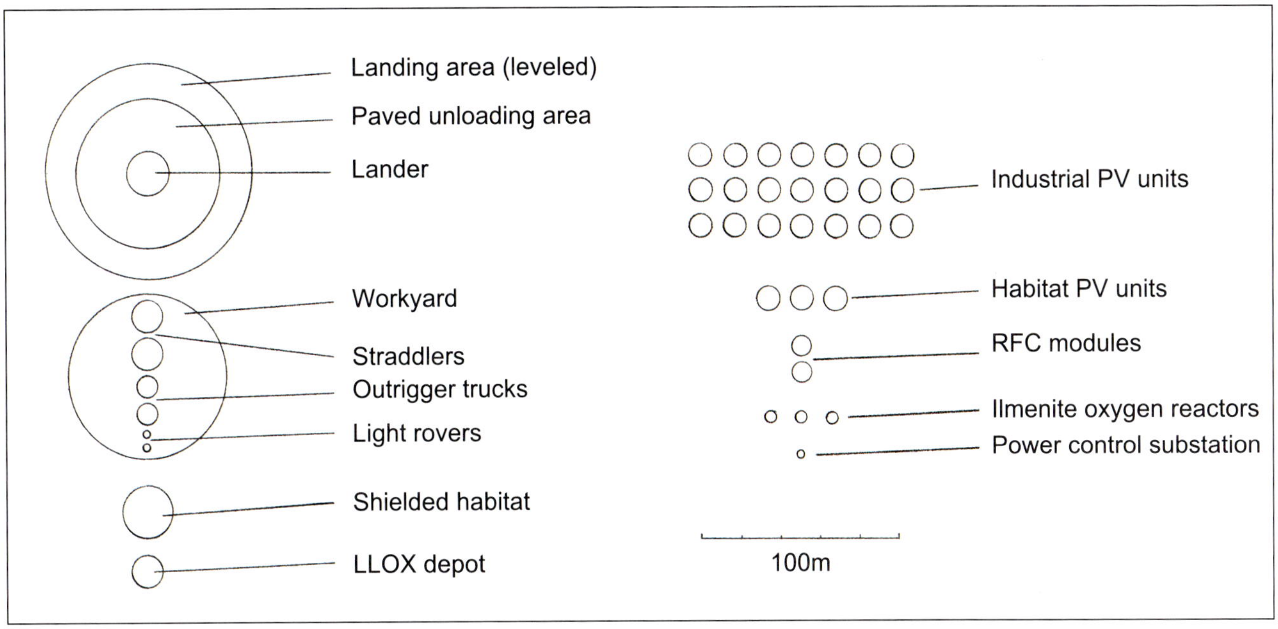

FIGURE 4 Scaled plan-area catalog of all major base elements is the first step to graphically depicting their interrelationships (courtesy of Boeing Co.).

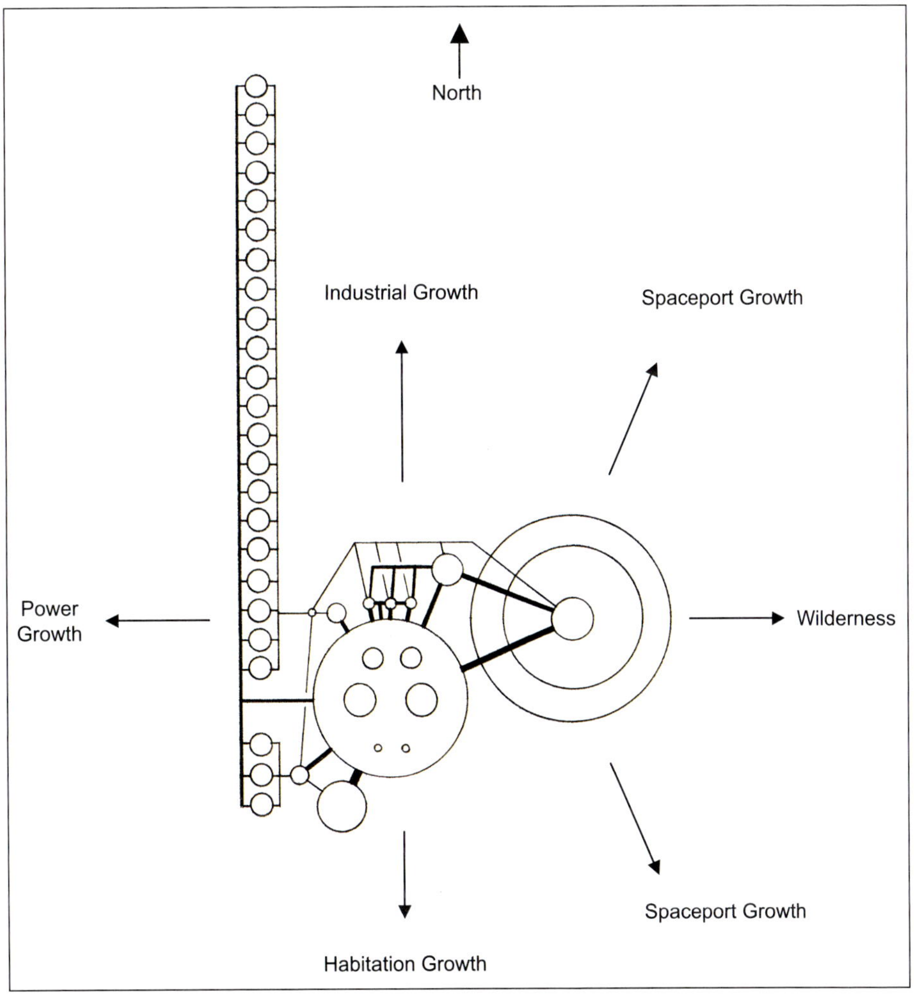

FIGURE 5 Element proximity diagram resolves the functional relationships among all major elements in a to-scale layout (courtesy of Boeing Co.).

that they appear even in the diagram's abstract language. The solar powerplant of the equatorial lunar base has such a requirement: none of the individual array panels can obstruct another's view of the Sun from dawn to dusk. Power calculations are based on a 300-hr lunar day (about 90% of the full lunar day, discounting the portion when the Sun is nominally within 9 deg of either horizon). The obstruction constraint leads to a linear, north–south installation, which appears clearly even in Figure 5. The solar-array panels could be arranged along the solar track (i.e., strung out east-west) instead, but given their dimensions this would require spacing them on 55-m centers in flat terrain to preclude mutual obstruction. Indeed, if the solar field were a few times larger than the one under study (e.g., if industrial operations were scaled up several times), parallel north–south lines spaced 55-m apart would be the proper configuration to minimize both driving distance and dc transmission losses. The proximity diagram immediately shows the importance of providing a single power bus connecting all of the array panels, rather than long separate cables.

Another particular requirement is for the oxygen reactors to be located in a straight line, so that hoppers removing the spent solids can be moved along a simple rail system. The final major orientation-specific requirement is for the spaceport to reflect astronautical constraints. Landers would typically approach the equatorial base from the east, along a terminal approach path elevated at only 15 deg, whereas launches would be almost vertical for a considerable distance before pitchover. During the landing approach, the most statistically common failure mode (albeit very rare, for a reliable, human-rated vehicle) is a short landing. Thus the safest place for the spaceport is east of the base, where the rare short landing will not crash on top of the base. The pad can, however, be adjacent to the base just past its east edge. Safety is not a driving issue; the man-rated lander hovers precisely upon touchdown, robotically targeted using beacons. Rather, debris impact on base equipment is the major concern; small particles travel ballistically as the lander's engines scour the ground. One design alternative is sheer separation, but this is impractical for a base with slow mobile robots because the required distance is many kilometers. Our solution is to deploy simple overlapping blast shields around the downrange side of the landing pad. The barrier height turns out to be reasonable, calculated as a function of the expected elevation distribution of particles given the lander rocket exhaust velocity. Such close proximity between the spaceport and base minimizes access road construction, minimizes access time, and keeps permanent utility lines short for the initial base. Later, as more frequent flights increase the cumulative probability of a landing failure, base growth naturally compensates by increasing the separation of active pads from other base elements.

Figure 5 also indicates practical zoning for base growth to avoid topological interferences in an open-ended future. Powerplants, supplying both industrial and human needs, would grow to the west, away from the polluting spaceport. Nuclear reactors would probably eventually supplant solar arrays for a large base and could be sited and shielded to the west. Industry could grow to the north, into the rich and accessible plains of the Sea of Tranquility. The pressurized habitation "village" could grow unimpeded to the south. If the base continues growing into a real lunar city, and if sited close enough to highlands at the Sea's edge, the human quarter might eventually spread into interesting foothill topography.

The spaceport could grow as needed to the east, with the region surrounding and beyond it remaining "wilderness." Preserving untouched lunar terrain within sight of the settlement might prove helpful for psychological relief if the base eventually grows into a densely populated city. In the enclosed, technological human environment of a lunar settlement, viewing barren wilderness would provide the inhabitants' only access to lunar nature. Accommodating wilderness preservation could become a major constraint eventually, albeit one easily foreclosed by short-sighted planning. It makes sense to combine an exclusion zone with the spaceport because the latter's function will keep it relatively development-free indefinitely into the future.

SITE PLAN

A fully developed proximity diagram can be trusted as a guide for laying out the actual site plan because element relationships have been accommodated according to their importance. Details of the elements are introduced, along with the physical infrastructure necessary to realize element relationships throughout the site. Progressively folding in constraints and details derived from a specific program and element set has enabled application of a simple method of developing site plans to the example of an early lunar base. The result is a to-scale site plan used to quantify operational activities in and around the base. In particular, phased buildup of the base, as constrained by flight, diurnal, and surface-vehicle schedules, can be engineered as a result (Buddington 1990). Figure 6 shows the final site plan for the robotic lunar base study, and Figure 7 shows a diorama overview.

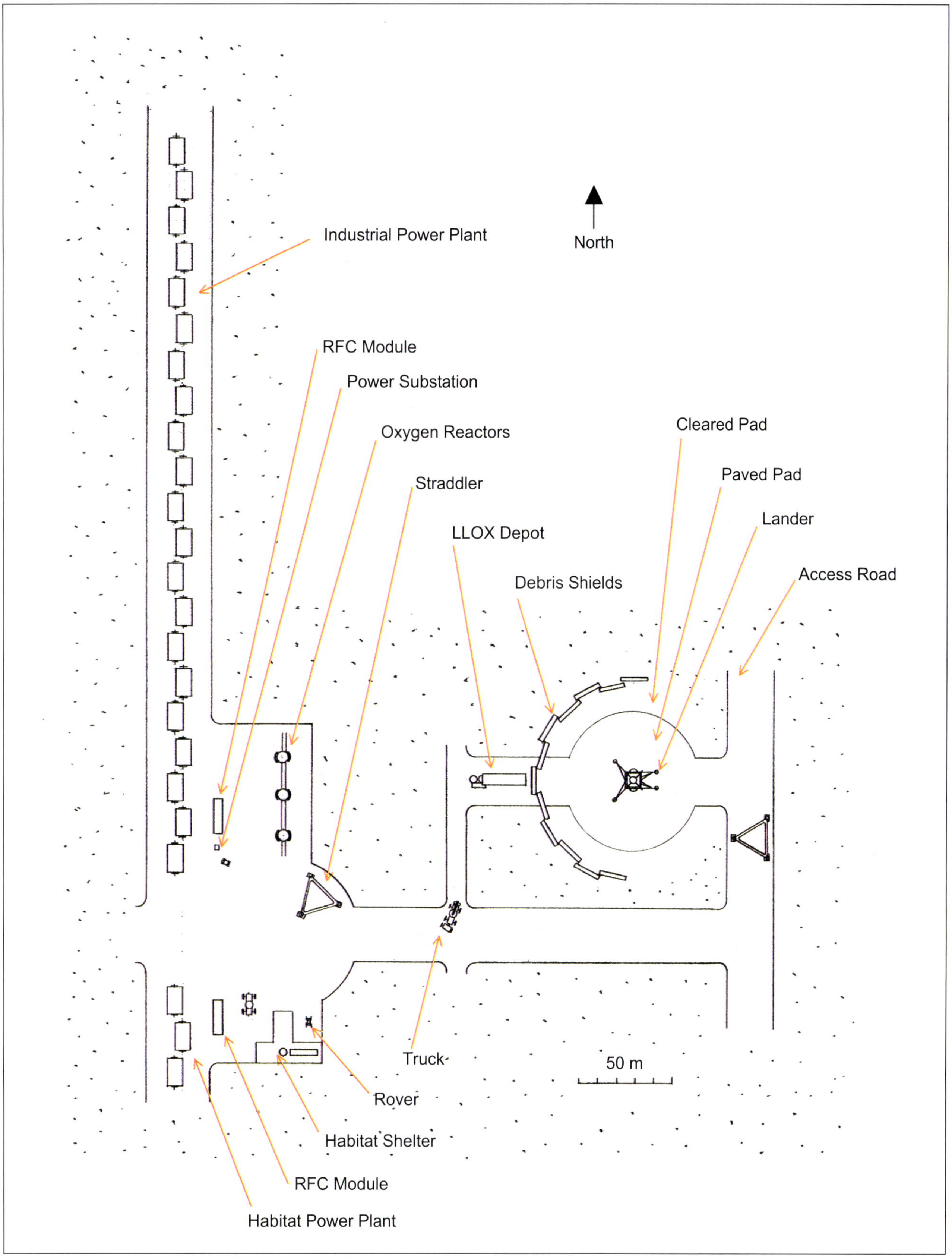

FIGURE 6 Completed lunar-base site plan fulfills requirements of the proximity diagram. Paved areas (nonstippled) are prepared by scraping the softest regolith layers down to a level depth of 20 cm and then depositing and compacting sieved gravel to control dust and provide predictable operations surfaces for the mobile robots. Road-building areas and material are byproducts of oxygen reactor feedstock mining; sitework construction takes half the time required for delivery of equipment from Earth at the 4/yr landing rate (courtesy of Boeing Co.).

FIGURE 7 Diorama of complete base design shows how the site plan comes to life (courtesy of Boeing Co.).

Power, grounding, and fluid lines are not shown, but reference to Figure 5 verifies that their layout is sensible. Lines are buried in 1-m-deep trenches, although serviceable connectors are above grade for inspection. The powerplant bus runs on the graded but unpaved surface beneath the solar arrays.

Maneuvering room for the large robotic vehicles has led to merging the shared workyard with small work zones around the three oxygen reactors, power storage modules, and power conditioning substation (Figure 8). This simplifies the grading, graveling, and compacting activities necessary to build these paved areas. Paving itself is necessary to limit dust contamination of critical systems and to provide a predictable surface for repeated robotic travel. Roads to accommodate the straddling mobile cranes regularly are 25 m wide; truck access roads are 10 m wide. At reasonable baseline excavation rates (governed by the oxygen production goal), it takes one year to prepare the substrate area indicated (Buddington 1990). Most areas are scraped 20 cm deep; excavating 1-m-deep foundations for the heavy elements (habitat shelter, oxygen reactors, and LLOX depot) allows reasonably small footings (Figure 9). Generating sufficient gravel and sand to pave most of this area to a depth of 5 cm requires almost an additional year. However, this two-year site preparation period amounts to only half the time required to bring all of the base elements from Earth anyway.

Most of the road construction is required for the spaceport. Two north–south roads establish the pattern. The west service road allows truck maintenance access to the LLOX depots associated with each landing pad developed; the depots need to be behind the debris barriers and connected by power and gas lines to the oxygen production area (Figure 10). The straddlers require easy access to landers for offloading operations, using the east road in front of the barriers. As more pads are developed, barriers would be erected between them to limit adjacent blast effects. An east–west arterial road links the spaceport complex with the rest of the base. A pair of north–south dust-control roads allows truck and straddler access to the solar powerplants. The arrays themselves are staggered to permit maintenance access to the tracking and control mechanisms located on their base structures despite the close-packed arrangement.

FIGURE 8 Operations analysis led to merging the workyard elements with smaller work zones required around the oxygen reactor and power storage elements (courtesy of Boeing Co.).

FIGURE 9 Half-meter-thick layer of regolith in the vaulted habitat shelter makes it one of the heaviest elements once assembled; footing size is kept modest by excavating down to 1 m depth, where naturally consolidated regolith makes a solid foundation (courtesy of Boeing Co.).

FIGURE 10 LLOX depot element needs to be close to the spaceport, close to the oxygen production area, protected from lander blast effects, and accessible for maintenance (courtesy of Boeing Co.).

The small separation between the base center and spaceport service road is the location of an eventual berm, made of gangue separated from the ilmenite reactor feedstock. Deposited and compacted incrementally, such a berm could reach 35 m in height, providing both a retrievable stockpile of these materials and a debris shield wall west of the spaceport. Five years of production at the baseline rate (100 mt/yr LLOX) yields a berm only 10 m high, 50 m wide and 110 m long; however, planning a consistent location for tailings seems advisable. Developing more landing pads fulfills the oxygen feedstock mining requirement for a few years. After that, growth of the habitation and power areas would provide more feedstock; ultimately mining would lead industrial growth to the north.

If such a base, established for research, demonstration, and production purposes, did continue to grow indefinitely, the zoning system outlined would lead naturally to a cruciform, cardinally oriented transportation and utility service infrastructure. The city center would grow into an "activity hub" at the intersection of these cardinal spines, which could grow to immense scales that absorb the initial base site while preserving the original regional separation. Evolved directly out of the original site constraints and site plan, such an orthogonal layout is immediately recognizable to city planners as the *decumanus, kardo,* and *forum* layout of all Roman cities—one of the most pervasive, persistent, and successful designs in the history of civilization. The simplicity of such a sectored scheme appears able to serve changing requirements in scale, element type, and even program emphasis as a permanent Tranquility Base grows.

References

Buddington, P. A. (1990), "Manifesting for a Lunar Robotic, Oxygen-Producing Base," SPACE 90, American Society of Civil Engineers.

Sherwood, B. (1990a), "Lunar Base Elements Designed for Robotic Operations," *Space 90: Engineering, Construction and Operations*, American Society of Civil Engineers.

Sherwood, B. (1990b), "Robotic Lunar Surface Operations: Engineering Analysis for the Design, Emplacement, Checkout and Performance of Robotic Lunar Surface Systems," NAS2-12108, Boeing Co. D615-11901, Jan.

Sherwood, B. (1990c), "Site Constraints for a Lunar Base," *Space 90: Engineering, Construction and Operations*, American Society of Civil Engineers.

Woodcock, G. R., Sherwood, B., Buddington, P. A., Bares, L. C., Folsom, R., Lousma, J., Mah, R., and Whittaker, W. (1990), "Application of Automation and Robotics to Lunar Surface Human Exploration Operations," *Space 90: Engineering, Construction and Operations*, American Society of Civil Engineers.

23

Mars habitat using locally produced materials

BAS LANSDORP
AND
KRISTIAN VON BENGTSON

INTRODUCTION

SUPPORTING PEOPLE WITH A LARGE ENOUGH, comfortable habitat during long-duration planetary missions is a major challenge. This chapter presents a concept for providing Mars surface crews with such a habitat, despite having launched only a small habitat core from Earth. Once developed, this concept could lead to a flexible way of constructing large and expandable habitats from local resources. A small habitat would be extended onsite by the crew, using *in situ* resource utilization (ISRU).

In this concept, habitat cores launched unmanned from Earth contain all necessary interfaces, subsystems, and airlocks but almost no living space. The living space is instead constructed on Mars from locally produced glass made from Martian regolith. Untreated Martian regolith results in black glass (Ray et al. 2004) but contains all of the ingredients for various types of clear glass.

A glass module has some advantages compared with conventional aluminum pressure vessels and inflatables. Greatest among them are external viewing and sunlight for psychological benefit and direct greenhouse insolation. Transparent glass results in habitat structures that can be used fully pressurized for people or plants.

An ISRU plant landed before the first human mission would collect regolith and treat it to obtain clear glass feedstock. Other major infrastructure elements required for constructing such a habitat are a nuclear powerplant and heavy rover for moving and positioning segments of the glass structure. The scale of regolith-processing and manufacturing operations required make this concept most appropriate for base construction to support large numbers of people.

CONCEPT DESCRIPTION

The habitat system starts with prefabricated aluminum core segments (Figure 1) launched from Earth to be extended by the glass construction. The middle segment contains four pressure doors, two of which are visible in the figure. Interior doors connect spaces within the habitat. Exterior doors connect to other habitat elements. All doors could be sealed in a depressurization contingency.

Tapered glass barrel segments constructed *in situ* can then be inserted between the middle core segment and hemispherical aluminum end caps (Figure 2) to yield large modules. The taper allows lap joints between the barrel segments (Figure 3), providing large contact surface areas for adhesive bonding. Filler caulk along the circumferential joint provides a reliable seal with the aid of internal habitat pressure.

Reference dimensions are shown in Figure 4. The habitat floor is located halfway between the center and the bottom of the 5-m-diameter segments. This results in a 4.33-m-wide floor on the inside and a ceiling height varying from 2.5 m on the edges to 3.75 m in the center. Each barrel segment is 1.5 m long, and the segments overlap 0.1 m.

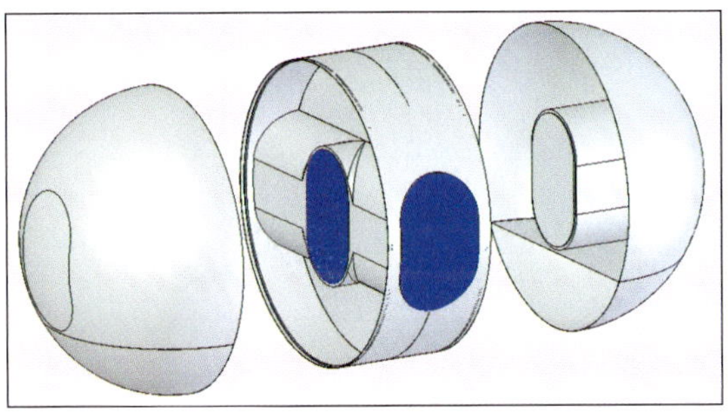

FIGURE 1 Prefabricated aluminum module core segments.

Copyright © 2008 by the authors. Published by the American Institute of Aeronautics and Astronautics, Inc., with permission.

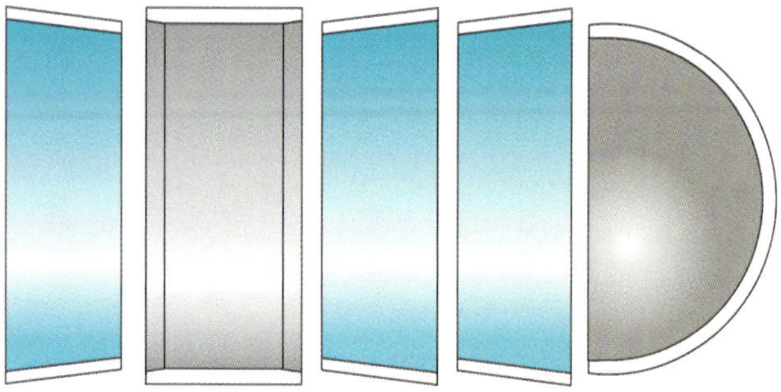

FIGURE 2 Glass segments inserted between prefab segments extend the module length as desired.

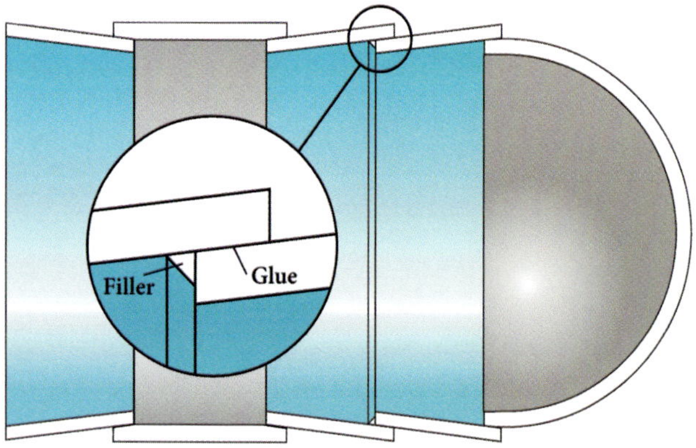

FIGURE 3 Assembled structure uses glued lap joints in shear, and pressure-sealed filler, to withstand the internal atmospheric pressure.

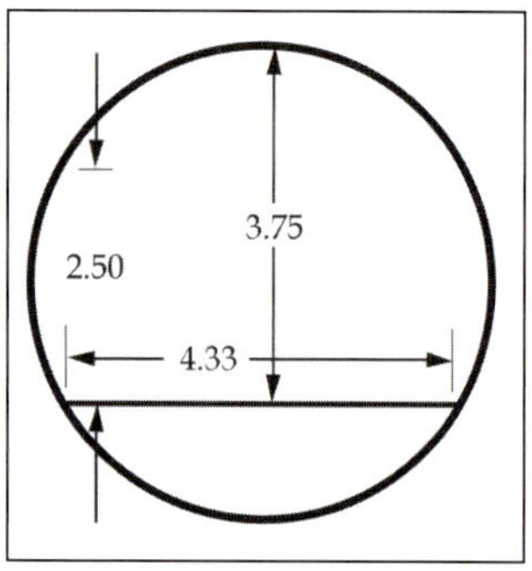

FIGURE 4 Reference module cross-sectional dimensions.

Internal design pressure is 0.75 bar, 25% lower than Earth sea-level pressure. This allows thinner glass and shorter prebreathe periods before EVA. The glass thickness is

$$t = \sqrt{t_a^2 + t_r^2}$$

where the thickness t_a required for pressure in the axial direction is

$$t_a = \frac{n\pi R^2 p}{2\sigma\pi R} = \frac{nRp}{2\sigma}$$

and the thickness t_r required for pressure in the radial direction is

$$t_r = \frac{nRp}{\sigma}$$

and where R is the segment radius, p is the internal pressure, n is the factor of safety, and σ is the tensile strength of the glass.

Therefore the total required thickness of the glass is

$$t = \sqrt{\frac{5}{4}}\frac{nRp}{\sigma}$$

For 2.5-m radius, 0.75-bar pressure, factor of safety of 5, and 50-MPa tensile strength of the glass, the resulting glass wall thickness is 2.1 cm, a promising result for transmission of light and view through the hull. Thus the volume of glass required for a single barrel segment is 0.49 m^3.

TECHNOLOGY APPROACH

Technologies required for *in situ* construction of glass habitat segments include soil treatment, glass melting, glass pouring, and connection of glass segments.

Martian regolith comprises the elements given in Table 1. If untreated regolith is melted, it forms a glaze, which, upon cooling, yields black glass (Ray et al. 2004). However, for some habitat applications clear glass is preferable. Many types of clear glass exist; this conceptual analysis assumes regular window glass, although further study would determine the optimum type for a Mars-environment

TABLE 1 Martian regolith composition, weight fractions.

CONSTITUENT	CONCENTRATION %
SiO_2	43.8
Fe_2O_3	18.2
Al_2O_3	7.2
SO_3	7.2
MgO	6.0
CaO	5.8
Na_2O	1.34
Cl	0.8
P_2O_5	0.68
TiO_2	0.6
MnO_2	0.45
Cr_2O	0.29
K_2O	0.10

TABLE 2 Ingredients of clear window glass.

INGREDIENT	Wt%
SiO_2	79
Na_2O	10
CaO	10
Al_2O_3	1

TABLE 3 Ingredient quantities required to produce 33 glass module segments.

INGREDIENT	Wt, %	MASS, kg	DENSITY, kg/m^3	V, m^3
SiO_2	79	32,196	2,200	15
Na_2O	10	4,076	2,270	1.8
CaO	10	4,076	3,340	1.2
Al_2O_3	1	408	2,700	0.15

application. The composition of typical window glass is tabulated in Table 2.

Thus all of the ingredients needed for window glass are available in the Martian regolith, which could be beneficiated to provide them. Such beneficiation is complex; for example, it could be done by decomposing the soil with hydrogen or fluorine gas. Although this is not a trivial undertaking, would be very energy intensive, and would require reagents, process monitoring, and machinery operating for a long time to meet the feedstock requirements shown next, it might make the best sense for the collection and beneficiation steps to take place before humans arrive so that the glass ingredients would be ready for them.

For 200 m^2 of habitable floor space, 33 glass segments are required. The mass of each segment is 1235 kg given a glass density of 2500 kg/m^3. The total glass mass is thus 41 mt. Table 3 gives the composition ratio of the ingredients, and Table 4 gives the amount of Martian regolith to be beneficiated to obtain the required quantities of each element. Sodium oxide is clearly the driving requirement for regolith beneficiation in making window glass. Sodium oxide dissolves in water, and so it would likely be recovered by flushing the regolith with water. However, because the reference design would require processing over 300 mt of regolith this way, the analysis incentivizes selection of a glass type that contains no sodium oxide.

TABLE 4 Feedstock requirements incentivize selection of appropriate glass recipes.

INGREDIENT	REQUIRED MASS, kg	TREATED MARTIAN SOIL MASS, kg
SiO_2	32,196	73,506
Na_2O	4,076	304,179
CaO	4,076	70,275
Al_2O_3	408	5,667

Production of the structure segments from the beneficiated and verified feedstock might likely be done with on-site crew providing oversight and quality assurance because the end product would be used in a vital application. Glass melting would be done in a large kiln. The specific heat capacity of glass is ~700 J/kgK. The glass must be heated to about

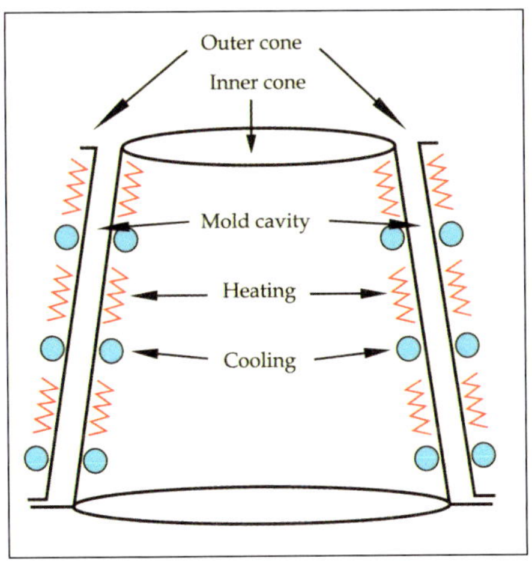

FIGURE 5 Double-cone mold directly yields net shape of the barrel segments and window-quality surface smoothness.

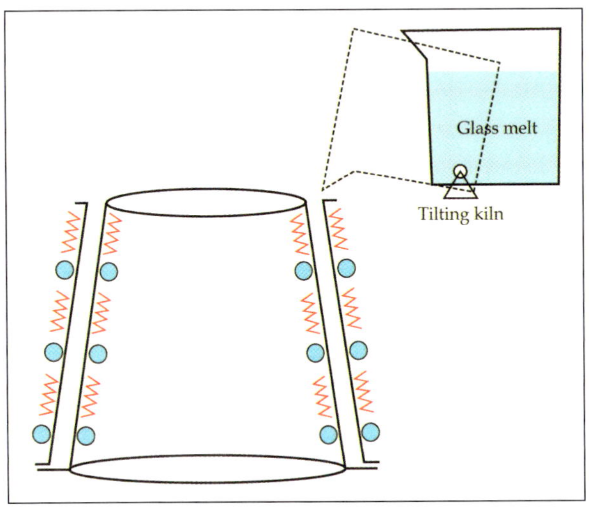

FIGURE 6 Molten glass is poured into the preheated mold.

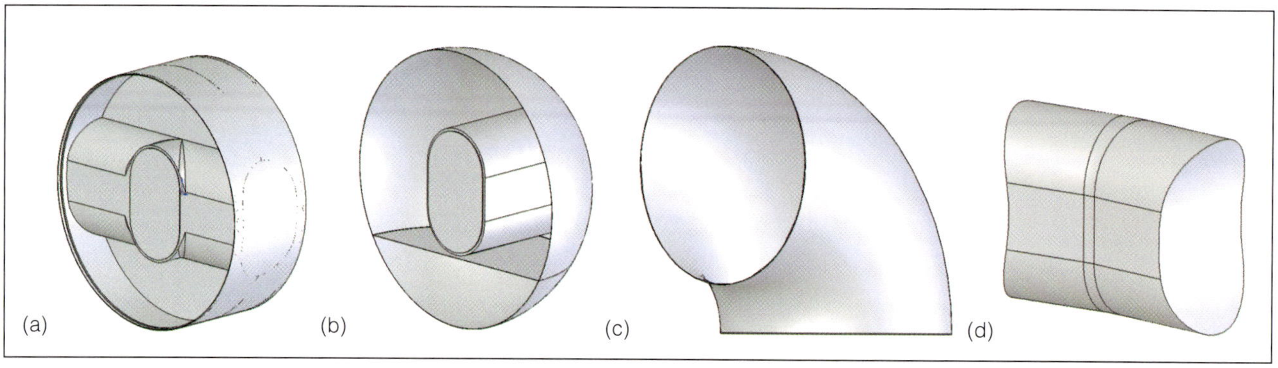

FIGURE 7 Prefabricated aluminum core elements connect wings made of glass barrel segments: (a) center node segment with airlock; (b) airlock/connector segment; (c) elbow transition segment; and (d) connecting node.

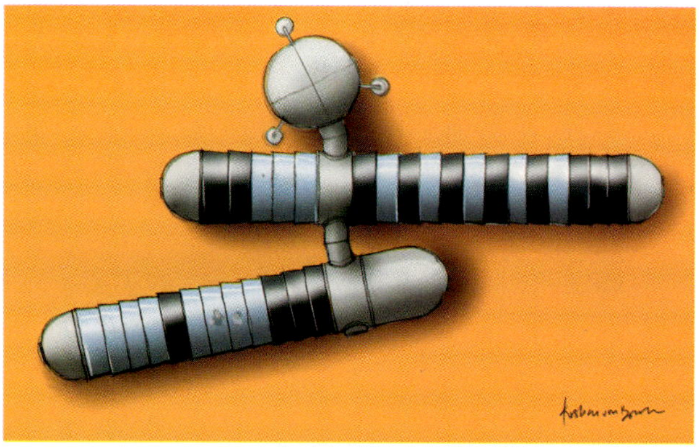

FIGURE 8 Example habitat system configuration, showing integration of aluminum segments with black-glass and clear-glass *in situ* sections.

1200°C from around 0°C. For each 1235-kg segment, this takes approximately $1{\cdot}10^9$ J. If 100-kW power is available from a nuclear plant, the heating process takes about three hours for each segment.

The mold is a rotationally symmetrical structure comprising two cones (Figure 5). Both cones have heating and cooling provisions. Glass is poured into the mold by tilting the kiln (Figure 6). Before the pour, the mold is preheated to about 600°C. During and after pouring, the mold is actively cooled. After pouring, the glass segment must be annealed for about a day, after which it can be removed from the mold. The mold's conical shape facilitates removal.

Habitat assembly proceeds one segment at a time. Laser-guided heavy equipment positions each segment for insertion. Prior to insertion, bonding lap-joint surfaces are treated with adhesive. Inside seams are sealed with vacuum-cured filler caulk. After all segments and the aluminum end cap are installed and sealed, that wing of the structure can be pressurized.

COMPLETE HABITAT SYSTEM

Four types of aluminum segment are required to construct a habitat system (Figure 7): a center node segment with airlock and four exits used to connect different sections of the complete system; an airlock segment that can also be used to connect to the airlock of another habitat system; a transition segment leading to an underground radiation shelter; and a hallway node that interconnects multiple center

FIGURE 9 Plan view of habitat system, also showing section cuts for Figure 10.

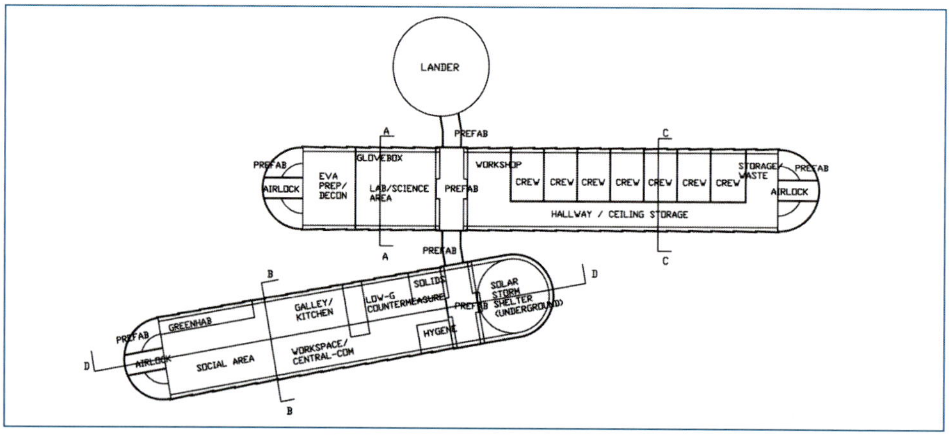

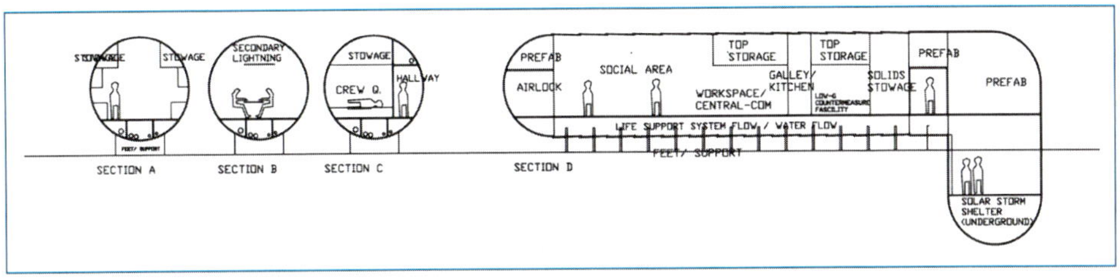

FIGURE 10 Section view of habitat system, showing underground radiation shelter.

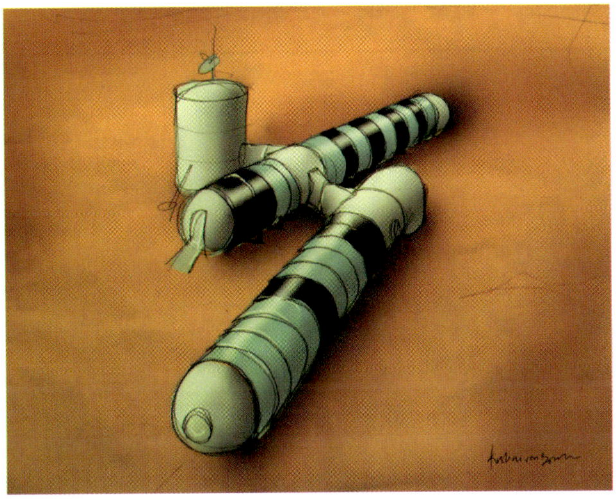

FIGURE 11 General view of exterior.

FIGURE 13 View to the outside.

FIGURE 12 View from the outside.

segments and links to external systems (e.g., a rover or grounded lander). These prefabricated aluminum segments would contain the life-support and other functional systems equipment. Distributed systems would be extended from these cores into the glass wings and installed as required.

Such a kit of parts could be used to assemble various habitat configurations (Figure 8). The construction of this configuration begins by attaching prefabricated core segments including airlocks to the habitat of a grounded human lander. Then assembly with glass segments begins. The upper-left wing, a working zone consisting of five glass segments, is built first because it contains the main airlock. This is followed by the lower-right wing, a radiation shelter for solar proton events. Then the upper-right (13 segments containing personal cabins) and lower-left (12 segments containing communal living space) wings are built. Each wing is roughly 30 m long. Figures 9 and 10 show plan and section views of the complete habitat system and its functional allocation, and Figure 11 shows an overall view.

Glass construction provides views of the surroundings, and orientation cues to differentiate the habitat wings. The pre-integrated, preverified airlock elements provide safety and a sense of security inside the glass structure because each section can be closed off independently. A slight cant in the connecting hallways allows optimal views from both crew quarters and communal spaces. Such angles also vary the natural lighting inside each habitat wing. Architecture based on a glass module structure offering significant transparent areas gives occupants quite a different perceptual experience than typically expected of cylindrical pressure modules (Figures 12 and 13).

References

Ray, C. S., Ramachandran, N., and Rogers, J. (2005), "Developing Glassy Magnets from Simulated Composition of Moon/Mars Regolith for Exploration Applications," *Materials Research Society Symposium Proceedings*, Vol. 851, edited by M. Chipara, D. L. Edwards, R. S. Benson, and S. Phillips, Materials Research Society, Warrendale, PA, p. 487.

INTRODUCTION

HUMAN CIVILIZATION AND ARCHITECTURE have defined each other for almost 10,000 years on Earth. As the space population grows over time, persistent issues of human urbanism will eclipse within a historically short time the technical challenges of space habitation that dominate current efforts. Although urban design teams will have to integrate many new disciplines (e.g., space-systems engineering, astronautics, solar-system science, life-support systems, and space resource utilization) into their already renaissance array of expertise, doing so will enable them to adapt ancient, proven solutions to startling opportunities afforded by expanding urbanism off-world. Inescapable facts about the Moon set boundaries within which tenable lunar urbanism and its component architecture will eventually develop.

24

lunar architecture and urbanism

BRENT SHERWOOD

LONG VIEW

Many decades still separate us from any kind of lunar development that could be considered urbanism. Indeed current planning puts deployment of even an initial lunar outpost around 2020, at the earliest. Why think about eventual lunar urbanism? Why examine a field so embryonic that its likely reality cannot yet be known? Three reasons motivate the analysis on which this chapter is based (Sherwood 1988, 2005).

First, Project *Apollo* proves that human expansion to other planets is possible. Indeed that vision has been central to U.S. space planning for five decades (Figure 1). As lunar activities ramp up, nonspecialists will anticipate lunar development by imagining concepts that imply the character of eventual lunar civilization. Such projections would be most relevant and useful if realistic, that is, grounded in inescapable facts that will constrain the eventual reality. Outlining such principles is the purpose of this chapter.

Second, those aspiring professionally to design the built lunar environment tend to originate either as space engineers who know little about urban development, architecture, and their history; or architects who know little about the quantitative characteristics of space environments or the way space systems are developed, tested, and operated. Eventually however, lunar planners will have to be well versed in both worlds. Preparing rigorously for that combined future will take time; the present analysis is intended as one small step to blend the two worlds.

Third, setting goals from the start can guide the paths that bridge present thinking to future history. Tangible ideas regarding what the far future must, might, or should be can constitute a sound compass for making the many incremental decisions along the way. So it is not premature to begin broad-based discussion of how humans will settle Earth's Moon. Recognizing likely end-states of off-world urbanism might help avoid wasted efforts as that urbanism develops.

To begin, we draw distinctions among three types of human activity, each with a special role to play

FIGURE 1 Earth's Moon, near side.

Copyright © 2008 by Brent Sherwood. Published by the American Institute of Aeronautics and Astronautics, Inc., with permission.

in the growth of space civilization: *traveling*, *staying*, and *living*. Space architecture so far has been entirely vehicular, based on components launched from Earth. This is true even for long-lived orbital space stations. Rocket launch governs their form from the outside in. Like trucks, vans, and the cargo containers they carry, this vehicular architecture only incidentally supports human activity: it is cramped, oddly shaped, noisy, and smelly. The interior human environment of such capsules, shuttles, and modules is adapted directly from methods and solutions optimized for atmospheric flight vehicles like high-performance airplanes and rockets.

Because vehicles themselves are optimized for traveling, longer-stay-time missions based on vehicular architecture require excrescent, retrofitted, or modular adaptation. The space shuttle exemplifies the excrescent type. Shuttle missions before the *Columbia* accident commonly carried "ab-ware" (*Spacelab*, *SpaceHab*) to extend their habitable volume so that seven workers could cohabitate for up to two weeks. This enabled, but also required, extensive ground support and preparation for every mission, and was ultimately limited by the capacity and operability of the vehicle itself.

The retrofitted and modular types characterize long-lived orbital stations. *Skylab* was a retrofit: built as an upper stage for the *Saturn V* rocket, but with the huge hydrogen tank modified to be a habitable research lab. Its capacious volume enabled greater spatial variety than found in vehicular architecture before or since, but its architectural sophistication was limited by what could be designed in, quickly and spartanly, prior to its launch. The modular approach is more common: *Salyut*, *Mir*, and the International Space Station (ISS) have all been of this type. Large habitable volume is obtained incrementally through the accretion of small modules. Contemporary modular space architecture does achieve operational efficiency by distilling staying from traveling (i.e., ISS for staying vs *Soyuz* and shuttle for traveling), but is ultimately limited by the dimensions of its units and the complexity of their multiple connections.

A space architecture of linked, pressurized cylinders, even one that sprouts appendages and enormous exterior structures, is still essentially vehicular. Such a habitat is like a train parked on a railroad siding—still a train even if immobile. Mission durations exceeding a year prove that, when specialized, such architecture can indeed enable individuals to work and stay in space. But although it is natural and common to envision even future space architecture based on this familiar vehicular vocabulary, only the first stages of permanent construction in orbit, or on planetary surfaces, can sensibly be vehicular. This realization drove the development of technology for larger, deployable modules like the inflatable TransHab originally developed by NASA, being matured by Bigelow Aerospace, and discussed in the Orbital Architecture section of this book.

ARCHITECTURE AND URBANISM

Submarine and Antarctic environments have been used as analogues for space. Remote and hostile, all three are intrinsically lethal and require artifice to sustain life, promote efficiency, avoid conflict and encourage conciliation, and prevent disaster. From these urgent needs emerged "human-factors engineering," the discipline that quantifies human behavior in response to the built environment and improves its design as a result. Human-factors engineering helps us understand human tolerance of key parameters in hostile settings. This enhances our ability to design an acceptable balance between minimizing resources and maximizing probability of mission success. But Earth's oceans and poles, from which people generally return within a year, can only model the remoteness and harshness of space to a certain point. Long interplanetary flights and planetary outposts start to blur the boundary between traveling and staying.

And finally, space cannot become a settled, economically viable human domain until people establish their lives there. Eventually travel time, expense, and risk will make it more practical for people to transform staying in space to living in space. Human living is an exceedingly complex activity that requires much more than passably engineered accommodation because it includes all we do: working, resting, playing, and growing (Figure 2). Designing for living is a vastly messy problem not tractable by supercomputers or with "human factors" alone. Human behavior cannot be captured in a numerical model of living. Living touches the sum of physical and abstract richness developed over all of human history, even as it continuously unfolds in the now.

The interactions between requirements for and effects of environments to support human living are subtle and continue to be honed over millennia as society evolves. Manipulating those environments with skill and grace demands a fine multivariate balance that only human experience, wisdom, and artistry can feasibly provide. This will be as true in space as it always has been on Earth, as far into the future as we can foresee (Figure 3). This artistry is the practice of architecture.

Architecture is the professional activity of coordinating a set of specialty industries and services

FIGURE 2 Richness of cities juxtaposes all facets of all kinds of human living: a) Barcelona, b) Santorini, c) New York, and d) Mumbai.

to make buildings that enable, enhance, and foster human living. Architecture addresses needs that range from the prosaic to the spiritual. Two millennia ago, the Roman architectural historian Vitruvius proffered a clear, concise, and complete statement of the qualities characterizing good architecture, translated since the 17th century as "firmness, commodity and delight" (Wotten 1624). This tripolar standard covers anything that architecture can do or be. *Firmness* refers to structural integrity, appropriate material qualities, proper fabrication, and safety. Firmness addresses the question: is it usable? *Commodity* subsumes all of the ways a work of architecture serves the programmatic purpose for which it is built, accommodating the physical and abstract needs of its occupants and environment. Commodity addresses the question: is it useful? *Delight* is often the diacritical signature of great architecture, frequently short-shrifted by modern commercial builders as a separable luxury. Delight addresses subtle but penetrating questions: would people rather use this than other solutions? Will it last? These three ancient principles apply to all ages, modes, and styles of architecture. They encapsulate complementary properties, without any one of which architecture cannot be simultaneously engineering, solution, and art.

At its best, architecture projects human values and aspirations; at the very least, it embodies human needs and behaviors. Because it depends on manipulating materials for human use, architecture has been whimsically called the "second oldest profession" (as have business management, taxation, and spying, incidentally). Architecture's purview is extensive and inclusive. All designed interfaces between human beings and their environment, from spoons to highways, gardens to sewers, and buildings too, are elements of architecture. Civic architecture, the largest-scale form, which services and embodies human community, is convolved inextricably with civilization itself.

FIGURE 3 Inca city at Machu Picchu, thought to have been a university town unique in pre-Columbian America.

Although known human cultural artifacts date back as far as 40,000 B.C., organized civilization arose 10 millennia ago as a result of several key inventions: agriculture, abstract writing, money, and urbanism. Natufian hunter–gatherers first began farming cereal when the climate in the Middle East warmed around 8000 B.C. (Bower 2005). Artifacts of early urban development appear in the Neolithic record at Jericho in 7500 B.C. and at Çatalhöyük from 7500–6500 B.C. (De la Croix and Tansey 1975) (Figure 4).

The earliest known applications of writing and urbanism were for commerce; they used formal design to facilitate densely efficient business intercourse. These permanent media (writing and building) became useful also for capturing and stimulating human sensibilities. By transcending mere functionality, these technologies spawned the recording arts of literature and architecture. Down through the millennia since then, civilization and its creative expressions have continued to define each other iteratively. We cannot even imagine "civilization" (from the Latin root *civis*, i.e., citizen) divorced from its created artifacts.

The city is architecture's grandest product, a built framework within which large numbers of people conduct individual but linked lives. As a tool to enable the evolution of increased social complexity, the city must first provide enduring organization and sustain the individual and collective needs of people living in it. By simultaneously accommodating most of the conflicting, singular services individual citizens need and desire, cities enable population density. The mixing caused by that density in turn animates a social organism much larger, more resourceful, and more consequential than any individual could be. The strength, capacity, and influence accessible to civic culture are what drive humans together to make cities wherever they live (Wright 2000).

An enduring and efficient civilization can achieve great things that advance the reach of the human spirit. But the extreme density encouraged by cities cannot alone guarantee greatness; urbanism often falls far short of both commodity and delight. Disease, exploitation, violence, environmental devastation, and spiritual impoverishment have historically accompanied high concentrations of people. As physical limits are approached, atavistic biological controls resurface in human populations. Certainly there is a significant gap between what is biologically tolerable

FIGURE 4 Çatalhöyük, one of the oldest known instances of human urbanism (based on a reconstruction by Orrin C. Shane, III). (Image courtesy of Time Books, Ltd.)

for the human species and what is spiritually desirable for human civilization. The practice of urban design tries to mitigate the negative aspects of dense populations while leveraging their special benefits.

Architecture integrates and reconciles disparate fields that only create a firm, commodious, and delightful environment when combined coherently. Traditional specialties contributing to modern terrestrial architecture include human activity programming, comparative historical analysis, abstract and representational modeling, psychology, structural engineering, law and regulation, materials testing and development, environmental control engineering, negotiation and contracting, construction management, engineering geology, economics, site engineering, landscaping, and art.

At the larger urban scale, designers of cities must in addition address mass transportation, civic logistics, waste management, industrial production, crime, commerce, power production and distribution, spectator events, communication networks and media, public recreation, law enforcement, resource conservation, death, park management, health maintenance, environmental protection, and defense.

Architects and urban planners work to satisfy the needs of all of these subjects simultaneously by manipulating the proportions, character, symbolism, and scale of material assemblages. In so doing they add incrementally to the long history of built human environments. Their core task—creative coordination and integration—remains invariant despite material and social features unique to time and place.

ANOTHER CHANCE

Let us now focus on a particular time (the late 21st century) and place (on and under the surface of Earth's dwarf planet companion, the Moon). Before that time and place, the vehicular nature of space vessels ensures that their design will continue to be influenced by only the barest skeleton—the human factors—of the tremendous array of architectural issues. Poised at the threshold of learning to inhabit the most novel environment since the dawn of man, and having only essayed tentatively into it, we are understandably preoccupied with technical challenges. Keeping people alive and physically healthy still dominates all other challenges of human space activity.

Landing a few people on the Moon, and learning how to keep a few people in orbit continuously, consumed the best engineering effort the 20th century could muster and is still beyond most nations. Developing a more open-ended, comprehensive, safe, and even commercial cis-lunar transportation infrastructure now occupies us. Later, growing this capability to accommodate greater numbers of people and greater distances will open a new level of technical problems by leaving behind the sustenance and protection of Earth. For example, sustaining large groups for long times swings the logistics trade in favor of life-support system closure. Long microgravity stays might require biochemical or inertial (rotating) prophylaxis against deconditioning. Protracted travel beyond low Earth orbit requires shielding against both continuous and acute radiation exposure. And living far away from Earth in confined, artificial environments will challenge psychological health in brand new ways. Solving just these four challenges reliably and elegantly will keep us busy well into this century.

And yet, once those problems are fundamentally solved, too, they will cease to pose the dominant design obstacle to space civilization. By the time multitudes of people can begin living in space, more ancient architectural issues will supersede the technical challenges of traveling and staying. We will have to establish an off-world urbanism that can provide the spectrum of amenities, stimulation, and cultural support that people require of cities anywhere. The urban complexities introduced by hundreds, thousands, or even millions of people living in space will come to dominate everything else. Technically on the verge of enabling astronauts to stay on the Moon, we have barely begun to prepare for solving the total architectural problem opened by doing so.

Unconcerted preparation takes three forms. First but least useful are the striking, utopian images that characterize 20th-century space colonization literature, both in studies and hotel-company promotions (O'Neill 1976; Ehricke 1985). These visions literally paint pictures of space civilization by projecting forward isolated details that might reveal more about their creators' interests than they do about life in space. They can be inspirational, but are of limited help for planning because they gloss over inconvenient practicalities.

Second and more provocative are the countless ideas explored in vignetted detail by science fiction writing, television, and cinema. As the 20th century progressed, the science-fiction audience was educated and sophisticated by developments in contemporary technology and a genre even developed rigorously based on physical feasibility. Both the realistic and the more fanciful stories have an important advantage: they are generally conceived by writers driven to explore implications and meaning, rather than ways and means. Everyone knows the "best" science fiction is about people responding to situations, not about special effects. Taken in aggregate, these fictional stories can both stimulate and caution our planning.

Third, the profession of terrestrial architecture is unwittingly well prepared for solving many eventually important problems of living in space. Trained and practiced architects and urban planners, supported by a heritage of millennia of experience, can guide us to focus on key issues, avoid the mannerist traps of simple visions, tap the wealth of futures concepts, and begin thinking seriously about viable and inspiring cities in space.

But to do so, architects need to learn about space. Engineering for space adds a large new set of tools and issues to the ancient panoply of architectural practice. Lunar urbanism will adapt to the human needs of its citizens according to principles that new technologies, new environments, and new ideas are unlikely to change deeply. Engineering realities of building on the Moon provide a new language for speaking an ancient message of urban accommodation.

Off-world urban design will require attention to all of the traditional architectural and planning subjects listed earlier, plus advanced and closed life support, radiation management, reduced-gravity biology, space mining, biomass production, and material recycling. These are all in addition to the full complement of disciplines specific to spacecraft engineering, including astrodynamics; propulsion; power production and distribution; structures and mechanisms; pressure containment; vibration and noise control; thermal management; guidance, navigation, and control; command and data handling; autonomy; reliability, safety and mission assurance; and development and testing of advanced technologies. Planet-surface architecture must further address launch and landing; alien engineering geology, weather, diurnal cycle, and gravity level; and alien wilderness management. To establish a mature and noble lunar urbanism, off-world urban designers will have to master many subjects.

That sudden technical and environmental enrichment of the architecture profession heralds a great step forward for human civilization. For the 10 millennia of its history, architecture has operated within a familiar, fixed range of conditions governed by the cradle of Earth. The space environment bursts that ancient design boundary, substituting a new set of freedoms and restrictions; traditional planetary constants become parameters. The shackles of gravity are released, but the easy dialogue between indoors and outdoors that humans have always enjoyed vanishes. Interior "exteriors" must arise because the actual exterior is lethal. The harsh rules of space, and its startling allowances, define a new relationship between people and the natural environment.

By being forced to rethink human living off Earth, we can remake urbanism. Starting afresh with the 10,000-year history of civilization as practice, we have a new opportunity to proceed thoughtfully. Space proffers an emphatic environmental transformation for our species, and the Moon provides a unique chance to experiment. A pristine realm affording utterly new opportunities fuels the designer's incessant hope: improving the human condition by creating a new standard of firm, commodious, and delightful urbanism. Perhaps the clarity with which we will have to treat human living on an alien world can even teach us how to live more lightly on Earth.

Given time and trial, the fuzzy problems of lunar human living can approximately sort themselves out, as they have done on Earth. But perhaps foresight can limit missteps through planning, even though space is an utterly novel domain. We should aim to design on the Moon an urbanism better than any yet created on Earth. And with today's accelerating rate of material progress, we must aim to do it hundreds of times quicker than the leisurely 10 millennia we had the first time.

LUNAR REALITY

Now let us look more closely at the necessary expression of lunar architecture and urbanism. The abundance of misleading images of lunar communities means that certain basic principles remain unobvious, so it might be helpful to outline the most probable rules that will constrain lunar architecture. The discussion that follows limns fundamental factual boundaries that likely contain the future possibilities. Not all of these principles will dominate lunar life until real urban growth supplants the first vehicular and outpost phases. Nor will they necessarily remain dominant for more than a few centuries, as they cannot account for unpredictable material progress. But they will likely circumscribe the first several generations of lunar architecture and urbanism.

Dense Population

Lunar urbanism will be densely populated at virtually all stages of its evolution. In most places on Earth, the many costs of spreading out in single-family dwellings are either low enough or external enough that homesteading appears natural, even when unnecessary for subsistence. On the Moon, however, every cubic meter of habitable volume must be imported, assembled, poured, or hewn, and sealed and sustained indefinitely—reverting otherwise to its native, lethal state. Resources for construction and life support would not commonly be consumed building any but the densest of city configurations. Lunar society will be almost fully urban.

Local Materials

The overwhelming majority of lunar civilization will depend on indigenous manufacturing. Off-world imports will inevitably be rate limited. Common objects will be made locally, not because supplying them from space is impossible, but because it is impractical by comparison. A specialized computer might be imported, but the chair in which the programmer sits, the room in which she works, the snack she munches, the scrap paper on which she doodles, and the light by which she sees must all somehow be produced on the Moon.

This pervasively local origin of the bulk artifacts of lunar culture, with its corollary need to fashion a human environment from the bottom up, will excite and occupy designers for generations. Incidentally, it prevents us from picturing it accurately yet. But some conclusions are unavoidable. For example, simplicity will favor human-powered interior transportation: lunar city dwellers are more likely to ride bicycles than electromagnetically levitated monorails. Elaborate, high-consumption, centralized transit systems are likely to be justified only for interurban traffic as long as a few kilograms of magnesium extracted from lunar soil can provide both mobility and exercise unobtrusively.

FIGURE 5 Oven dish made from molten unrefined basalt. Corning used a familiar mold to cast this prototype in the 1960s. Lunar basalt will easily yield this same kind of black glass.

Ubiquitous products will be made as quickly, simply, and cheaply as possible from available resources. We can expect most surface buildings to be made primarily of lunar concrete reinforced with indigenous metal rebar or glass fibers, serving both structural and shielding needs with minimal industry. We can expect iron and alloys of titanium and aluminum to be used as commonly as are steel and plastic on Earth. And we can expect glass to be everywhere—among the easiest materials fabricated from lunar regolith, glasses of varying purities will comprise everything from tunneled cavern linings and architectural elements, to structural and optical fibers, to utilitarian objects (Figure 5). This might well mean a built landscape dominated by poured, masonry, fired, and vitreous materials. Again, these are not all the moon makes possible, but they will be the most expedient and are therefore likely to be the most common.

Hermetic Interior Environment

Lunar architecture must be an interior architecture. Heavily shielded havens are required during solar proton events (SPE, or flares). And cosmic rays (which Earth's atmosphere attenuates) irradiate the lunar surface semi-isotropically and continuously; the best long-term countermeasures are not yet known. It might turn out that, when not actually working in the space environment, people living there will voluntarily limit their unshielded exposure as much as possible. The image of miraculous, crystalline pressure domes scattered about planetary surfaces, affording a suburban populace with magnificent views of raw space, is a baseless, albeit persistent, modern icon. Such architecture would bake the inhabitants and their parklands in strong sunlight while poisoning them with space radiation at the same time.

However, the natural landscapes of the Moon's surface (Figure 6), home (Figure 7), and the nighttime

FIGURE 6 The Moon has its own, alien natural aesthetic. Up close, the lunar landscape is all about subtle shades of gray (from *Apollo 17*, EVA-3) (courtesy of NASA).

FIGURE 7 The most poignant sight in the near-side lunar landscape is Earth; it is small enough in the sky to be blocked by holding up one's gloved thumb (from *Apollo 8*) (courtesy of NASA).

antisolar sky full of bright stars will be especially attractive to human sensibility, just because they are natural. A lunar lifestyle that restricts recreational viewing to special times might evolve, perhaps spurring ritual behavior and special surface architectures for that purpose. Primarily or effectively subterranean then, lunar cities would be heavily top-shielded by concrete superstructures, by regolith overburden, or perhaps even by areas of untouched wilderness overlying tunneled city caverns. The natural and engineered planetary surface will be the single most important architectural interface on the Moon.

That interface must also include a continuous hermetic boundary capable of containing atmospheric pressure. Although the enclosures inside lunar cities can be structurally rather conventional, every square meter of the actual city wall must withstand over 100,000 newtons of force exerted by the air within it. Indeed, a regolith overburden with sufficient weight to counteract this pressure would exceed by many times the thickness required for safe radiation shielding alone. Thus pressurized, lunar cities will in effect be grounded spaceships; no other single feature argues more strongly for an economical, underground urbanism there.

Lunar life need not be troglodytic, though. Many ages of architecture, three of which provide contrasting programmatic examples, have been conceptually or explicitly interior. Roman urbanism was conceived and executed as a sequence of controlled volumes and views that regarded all of the natural landscapes it overran, from the Middle East to the British Isles, as alien. Turning inward away from natural features, Romans imposed the same planning schemes everywhere, creating their own universe around themselves, civilizing it with gods of their convenience, and arranging in it an ordered landscape of their choosing (Scully 1991) (Ward-Perkins 1974). Virtually all outdoor spaces in Roman cities functioned as urban "rooms" within which the public rituals of Roman society could be played out (Figure 8). The Roman invention of concrete allowed enclosed volumes of a truly public scale never before seen, and the legacy both of those volumes and of the street facades that surfaced and announced them remains alive today (Brown 1975; Ward-Perkins 1977). From Roman urbanism, we learn that necessarily interior lunar urbanism can nonetheless be grand and theatrical and promote civic life.

In the western medieval millennium following the Roman Empire, northern cold and frequent, local warfare among independent fiefdoms conspired to produce a genuinely interior environment. Often little more from the outside than a densely shielded pile, medieval architecture peered out of halls and chambers through tiny slits recessed in thick masonry walls (Stierlin 1966; Backes and Dolling 1969) (Figure 9). The intellectualism of Christianity encouraged introspection, and even ornament shrank largely off the stone architecture to cloak the people instead (Toy 1984; Stoddard 1972). To the east Byzantine piety and Islamic wonder developed enormous and lavishly ornamented interior spaces, now in the service of religious mystery rather than a secular civic public (MacDonald 1977) (Figure 10). Eventually belief inspired the West to refine its masonry construction technology to recover volume, stretching the old Roman basilica form upward and flooding it with light from above. Gothic religion came to sustain an interior architecture as potent, grand, and influential as anything Roman (Scully 1991; Grodecki 1977). From medieval architecture, we learn that interior lunar urbanism can use precious but dangerous external views sparingly (Figure 11), yet still be emotionally and spiritually inspiring.

Most familiar to modern citizens, 20th-century North American business evolved multiple types of "interior outdoors," ranging from the grand (Figure 12) to the mundane. The inclusive interior shopping mall in particular defines an architecture that maintains economic concentration despite the automotive dispersion of population. Wrapped within parking lots and structures, the mall's manufactured interior landscape entertains and stimulates temporary pedestrians along intersecting, faux-outdoor streets of retail facades (Figure 13). Roman-like consistency of style makes Toronto and San Francisco essentially the same (Portman and Barnett 1976). From capitalism-driven mall architecture we learn that interior lunar

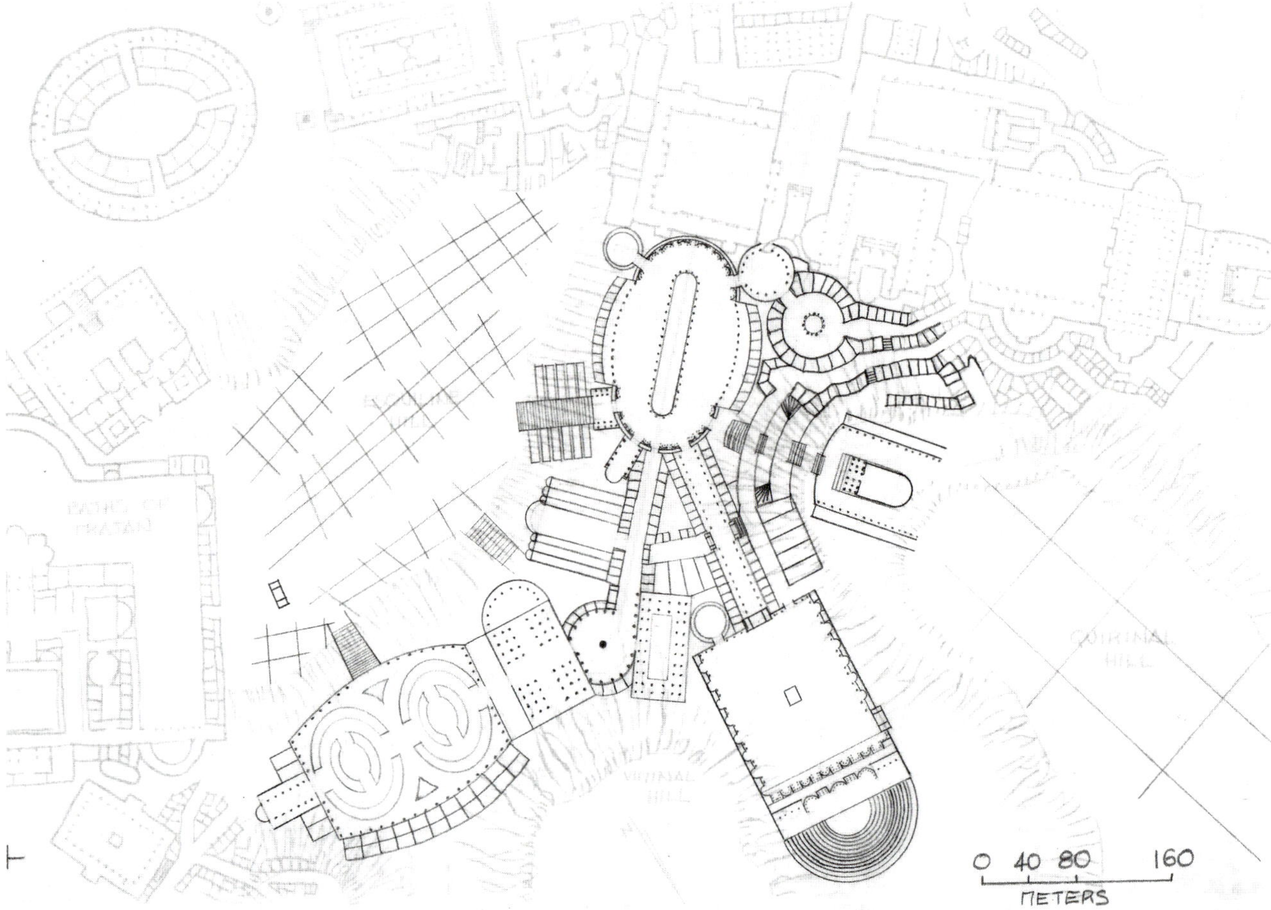

FIGURE 8 Roman architecture controlled the human environment so that even outdoor urban spaces became interiors. Study by the author for a forum complex in ancient Rome as it might have been built by Hadrian.

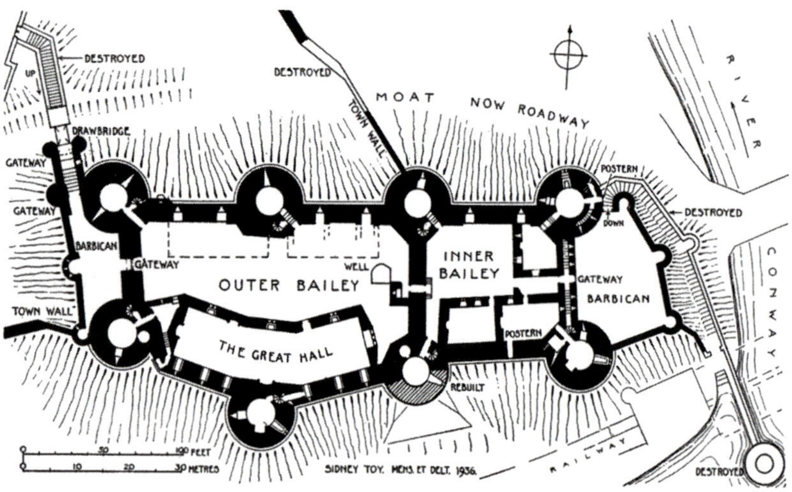

FIGURE 9 Medieval architecture interiors evoked protectionism. Conway Castle, built 1283–1287, view from the southwest and plan. (courtesy of Dover Publications, Inc., used with permission.)

urbanism can create an interior outdoors that is transient, adaptable, cheerful, and familiar.

Civic pride, protectionism, spiritualism, and commercialism, embodied in the well-known built expressions just briefly reviewed, will be among the old and new motives guiding lunar civic building. Referring eclectically to the rich human past, a pluralistic 21st-century lunar culture will embody its own aspirations in the public interiors it builds. But all types of lunar interiors will share uniquely lunar proportions.

Reduced-Gravity Proportions

Lunar architecture must accommodate a larger scale of human movement than we are used to on Earth. Although details await experience, the stride of a natural human gait in lunar gravity will be longer and rise higher. (Note that the side-to-side, bounding gait used by *Apollo* astronauts was likely caused by their rigid, bulky spacesuits rather than lunar gravity, and so old videos are not likely a helpful design guide.) Human

FIGURE 10 Blue Mosque in Istanbul converts interior space into a vision of heaven.

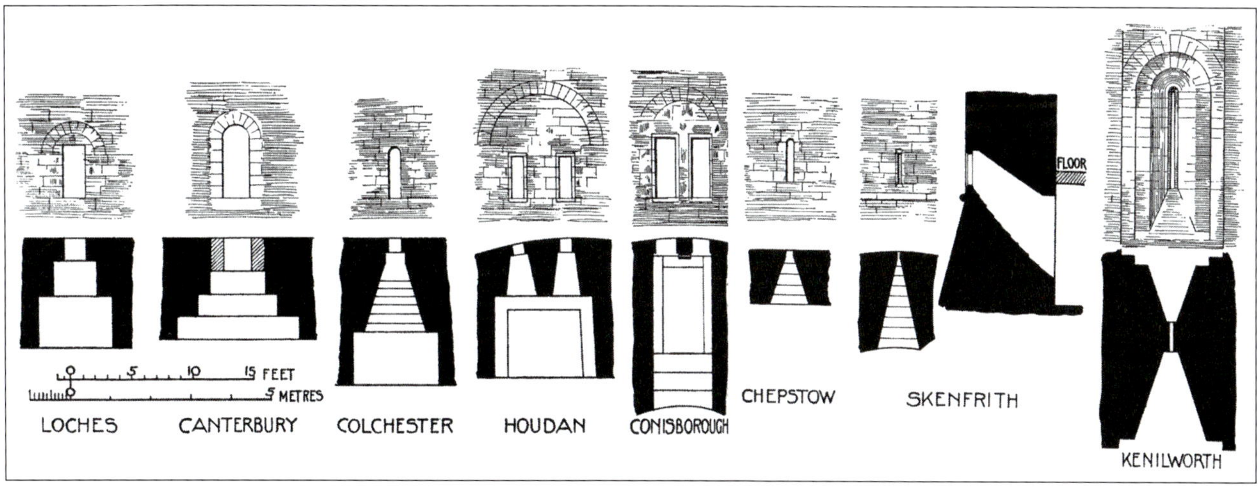

FIGURE 11 Hallway and basement windows from various medieval castles—a paradigm for radiation-limiting lunar fenestration? (courtesy of Dover Publications, Inc., used with permission.)

factors will have a new problem to solve: the proper dimensions of standard ceiling heights, doorways, and corridor widths.

In addition, supporting structures are always governed by economy, therefore, interior structures will be much more slender on the Moon than on Earth. For example, concrete columns, beams, and ribs can be less massive than we are used to. Lunar architecture will tend to be lighter and seem more expansive despite its pressurized closure, its exterior shielding, and its urban crowdedness. The reduced weight-bearing requirement, when coupled with the

FIGURE 12 Hyatt Regency in San Francisco: spare roof fenestration and artificial plants make soaring space an almost rain-forest-like faux outdoors.

FIGURE 13 Faux outdoors of the late-20th-century shopping mall offers climate-controlled streetscape. Underground lunar architecture need not be troglodytic.

easy availability of reduced iron in lunar regolith, might lead to a renaissance of cast-iron structural members.

Nonsterile Environment

Like terrestrial life, lunar life will be nonsterile. Human beings are elaborate ecological hosts, having evolved in the septic biosphere of Earth a vast web of symbiotic, commensal, and parasitic interactions with microscopic organisms. We cannot seriously plan a sterile off-world human ecology. Indeed, the biological "dirtiness" of humans poses a serious challenge to planetary protection considerations for places like Mars that might be hospitable to native or imported life (Sherwood 2004). Inside the hermetic confines of a lunar city, pathogen management will be a difficult but real problem, both for preventing hazardous infections and ensuring beneficial inoculations (Seppa 2005).

Lunar cities themselves will host life as well. For example, bacteria that metabolize by corroding metal, and can live in environments extreme in temperature, pressure, radiation, and toxics, will exploit niches in space. Healthy plant ecologies require rich soil cultures in any case, and so Earth life will accompany people to the Moon. Similarly, feral pets and research animals will eventually co-inhabit lunar cities. Although Norwegian rats, pigeons, and sparrows should be avoidable, it is unlikely that a secret, intentional release of fertile lab rodents will never occur. And it appears highly unlikely that off-world urbanism could grow without bringing along hardy stowaway organisms like the cockroach. Expansion will be too fast, and quarantine too porous, to preclude vermin from colonizing the Moon with us.

Irreplaceable Wilderness

Finally, the Moon must be a place of unprecedented demarcation between wilderness and human use. The ancient architectural form of the "town wall" will recur on the Moon—not to protect inhabitants from outside dangers, but rather to keep routine human activity from inexorably overrunning the native lunar landscape. Fragile though the Earth's biosphere might be in the face of modern development, its ultimate resilience has spoiled us. The encroachment of living things, relentless weathering, and finally even the implacable tectonics of Earth's

FIGURE 14 Once touched, lunar wilderness bears this mark forever, and so its character—and perhaps value—is irrevocably changed (courtesy of NASA).

geology render most signs of human action here into transience. Left alone, even denuded forests and ravaged desert ecostructures can eventually recover.

The inanimate lunar wilderness, however, is truly fragile and effectively irrecoverable. At least millions of years are required for micrometeorite "gardening" to remake just centimeters of regolith. The forces that reclaim strip mines and ruins on Earth simply do not exist on the Moon; the first trek through a pristine region of the Moon's unique "magnificent desolation" ruins its ineffable wilderness value practically forever (Figure 14). Surface exploration, strip-mining, construction, and recreation will be facts of human activity on the Moon. But so, sooner or later, will be human demands for utter preservation of untouched wild regions. Wilderness appreciation cannot be participatory on the Moon the same way it is on Earth. The solace and emotional renewal afforded by passively contemplating wilderness will induce radically new forms of urban design, specialized architectures, and art, to accommodate that human need on the Moon. The Moon's small size (about the same area as Africa) increases the urgency of preventing total surface development; however, it also creates a close horizon that can help to isolate areas visually.

CONCLUSION

A few salient characteristics of lunar architecture and urbanism grow directly out of facts as intrinsic to the Moon as weather is to the Earth. By accepting them as boundary conditions, we can project the built human lunar environment more aptly. Many types of people—including designers, authors, illustrators, engineers, explorers, leaders, and planners—will continue to be inspired by thinking about living in space and on the Moon. Space architects should inject as much realism as possible into their thoughts. Rigorous designs can be even more exciting and romantic than specious fantasies and are more helpful for shaping our collective vision of the future. By starting from a few accurate principles—that lunar urbanism will be densely populated, hermetic, interior, kinesthetically expansive, visually lightweight, and based on indigenous materials; that it will be nonsterile; and that lunar wilderness will become irreplaceably precious—those who plan can contribute to the responsible realization of one of the grandest projects ever imagined in human history.

References

Backes, M., and Dolling, R. (1969), *Art of the Dark Ages*, Abrams, New York, pp. 13–245.

Bower, B. (2005), "Cultivating Revolutions," *Science News*, Vol. 167, Feb. p. 88.

Brown, F. E. (1975), *Roman Architecture*, Braziller, New York, pp. 9–11.

De la Croix, H., and Tansey, R. G. (1975), *Gardner's Art Through the Ages*, 6th ed, Harcourt Brace Jovanovich, New York, pp. 43–47.

Ehricke, K. A. (1985), "Lunar Industrialization and Settlement—Birth of Polyglobal Civilization," *Lunar Bases and Space Activities of the 21st Century*, Lunar and Planetary Inst., Houston.

Grodecki, L. (1977), *Gothic Architecture*, Abrams, New York.

MacDonald, W. (1977), *Early Christian and Byzantine Architecture*, Braziller, New York, pp. 11–18; plates, 33–96.

O'Neill, G. (1976), *The High Frontier*, Bantam Books, New York, pp. 4–304.

Portman, J., and Barnett, J. (1976), *The Architect as Developer*, McGraw–Hill, New York, pp.4–191.

Scully, V. (1991), *Architecture: The Natural and the Manmade,* St. Martin's Press, New York, Chap. 5 and 7.

Seppa, N. (2005), "Good Exposure," *Science News*, Vol. 167, Jan., p. 68.

Sherwood, B. (1988), "Lunar Architecture and Urbanism," Second Conference on Lunar Bases and Space Activities of the 21st Century, Houston, TX, April; also NASA CP 3166, 1992.

Sherwood, B. (2004), "Progressive Protocol for Planetary Protection During Joint Human and Robotic Exploration of Mars," International Astronautical Congress, Paper IAC-04-IAA.3.7.2.10, Oct.

Sherwood, B. (2005), "Lunar Architecture and Urbanism," Society of Automotive Engineers, Paper 2005–01–2914; also *Transactions of the Journal of Aerospace*, Vol. 114-1.

Stierlin, H. (ed.) (1966), *Architecture of the World: Romanesque*, Benedikt Taschen Verlag, Lausanne, pp. 7–188.

Stoddard, W. S. (1972), *Art and Architecture in Medieval France*, Harper & Row, New York, pp. 354–360.

Toy, S. (1984), *Castles: Their Construction and History*, Dover, New York, pp. 50–140, 151, 153–229.

Ward-Perkins, J. B. (1974), *Cities of Ancient Greece and Italy: Planning in Classical Antiquity*, Braziller, New York, pp. 27–36; plates 39–86.

Ward-Perkins, J. B. (1977), *Roman Architecture*, Abrams, New York.

Wotten, H. (1624), *Principles of Architecture*, translated from the original by Marcus Vitruvius Pollio, *The Ten Books of Architecture*, *Royal Inst. of British Architects,* http://www.architecture.com [accessed 30 May 2009]. comBook 1, Chap. 3.

Wright, R. (2000), *NonZero*, Random House, New York, p. 47.

PART 4
Earth-Based Space Architecture

Not all space architecture is in space. Hostile-environment analogues, support facilities, and architecture inspired by space are all important aspects of the new field of space architecture.

Part 4 is divided into three sections: 1) simulators and analogues, two kinds of research or test facilities that approximate some of the conditions we have to deal with when designing for space and that offer important lessons and preparation; 2) support infrastructure, which is the Earth-based part of integrated space missions; and 3) "spin-off" architecture that brings the benefits of space systems, technologies, and perspectives directly back home to Earth.

SIMULATORS AND ANALOGUES

Space architects seek applicable lessons wherever they may be found. Two major sources are space mission simulators and terrestrial analogue environments like polar research stations and submarine facilities (extreme environments where psychological stress drives human behavior and high technology thrives).

In *Chapter 25*, "Human-Space-Mission Simulators," Susmita Mohanty et al. catalog 27 space mission simulators going back to the 1970s, describing in detail three major research campaigns by European, American, and Russian investigators to study system and human performance in simulated space mission conditions." Larry Toups et al., in "Antarctic Habitat Analogue" (*Chapter 26*), describe the design and rapid deployment of a prototype inflatable shelter at McMurdo Station in the antarctic. The technology demonstration informs lunar habitat concepts being developed now and new types of antarctic remote science stations. *Chapter 27*, "Halley VI Antarctic Research Station," concludes this section with Hugh Broughton's description of the design, construction, and deployment of the world's most advanced permanent polar research facility, analogous in many ways to a modular, relocatable, remotely assembled lunar base.

SUPPORT INFRASTRUCTURE

For a long time, space architecture will depend on Earth-based elements: prototype, factory, and test facilities; logistics and launch support facilities; and control and communications facilities. Installations that interface physically with space flight systems require special architectural consideration.

In *Chapter 28*, "Planetary and Lunar Surface Simulator," David Nixon et al. describe a European Space Agency research and test facility designed to support developmental testing of flight systems, like the robotic ExoMars rover, and to enhance public engagement in space exploration. Constance Adams and Georgi Petrov, in "Spaceport Design" (*Chapter 29*), analyze how spaceport facilities combine features of airports with unique considerations for orbital ascent and landing, including range safety and ground-support operations.

SPIN-OFF ARCHITECTURE

Learning how to meet the extreme requirements found in space architecture can teach us techniques for directly improving the quality of life on Earth. In space, the stuff of life cannot be taken for granted. Hence, we develop a "spaceman" mentality where our habitable world is a closed, tightly coupled system in which everything we need is valued and everything we do has immediate ramifications. This approach allows us to transcend the "cowboy" state of mind, where our world appears limitless and insensitive to our own actions. In "Space Architecture for the Mothership: Bringing It Home" (*Chapter 30*), Andreas Vogler and Arturo Vittori wrap up Part 4, and the book, as they show how architecture using technologies developed to take us to the planets can transform the human condition now.

25

human-space-mission simulators

SUSMITA MOHANTY
SUSAN M. FAIRBURN
BARBARA IMHOF
STEPHEN RANSOM
AND
ANDREAS VOGLER

INTRODUCTION

THIS CHAPTER SURVEYS THE HISTORY, current state, and known plans for ground-based facilities that simulate human space missions. Such simulators have been shown to have great value for medical, physiological, psychological, biological, and exobiological research, and for subsystem test and development, particularly of closed-loop life-support systems.

The survey is based on a detailed study conducted in 2006 for the European Space Agency (ESA) to support design of a Facility for Integrated Planetary Exploration Simulation (FIPES) (LIQUIFER 2006). The survey data served as reference material for development of the FIPES system architecture and ensured that FIPES system requirements reflected the design, experience, and lessons learned from the use of such facilities.

The survey provides a comparative survey of 27 human-space-mission simulators, essentially all of the simulators built and operated to date. It presents consolidated descriptions of each, classified into 1) site and purpose, 2) key technical data, 3) scientific and medical research functions, and 4) technology test and development functions. Then, selected summaries are presented of three sets of relatively recent simulation campaigns: one European, one American, and the other Russian-International. The chapter concludes with some key lessons learned from these campaigns.

USE OF SIMULATORS

Medical and psychological aspects are major issues for long-duration space missions—those beyond the half-year range baselined for the International Space Station (ISS). The isolated and confined nature of spaceflight, in particular for missions beyond low Earth orbit (LEO), along with its potential hazards, poses great challenges and risks related to human performance. These risks are influenced by boredom, crew autonomy and increased mutual reliance, extreme crowding, duration of flight, interpersonal tensions, mechanical breakdowns, poor communications, scheduling constraints and requirements, and sleep disturbances. To prepare for crewed exploratory missions to the Moon and Mars, studying these risk factors in a ground-based, simulated mission environment is an important step to minimizing eventual mission risk. Simulators provide a platform to conduct research in psychology, physiology, medicine, mission operations, human factors, and habitability. In addition, human-space-mission simulators play an important role in developing and testing mission system hardware and software. The surveyed simulators are listed next.

The *early simulators* are the following: Regenerative Life-Support Study, Apollo ground-based tests, Skylab Medical Experiments Altitude Test (SMEAT), Skylab Mobile Laboratory (SML), BIOS-1 and BIOS-2, Ben Franklin Underwater Research Laboratory, and Tektite-I and II Underwater Research Laboratory.

The *recent simulators* are the following: Isolation Study of European manned Space Infrastructure (ISEMSI-90), Experimental Campaign for European Manned Space Infrastructure (EXEMSI-92), Lunar Mars Life Support Test Project (LMLSTP), Biosphere-2, Canadian Astronaut Program Space Unit Life Simulation (CAPSULS), Human Behavior in Extended Spaceflight (HUBES-94), and Simulation of Flight of International Crew on Space Station (SFINCSS-99).

Next are the *present simulators*: Flashline Mars Arctic Research Station (FMARS), Mars Desert Research Station (MDRS), Bioregenerative Planetary Life Support Systems Test Complex (BIO-Plex), BIOS-3, Aquarius and NASA Extreme Environment Mission

Copyright © 2008 by the authors. Published by the American Institute of Aeronautics and Astronautics, Inc., with permission.

Operators (NEEMO), Concordia, and Closed Ecology Experiment Facilities (CEEF).

The only virtual simulator is the Interactive Mars Habitat.

Finally the *planned simulators* are the Integrated Human Exploration Mission Simulation Facility (INTEGRITY), Environmental Habitat (EnviHab), European Mars Analogue Research Station (EuroMARS), Australian Mars Analogue Research Station (MARS-Oz), and Integrated Planetary Simulator Studies.

These simulators were chosen because salient technical information was accessible for comparative study:

- Purpose and mode of use
- Development and adaptation for extended use
- Cost and method of funding and operation
- Public outreach
- Crew size
- Length of simulated missions
- Size of facility
- Facility power requirements

COMPARATIVE ANALYSIS

The comparative analysis is presented in a set of four tables. The first (Table 1) provides basic descriptive information and citations for each simulator. Table 2 summarizes research functions. Table 3 summarizes technological tests and developments. Table 4 allows further comparative analysis via synthesis of information into useful architectural and environmental metrics including pressurized volume, habitable volume, private volume, atmosphere, temperature, noise level, number of airlocks, number of viewports, and other valuable factors for assessing simulator conditions.

TABLE 1 Summary description of Earth-based human-space-mission simulators.

NAME	AGENCY	SITE	PURPOSE	REFERENCE
EARLY SIMULATORS				
Regenerative Life-Support Study	NASA	McDonnell Douglas, Huntington Beach, stationary, surface	Manned, 60- and 90-day closed environment test of ECLS system (4 crew: male)	NASA (year)
Apollo ground-based tests	NASA	NASA, stationary, surface	*Apollo* mission flight preparation in realistic environment (3 crew: male)	NASA (2008a)
Skylab Medical Experiments Altitude Test (SMEAT)	NASA	NASA Johnson Space Center, transportable/mobile, surface, seaborne	56-day, biomedical flight preparation tests for *Skylab* missions (3 crew: male)	NASA (1972); Skylab program office (1974); Belew (1977); Primeaux and Larue (1975); Biotechnology Jaco, (1973)
Skylab Mobile Laboratory (SML)	NASA	Mobile, transportable by Lockheed C-5A Galaxy	Facility to obtain data on *Skylab* crewmen 30 days before liftoff, within 1 hour after recovery, and for preflight physiological baseline	Primeaux and Larue (1975)
BIOS-1 and BIOS-2	Russia	Russia	Facility to assess the bioregenerative abilities of algae and long-term closed-system food production mission—365 days (BIOS-2: 3 crew)	Ash et al. (2006)

(Continued)

TABLE 1 (continued) Summary description of Earth-based human-space-mission simulators.

NAME	AGENCY	SITE	PURPOSE	REFERENCE
EARLY SIMULATORS				
Ben Franklin Underwater Research Laboratory	Private	Mobile (submarine)	Summer 1969 mission to monitor living in closed, confined environment (6 crew)	Belew (1977)
Tektite I and II Underwater Research Laboratories	NASA, NOAA, and private	U.S. Virgin islands stationary, underwater	Tektite II: 14-day mission to monitor physiological and psychological aspects of living at depth in relative isolation; first all-female crew (5 crew)	IMBP (2008); NASA (2008d)
RECENT SIMULATORS				
Isolation Study for the European Manned Space Infrastructure (ISEMSI-90)	ESA	NUTEC (Norwegian Underwater Technology Center), Norway, stationary, surface	28-day test of closed-loop system to examine psychophysiological themes, in addition to contamination and teleoperation factors (1990) (6 crew, male)	Vaernes et al. (1993); Vaernes (1996)
Experimental Campaign for the European Manned Space Infrastructure (EXEMSI-92)	ESA	DLR Cologne, Germany, NUTEC as project manager/ contractor, stationary, surface	Long-duration test of closed-loops system—60-day test; main aims were organization of the management of the "flight" and psychological/physiological experiments (1992) (4 crew: 3 male, 1 female)	Vaernes (1996); Bilchi (1995); Cazes et al. (1996); Gushin et al. (1996); Kasse et al. (1996); Milon et al. (1996)
Lunar–Mars Life Support Test Project (LMLSTP)	NASA	NASA Johnson Space Center, stationary, surface	Long-duration test of closed-loop system employing biological and physicochemical techniques (4 crew: 3 male, 1 female)	Permanent.com (2008a); NASA (2006); Lane et al. (2002)
Biosphere-2	Private	Arizona, USA, stationary, surface	Very long-duration ecological environment tests (8 crew: 4 male, 4 female)	Wikipedia (2008a); Biospheres.com (2008); MacCallum et al. (2004) Cohen and Titman (1996); Schrunk et al. (1999)
Canadian Astronaut Program Space Unit Life Simulation (CAPSULS)	CSA	DCIEM (Defense and Civil Institute of Environmental Medicine, Toronto, Canada, stationary, surface	Short-duration (7-day) test to gain firsthand experience on operational aspects of a typical space mission, with psychological objectives (6 crew, male) (1994)	Kass and Kass (1995); Canadian Space Agency (2008)
Human Behavior in Extended Spaceflight (HUBES)	IBMP, ESA, NUTEC, EAC	*Mir* Simulator, IBMP, Moscow, Russia, stationary, surface	Long-duration test of closed-loop systems, high-fidelity simulator; 135-day research into human-related effects (psychophysiological) of manned spaceflight (1994–95) (3 crew, male)	Vaernes (1996); Vaernes and Bichi (1995); ESA (1994)

(Continued)

TABLE 1 (continued) Summary description of Earth-based human-space-mission simulators.

NAME	AGENCY	SITE	PURPOSE	REFERENCE
Simulation of Flight of International Crew on Space Station (SFNICSS)	IBMP, NASDA, CSA	IBMP, Moscow, Russia, stationary, surface	Long-duration test of closed-loop systems to observe effects of multiple crews of mixed gender, multicultural, living and working in a simulated space environment for extended periods (4 weeks to 240 days), 1999 (multiple crews: male-female, multicultural)	Baranov (2001); Baranov et al. (2008); IBMP (2008); Space.com (2008b)
PRESENT SIMULATORS				
Flashline Mars Arctic Research Station (FMARS)	Mars Society	Devon Island, Canadian Arctic, stationary, surface	Long-duration isolation tests and geological field work (6 crew: male-female)	Mars Society (2008b); Wikipedia (2008c)
Mars Desert Research Station (MDRS)	Mars Society	Hanksville, Utah, USA, stationary, surface	Long-duration isolation tests and geological field work (6 crew: male-female)	Mars Society (2008c); Wikipedia (2008d)
Bioregenerative Planetary Life Support Systems Test Complex (BIO-Plex)	NASA	NASA Johnson Space Center, stationary, surface	Long-duration, biological and physicochemical life support technologies testing (4–6 crew)	Imhof (1997); Villarreal and Tvi (2001)
BIOS-3	Institute of Biophysics	Krasnoyarsk, Siberia, stationary, surface	Development of bioregenerative life-support systems for Moon/Mars missions (1–3 crew: male)	Wikipedia (2008c); Wheeler (1997); Allen (2008); Permanent.Com (2008b); Salishery et al. (1997)
Aquarius and NASA Extreme Environment Mission Operations (NEEMO)	NOAA, NASA	Stationary, underwater	Extreme environment, research habitat designed as a multi-objective mission analogue for long-duration spaceflight (6 crew: male-female)	NASA (2008b); NASA (2008c); Todd and Reagan (2008)
Concordia	ESA, IPEV, PNRA	Antarctic, stationary, surface	Medical, physiological, and psychological research for long-duration space missions (16 crew: male-female)	ESA (2008b, 2008c, 2008d); French Polar Inst; Wikipedia (2008b)
Closed Ecology Experiment Facilities (CEEF)	Institute for Environmental Sciences	Rokkasho, Japan, stationary, surface	Ecological tests (2 crew: male + animals)	Kornatsubara et al. (2005); Shinohara et al. (2005); ESTEC (2008)
VIRTUAL SIMULATORS				
Interactive Mars Habitat	Nexterra	Interactive Web site	Platform for design development offering virtual walk-throughs of a highly detailed 3D computer model of a Mars habitat, a pressurized rover and a greenhouse (no crew)	Interactive Mars Habitat (2008)

(Continued)

TABLE 1 (continued) Summary description of Earth-based human-space-mission simulators.

NAME	AGENCY	SITE	PURPOSE	REFERENCE
PLANNED SIMULATORS				
Integrated Human Exploration Mission Simulation Facility (INTEGRITY)	NASA	NASA Johnson Space Center, surface, stationary	Life support and habitat research, including simulated Moon or Mars terrain where astronauts can evaluate extraterrestrial surface tasks	Interactive Mars Habitat (2008); John Frassanito and Assoc. (2006); David (2008)
Environment Habitat (EnviHab)	DLR	DLR Cologne, stationary, surface	Space analogue for interdisciplinary and international research in medicine, psychology, and environmental sciences (8 crew)	DLR (2004)
European Mars Analogue Research Station (EuroMARS)	Mars Society	Iceland, stationary, surface	Geological and exobiological field work (6 crew)	Mars Society (2008b); Wikipedia (2008c)
MARS-Oz	Mars Society	Australia, transportable, surface	Geological field work (multiple crew)	Clarke and Wilson (2008); Mars Society (2008a)
Integrated planetary simulator studies	NASA	NASA Johnson Space Center, stationary, surface	Habitat development to support lunar exploration	—

TABLE 2 Scientific and medical functions targeted by the simulators.

SIMULATORS	SCIENTIFIC / MEDICAL INVESTIGATION									
	UNSPECIFIED PHYSIOLOGY	CARDIOVASCULAR	MUSCULOSKELETAL/ METABOLIC	NEUROPHYSIOLOGY	NUTRITION AND ENDOCRINOLOGY	HEMATOLOGY	MICROBIOLOGY	EXOBIOLOGY	PSYCHOLOGY	HUMAN FACTORS
EARLY SIMULATORS										
Regenerative Life-Support Study	✓	✓	✓	✓	✓	—	✓	—	✓	✓
Apollo ground-based tests	✓	✓	✓	—	—	✓	✓	—	—	—
Skylab Medical Experiments Altitude Test (SMEAT)	—	—	✓	✓	✓	✓	—	✓	—	—
Skylab Mobile Laboratory (SML)	✓	✓	✓	✓	✓	✓	✓	—	—	✓
BIOS-1 and BIOS-2	✓	—	✓	—	✓	—	—	—	—	—
Ben Franklin Underwater Research Laboratory	✓	✓	✓	—	✓	✓	✓	—	✓	✓
Tektite I and II Underwater Research Laboratories	✓	✓	✓	—	—	—	✓	—	✓	✓

(Continued)

TABLE 2 (continued) Scientific and medical functions targeted by the simulators.

SIMULATORS	SCIENTIFIC / MEDICAL INVESTIGATION									
	UNSPECIFIED PHYSIOLOGY	CARDIOVASCULAR	MUSCULOSKELETAL/ METABOLIC	NEUROPHYSIOLOGY	NUTRITION AND ENDOCRINOLOGY	HEMATOLOGY	MICROBIOLOGY	EXOBIOLOGY	PSYCHOLOGY	HUMAN FACTORS
RECENT SIMULATORS										
Isolation Study for the European Manned Space infrastructure (ISEMSI-90)	✓	✓	✓	—	✓	—	✓	—	✓	—
Experimental Campaign for the European Manned Space Infrastructure (EXEMSI-92)	✓	✓	✓	—	✓	✓	✓	—	✓	—
Lunar Mars Life Support Test Project (LMLSTP)	✓	—	✓	—	✓	—	✓	—	✓	✓
Biosphere-2	✓	—	✓	—	✓	—	✓	—	✓	✓
Canadian Astronaut Program Space Unit Life Simulation (CAPSULS)	✓	—	—	—	—	—	—	—	✓	✓
Human Behavior in Extended Spaceflight (HUBES)	✓	✓	✓	—	✓	—	—	—	✓	✓
Simulation of Flight of International Crew on Space Station (SFNICSS)	✓	✓	✓	—	✓	✓	✓	—	✓	✓
PRESENT SIMULATORS										
Flashline Mars Arctic Research Station (FMARS)	✓	—	—	—	—	—	—	—	✓	✓
Mars Desert Research Station (MDRS)	✓	—	—	—	—	—	—	—	✓	✓
Bioregenerative Planetary Life Support Systems Test Complex (BIO-Plex)	✓	—	✓	—	✓	—	✓	—	—	—
BIOS-3	✓	—	✓	—	✓	—	—	—	—	—
Aquarius and NASA Extreme Environment Mission Operations (NEEMO)	✓	✓	✓	—	—	—	—	—	✓	✓
Concordia	✓	—	—	✓	—	—	—	—	✓	—
Closed Ecology Experiment Facilities (CEEF)	✓	—	—	—	✓	—	—	—	—	—
VIRTUAL SIMULATORS										
Interactive Mars Habitat	—	—	—	—	—	—	—	—	—	✓
PLANNED SIMULATORS										
Integrated Human Exploration Mission Simulation Facility (INTEGRITY)	✓	—	✓	—	✓	—	✓	—	✓	✓
Environmental Habitat (EnviHab)	✓	—	—	—	—	—	—	✓	✓	—
European Mars Analogue Research Station (EuroMARS)	—	—	—	—	—	—	—	✓	—	—
MARS-Oz	✓	✓	✓	—	✓	✓	✓	✓	✓	✓
Integrated planetary simulator studies	—	—	—	—	—	—	—	—	✓	✓

TABLE 3 Technology test and development functions—Darker shading means greater certainty regarding investigations conducted.

SIMULATOR	TECHNOLOGY, TEST, AND DEVELOPMENT: PLANETARY ENVIRONMENT	ENVIRONMENTAL CONTROL AND LIFE SUPPORT	HYGIENE	WASTE MANAGEMENT	FOOD PREP AND STORAGE	SURGERY/ DENTISTRY	EVA	INTERIOR HABITAT ARCHITECTURE	INFOTAINMENT
EARLY SIMULATORS									
Regenerative Life-Support Study	—	✓	—	✓	✓	—	—	—	—
Apollo ground-based tests	—	✓	✓	✓	—	—	—	—	—
Skylab Medical Experiments Altitude Test (SMEAT)	—	✓	✓	✓	✓	—	—	—	(✓)
Skylab Mobile Laboratory (SML)	—	✓	—	—	—	✓	—	—	—
BIOS-1 and BIOS-2	—	✓	—	✓	—	—	—	—	—
Ben Franklin Underwater Research Laboratory	—	—	—	—	—	—	—	—	—
Tektite I and II Underwater Research Laboratories	—	✓	✓	✓	—	—	(✓)	—	(✓)
RECENT SIMULATORS									
Isolation Study for the European Manned Space Infrastructure (ISEMSI-90)	—	✓	✓	✓	—	—	—	—	—
Experimental Campaign for the European Manned Space Infrastructure (EXEMSI-92)	—	✓	✓	—	✓	—	—	—	—
Lunar Mars Life Support Test Project (LMLSTP)	—	✓	✓	✓	(✓)	—	—	—	—
Biosphere-2	—	✓	—	✓	✓	—	—	—	✓
Canadian Astronaut Program Space Unit Life Simulation (CAPSULS)	✓	—	—	—	—	✓	—	—	—
Human Behavior in Extended Space-Flight (HUBES)	—	✓	—	—	✓	—	—	✓	✓
Simulation of Flight of International Crew on Space Station (SFNICSS)	—	✓	✓	✓	✓	—	—	—	✓
PRESENT SIMULATORS									
Flashline Mars Arctic Research Station (FMARS)	—	—	—	—	—	—	(✓)	—	✓
Mars Desert Research Station (MDRS)	—	—	—	—	—	—	(✓)	—	✓
Bioregenerative Planetary Life Support Systems Test Complex (BIO-Plex)	—	✓	✓	✓	✓	—	—	✓	—
BIOS-3	—	✓	✓	✓	✓	—	—	—	—

(Continued)

TABLE 3 (continued) Technology test and development functions—Darker shading means greater certainty regarding investigations conducted.

SIMULATOR	TECHNOLOGY, TEST, AND DEVELOPMENT								
	PLANETARY ENVIRONMENT	ENVIRONMENTAL CONTROL AND LIFE SUPPORT	HYGIENE	WASTE MANAGEMENT	FOOD PREP AND STORAGE	SURGERY/ DENTISTRY	EVA	INTERIOR HABITAT ARCHITECTURE	INFOTAINMENT
PRESENT SIMULATORS									
Aquarius and NASA Extreme Environment Mission Operations (NEEMO)	—	✓	—	—	—	—	(✓)	—	—
Concordia	—	✓	✓	✓	—	—	—	—	—
Closed Ecology Experiment Facilities (CEEF)	—	✓	—	✓	✓	—	—	—	—
VIRTUAL SIMULATORS									
Interactive Mars habitat	—	—	—	—	—	—	—	✓	✓
PLANNED SIMULATORS									
Integrated Human Exploration Mission Simulation Facility (INTEGRITY)	—	✓	✓	✓	—	—	✓	—	—
Environmental Habitat (EnviHab)	—	✓	—	✓	—	—	—	—	—
European Mars Analogue Research Station (EuroMARS)	—	—	—	—	—	—	(✓)	—	—
MARS-Oz	—	—	—	—	—	—	(✓)	—	✓
Integrated planetary simulator studies	—	—	—	—	—	—	✓	—	—

SELECTED SIMULATOR HIGHLIGHTS

To explore differences in research focus, operations, and overall architecture, we now discuss three sets of relatively recent (1990–2000) simulation campaigns involving European, American, and Russian agencies and researchers:

- European: ISEMSI-90, EXEMSI-92, HUBES-94
- American: LMLSTP
- Russian/International: SFINCSS-99

The first campaign is a series of three isolation tests (ISEMSI-90, EXEMSI-92, HUBES-94) commissioned by ESA. The second is a series of isolation chamber tests (LMLSTP, 1995–1997) conducted by NASA Johnson Space Center (JSC). The third is a series of isolation studies (SFINCSS-99) managed by the Russian Federation State Research Center (SRC) and Institute for Biomedical Problems (IBMP) of the Russian Academy of Sciences, in collaboration with several ISS partner space agencies: National Aerospace Development Agency of Japan (NASDA), Canadian Space Agency (CSA), and the European Space Agency (ESA). Together these examples highlight 1) how useful space mission simulators are for understanding psychological, sociological, equipment, and operational issues before a mission is designed and 2) how critical the simulator architecture is for driving these issues and for representing actual mission conditions.

European Isolation Tests

ESA's Long Term Program Office conducted the ISEMSI-90, EXEMSI-92, and HUBES-94 experimental campaigns. These campaigns aimed to obtain information on psychological and physiological effects of long-term isolation and confinement of a small crew under conditions simulating those expected to exist in a space station. Details of each campaign's duration, location, crew size, simulator type, volume, and experiments are in Table 5.

TABLE 4 Technical data—Numbers in italics are estimates by the authors.

SIMULATOR	KEY TECHNICAL DATA														
	SIMULATION DURATION, DAYS	CREW SIZE	GENDER	DIMENSIONS, m	PRESSURIZED VOLUME, m^3	HABITABLE VOLUME, m^3	PRESSURIZED VOLUME PER CREW	HABITABLE VOLUME PER CREW	AREA, m^2	AREA PER CREW, m^2	INTERNAL PRESSURE, kPa	EXTERNAL PRESSURE, kPa	MAIN AIRLOCK VOLUME, M^3	NUMBER OF AIRLOCKS	NUMBER OF VIEWPORTS	COST
EARLY SIMULATORS																
Regenerative Life-Support Study	60	4	M	3.66 × 12.19	116	N/A	29	N/A	36	9	70	100	4.5	1 + 2 small (46 cm)	N/A	N/A
Apollo ground-based tests	7.6	3	M	—	8.8	N/A	2.9	N/A	—	0	34.5	>10	N/A	1	N/A	N/A
Skylab Medical Experiments Altitude Test (SMEAT)	56	3	M	6.1 × 2.3 (8.4 total height)	*81*	N/A	27	N/A	35.3	11.8	34.5	100	20	1 + 1 small (46 cm)	15	N/A
RECENT SIMULATORS																
Lunar Mars Life Support Test Project (LMLSTP)	91	4	3M + 1F	6.1 × 8.4	226.5	N/A	56.6	N/A	87.3	21.8	N/A	100	20	1	15	N/A
ISEMSI-90	28	6	M	—		118	N/A	19.7	—	—	—	—	—	—	—	—
EXEMSI-92	60	4	3M + 1F	2.2 × 6.6	28.7	94.4	7.2	23.6	—	—	—	—	—	—	—	—
HUBES-94	135	3	M	—		100	N/A	33.3	—	—	—	—	—	—	—	—
SFINCSS-99	240	4	M	—	100	—	25.0	—	—	—	100	100	—	—	—	—
Biosphere-2	700	8	4M + 4F	154 × 110 × 28	203,880	N/A	25,485	N/A	13,000	1,625	100	100	N/A	N/A	glass front	$150M USD

TABLE 4 (continued) Technical data—Numbers in italics are estimates by the authors.

SIMULATOR	KEY TECHNICAL DATA															
	SIMULATION DURATION, DAYS	CREW SIZE	GENDER	DIMENSIONS, m	PRESSURIZED VOLUME, m^3	HABITABLE VOLUME, m^3	PRESSURIZED VOLUME PER CREW	HABITABLE VOLUME PER CREW	AREA, m^2	AREA PER CREW, m^2	INTERNAL PRESSURE, kPa	EXTERNAL PRESSURE, kPa	MAIN AIRLOCK VOLUME, M^3	NUMBER OF AIRLOCKS	NUMBER OF VIEWPORTS	COST
PRESENT SIMULATORS																
Flashline Mars Arctic Research Station (FMARS)	14	6	M + F	8.2 × 7.3	352	N/A	58.7	N/A	100	16.7	100	100	*3.2*	1?	6?	$680k USD
Mars Desert Research Station (MDRS)	14	6	M + F	8.2 × 7.3	352	N/A	58.7	N/A	100	16.7	100	100	*3.2*	2 + 1 small	6	$573k USD
Bioregenerative Planetary Life Support Systems Test Complex (BIO-Plex)	425 (planned)	6		4.61 × 11.38 (5 pcs)	*900*	N/A	150	N/A	*400*	*66.7*	N/A	100	N/A	N/A	N/A	N/A
BIOS-3	180	3	M	15 × 8.4 × 2.5	315	N/A	105	N/A	126	42	N/A	100	N/A	N/A	N/A	N/A
Aquarius and NASA Extreme Environment Mission Operations (NEEMO)	10	6	M + F	4 × 13.7	172	11	28.7	1.8	*30*	5	250	250	N/A	1	*3*	N/A
Concordia	420	16	M + F	17 × 17 (2 pcs)	6,800	N/A	425	N/A	*1,200*	75	64.5	64.5	N/A	N/A	*40*	N/A
Closed Ecology Experiment Facilities (CEEF)	120	2	M	—	*1000*	*N/A*	*500*	*N/A*	*333*	*166.5*	*102*	*100*	*N/A*	*N/A*	*N/A*	*N/A*

TABLE 5 Comparison of ISEMSI, EXEMSI, and HUBES campaigns and experimental plans.

CAMPAIGN	ISEMSI-90	EXEMSI-92	HUBES-94
Duration (days)	28	60	135
Crew size (number, gender)	6 (all male)	4 (3 male, 1 female)	3 (all male)
Crew makeup/language	European/English	European/English	Russian/Russian
Habitable volume	118 m^3	94.4 m^3	100 m^3
No. of chambers	Four main chambers for living and working	Two main chambers allowed separate living and working	One chamber for both living and working
No. of psychological experiments	7	10	20
No. of physiological experiments	5	9	14
No. of operational experiments	11	12	1
Location	Norwegian Underwater Technology Center (NUTEC), Bergen, Norway	Institut für Flugmedizin (Institute for Flight Medicine), Deutsche Luft- und Raumfahrt (DLR), Cologne, Germany	Institute for Biomedical Problems (IBMP), Moscow

ISEMSI-90 During the four-week isolation period, a crew of six male scientists and engineers was asked to perform tasks accepted as real, meaningful, and similar to those performed on a space station. These tasks were selected to require collaboration among crew members, which was observed and measured according to performance criteria. The experimental purpose of ISEMSI was based on psychological and physiological experiments, as well as operationally oriented studies in contamination and teleoperation. The contamination studies included experiments in microbiological and chemical contamination, long-term evolution of the chemical composition of the chamber environment, and operational calibrations of the monitoring equipment. Noteworthy was a set of teleoperation experiments focused on telemedicine, teletraining, and telescience.

Psychological themes investigated during the experiment included 1) social interaction/communication, 2) autonomic nervous system, 3) crew performance, 4) cognitive demand, 5) subjective status, and 6) sleep and rhythmicity.

Physiological investigations included 1) psychoendocrinology, 2) immunology, 3) blood volume regulation, 4) body fluid balance, 5) lower-body negative pressure, and 6) heart rate and heart-rate variability.

ISEMSI was a scientific success, but identified several shortcomings in operational management, crew selection and training, and data management and storage. The next step was the EXEMSI-92 campaign.

EXEMSI-92 The reference mission scenario for EXEMSI was a space laboratory in LEO, where a visiting crew would service the orbital complex and perform payload operations. Similar to astronauts in a space station, four "emsinauts" lived and worked for 60 days in this closed-system simulation scenario. They had to provide for themselves and to work on several tasks specified by a team of scientists. The multinational crew comprised three men and one woman. The ground crew comprised one man and two women.

One important aim of the EXEMSI investigation was organization of the management of the "flight" as a "mini space mission." One of its main achievements was demonstrating that a major, useful project could be planned and executed in less than a year's time and on a moderate budget. In addition to achieving its scientific (physiological and psychological) objectives, EXEMSI provided valuable experience training chamber crew and ground control crew. It covered all phases of a mission, from the call for experiment proposals, through crew selection and training, integration and testing of the facility and its equipment, to daily monitoring and managing of the mission, and finally to post-isolation data collection and evaluation.

The main part of the study was carried out by four crew members in a vessel 2.2 m in diameter and 6.6 m in length (Figure 1). The complex served habitat, sanitary, galley, and stowage functions. A main lock in the front end served as entrance and experimentation area. Both chambers could be operated independently. The complex was augmented by a

(a) 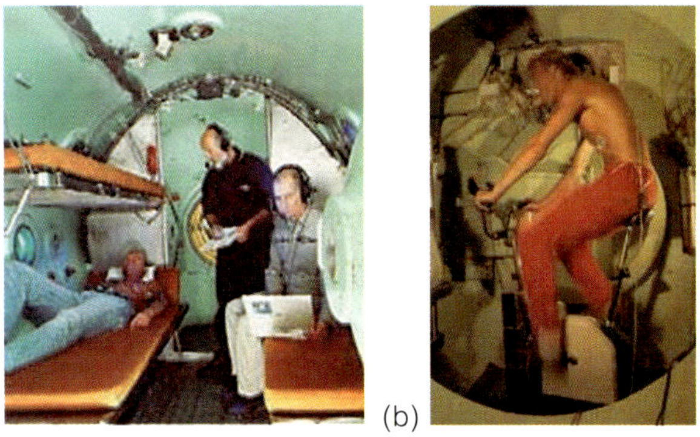 (b)

FIGURE 1 Titan Hyperbaric Chamber used for EXEMSI: a) interior view of living chamber and b) test subject exercising on a bicycle ergometer (courtesy of DLR).

custom-built steel container that served as the main laboratory facility. The life-support system regulated temperature, moisture, and percentage of oxygen in the air, analogous to spacecraft. Overall the environment was considered representative of a space station. The Campaign Control Center controlled and monitored the facility and experiment operations.

The numerous psychological objectives of the study included investigations into the crew's social behavior, interrelations, cohesion, efficiency, and team formation. The focus was a critical comparison of various test methods, leading to recommendations for their application in selection, training, and support for future studies of this kind. The study consisted of three phases:

- *Pre-isolation phase*: Initial individual and group assessments to understand motivation, characteristics, and styles of the crew members and the state of the crew and to make a prognosis for the behavior of the group and its members
- *Isolation phase*: Tests and observations to analyze crew behavior and group dynamics and to detect manifestations of stress
- *Post-isolation phase*: Final assessments and debriefings

Quantitative and qualitative individual and group tests were carried out. Direct methods (e.g., questionnaires and tests) as well as indirect methods (e.g., observations of behavior) were used. These had cognitive, affective-emotional, and social components.

Before isolation the crew members expressed strong confidence in the team and in their own personal capabilities, and the leadership of the commander seemed uncontested. Noting the relatively short experimental period and the absence of real risk, there was some question as to whether the crew's behavior (isolation effects and denial of anxiety) would emulate a real spaceflight situation or might even be dangerous. The confinement and isolation were experienced as the major stress factors, but during isolation there were no clear manifestations of stress. Crew members described themselves as a heterogeneous but harmonious group that was successful in their mission, maintaining cohesion by opposing external authority (management and ground crew). It was observed that the woman in the mixed-gender crew was never involved in conflicts and acted as a peacemaker. Again, it was questioned whether group cohesion would have persisted in a life-threatening crisis or even in a prolongation of the experiment.

The psychological state of the crew and their need for psychological support during prolonged isolation were observed using well-established methods employed during Soviet spaceflights. Communications between the commander and the crew interface coordinator were analyzed, and crew disposition was observed and analyzed for information about the process of group formation and the role of each crew member in this process. The key findings were adaptive changes in communication:

- Use of unplanned contacts and intensive contacts with a preferred ground crew member
- Resistance to penetration into crew life (increase in aggressive statements and self-justifications, reduction of report length and claims)
- Closing communication to "outsiders" by using a special code and decreasing discussion of problems

Daily monitoring also included a range of physiological variables. The study team developed a food and nutritional management system that provided online analysis of all available foods (macronutrients, water, minerals, and vitamins) and allowed for an accurate record of daily food intake. The findings revealed that eating and nutrition during the 60-day study were not problematic, although vitamins B_1 and B_6 were rather low and warranted supplements. Crew members rated food appreciation on a daily basis via questionnaires. They conveyed that food offered daily pleasure and social activity, which in turn was seen to potentially decrease stress caused by confinement and isolation. The crew rated food quality very high. Their satisfaction was attributed in part to having involved the crew directly in selection of the menu prior to the mission; in addition, there was large menu variety and abundance, which allowed crew members choice in food intake. The third step of the campaign was HUBES-94.

HUBES-94 The HUBES mission endeavored to extend findings through comparison and validation of psycho-sociological methods and tools for use in crew selection, training, monitoring, and in-orbit flight support. The primary purpose was to improve understanding of human-related effects of long-duration manned spaceflight. Thirty-one studies were carried out in the following areas: individual performance, group behavior, medicine, immunology, chronobiology, and nutrition. The experiment was preceded by a two-part training phase, with the second part included to minimize problems experienced on ISEMSI and EXEMSI with crew assignment procedures. The main study ran the complete 135-day period as planned, and was followed by a two-week post-testing period involving a range of psychological and physiological evaluations.

Prior relevant campaigns, including ISEMSI-90 and EXEMSI-92, were taken into account, particularly their findings on the psychology of group dynamics and on individual performance under isolation and confinement. HUBES focused on the process of how to select crew for a real, long-duration spaceflight (e.g., EUROMIR 95). Through the combined objectives of psychological methodology and crew selection, the mission aimed to improve knowledge about human requirements for extended space missions. EUROMIR 95 was selected as a model for the HUBES experiment and thus informed the duration, crew, schedule organization, workload, mission control, setup for communications and data processing, and facility layout. Modeling HUBES (surface) against EUROMIR 95 (space) allowed subsequent evaluation of the additional stress induced by microgravity, with respect to isolation and confinement.

The facility was set up like a small space station; the crew living quarters provided limited living space and comforts. The fidelity of the experiment to a *Mir* mission was maximized to the greatest degree possible. For example, organization of workstations into racks, division into working zones and leisure zones, and dimensions of the various zones were the same as *Mir*. Total internal volume of the simulator was 100 m^3, similar to *Mir*. Consideration was given to many aspects of habitability: traffic flow, privacy, meals and food preparation zone, workstation setup, lighting, sleeping area, and stowage. Operations and communication aspects were also configured to resemble a *Mir* mission to the greatest degree possible. Similarly, crew members were selected per the same medical, psychological, and professional qualifications as Russian and ESA astronauts. Because of the requirement to speak Russian, no European Union member state candidates were selected.

GLOBEMSI-96 DATABASE The three successive ESA campaigns involved a wide range of experiments in three broad categories: psychology, physiology, and operations. ESA implemented an elaborate database management system to document and analyze these experiments through a well-planned system called Global Analysis of Scientific Data from the European Manned Space Infrastructure (GLOBEMSI) (Tafforin and Bichi 1996). Measurements of psychological and physiological dependent variables were made via different methods including 1) individual questionnaires, computer tests, video recordings, audio recordings, etc. and 2) collection of blood, saliva and urine samples, echocardiography, body weight, skin temperatures, etc. Independent variables included experiment time, date, subject identification, and other data from operational experiments such as gas contamination, water composition, subjects' nutrition, etc.

The GLOBEMSI-96 database integrated all of these data to support analysis of interrelations among all of the variables. It was specifically formatted for global, multidisciplinary data analysis; designed to be fully compatible with both PC and Macintosh; ran on Microsoft Excel 5.0; and was developed with Visual Basic Applications. Data management was planned from the outset to be relational through the use of a conceptual data model, allowing navigation of the database. Based on the three campaigns, GLOBEMSI accumulated 89 experiments: 37 psychological, 28 physiological, and 24 on operational aspects. Sixty-five principal investigators performed experiments for 38 parameter units (metric, reaction time, pressure, etc.) from 31 procedures (video, computer test, monitoring, and telemedicine), and used 120 key words for data coding (e.g., cardiovascular, interaction, nutrition, interpersonal, etc.).

GLOBEMSI shows how a series of study protocols can be developed and carried out to achieve multiple multidisciplinary investigations. One of its key features was adoption of an ethological approach (a combination of laboratory and field science concerned with the connectedness of physiological and psychological systems involved in the adaptive process). This approach required the team to find ways to carry out the research while still preserving integral characteristics of living and working conditions (communications, crew tasks, volume, etc.) to allow analysis of spontaneous behaviors.

GLOBEMSI fulfilled its experimental design by varying the architecture of the simulation facilities, the duration of the isolation and confinement period, crew size, and cultural differences. From an experimental design perspective, GLOBEMSI involved a complex set of methods and approaches to evaluate a range of psychological and physiological parameters. GLOBEMSI parameters can be used as social or individual adaptive indexes in a space mission simulation to forecast less evident behavioral and psychophysiological disturbances during real space missions.

NASA Isolation Tests

The primary goal of the Lunar-Mars Life Support Test Project (LMLSTP), conducted from 1995 through 1997 at NASA JSC, was to test an integrated, closed-loop system that employed biological and physicochemical techniques for water recycling, waste processing, and air revitalization for human habitation (Lane et al. 2002). NASA evolved and upgraded the previous *Skylab* Medical Experiments Altitude Test (SMEAT) simulator project into the LMLSTP, which was built in the same vacuum chamber SMEAT used (Figure 2). As an analogue environment for long-duration missions, the conditions of isolation and confinement enabled studies of human factors, medical sciences (both physiology and psychology), and crew training.

LMLSTP was planned, designed, and operated by the Advanced Life Support Group at JSC (Table 6). It was based on the advanced life-support system concept that a human life-support system supplying food, water, and oxygen, open with respect to energy but closed with respect to mass, could operate indefinitely in space without resupply from Earth. This requires regenerative or recycling technologies, so as part of the technology development effort a series of tests was conducted.

FIGURE 2 LMLSTP test chamber at NASA Johnson Space Center (courtesy of NASA).

PHASE I The test was performed in August 1995. Its purpose was to obtain engineering and scientific data to demonstrate ability of a wheat crop to provide air revitalization for a human test subject for a 15-day period. The test also characterized crop growth and testbed performance of wheat grown from seed to harvest in the closed, controlled atmosphere of the growth chamber section of the VPGC using hydroponics and high-intensity light. The test chamber was divided into two sections: plant growth chamber and airlock, which was used as the human habitation chamber (Figure 3).

PHASE II The second human test to validate regenerative life-support technologies began on 12 June 1996. The ground-based LSSIF testbed facility was constructed using an existing 6-m vacuum chamber outfitted with life-support systems for four test subjects in a closed environment.

PHASE IIA The third human test began on 13 January 1997. It used hardware representative of the ISS, then scheduled for first element launch in 1998. As in phase II, this was an integrated test recycling the air and water required for four crew members. Results were combined with results from NASA Marshall Space Flight Center (MSFC) tests to evaluate and compare advanced life-support system technologies. The test was successfully completed on 14 March 1997.

PHASE III This was a 90-day test with four crew members and two test chambers interconnected by gaseous air exchange. The crew lived in the test facility; oxygen was augmented and carbon dioxide consumed by wheat growing in the VPGC. The phase III systems comprised plant growth systems, solid waste

TABLE 6 LMLSTP simulation campaigns ran in four phases.

CAMPAIGN	PHASE I	PHASE II	PHASE IIA	PHASE III
Year	1995	1996	Early 1997	Late 1997
Duration	15 days	30 days	60 days	90 days
Crew size	1	4	4	4
Test focus	Air revitalization system (ARS)	integrated physicochemical ARS, water recovery system (WRS), and thermal control system (TCS)	ISS integrated environmental control and life-support system (ECLSS)	Integrated physicochemical and biological ARS, WRS, and TCS test
Facilities	Variable-pressure growth chamber (VPGC)	Life-Support Systems Integration Facility (LSSIF)	LSSIF	LSSIF and VPGC

FIGURE 3 Montage of interior and exterior LMLSTP test images showing crew members performing various tasks (courtesy of NASA).

incineration system, ARS, product gas-transfer system, water supply and WRS, TCS, energy balance instrumentation, food system, crew accommodations, and facility support systems. A unique, microbe-based bioreactor designed and built at JSC was the primary water-recycling component. For the first time in this series of tests, an incinerator was used in the solid waste processing system to turn crew fecal matter into ash and gaseous carbon dioxide for uptake by the wheat.

A product-gas-transfer-system interface between the LSSTF and VPGC chambers was responsible for gaseous exchange between them, including correctly balancing oxygen and carbon dioxide for the crew in the LSSTF and the wheat crops in the VPGC. A side loop in this exchange was the use of carbon dioxide collected in the VPGC airlock from the incinerator for the wheat crop and the use of oxygen generated by the wheat for the incinerator.

Russian/International Isolation Tests

The main purpose of SFINCSS-99 was to obtain experimental data on implications of long-term isolation and confinement in simulated ISS conditions (Baranov et al. 2001). It focused on the ISS assembly phase, which required a highly demanding schedule that would impact both mental and physical crew-member health. Instruments used for health monitoring and prediction, work-ups, and medical care in flight would need to be substantially upgraded. The associated duties would require more flexible scheduling of operations and procedures, whereas *Mir* in-flight work/rest schedules were ironclad. This difference motivated keen scientific interest in learning how rigid and flexible work/rest schedules could affect the postflight status of crew members enduring long-term isolation and confinement.

The Institute for Biomedical Problems (IBMP) in Moscow was commissioned by the Russian Space Agency to research and develop methods, tools, and prototypes of biomedical equipment to enhance existing medical care for space crews. Integrated verification of these methods during a simulation of the most challenging period of the ISS program would allow a definitive conclusion regarding their potential for crew mental and physical health maintenance. Prior spaceflight simulation studies led to the hypothesis that adaptation of a group of humans subjected to closed, controlled environments would proceed in phases. Each phase (after approximately two months into the isolation and confinement period) would culminate in a transition where the body's functional controls readjusted. The associated strain in body functions would damage professional efficiency. Testing this hypothesis would significantly increase spaceflight safety.

The following objectives were set:

- Determine the effects of monotony from long-term isolation and confinement on space crew performance and human body functioning
- Compare the psychophysiological status of test subjects on fixed vs flexible work/rest schedules
- Identify regular patterns of bodily adaptation to the artificial climate of pressurized modules
- Observe the behavior of simultaneous crews during "rendezvous" missions as they tend to their assigned tasks while also interacting with each other
- Evaluate robustness and efficacy of the flight control systems and research facilities and procedures proposed for ISS utilization; framing of relationships among investigators

Other aspects important for ISS included collaboration and interaction of groups differing in culture, length of tour, and adaptation to spaceflight conditions in relatively isolated ISS modules. Experience from the ISS phase I program (joint missions on *Mir*) revealed national, cultural, professional, and other differences serving as a substrate for establishment of psychological subgroups. Such conflicts could lead to negative consequences for mission success and safety. Until SFINCSS-99, there had been no reports from simulating or investigating these phenomena. Thus the experiment investigated the psychology of interaction between several crews assigned independent mission tasks. Sources of psychological information included videotapes of crews, observations of relationships within and between the groups, voice communications with controllers, e-mails, checklists and questionnaires, and other systematized methods.

To determine the extent to which the monotony of isolation and confinement might impair space-crew performance and physical state, the SFINCSS-99 configuration monitored parameters that characterize efficiency and proficiency of operators on a space mission including their psycho-emotional, cardio-respiratory, biochemical, immunologic, hematological, morphobiochemical, and metabolic status, and immunity-microflora system.

The project included the following fields of research: intergroup and group behavior; individual and operator's psychology, behavior; clinical/physiological investigations; biochemical and immunologic assays; sanitary/hygiene and microbiological survey; biological studies; and engineering and technological experiments (included database verification). Strict medical criteria were used to select test subjects. A total of 21 test subjects (15 Russians, 3 Japanese, 1 German, 1 Canadian, and 1 French subject) participated. The test subjects were divided into two groups: long-duration *primary crews* and short-duration *visiting crews*:

- Three primary crews (Table 7): Group 1 (240 days), Group 2 (110 days), Group 3 (110 days)
- Four visiting crews: Groups 4 and 5 (7 days each), Group 6 (4 days), Group 7 (28 days)

The SFINCSS isolation campaign began with Group 1 entering EU-100 on 2 June 1999 for their 240-day isolation and ended when Group 7 completed their 28-day visiting confinement on 14 April 2000.

There were four visiting crews comprising three to four members of mixed sex and nationality and including experienced cosmonauts. Their visits ranged from 4–28 days. Groups 4 and 5 (three and four members, respectively) visited for seven days; the international Group 5 comprised members from France, Germany, Japan, and Russia. Group 6 comprised entirely medical professionals who devoted four days to comprehensive medical examinations of Group 1 to evaluate their fitness to continue the experiment. Group 7 was intended to test methods of counteracting adverse effects of underuse of the musculoskeletal apparatus. It consisted of four male subjects (three physicians and one engineer). Group 7 entered EU-100 for a 28-day stay, and for five days they interacted with members of Group 3.

Equipment and food supplies were provided once a month to each module. Crews were required to follow the work-rest schedule determined by mission managers and underwent medical checkups by a physician. Inside the modules, they exercised per a training program and administered other countermeasures utilized on long-term space missions. Psychological care was analogous to what had been typically provided to crews on *Mir*. The hermetic isolation chamber consisted of three interconnected modules with a total volume of 350 m^3 (about the size of three large, interconnected trailers) (Figures 4 and 5).

TABLE 7 Three primary crews of SFINCSS-99 incrementally tested alternative operations scenarios against a familiar baseline.

PRIMARY CREWS	DAYS	CHAMBER	NUMBER, GENDER, AGE	NATIONALITY	PROFESSION	REMARKS
Group 1 *Mir* schedule	240	EU-100 (100 m^3)	Four males 37–48 years	Russian	Three physicians, one engineer	Group 1 performed scheduled operations during eight working hours maximum, with energy expenditure of 2400–2600 kcal. They lived on a schedule typical for *Mir* crews. Physical exercise adhered to the *Mir* protocol; i.e., a four-day cycle.
Group 2 ISS assembly phase: fixed work–rest schedule	110	EU-37 (200 m^3) Mars Flyer module	Four males 27–45 years	One German (commander) Three Russians	Three physicians, one engineer	Group 2 entered EU-37 on day 28 after the beginning of the experiment and remained until day 138 (total: 110 days). The crew had a *fixed* work–rest schedule and was charged with more demanding and tedious tasks that cost a daily average of 3200–3600 kcal. The increased energy expenditure mimicked ISS assembly operations. This was achieved via an EVA simulator and exercise machines. The work day was up to 12 hours.
Group 3 ISS assembly phase: flexible work–rest schedule	110	EU-37 (200 m^3) Mars Flyer module	Three males, one female 27–37 years	Japanese, Austrian, Russian, and Canadian	Three physicians, one sports expert	Group 3 began in EU-37 22 days after the return of Group 2, and remained in the module for 110 days. The crew faced the same challenges and difficulties as its predecessor, but with a *flexible* work–rest schedule. The work–rest schedule was the result of two-step planning. The crew was given just a task list for each day, with the emphasis on chief duties, job quality, and time limit, and they were free to decide their own task approaches, priority setting, task timing and sacrifice of leisure time or other activities.

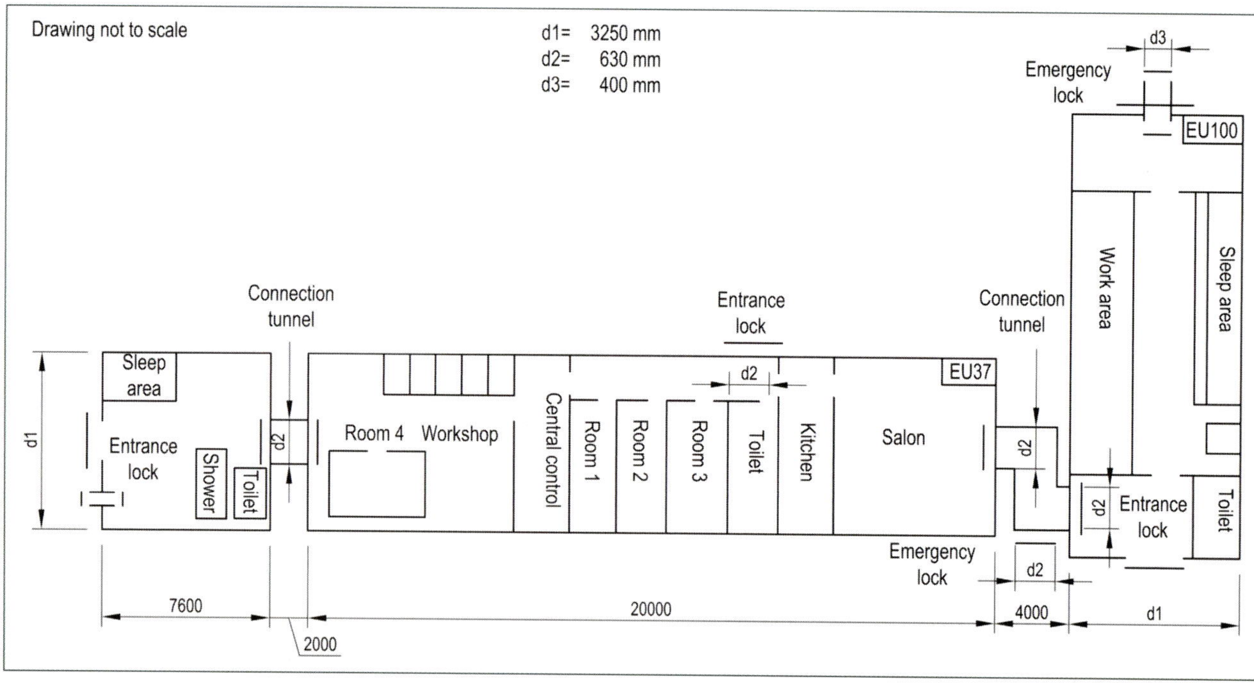

| **FIGURE 4** General layout of the SFINCSS habitat (courtesy of IBMP).

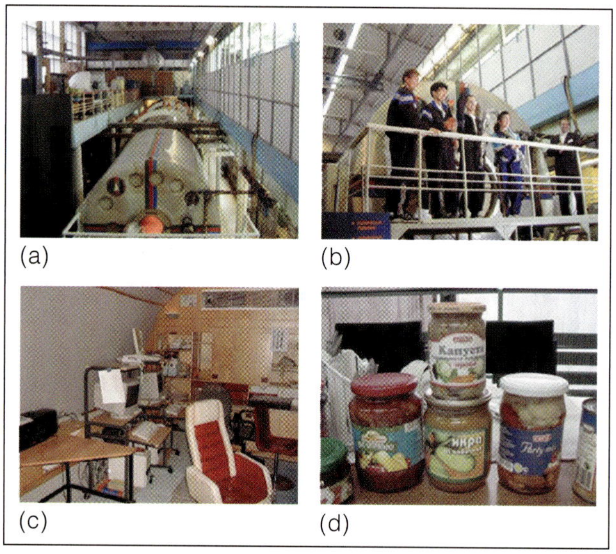

FIGURE 5 SFINCSS simulator: a) hyperbaric chamber exterior, b) international crew, c) interior working area, and d) representative food (courtesy of NASDA).

Group 1 (a Russian core crew) occupied a 100-m^3 module, and their sleeping areas were like those found on Mir, separated from the work area and each other by only a curtain. Groups 2 and 3 stayed in a 200-m^3 module that had individual sleeping cabinets and a large living area. During their isolation, crew members kept in contact with the outside world through telephone, Internet, and video.

The SFINCSS environmental parameters were

- Temperature: from 18–24 °C
- Humidity: 40–75%
- Airflow rate: 0.08–0.2 m/s
- Barometric pressure: 660–860 mm Hg
- Oxygen partial pressure: 140–200 mm Hg
- Carbon-dioxide partial pressure: up to 8 mm Hg
- Acoustic limit: 60 dBA
- Dust loading: 0.15 mg/m (daily average), maximum 0.5 mg/m^3
- Lighting: 50–300 lux (10 lux under a blue filter at night)

SFINCSS crews were supplied with thermostabilized, freeze-dried, sublimated, and partially dried and fresh foods, consistent with the diet of *Mir* crews. Personal hygiene items and techniques were provided to the crew to prevent skin or dental problems. A shower cabin was provided for full-body bathing. A washbasin and tap water were available for hygiene needs and dishwashing. Crew members were allowed personal hygiene items of their choice.

The experimental flow proceeded without incident until part way through Group 3's stay, when conflict arose between Groups 1 and 3 that resulted in the transfer hatch between two of the modules being locked. The situation persisted for just over one month and was only resolved through intercession by the project management team. Once it was resolved, the crews reopened the hatch and renewed contact.

LESSONS LEARNED FROM THE THREE CAMPAIGNS

This section integrates a few lessons from the three simulation campaigns just described (Vaernes 1996; Lane et al. 2002; Baranov 2001; Baranov et al. 2001). These lessons are based directly on findings, observations, commentaries, or conclusions documented in the publications cited and do not attribute causality. They are grouped into four categories that drive simulator planning, architecture, and design, and each is annotated with the simulation experiment where it arose. These lists indicate the nature and diversity of issues involved in simulator architecture.

Operations and Logistics

Simulation Planning, Operations, Data Management, Crew Selection, and Training Issues:

- Earth simulations permit study of scientific and operational aspects of space missions at a fraction of the cost of an in-orbit precursor mission (EXEMSI-92).
- Careful simulation planning is required to yield useful information and to maximize mission-relevant experience (EXEMSI-92).
- Large, representative, multidisciplinary campaigns require extensive planning for crew selection and training (ISEMSI-90).
- Further improvements are indicated in operational management and mission and simulation control (GLOBEMSI-96).
- Technologies, procedures, and protocols that enable crew autonomy are indicated for long-duration mission simulations (GLOBEMSI-96).

Human Factors

Physiological, Psychological, and Sociological Issues:

- Aerobic and resistive countermeasures provide a training stimulus when performed on separate days. Further work is needed to explore possible negative effects on strength training when aerobic exercise is performed on the same day as resistive training (LMLSTP).
- Increasing variety of exercise protocols, exercise devices, and addition of virtual reality headgear or other forms of entertainment during exercise can improve exercise compliance (LMLSTP).
- Physical exercise in the course of extended isolation and confinement is more than a countermeasure; it also provides psychological support, diluting the monotony inherent in this environment (LMLSTP).
- The most reliable instruments for psychological survey include group methods, nonobstructive tests, indirect instruments, and qualitative tools. The least reliable include strictly quantitative methods, self-evaluations, and standard debriefing techniques (EXEMSI-92).
- Role and responsibility clarification must be made clear and specific to an extraordinary degree to help mitigate misunderstandings and erroneous assumptions that naturally arise between groups that are physically and visually separated (EXEMSI-92).
- LMLSTP adopted lessons from other environments, which fell within three general themes: 1) further inclusion of the control room (CR) group as an integral part of the team and acknowledgment of their contribution; 2) mutual understanding of the daily issues facing the crew and CR groups and strategies for managing that interface; and 3) a reasonable work-rest schedule for individuals in the CR (LMLSTP).
- It is essential to allow dialogue, to take crew opinions and suggestions seriously, and to establish clear rules of confidentiality (EXEMSI-92).
- Taking data from past simulations and educating crews about teamwork styles for extended missions helped with familiarization and integration of team members (LMLSTP).
- In shorter-duration simulations (ISEMSI-90), interpersonal distances increased. On longer-duration simulations (EXEMSI-92), crew spent more time in private quarters (GLOBEMSI-96).
- Inclusion of family members is a powerful crew-support method for many reasons and a source of strength for the crew-member families (LMLSTP).
- Several problems are of a type documented earlier in spaceflights, including cross-cultural interactions between carriers of various national traditions. According to investigators from different participating countries, these problems stem from objective nuances in mentality, customs, and treatment of various situations (SFINCSS).
- Arrival of the visiting crew is a mighty stress and poses additional challenges to relations between groups (SFINCSS).
- Cowork of several crews very much alike by their composition but with different spaceflight or/and ground-based test experience begins with interadaptation. The presence of another crew is perceived as an external factor to fit to. On average, it takes three weeks to make relations between the crews comfortable (SFINCSS).
- Joint leisure hours and off-duty communication considerably expedite mutual adjustment, whereas cultural differences or poor knowledge

of a partner's language complicate informal communication and appreciably impede the process (SFINCSS).

Habitability

Habitat Design Issues (e.g., Configuration and Function, Stowage, Privacy, Group Interaction, Noise, Temperature, Lighting, Windows, Housekeeping):

- Internal simulator facility configurations and their evaluation should be developed in tandem with human performance disciplines (LMLSTP).
- Provide more acoustic insulation between low-noise (e.g., sleep, relaxation) and high-noise areas (e.g., equipment bay). Provide crew members with earplugs (LMLSTP).
- Loud equipment should be run at night and away from the sleep quarters. Equipment should be tested prior to a mission in an integrated operational setting, and predetermined noise level limits should be identified (LMLSTP).
- Locate hygiene facilities and trash away from dining and public gathering areas (LMLSTP).
- Include an operations panel inside the simulator to display problem and source so that the cause and seriousness can be ascertained by the crew. Audible alarm notifications on computer screens (individual and shared) could be very useful (e.g., when something goes wrong with a piece of equipment) (LMLSTP).
- Provide both dedicated use and multifunctional areas (LMLSTP).
- Ensure gradation from public to private areas (LMLSTP).
- For future crew quarters designs, the types of crew activities should be traded with the amount of space required to support those functions. In addition, the crew should be able to reconfigure their personal spaces within the limitations of exterior geometry (LMLSTP).
- Before the mission, allow each crew to thoroughly plan where to stow supplies and hardware not being routinely passed in and out. Establish a dedicated labeling system for these items (LMLSTP).
- Investigate and test a palette of materials, colors, and textures to determine their viability for various applications and locations (LMLSTP).

Environmental Control and Life Support

Issues Related to Air, Water, Food and Waste Management:

- Closed-loop life-support system development goals can only be achieved in a controlled test chamber because of the complex interactions between the sources and sinks. The complexity of these interactions rises as food preparation and waste processing systems are integrated into habitats (LMLSTP).
- Comprehensive air quality analysis is needed to determine whether preventative measures to limit pollution are effective, to ascertain if the ARS is capable of dealing with the pollutant load on a sustained basis, to detect any new sources of air pollution, and to judge whether the air is acceptable for crew health. Future research should focus on understanding risks that specific pollutants pose to crew health and then developing analyzers capable of addressing those risks using a minimum of resources (LMLSTP).
- A food and nutritional management system is a powerful tool to permit optimal management of food and eating onboard, while also allowing online analysis of crew-member nutritional status so that food intake and supplements can be adjusted as needed (EXEMSI-92).
- Direct involvement of the crew in menu planning and selection, as well as menu variety and surplus, allows crew members choice of food intake and contributes to overall food and mission satisfaction (EXEMSI-92).
- Comprehensive research is needed in food processing and preparation in an enclosed environment. A menu developed from a basic crop list can be acceptable for a crew for 10 days and meet most of the nutritional requirements; however, it will be a very labor-intensive diet with excessive waste (LMLSTP).
- Contamination study findings for one simulator experiment revealed that although there were established rules for hygiene, the rules were not followed. Findings showed that the disinfectants used were not effective for eliminating microbial growth, the crew lacked training in environmental hygiene, and the crew did not appoint anyone as responsible for hygiene matters (EXEMSI-92).
- There is a need for clear definition of wet and dry trash. It is also important that each crew member understands the difference and be trained on how to handle both kinds of trash (LMLSTP).
- The galley should be supplied with a trashcan with a foot control for its lid. A lid that pivots open easily would also be acceptable. This helps ensure sanitary conditions when the trashcan is used during meal preparation (LMLSTP).
- Address and eliminate odor issues resulting from trash and crew-member hygiene (EXEMSI-92).

FUTURE PLANS

ESA and IBMP plan to conduct a new series of long-duration space mission simulations called Mars500 (ESA 2008a; space.com 2008a), leading up to a final, unprecedented long-duration Mars mission simulation of nearly 520 days. This involves a crew of six living and working in a lattice of six interconnected, hermetically sealed modules in a Moscow laboratory. The Mars500 campaign attempts to reflect the major phases of a mission to Mars (e.g., transit time, surface stay and exploration, communication lag, autonomous decision making, limited consumables). The mission scenario simulates an opposition-class mission profile: outbound flight (250 days), Mars surface operations (30 days), and return flight (240 days). This is the most ambitious duration ever attempted in a ground-based simulator and could set a new benchmark for fidelity, extremity, and insight into human space missions. |

References

Allen, J. (2008), Russian Bios Project," http://www.biospherics.org/russia.html [retrieved April 2008].

Ash, S., et al. (2006), "Technical Proposal for Commercialisation and Space Tourism," International Space University Strasbourg, http://mss02.isunet.edu/projects/tp1/TP1_LiteraryReview.pdf, [retrieved March 2006].

Baranov, V. M. (ed.) (2001), *Simulation of Extended Isolation: Advances and Problems,* Firm SLOVO, Moscow.

Baranov, V. M., et al. (2001), "Project SFINCSS-99—Simulation of a Flight of International Crew on Space Station," Russian Federation State Research Center—Inst. for Biomedical Problems, Report, Moscow, Russia. *Simulation of Extended Isolation: Advances and Problems,* Moscow, Firm SLOVO, Moscow, pp. 17–24.

Belew, L. F. (ed.) (1977), "Skylab, Our First Space Station," NASA SP-400, Feb. p. 19.

Bichi, A. (1995), "The Recent ESA Activity in Human Requirements Investigations," Society of Automotive Engineers, Paper 951511.

Biosphers.com (2008), "Biosphere 2: The Experiment," http://www.biospheres.com/experimentchrono1.html [retrieved April 2008].

BioTechnology, Inc. (ed.) (1973), "Skylab Medical Experiments Altitude Test (SMEAT)," NASA TM X-58115, Oct.

Canadian Space Agency (2008) "Canadian Astronaut Program Space Unit Life Simulation (CAPSULS)," http://www.space.gc.ca/asc/eng/astronauts/osm_capsuls.asp [retrieved April 2008].

Cazes, C., Rosnet, E., Bachelard, C., Le Scanff, C., and Rivolier, J. (1996), "Group Dynamics During the EXEMSI Isolation Study. Experimental Campaign for the European Manned Space Infrastructure," *Adv. Space Biol. Med.*, Vol. 5, pp. 245–262.

Clarke, J., and Willson, D. (2008), "The MARS-OZ Analogue Mars Research Station: a Status Report," http://www.marssociety.org.au/amec2003/proceedings/08-Jon_Clarke_1.htm [retrieved April 2008].

Cohen, J. E. and Tilman, D. (1996), "Biosphere 2 and Biodiversity: The Lessons So Far," *Science*, Vol. 274, Nov. 15 pp. 1150, 1151.

Concordia Base (2008), "Concordia: A New Permanent Continental Station in Antarctica," http://www.concordiastation.org/ [retrieved April 2008].

David, L. (2008), "World Space Congress: Daily Updates: NASA JSC looking for INTEGRITY," Space.com http://www.space.com/news/wsc_astronotes-1.html [retrieved April 2008].

DLR, (2004), EnviHab Workshop, DLR, Cologne-Porz, Sept.

ESA (1994), Press Release," No. 24-1994, 22 Aug.

ESA (2008a), "Announcement of Opportunity for Research in the Context of ESA's Participation in the Mars500 Programme AO-07-Mars500," http://www.spaceflight.esa.int/users/downloads/ao2007/AO-07-Mars500.pdf [retrieved April 2008].

ESA (2008b), "Concordia Station," www.spaceflight.esa.int/concordia/ [retrieved April 2008].

ESA (2008c), "Mission to Mars via Antarctica," http://www.esa.int/esaCP/SEMBZA8A9HE_Life_0.html [retrieved April 2008].

ESA (2008d), "Preparing a Human Mission to Mars via Antarctica and Toulouse," http://www.esa.int/esaCP/SEMOS4T1VED_index_0.html [retrieved April 2008].

ESTEC (2008), Study on the Survivability and Adaptation of Humans to Long-Duration Interplanetary and Planetary Environments," ESTEC/Contract No. 14056/99/NL/PA; also "European Initiatives in Advanced Life Support Developments for Humans in Interplanetary and Planetary Environments," HUMEX-TN-003, http://www.ecls.esa.int/ecls/attachments/ECLS/Perspectives/humex/tn3.pdf [retrieved April 2008].

Interactive Mars Habitat (2008), www.exploremarsnow.org [retrieved April 2008].

French Polar Inst., "Concordia Project Information on the Surface Transport System Set up for Servicing the Dome C Site," Plouzane, France.

Gushin, V. I., Kolinitchenko, T. B., Efimov, V. A., and Davies, C. (1996), "Psychological Evaluation and Support During EXEMSI. Experimental Campaign for the European Manned Space Infrastructure," *Adv. Space Biol. Med.*, Vol. 5, pp. 283–295.

IMBP (2008) "SFINCSS-99: Isolation 240 Days," http://www.imbp.ru/webpages/engl/SFINCSS-9/sf_pr03_e.html [retrieved April 2008].

Imhof, B. (1998), "Bioplex, A Future Life on Extra-Terrestrial Planets," Third International Conference of Biosphere Science and Life Support Systems, Jan.

John Frassanito and Assoc. (2006), "Computer Renderings for NASA by John Frassanito & Associates," http://www.frassanito.com/integrity/ [retrieved March 2006].

Kass, J. R., Ellmers, F., and Schiemann, J. (1996), "Operational Evaluation of the EXEMSI Project. Experimental Campaign for the European Manned Space Infrastructure," *Adv. Space Biol. Med.*, Vol. 5, pp. 357–373.

Kass, R., and Kass, J. (1995), "Group Dynamics Training for Manned Spaceflight and the CAPSULS Mission: Prophylactic Against Incompatibility and Its Consequences?" *Acta Astronautica,* Vol. 36, No. 8–12, Oct.–Dec., pp. 567–573.

Komatsubara, O., et al. (2005), "Estimation of Energy Requirements of Eco-nauts in the Closed Ecology Experiment Facilities (CEEF)," Society of Automotive Engineers, Paper 2005-01-3004, July.

Lane, H. W., Sauer, R. L., and Feeback, D. L. (eds.) (2002), *Isolation: NASA Experiments in Closed-Environment Living: Advanced Human Life Support Enclosed System,* American Astronomical Society, San Diego, CA. Vol. 104, Science and Technology Series, pp. 131–139.

LIQUIFER Systems Group, Architecture and Vision, M+W Zander, Fibre Design, Inc. (2006), "Summary Report – LSG-FIPES-SR-2006," ESA/ESTEC Contract No. 19397/05/NL/Sfe, Noordjwik, The Netherlands, Nov.

MacCallum, T., Poynter, J., and Bearden, D. (2004), "Lessons Learned from Biosphere 2: When Viewed as a Ground Simulation/Analogue for Long Duration Human Space Exploration and Settlement," Society of Automotive Engineers, Paper 2004-01-2473, July.

Mars Society (2008a), "Mars-Oz: A Simulated Mars Base in the Arkaroola Region," http://www.marssociety.org.au/marsoz.php [retrieved April 2008].

Mars Society (2008b), "Mars Society Flashline Mars Arctic Research Station," www.marssociety.org/arctic/index.asp [retrieved April 2008].

Mars Society (2008c), "Mars Society Mars Desert Research Station (MDRS)", http://www.marssociety.org/mdrs/ [retrieved April 2008].

Miller, J. W. and Koblick, I. G. (1995), *Living and Working in the Sea,* Second ed., Five Corners Publications.

Milon, H., Decarli, B., Adine, A. M., and Kihm, E. (1996), "Food Intake and Nutritional Status During EXEMSI. Experimental Campaign for the European Manned Space Infrastructure," *Adv. Space Biol. Med.*, Vol. 5, pp. 79–91.

NASA (1972), "Skylab Medical Experiments Altitude Test (SMEAT)," NASA, Rep. TMX-58115, Oct.

NASA (2002), "Regenerative Life Support Study," *Isolation: NASA Experiments in Closed-Environment Living*, http://lsda.jsc.nasa.gov/books/ground/chambers.pdf.

NASA (2006), "NASA's Advanced Life Support Program," http://advlifesupport.jsc.nasa.gov/lmlstp.html, [retrieved March 2006].

NASA (2008a), "Apollo Ground Based Tests," *Isolation: NASA Experiments in Closed-Environment Living*, http://lsda.jsc.nasa.gov/books/ground/1.2Overview.pdf [retrieved April 2008].

NASA (2008b), "From the Ocean Depths to Deep Space," www.nasa.gov/missions/shuttle/neemo.html [retrieved April 2008].

NASA (2008c), "NASA NEEMO: Behind the Scenes," http://spaceflight1.nasa.gov/shuttle/support/training/neemo/facilities.html [retrieved April 2008].

NASA (2008d), "Tektite-1 Food Developments," http://history.nasa.gov/SP-202/sess2.3.htm [retrieved April 2008].

Permanent.com (2008a); "Biosphere 2, Mission 1 CELSS Project," http://www.permanent.com/s-bio2m1.htm [retrieved April 2008].

Permanent.com (2008b), "Russian CELSS Studies," http://www.permanent.com/s-bios3.htm [retrieved April 2008].

Primeaux, G. R., and LaRue, M. A. (1975), "Skylab Mobile Laboratory," NASA TN D-8028, July.

Salisbury, F. B., Gitelson, J. I., and Lisovsky, G. M. (1997), "Bios-3: Siberian Experiments in Bioregenerative Life Support," *BioScience*, Vol. 47, No. 9, pp. 575–585.

Schrunk, D., Sharpe, B., Cooper, B., Thangevelu, M., and Bond, P. (1999), *The Moon. Resources, Future Development and Colonization,* Wiley, Chichester, England, U.K., Appendix J, pp. 316, 317.

Shinohara, M., et al. (2005), "Workloads and Environment of Closed Habitation Experiments in CEEF (Closed Ecology Experiment Facilities) and Physio-Psychological Changes in Habitants (Econauts) During the Experiments, Institute for Environmental Sciences," Society of Automotive Engineers, Paper 2005-01-3005, July.

Skylab Program Office (1974), "Skylab Lessons Learned," NASA TM X-64860.

Space.com (2008a), "Europe to Join 500-Day Mock Mission to Mars," http://www.space.com/news/070402_mars500_esa.html [retrieved April 2008].

Space.com (2008b), "The Real World—Moscow Style," http://www.space.com/news/spacestation/isolation_russia_000412.html [retrieved April 2008].

Tafforin, C., and Bichi, A. (1996), "Global Analysis of Scientific Data from the Three Experimental Campaigns for European Manned Space Infrastructure," Society of Automotive Engineers, Paper 961442.

Todd, B., and Reagan, M. (2004), "The NEEMO Project: A Report on How NASA Utilizes the 'Aquarius' Undersea Habitat as an Analogue for Long-Duration Space Flight," American Society of Civil Engineers. p. 103.

Vaernes, R. J. (1996), "Lessons Learned from ISEMSI and EXEMSI. Isolation Study for the European Manned Space Infrastructure. Experimental Campaign for the European Manned Space Infrastructure," *Adv. Space Biol. Med.*, Vol. 5, pp. 375–396.

Vaernes, R. J., and Bichi, A. F. (1995), "135 Days in Isolation and Confinement: The HUBES Simulation," Society of Automotive Engineers, Paper 951512, July.

Vaernes, R. J., Schonhardt, A., Sundland, H., and Thorsen, E. (1993), "General Description of ISEMSI (Isolation Study for European Manned Space Infrastructures: Technical Scenario, Selection of Candidates, Operational Aspects, and Organization," *Adv. Space Biol. Med.*, Vol. 3, pp. 35–58.

Villarreal, J. D., and Tri, T. O. (2001), "Bioregenerative Planetary Life Support Systems Test Complex (BIO-Plex): Progress to Date," Society of Automotive Engineers, Paper 2001-01-2317, July.

Wisconsin Center for Space Automation and Robotics (2008), "Integrated Human Exploration Mission Simulation Facility (INTEGRITY)," http://wcsar.engr.wisc.edu/integrity.html [retrieved April 2008].

Wheeler, R. M. (1997), "Bios-3 Project in Krasnoyarsk, Russia," *BioScience*, available online through PubMed, http://www.ncbi.nlm.nih.gov/pubmed//11538718.

Wikipedia (2008a), "Biosphere 2 Wikipedia," http://en.wikipedia.org/wiki/Biosphere_2 [retrieved April 2008].

Wikipedia (2008b), "Concordia Wikipedia," http://en.wikipedia.org/wiki/Concordia_Station [retrieved April 2008].

Wikipedia (2008c), "FMARS Wikipedia," http://en.wikipedia.org/wiki/Flashline_Mars_Arctic_Research_Station [retrieved April 2008].

Wikipedia (2008d), "MDRS Wikipedia", http://en.wikipedia.org/wiki/Mars_Desert_Research_Station [retrieved April 2008].

Wikipedia (2008e), "Russian BIOS-3," http://en.wikipedia.org/wiki/BIOS-3 [retrieved April 2008].

Bibliography

Bluth, B. J., and Helppie, M. (1987), "Soviet Space Stations as Analogues, Second Edition, with MIR Update," NASA Grant NAGW-659, May.

Dudley-Rowley, M., Gushin, V., and Gorry, T. (1999), "A Social States Index for Multi-National Crews Co-Contained in the ISS Simulator, Moscow, Russia," Society of Automotive Engineers, Paper 1999-01-2101, July.

Dudley-Rowley, M., Nolan, P., Bishop, S., Farry, K., and Gangale, T. (2000), "Ten Missions, Two Studies: Crew Composition, Time, and Subjective Experience in Mars-Analogue Expeditions" *On to Mars: Colonizing a New World*, edited by R. Zubrin and F. Crossman, Apogee Books, Burlington, Ontario, Canada, p. 1–18.

Micheels, K. A. (2004), "Lessons Learned: The Design, Fabrication and Deployment of the Flashline Mars Arctic Research Station," Society of Automotive Engineers, Paper 2004-01-2271, July.

Micheels, K. A. (1999), "The Mars Surface Habitat: Issues Derived from Design of a Terrestrial Polar Analogue," Society of Automotive Engineers, Paper 1999-01-2140, July.

Mohanty, S., Fairburn, S., Imhof, B., Ransom, S., and Vogler, A. (2008), "Survey of Past, Present and Planned Human Space Mission Simulators," Society of Automotive Engineers, Paper 2008-01-2020, July.

Stewart, R. A. (1988), "Habitability and Behavioral Issues of Space Flight," *Journal: Small Group Research*, Vol. 19, No. 4, pp. 434–451. http://sgr.sagepub.com/cgi/content/abstract/19/4/434 [retrieved April 2008].

Thirsk, R., Williams, D., and Anvari, M. (2007), "NEEMO 7 Undersea Mission," *Acta Astronautica*, Vol. 60, No. 4–7, pp. 512–517, http://adsabs.harvard.edu/abs/2007AcAau..60..512T [retrieved April 2008].

Vaernes, R. J., Baranov, V. M., Demin, Y. P., and Stepanov, V. A. (1995), "HUBES Executive Summary," Norwegiean Underwater Technology Center (NUTEC), Inst. for Biomedical Problems (IBMP), NUTEC Rept. 10-95, Moscow.

26

antarctic habitat analogue

LARRY TOUPS
DAVE CADOGAN
AND
CRAIG SCHEIR

INTRODUCTION

ANALOGUES AND MOCK-UPS ARE ESSENTIAL INFORMATION tools for designing long-duration habitats for the Moon and Mars. They provide a way to demonstrate structural concepts, concepts of operations, and internal layouts. This unique approach to architecture makes sense where there is no precedent. By building a developmental series of full-scale prototypes subjected to field conditions, features and functions can be validated before incorporation into final designs and construction methods.

Many NASA planet surface system technology demonstrations (e.g., unloading systems, rovers, and power systems) take place for a short period of time (usually less than two weeks) in a specific analogue location that simulates expected terrain or environments. Antarctic analogues add layers of relevance to such testing: simulated operations under psychological and physical stress conditions imposed by remote location, extreme environment, and long mission duration.

For example, consider lunar habitats that are densely packaged, violently transported half a million kilometers, self-deployed and -verified long before crew arrival, survive long periods of unoccupied dormancy, and remain life-critical structures for decades. Systems that monitor health and habitability of such habitats and relay that information back to mission control are vital, and so the remote and extreme conditions of Antarctic analogues can test their performance and reliability during development.

Albeit an incomplete analogue for the lunar environment, Antarctica allows the study of several specific factors: advanced mission planning, system packing, transport survivability, deployment in a gravitational environment while in harsh conditions, human interface while wearing protective equipment, and long-term survival in extreme environments.

This chapter describes an antarctic analogue project that NASA, the National Science Foundation (NSF), and ILC Dover began in 2007. The project designed and built an expandable-technology, lunar-analogue habitat unit now undergoing demonstration testing over several years. Principal project objectives are demonstration of expandable structures and use of an integrated health monitoring system (IHMS). The structure continuously relays its health status (collected by wireless sensors and video) from its location in McMurdo Station to NASA Johnson Space Center (JSC) in Houston.

The habitat has already yielded information regarding transport and deployment, sensor integration, reconfigurability, habitability, performance in harsh environments, radiation shielding, and dust mitigation. These results inform ongoing NASA lunar architecture studies. In parallel, performance data are being studied by the NSF Office of Polar Programs (OPP) to determine if this class of structures can improve mission efficiency for polar exploration.

Copyright © 2008 by SAE International, published by the American Institute of Aeronautics and Astronautics Inc., with permission. This paper was originally presented as "Expandable Habitat Technology Demonstration for Lunar and Antarctic Applications," SAE Paper 2008-01-2024.

PROJECT BACKGROUND

NSF Antarctic Program objectives are analogous to NASA lunar exploration objectives. Small teams performing scientific research in the antarctic environment require protection from its harsh and unpredictable climate. While conducting their work away from permanent bases, they need shelters that are easily transportable and mass efficient. Remote research sites typically employ tents (e.g., Scott, dome) or rigid-framed, deployable shelters (e.g., Jamesway, Rac-Tent) consisting of plywood floors, wooden supports, and fiberglass/wool insulation. Such structures, balancing transportability with internal volume and habitability, represent the range of habitat options used.

The NASA Innovative Partnership Program (IPP) provided an opportunity for the NSF, NASA, and ILC Dover to study inflatable, deployable structures to enlarge the existing menu of structures approaches. NASA JSC led the project and developed the sensor and monitoring systems. NSF provided the test facility, transportation to the site, and contractor personnel to monitor the system in Antarctica. ILC Dover designed and manufactured the structure, electrical system, and pressurization system. Project duration was approximately one year, with the bulk of manufacturing occurring over a two-month span.

Program goals and objectives centered on building a knowledge base regarding large, expandable structures for lunar and Earth polar environments.

The *NASA study objectives* are as follows:

- Packing efficiency and packing methods
- Shipping/handing survival (vibration/environmental)
- Deployment operability in a gravitational environment and in polar gear (emulating spacesuits)
- Adaptability to uneven and rugged surfaces representing the lunar surface and guying practices
- Reusability and reconfigurability through joining of large components (habitats and airlocks)
- Performance in a harsh environment (cold, UV, flex, crew interface)
- Internal suspension/attachment of components (electrical, partitions, equipment, etc.)
- Deployment with integrated electronics (power, lighting, sensors, etc.)
- Remote structural health monitoring over long periods of time
- Use of *in situ* materials for radiation shielding
- Lunar dust mitigation practices
- Integration and function of windows

The *NSF study objectives* are as follows:

- Performance of high packing-efficiency deployable structures
- Transportability and setup under harsh conditions (wind, cold) and while wearing extreme cold-weather gear
- Simplicity of packing and deployment to reduce personnel required
- Damage tolerance and safety
- Modularity and reconfigurability
- Long-term survivability
- Power consumption in the Antarctic environment
- Multiple-use performance

The expandable-structure technology demonstrator was deployed at McMurdo Station in Antarctica by members of the IPP team in January 2008 for a year of continual monitoring by NSF personnel and remote study via an integrated sensor system.

SYSTEM OVERVIEW

The demonstration system comprises two inflatable habitat halves, inflatable airlock, doors, windows, insulation package, sensors and instrumentation, and inflation system. The structure is built up from a series of intersected tubular sections of thermally welded, coated fabric that form faceted, inflatable arches. Each element is an independent inflated volume that can be connected to adjacent elements via inflation port and zippers, making the system footprint easily expandable. The inflatable structure is pressurized to 6.9 kPa. Internal footprint of the completed habitat is 4.9 × 7.3 m. A flexible insulation package wraps the exterior, including under the floor. Guy lines and ground anchors are used to stabilize the structure for high-wind conditions (up to 44.7-m/s wind load).

The 453-kg system has a 15:1 packaging efficiency; it is transported in two 1.21 × 2.43 × 0.76 m packages (4.53 m^3), yet provides 70.8-m^3 living space when deployed (Figure 1). The flexible nature of the materials allows the shape of the transportation package

FIGURE 1 Expandable habitat fully deployed and half-unit packed for transport.

to be fitted to the transport vehicle (e.g., Twin-Otter aircraft) to simplify logistics. It is deployed by a three-man crew in 50 minutes (the actual inflation takes 10.2 minutes using a standard blower). A low-power, compact pressure-compensation system maintains pressure over long periods to compensate for pressure decay by permeation or variable barometric pressure. The system has two 1.46-kW quartz convection electric heaters, LED light strings, electrical outlets, and interfaces for attaching equipment to the walls.

Built-in exterior pockets on the lateral sides of the unit allowed feasibility testing of a radiation-shielding concept during initial deployment. The pockets were shovel-filled with snow to simulate addition of a prescribed thickness of lunar regolith. The pockets performed well for the intended test purpose, but also demonstrated utility for Antarctic applications like exterior storage space, guy line replacement features, and water production from ice.

The IHMS sensor system uses wireless and wired sensors, a central data-acquisition system, and Internet connection to enable remote monitoring and tracking of system performance during the harsh Antarctic winter. Monitored parameters include temperature, pressure, humidity, CO_2 concentration, power consumption, and light impingement. The system is also equipped with internal and external web-based cameras to document usage patterns and allow remote inspection (Figures 2).

EXPANDABLE STRUCTURES

The earliest credible work on expandable (i.e., inflatable) space habitats was published by Werner von Braun in 1946 and enhanced throughout his career (von Braun 1951). Von Braun recognized early the packaging advantage of collapsible and expandable structures. His work directly influenced habitat, tunnel, and airlock developments at NASA Langley Research Center (LaRC) and Goodyear Aerospace in the 1960s. The first demonstration of human-rated inflatable structures in space occurred in March 1962 when Alexei Leonov performed the first spacewalk by exiting Voskhod 2 through an expandable airlock (Cambridge 1990). Since then, many single-person inflatable habitats (i.e., spacesuits) have flown in space.

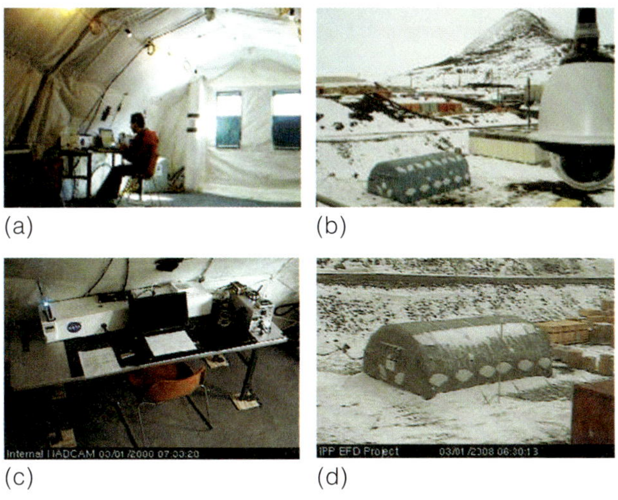

FIGURE 2 For documenting usage patterns and allowing remote inspecting, note the a, b) internal and external web cameras and c, d) pictures taken from them.

Modern-day development of deployable habitat structures began in the late 1980s at Lawrence Livermore Laboratory (Hyde et al. 1990). This work yielded the technology basis adopted for NASA's Transhab (see Chapter 8) (Kennedy 2002). TransHab in turn became the design basis for Bigelow Aerospace development of systems for commercial microgravity applications (see Chapter 9). After the Vision for Space Exploration was announced, NASA LaRC and ILC Dover manufactured and tested a prototype habitat structure for planetary environments (Cadogan et al. 2006).

Much of the technology used to design and fabricate expandable space habitats is derived from terrestrial applications. Many expandable structures have been fielded to provide battlefield protection from chemical and biological agents over the past half-century. The M51 and more recent M28 systems are two examples designed to operate in harsh environments per MIL-STD-810E. These structures are designed to be easily transportable in a packed state and robust enough for military use. Numerous expandable military shelters have been developed over the past several decades. A very large expandable structure is the mobile-platform-based, 30.5-m-diameter SBX Radome used in the Pacific Ocean for missile defense. This inflatable structure can withstand category 5 hurricane winds and has a 20-year service life in the harsh marine environment. Figure 3 shows examples from the history of inflatable/expandable structures.

SYSTEM REQUIREMENTS

Requirements for the Antarctic habitat demonstration project derived from the combined objectives of NASA and NSF already listed (Dover 2007). Broadly summarized, NASA's lunar-analogue objectives included field demonstration of structure offloading, positioning, and setup; dual ingress and egress; practicality of using local materials for radiation shielding; habitat element leveling, alignment, and connection;

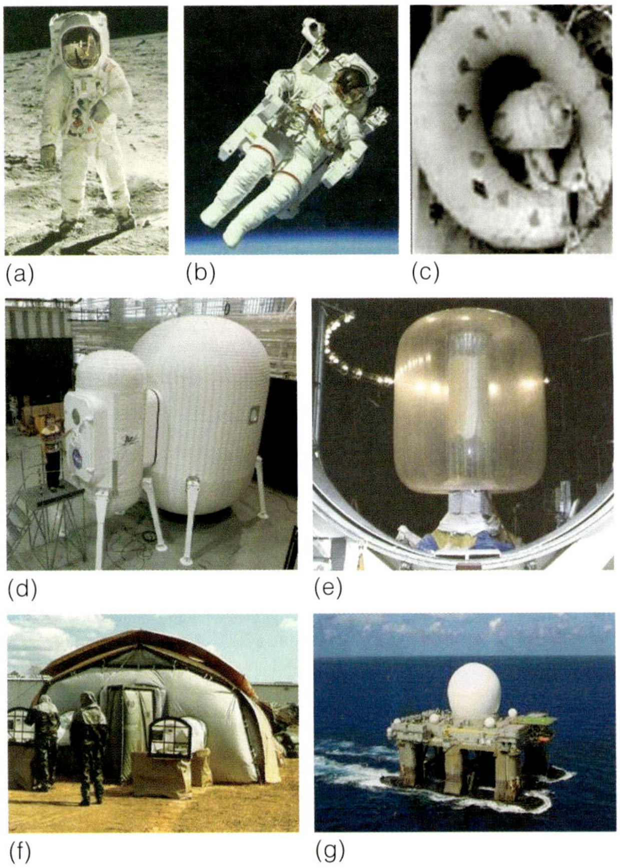

FIGURE 3 Inflatable/expandable structures: a) *Apollo* and b) shuttle "single-person-habitat" spacesuits, c) NASA LaRC Toroidal Habitat, d) ILC InFlex Habitat, e) NASA Transhab, f) ILC M28, and (g) ILC SBX Radome.

dust mitigation; and integration and function of windows in the structure (Toups 2007). Four categories of system requirements are given here:

1) Structural
   - Deployable habitat
   - Airlock
   - Rear door for alternative ingress/egress
   - Structural feature to enable *in situ* resource utilization radiation protection demonstration and/or structural stabilization
   - Viable design for incorporating window(s) in the deployable structure
   - Easily repairable materials
   - Consistent intrinsic insulation (walls and floor) with a minimum *R* value of 7 (consistent with the Jamesway habitat)

2) Loading
   - Withstand dynamic loading caused by 44.7-m/s wind and loads imposed by snow and ice
   - Withstand kick load of 57 kg over 25.8-cm^2 area
   - Internal walls equipped with features to support localized loads of at least 1.36 kg

3) Environment
   - Survive −50 to 8°C for approximately one year
   - No deterioration of structural performance during period of operation
   - Fire-retardant materials whenever applicable
   - No damage from vibration during shipping or shock from reasonable handling

4) Operations and logistics
   - Easy assembly in the field
   - Disassembly and reassembly to accommodate relocation
   - Survive up to three deactivations and redeployments in the field after initial setup
   - Deployment time ≤4 hours after removal from shipping containers
   - Sectioned so ≤4 personnel trained in deployment ops can move and deploy it
   - Sections provisioned for inflation system either through pneumatic interconnects between sections or with separate lines from inflation source
   - Structural feature to enable electrical outfitting

DESIGN DESCRIPTION

There are two volume elements: main habitat structure and airlock. The airlock was originally intended to be relocatable to either end of the habitat, but system requirements were changed late in the program to instead include a larger entry door at one end for ingress of a large submersible robot (the size of a small car) for storage. The main habitat volume is designed as a two-module structure conjoined using a multiple-zipper attachment. The doors and larger elements are replaceable and reconfigurable to allow changing the system footprint. Figure 4 shows the general arrangement. The entire structure is normally surrounded by a snow skirt extending 1.2–1.8 m from the base. (This was not demonstrated at McMurdo because of limited snow cover during project deployment.) The habitat structure provides 2.43-m maximum headroom and 35.7-m^2 floor space. The airlock provides 1.98-m maximum headroom and 2.23-m^2 floor area.

The main structure wall is built up of conjoined, cylindrical tubes inflated for rigidity. The exterior is covered by a flexible insulation laminate and the interior by a fire-retardant liner layer that protects the inflatable wall from inadvertent damage. The floor is the same material used in the inflatable wall, but is sandwiched by a full insulation layer underneath

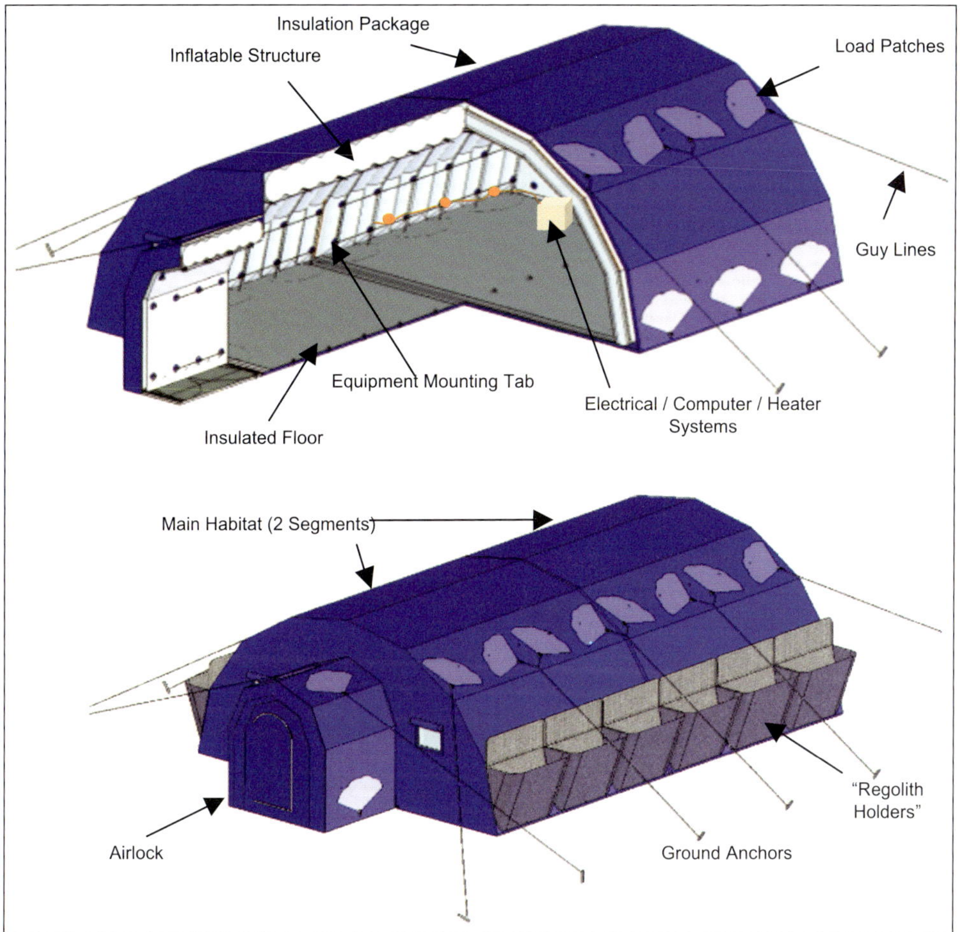

FIGURE 4 Inflatable habitat system general arrangement.

load-stabilizing capacity. The final selection was also influenced by contextual analysis of the configurations.

The structure with inflatable walls was selected because of unique mass and volume advantages absent in some of the rigid or pressurized configurations. Deployment by wall inflation facilitates floor and door designs that further augment high packing efficiency, light mass, and deployment-ease advantages. Several options were considered for the floor: coated fabrics, rigidizable floor, wood panels, inflatable floor made from drop-stitch fabric, and flexible laminate of coated fabric over insulation material. The insulation laminate was selected because of its lower mass and higher packing efficiency. A similar trade study was conducted to select the door design. The configurations evaluated included sprung overlap flaps, wooden door on hinges, inflatable-frame fabric door, hook-to-close flap, zipper door, and drop-thread door on hinges. The zipper door was selected for design and operational simplicity and its minimal mass and packing burden. Figure 5 shows how some of these details are integrated into the overall configuration.

and modular foam floor on top. The foam floor is an industrial flooring material that provides more insulation. The outer insulation blanket and the interior liner are indexed to the inflatable wall to preclude shifting of the three individual layers prior to and during deployment.

The structure also includes several fan patches to tether guy wires and anchors that serve to stabilize the structure under dynamic loading. The habitat is outfitted with windows that can be covered with zippered flaps made from flexible insulation laminate. Numerous mounting tabs are integrated into the interior surface for equipment, electrical, and sensor outfitting.

Configuration Development

Several design configurations were considered for the habitat structure, and a trade study was conducted to guide selection. The options included structures with inflatable walls, structures stabilized by air pressure, and structures with rigid but modular wall construction. The critical trade-study parameters were system mass, packing efficiency, thermal regulation, power consumption, and

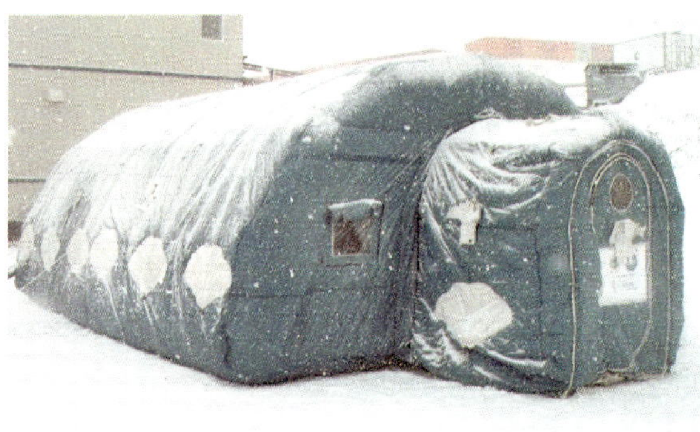

FIGURE 5 Airlock, door, and window details.

Insulation

The insulation blanket is analogous to the thermal cover and micrometeoroid and orbital debris shield of a planetary exploration habitat structure. Hence the Antarctic analogue demonstration supports feasibility of deploying such a structure in harsh planetary conditions. The insulation cover provides thermal and environmental protection, consistent *R* value, and constant temperature inside the habitat. The collapsible/expandable structure design and operational environment impose stringent requirements on thermal cover design. Critical requirements include low mass, high flexibility and compressibility for packaging, complete recovery of thickness upon deployment, and retention of thermal conductivity and structural integrity throughout the temperature range of −50 to 8°C.

Conventional insulation materials for polar environments include foam and fiberglass, which could not meet the expandable-structure requirements. Other state-of-the-art materials such as microfibers and aerogels were also considered. These materials have a very high *R* value compared to fiberglass and polyurethane foams of similar thickness. However, their high density (e.g., Aspen Spaceloft™ 6200 aerogel at 0.13 g/cc) would significantly increase total system mass when compared to other insulation materials (e.g., EPS P2000 Foam at 0.023 g/cc, Thinsulate™ G200 at 0.01g/cc, and fiberglass at 0.016 g/cc). Their incompressible nature is not conducive to efficient packing, they have high cost, and particulate contamination is a concern. fiberglass would yield a 10.2-cm thermal blanket and shed particulates, and so it was dismissed also. Thinsulate™ type G insulation was selected for availability, lightweight, high packing efficiency, and retention of insulation properties under damp conditions. The blanket is fabricated by sandwiching two layers of Thinsulate™ between coated fabric layers to provide the final laminate (Figure 6). The coated-fabric exterior is used to heat-seal indexing tie tabs and load-distribution guy patches.

The total thickness of the insulation was 3.81 cm, with a calculated *R* value of 6. The *R* value of the air gap in the inflatable wall and trapped air gaps under the insulation is approximately 1–2, so that the cumulative insulation during operation is concluded to be an *R* value of 7–8. Figure 7 shows factory integration of the insulation over the inflatable structure.

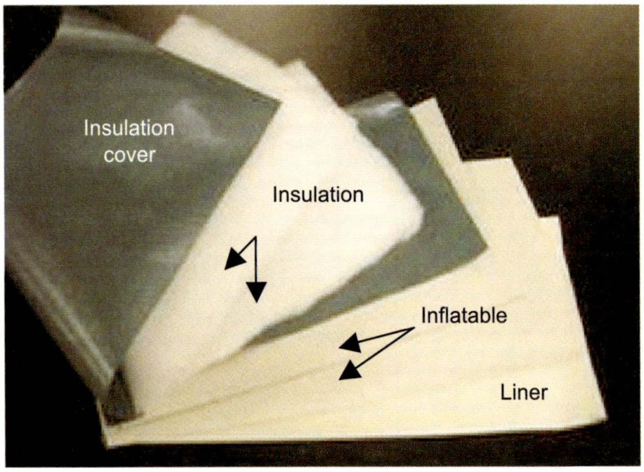

FIGURE 6 Laminate wall with coated-fabric, insulation, inflatable-tube, and liner layers.

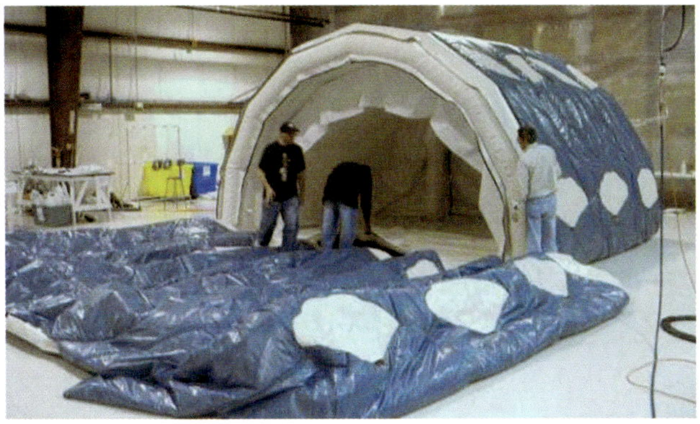

FIGURE 7 Integration of insulation blanket to main inflatable structure.

Electrical Systems

The habitat is outfitted with sensor packages to monitor structure health and internal and external environments (Figure 8). NASA used both technologically mature [high technology readiness level (TRL)] and experimental (lower TRL) sensor packages. The sensor packages are given here:

External
- Light impingement
- Surface temperature (RuBee Tags)
- Weather station (wind speed, direction, temperature)
- Web camera

Internal
- Temperature (RuBee Tags)
- CO_2 monitor
- Internet-controlled web camera
- System, heater, and air pump power consumption

Embedded within the inflatable structure
- Temperature
- Pressure

The external and embedded sensors, and some of the internal temperature sensors, are wireless devices. Embedded sensors have dedicated antennas, whereas surface-mounted temperature "RuBee" tags

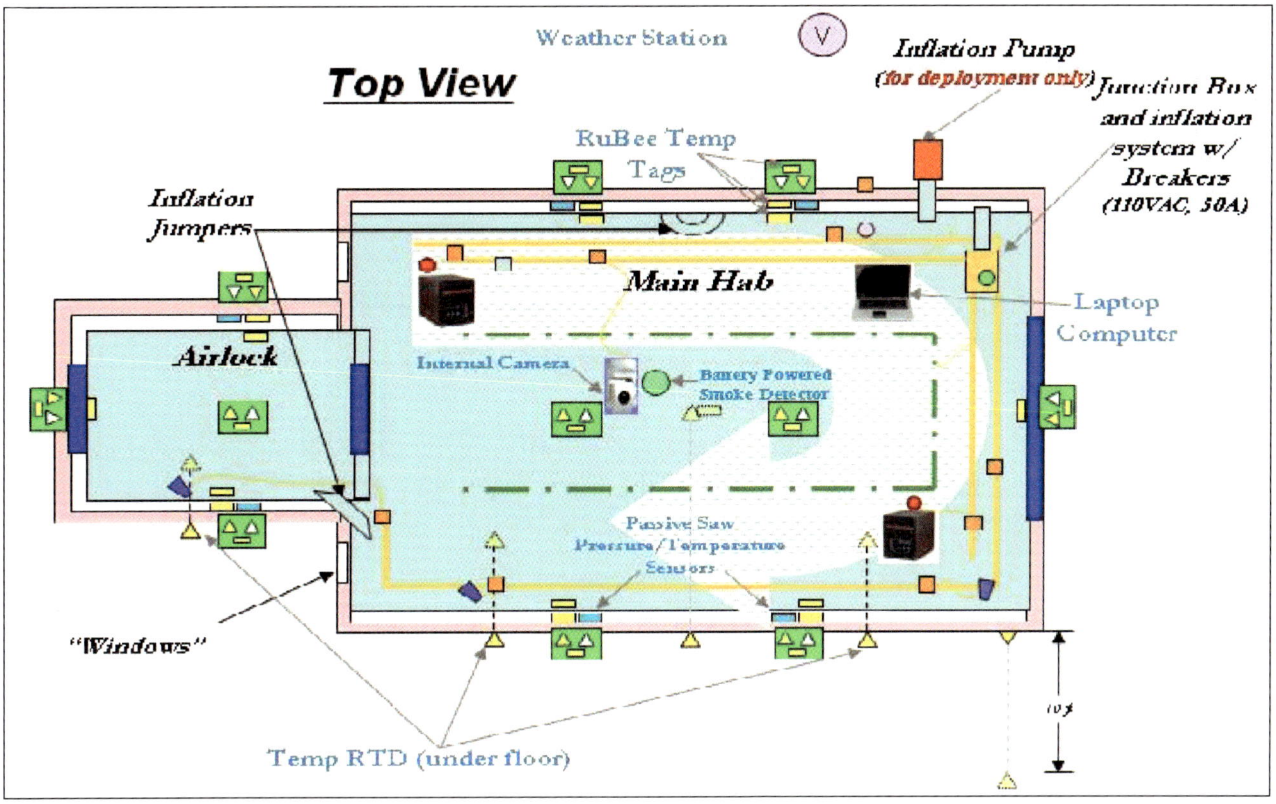

FIGURE 8 Sensor and electrical systems.

are RFID type tags linked to the monitoring system via a single-loop antenna routed around the inside perimeter of the habitat.

Several of the power monitors are hardwired to the data system; the remainder are connected through a wireless USB hub. All data are logged to a local computer and accessible with a local laptop. The computers are on the McMurdo local area network and accessed via Internet from NASA JSC for monitoring and review.

Regolith Holders

The habitat is equipped with storage bags on the exterior sidewalls (Figure 9). These pockets could be useful for NSF antarctic habitats as storage space, water melters, and stabilizing aids when filled with snow. For NASA, such pockets could be used for holding large quantities of regolith to provide radiation protection to inhabitants and equipment during long-duration missions. One possible advantage is a simple filling operation using primitive tools (Figure 9).

The simple design of the pockets includes attachment features that enable the contents to be removed easily. The demonstration unit is equipped with 10 regolith containment bags, each with 1.2-m-wide inlet opening and 0.9-m pocket depth. A closing flap is attached to each bag with webbing and D-rings, allowing removal and draining.

FIGURE 9 "Regolith pocket" detail (top) and analogue filling operation.

CONCLUSION

The demonstration unit described in this chapter was tested at the material level, component level, and system level, demonstrating all project objectives including survivability of the system throughout its anticipated life cycle. The sensor systems performed well after installation and in remote operation. Sensors embedded within the structure during manufacture survived installation, shipping, and deployment. The internal camera is operated from NASA JSC to monitor the inflation-system pressure gauge and verify pressure maintenance. All weather station functions have also been demonstrated. Data are continuously collected from McMurdo and studied to assess system performance.

This analogue project demonstrated to the NSF that expandable systems can help achieve OPP missions because of transportability, deployment rapidity, and structural stability. Expandable polar systems were found to "pack like a tent" but "act like a building" once deployed. The project demonstrated to NASA that such systems can have high packing efficiencies, be reconfigurable, be rugged and durable, and withstand extreme-environment conditions. Deployment and testing of habitat analogues in Antarctica can provide information valuable to space architects as they develop system design and operations concepts for lunar surface missions. |

References

Cadogan, D. P., Scheir, C., and Dixit, A. (2006), "Intelligent Materials for Deployable Space Structures (InFlex)," Paper 2006-01-2065, July.

Cambridge (1990), *Voskhod 2 Inflatable Airlock*, The Cambridge Encyclopedia of Space, Cambridge Univ. Press, Cambridge, England, U.K., pp. 52, 53.

Dover (2007), "Design Verification Compliance Matrix," ILC Dover, Frederica, DE, internal document, June.

Hyde, R., Ishikawa, M., and Wood, L. (1990), "Mars in this Decade: The Great Exploration," Sixth National Space Symposium of the U.S. Space Foundation, Colorado Springs, Co, April.

Kennedy, K. (2002), "Lessons From Transhab," AIAA Paper 2002-6105, Oct.

Toups, L. (2007), "NASA Advanced Programs Office Objectives for the Antarctic Habitat," NASA internal program document, Jan. 18.

Von Braun, W. (1951), "Inflatable Space Station," First Symposium on Space Flight, Hayden Planetarium, New York, Oct. 12.

Bibliography

Cadogan, D. P., Ferl, J., and Dixit, A. (2007), "Dust Mitigation Solutions for Lunar and Mars Surface Systems," Society of Automotive Engineers, Paper 2007-01-3213, July.

Dietz, A. E., Profitt, R. B., and Chabot, R. S. (1969), *Design Manual for Ground Mounted Air Supported Structures (Single and Double Wall)*, U.S. Army Natick Lab., Natick, MA, Technical Rept. 69-59-GP, Jan.

27

Halley VI antarctic research station

HUGH BROUGHTON

INTRODUCTION

ANTARCTICA—THE COLDEST, WINDIEST, DRIEST, and most remote continent on Earth—provides a unique environment for both the study of Earth-system science and parallel analysis of habitat regimes for interplanetary exploration. The British Antarctic Survey (BAS) operates two research stations on the continent. Halley is the most southerly of these at 75° 35' S 26° 39' W and is located 16,000 km from the United Kingdom on the 150-m-thick floating Brunt Ice Shelf, which moves 400 m/year toward the sea. Snow accumulation means the "ground level" rises by over 1 m every year. The sun does not rise above the horizon for 105 days during Austral winter and does not set for a similar period in the summer. Temperatures drop to −56°C, and the site can be buffeted by winds in excess of 160 km/hr. Access by ship and plane is limited to a three-month window from December to February. All materials and components required to sustain the existing base or construct a new base have to be dragged from a ship across fragile sea ice with a limited bearing capacity of 9.5 mt.

The base at Halley was established in 1956, the International Geophysical Year (IGY). It filled an important gap in the IGY antarctic network with studies in meteorology, glaciology, seismology, radio astronomy, and geospace science. Many of these studies have continued uninterrupted since then. Studies at Halley are crucial for a global perspective on ozone depletion, atmospheric pollution, sea-level rise, and climate change. Ozone has been measured at Halley since 1956. A springtime depletion in stratospheric ozone was discovered by BAS in 1985, and this led to the international response to curtail production of chlorofluorocarbons. Lying within the spectacular auroral zone, the station is ideally situated for geospace research. Halley is an apt analogue for a planetary base.

PRIOR ARCHITECTURE

The first two stations built on the site were little more than timber huts allowed to be slowly buried under the snow and ice. Although 50 years had passed since Scott built his hut at Cape Evans, there had been little development in antarctic station design in the intervening period. Change did occur with the third and fourth stations, where the accommodation was placed within tunnels that were purposefully allowed to become buried. These stations had only limited life spans before the inexorable creep of the ice sheet crushed them.

By the time Halley V (Figures 1 and 2) was commissioned, it was clear that a new concept was needed. The current station, completed in 1992, was designed as a single-story accommodation raised above the ice on jackable steel legs. By raising the buildings on stilts, wind is forced to accelerate underneath, creating a wind scoop directly under the platform and depositing a significant wind tail drift on the leeward side. Throughout the year this keeps the building out of the ice as the snow level rises all around. Living above the ice has proved psychologically beneficial for residents, with good natural light and views onto the ice in summer and of the Aurora Australis in winter. This design approach has also improved operational efficiency, albeit with one or two shortcomings. Plastic movement of the ice distorts the legs supporting the platform. Year on year a team of steelworkers is therefore shipped to the Antarctic to cut off the legs at snow level, realign the steelwork above and splice the two parts back together. Once this task is completed, a team of 40 people is needed to slowly jack the building higher above the ice in preparation for the next year's snow accumulation.

Copyright © 1998 by Hugh Broughton. Published by the American Institute of Aeronautics and Astronautics, Inc., with permission.

FIGURE 1 Existing accommodation building at the Halley V Antarctic Research Station.

FIGURE 2 The author at Halley V.

Although the current base continues to operate effectively, a significant calving event on the ice shelf is predicted within the next decade, which will send Halley V floating out to sea. As a result, in 2004 the BAS teamed up with the Royal Institute of British Architects to launch a three-stage international competition for the design of a new and fully relocatable base to be built 16 km further inshore. The new base would provide a home and workplace for 16 people in winter and 52 in summer. The competition attracted 86 entries from around the world. In July 2005 the British team of Hugh Broughton Architects and engineers Faber Maunsell were selected. In December 2007 construction commenced on site. By February 2008, the first Halley VI module was fully clad, ready to endure its first winter on the ice.

This chapter describes the design concept for Halley VI as an analogue for space habitation facilities that will be optimized for extreme environments and modular construction on site in remote locations.

DESIGN

The Halley VI design was developed in direct response to the demands of science, comfort of the residents, buildability, and operations inherent in the life of an Antarctic research station. The designers recognized from the outset it was crucial to fully understand the processes involved at Halley and to use this knowledge to ensure the design best reflected the needs of the science and the aspirations of the people who would live and work at the station.

Modularity

To meet these demands, it was crucial to create a design that maximized flexibility. This was achieved with a modular approach. Modularity has significant benefits: flexibility, easier construction and maintenance through repetition, easier relocation, enhanced fire-safety and acoustic performance through separation, and greater overall robustness. Connected together, multiple modules form the new station and are used for a wide variety of activities ranging from laboratories and bedrooms to recreation areas and energy centers.

Habitability

The plan form of the modules responds to BAS requirements. A standard module (Figure 3) accommodates eight ergonomically planned bedrooms within four structural bays arranged on a 2.6-m grid. This sets the size of the standard module. The compact bedrooms measure 2.5 × 3.6 m internally and are homey, comfortable spaces that promote emotional well-being without being so comfortable that residents hide away from the community. Each has a window for natural light and views and includes daylight-simulation lamps, long beds to accommodate taller residents, data connections, and excellent storage. Walls are fitted with pin boards that allow personalizing the space, and behind each bed is an alcove for personal possessions. A warm color palette selected with the help of a color psychologist helps assist sleep patterns both during the 24-hour

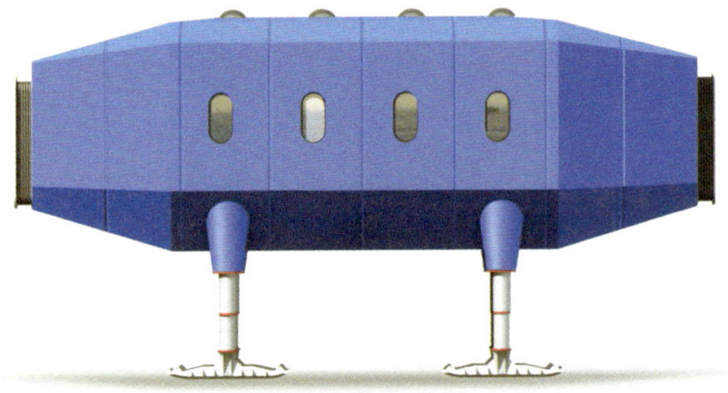

FIGURE 3 Standard module is the building block for Halley VI.

areas are enriched through strong colors to suit and reflect uses and activities, for example, a reassuring green in sleeping module corridors and a welcoming yellow in the entry module. The colors provide location identity and emphasize contrast with the broad, top-lit areas. In many of the modules the corridor unexpectedly opens out at the ends to give views onto the Antarctic landscape, punctuating the journey through the station and providing spaces for chance encounters with other residents. Through manipulation of the cross section and plan, each module becomes a destination in its own right, creating spatial variety to help stimulate residents (Figure 4).

The needs of science and scientists are at the forefront of the design. Science modules are placed at the southern end of the station, close to the clean air sector, optical dark zone, and geospace radar arrays. At Halley V science is conducted from two separate platforms. Colocation of scientific activities at Halley VI brings benefits through cross-fertilization of ideas and team building. The ever-changing nature of scientific work carried out at Halley means that the working environment of the scientists needs to be as flexible as possible. The modular solution allows extra modules to be added to the station to suit changing science programs. The design therefore meets BAS' current requirements, with extra built-in flexibility for minimum alterations despite future changes.

summer sunlight and winter darkness, when seasonal affective disorder can be a particular problem.

As there is no internal structure—roof support is provided by the perimeter—the modules are extremely versatile, allowing layouts to be reconfigured to suit changing science programs or operational drivers. Within the science modules some partitions are omitted altogether. Offices, for instance, are open to circulation to promote social interaction as well as a light, spacious working environment. Windows in all rooms allow views of the snowscape, providing a connection between the internal, protected space and sometimes brutal conditions outside.

Particular attention has been paid to connections between spaces. The central passageway opens out in plan to a generous 1800 mm wide, reflected in section by raising the ceiling in these toplit areas to give a sense of openness. These wider, taller areas break down the apparent length of the passage and allow for incidental meetings and places to enjoy the station's large collection of photos. The link

Central Module

The majority of activities that take place at Halley can be accommodated using a repeated standard module. However, some activities are constant and require a distinct approach. These activities are housed in a special central module (Figure 5). This is the principal space for eating, drinking, and recreation and is the major "destination" of the station.

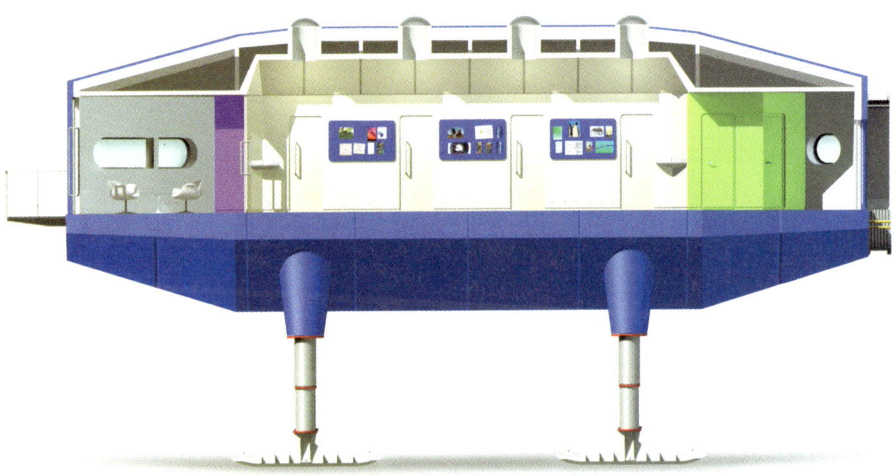

FIGURE 4 Each module is a destination in its own right.

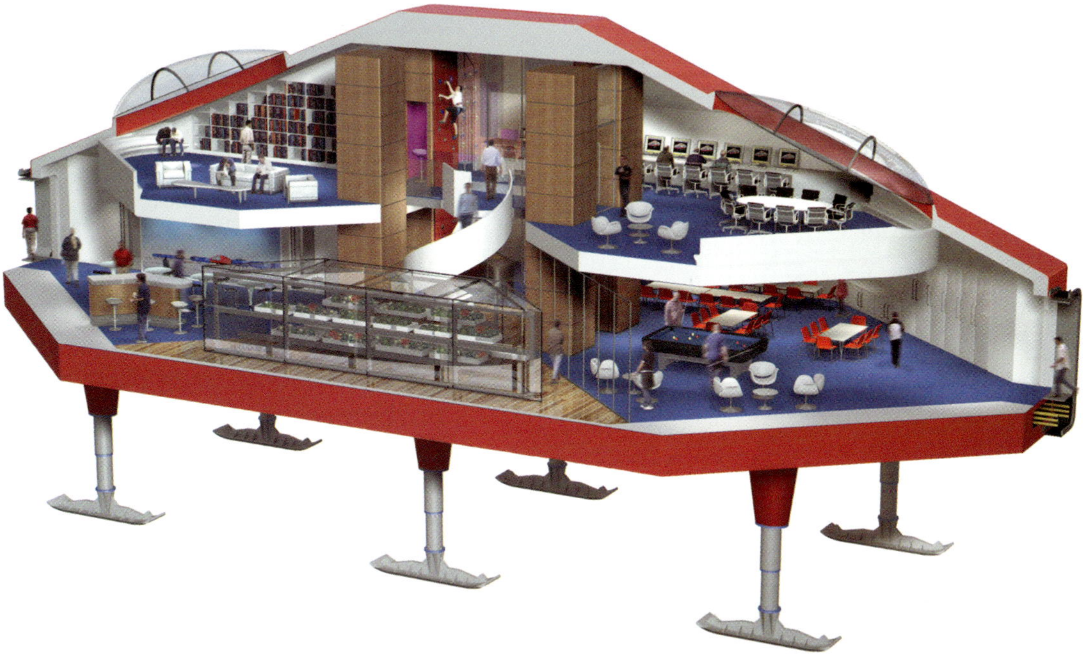

FIGURE 5 Central module is the heart of Halley VI.

Movable acoustic partitions on the lower deck of the central module maximize flexibility. The lower floor can be divided into a number of cellular spaces or converted into a large open-plan area. Combinations of open and closed partitions allow a wide range of activities to take place simultaneously. For example, in winter the dining room needs to be kept at an intimate scale to suit the low population of 16 people, and a significant amount of space is needed for recreation to sustain the crew through the dark months. Within the central module there are areas for darts, table tennis, pool, and computer games. A sauna, gym, and music room are included in adjacent modules. In summer there is less need for recreational space, and the dining room needs to expand to accommodate up to 60 people. It is particularly important that the entire summer complement be able to dine together to promote a sense of community.

In combining the key social functions of the station, the design of the central module also recognizes the difference between quiet and noisier activities. An upper deck houses video lounge, library, and office area. Within these upper areas "cockpit" roof lights offer great views of summer snowscapes and winter displays of the Aurora Australis. Access to the upper level is via a helical stair feature that climbs through an atrium glazed with a high-performance, translucent curtain wall. In summer this space is filled with diffuse light. In winter, the translucent panels act as a screen for color-change LEDs to simulate the changing daytime light of a more temperate climate. A hydroponics installation at the center of the atrium nourishes the crew and provides the refreshing sight of greenery at the heart of the station—an unexpected moment of spatial drama.

The interior design of the central module is characterized by a warm palette of materials: timber paneling, carpets, comfortable furniture, and a Halley history display all help stimulate the well-being of a crew living far from home for extended periods.

Site Layout

The 1970-m^2 station is centered on two modular platforms (Figure 6). The northern platform provides the principal habitat. The southern platform contains the science modules. Separation into two platforms creates a refuge in case of catastrophic failure of one platform or plant module. Key features of the site layout are the following:

- The platforms are perpendicular to the prevailing wind to minimize snow management.
- Entrances to the platforms are placed in the center to reduce circulation distances and offer a choice of journey once off the ice.
- The physical plant modules on each platform face each other. This removes the plant for the science modules to the north of the platform, limiting its impact on the clean air sector.
- A pedestrian and services bridge connects the plant modules. Two melt tanks are placed under the bridge and provide potable water. Water treatment, fuel, electrical, data, and communications are linked between the platforms.

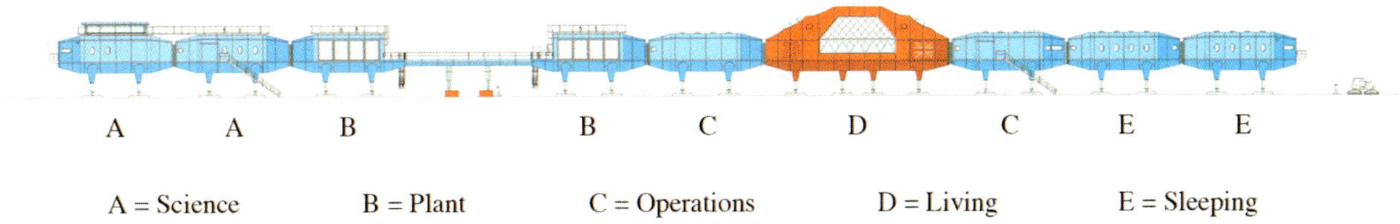

FIGURE 6 Halley VI elevation with science platform (left) bridged to living platform (right).

- The bridge provides an above-ice link for scientists with the main platform for use in poor weather.
- The southern-most science module has an upper-level observatory with a clear panoramic view of the southern skies for the meteorology operations room and ozone laboratory.
- The northern-most winter sleeping module includes a quiet lounge with panoramic views where residents can escape from the throng of the main social spaces.

Structure and Envelope

The structural design is based on simplicity, standardization, and buildability. The modular structure is designed for easy transport to the site, to minimize the annual maintenance effort to overcome snow accumulation, and for easy relocation to another site.

The structural subframe of the modules is formed from a lightweight steel space frame (Figure 7). The frame is supported on steel legs supported on special skis. The assembly mass is 9.5 mt, which is within the expected bearing capacity of the sea ice. The skis, integral to delivery and relocation of the modules, also act as spreader-beam foundations. In operation, each ski is secured with removable pegs that provide a shear key to the ice to avoid unwanted sliding of the modules.

A braced steel goalpost frame forms the superstructure of each standard module. Floors are formed from prefabricated composite timber cassettes that can be quickly slotted together on site. Preformed floor hatches give access to services within the belly of the modules.

The cladding needs to be erected quickly, safely, and efficiently and needs to provide a robust envelope that resists moisture, snow spindrift, the passage of cold air, electrostatic conduction, thermal shock, and high levels of ultraviolet (UV) light. It is formed from relatively lightweight, glass-reinforced-plastic (GRP) panels fixed back to the structure with flexible shock mounts and to each other with sealed rigid pressure plates, forming a semimonocoque enclosure. The panels consist of closed-cell polyisocyanurate foam insulation encapsulated within the GRP, finished with a UV-resistant paint designed to resist the abrasive impact of wind-driven snow and ice. The upper section of the leg is formed by a pair of rectangular tubes integral to the space frame. The lower leg consists of a large circular tube with integral jacking mechanism. The jacking mechanisms overcome annual snow accumulation. The tube is fabricated as a separate cassette, slotted into the upper part of the leg onsite. The jacking system is hydraulic, as this technology is well understood by the vehicle team operating at Halley.

The jacking system yields a significant reduction in the overall maintenance burden for Halley VI compared to Halley V. The main effort is to push snow beneath the platform feet using D5 Caterpillar bulldozers, the vehicles used for the majority of tasks at the station. The whole operation for the entire station can be completed in six days, with just two vehicle operators and one supervisor to control the leg movements.

FIGURE 7 Standard module under construction at Halley.

Despite its special configuration, the central module is technically the same as the standard modules. The central module's superstructure was developed to minimize beam weights, thereby easing construction. The cladding uses the same system as the standard modules.

Construction and Logistics

Repetitive modules use fewer different construction components, reducing the time required to build the station. Many of the rooms were prefabricated as pods to maximize on-site installation efficiency and quality upon completion. An assembly-line approach was applied to the construction of the modular units, thereby minimizing the risk of program delays. A test module was built in 2007 in Cape Town to ensure buildability and to carry out crucial checks on structure, legs, and cladding. Separate material tests were conducted in laboratory conditions to check suitability for use in cold climates and to prove fire performance. On arrival in Antarctica in December 2007, the space frame, legs, and transit skis were offloaded from the cargo ship and skied directly to the site, offering an excellent start to the construction process.

Relocation

Completed modules have a mass of ~80 mt. This allows them to be easily moved to a new site during their projected 20-year lifetime to escape future calving events. Because the station is being constructed at Halley V and then moved to its new site, this relocation strategy will be proved early in the life of Halley VI.

Extensive vehicle testing conducted at Halley in the 2005–2006 season demonstrated that the standard modules could be pulled using a single D5 Caterpillar bulldozer across a prepared ice surface (Figure 8). The central module weighs 120 mt. Further testing in the 2006–2007 season was conducted to demonstrate that this module can be relocated using either bulldozers or CAT Challengers, which are being used for the construction.

FIGURE 8 Proof-of-concept vehicle tow testing at Halley.

Services

The modular approach to building services, a key component in achieving resilience and ease of maintenance, reinforces the modular concept of the station. The design responds to the needs for sustainability and energy efficiency. An energy center at the end of each platform centralizes generators, sewage treatment, and fuel storage. Separate energy modules add flexibility, while ensuring good acoustic, vibration, odor, and fire isolation from habitable areas (Figure 9). Oriented to face each other, the modules enjoy an efficient arrangement for sharing water generation and refueling. The modules are connected with a utility bridge that also provides a walkway between the platforms.

Some services, such as ventilation, heating, and humidification, are distributed within each module because both local control and modular adaptability are of critical importance. Components are standardized to maximize interchangeability of parts and reduce the number of onsite spares.

Environment and Sustainability

Sustainability drives energy use at all levels of the design. The combination of a well-insulated, sealed enclosure and good control ensures efficient energy consumption around the clock. Allowance is made for bolt-on integration of sustainable sources of energy such as wind turbines and photovoltaic arrays. For the present, technology and environmental conditions at Halley do not allow reliance on these sources. The energy modules include solar thermal panels to supplement waste heat collected from CHP generator engines. The new station makes use of low-energy equipment from lighting to domestic white goods to fans and pumps. A vacuum drainage system with low-water-use equipment yields a 50% reduction in potable water use compared to Halley V.

Halley VI is the most environmentally friendly and sustainable facility BAS has ever built. Low on environmental impact during construction, with an extremely efficient, environmentally aware performance life cycle, it can be easily moved and eventually taken apart when the time comes for it to leave the ice and be decommissioned. Simplified processes dramatically improve operation and maintenance procedures, reducing the numbers of staff needed to look after the station.

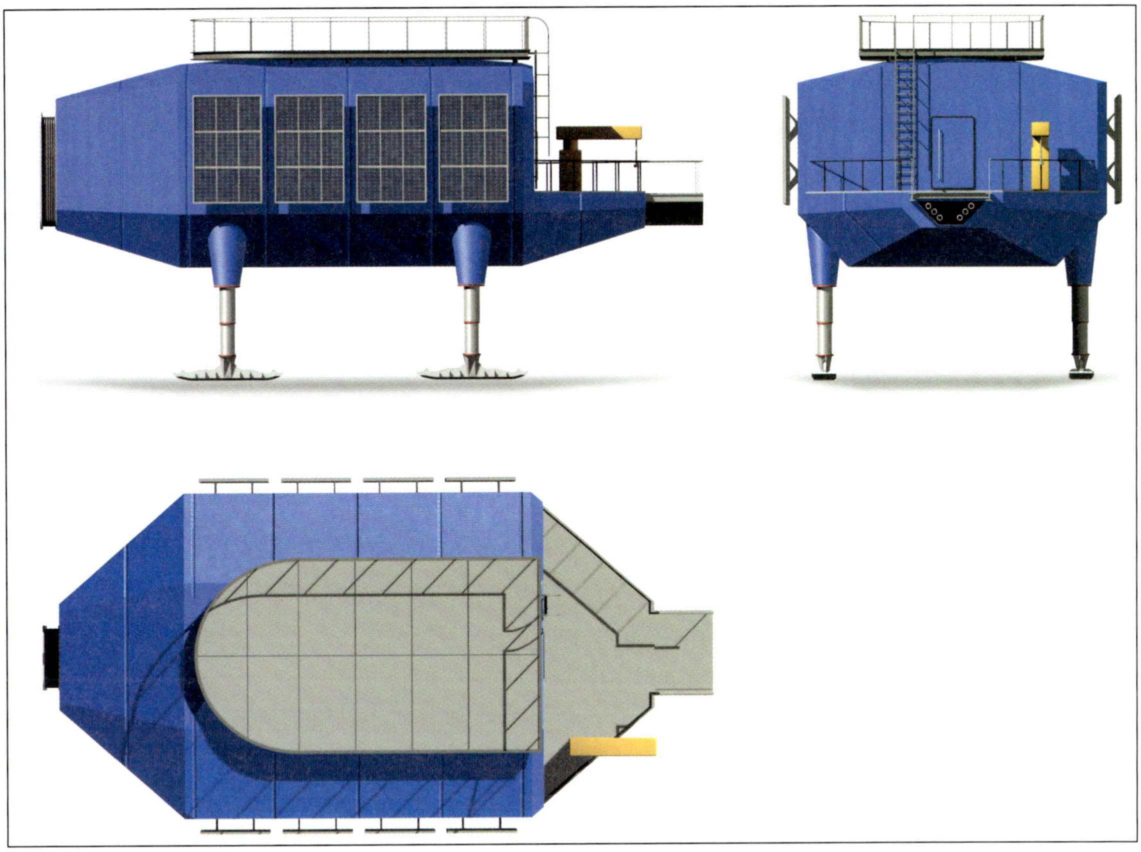

| **FIGURE 9** Energy module consolidates building services.

The right energy balance ensures low use of primary energy through better insulation, better building management, and better equipment.

We consider Halley VI to be a visitor to Antarctica, not a resident (Figures 10). The buildings rest entirely on the surface of the ice shelf. This mobility and flexibility means that the new station will survive and perform on the ice for far longer than any of its distinguished predecessors. Historically, the life expectancy for ice stations at Halley has been about 10 years. The design life for this project is 20 years. Our design admits the opportunity to add (or reduce) the number of modules. It provides flexibility to be adapted, rearranged, and relocated. The new Halley VI can therefore continue to respond to the changing needs of Antarctic science for many years more than its projected design life.

Linked together, the ski-based, jackable modules create a dramatic new station to propel Antarctic design into the 21st century (Figure 11). One hundred years after Scott and Shackleton built timber huts on this frozen continent, Halley VI introduces the first relocatable, modular research station, with optimized accommodations for both living and working. The station is packed with stimulating areas for recreation and relaxation and allows total flexibility for growth and change. It will be the envy of Antarctica, a beacon for sustainable living, and, above all, an icon to draw attention to some of the most significant and influential science conducted on our planet today.

FUTURE APPLICATIONS

A modular, relocatable solution to science research facilities in remote locations yields advantages of flexibility, versatility, easier construction, easier maintenance, and overall robustness. Halley VI modules are sized to match the needs of its residents and their science. Materials were selected to withstand prevailing environmental conditions and to suit the planned construction process. Many of the approaches and concepts explored for Halley can be adapted to suit other remote research stations for polar regions or planetary missions. Modules could be even smaller in size and constructed using sophisticated lightweight materials, while still providing ergonomic and psychologically supportive living conditions. Smaller modules could be prefabricated in larger subsections or even as complete units within the dimensional constraints of the logistics chain, designed to fit within the cargo hold of a flight vehicle or deck of a landing barge, and be configured for relocation on skis across ice or tracks across drier ground (Figure 12).

(a)

(b)

FIGURE 10 Halley VI is conceived as a visitor in the antarctic environment: a) model and b) emerging reality.

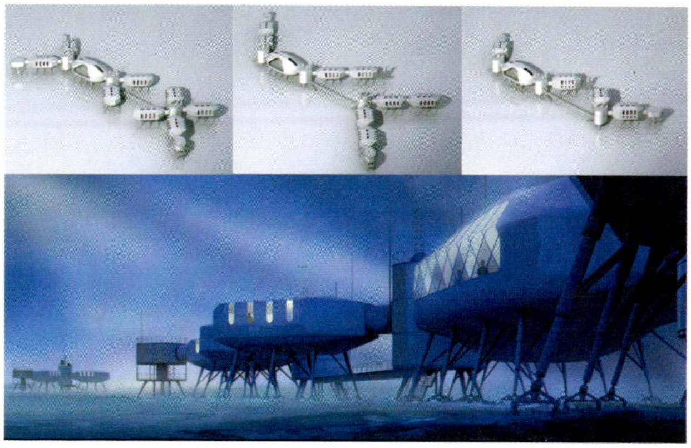

FIGURE 12 Small, pre-integrated modules enable temporary research stations at areas of special interest.

FIGURE 13 Modular approach maximizes flexibility, providing an excellent analogue for space architecture.

The application of modular solutions to space architecture is inevitable given logistical limitations. An embryonic planetary version of the Halley model could encompass one service and one habitat module, and host a crew of four to six people. A service module could house habitat systems and stowage. A habitat module could provide science facilities, sleeping, washing, eating, and social space for the crew and use rack systems for outfitting flexibility. From this basic architecture there are limitless opportunities for expansion to incorporate operational, scientific, and additional social modules.

A simple kit of modules could be developed for living and working, for services, for entry and connection, and for the primary social space. These modules could be connected together in various ways to minimize circulation and maximize useable space (Figure 13). Cluster size would depend on location, environment, use, logistics, budget, and the extent to which human factors drive the decision making.

The flexible concept adopted for the Halley VI Antarctic Research Station is analogous to likely requirements for space architecture in orbital and planetary environments, as is evident in precedents such as the International Space Station. It also demonstrates that an ergonomic, modular, relocatable design not only works in Earth's antarctic, but offers an excellent model for research stations on the Moon and Mars.

In the Halley VI project human-factors design was given high priority both at strategic and detailed levels, recognizing that inhabitants would be confined to the station for long periods, often living in 24-hour darkness. In any remote or space location, the interior design must be underpinned by thorough investigation of architectural first principles, careful manipulation of daylight and volume, and a desire to create a home for residents. In this respect the lessons learned from Halley VI and other recently constructed Antarctic research facilities offer significant precedents and guide for space architecture.

FIGURE 11 Halley VI is an icon for the British Antarctic Survey's world-class Earth system science.

28

planetary and lunar surface simulator

DAVID NIXON
TRULS OVRUM
AND
PAUL CLANCY

INTRODUCTION

SINCE THE BIRTH OF THE SPACE AGE, interest in exploring planetary surfaces has been an important aspect of global space activities. Since the 1960s, the United States, Russia, and Japan have deployed planetary, lunar, and asteroidal robotic surface probes. The European Space Agency (ESA) achieved a successful landing of the *Huygens* probe on Titan as part of NASA's *Cassini* mission, and ESA's *Rosetta* mission is on its way to land a probe on the surface of a comet. Planetary and lunar exploration figures increasingly in the space activities of all global players capable of such achievements. In the case of ESA, the ExoMars mission will look for past and/or present life on Mars, leading to additional studies of Mars sample return and eventually human Mars missions.

Experience to date, particularly on the Moon and Mars, has shown that rover and lander interactions with their environments are complex, both strongly operationally constrained and subject to unplanned contingencies. This means that, in addition to sophisticated design and analysis, landers and rovers need extensive field testing in simulated surface environments before launch to verify their requirements and validate their performance.

This chapter describes an ESA project to develop a simulation facility to support all aspects of testing European landers and rovers. This first-of-a-kind simulation facility is purposely designed to enable specialized, simulations-based research and testing on lander and rover surface operations for Mars, Moon, and other locations such as comets and near-Earth objects (NEOs). The project also has international value and provides a means of public observation of rover and lander development and testing work. The first intended use of the facility is developmental testing of the ExoMars rover.

The ESA planet surface simulator project was designed by a team comprising 4SPACE s.a.r.l., Paris, France; Altus Associates architects, Los Angeles, California; Arup, Cardiff, United Kingdom, and Amsterdam, The Netherlands; Davis Langdon, London, United Kingdom; and Architecture + Vision, Munich, Germany.

REQUIREMENTS

Study began with preparation of requirements to define the simulated environments, equipment, and materials necessary to support rover and lander testing. Figure 1 illustrates the ExoMars rover. It must undergo realistic testing in the types of terrain it will encounter on Mars, which can vary considerably: from flat to sloped to undulating terrain, fine soil to granular soil, and outcrops, rocks, craters and other features of varying shape and size. Terramechanical performance and autonomous navigation functions are particularly important for ExoMars rover testing. ESA estimated it would need a 30-m-diameter circular area of simulated terrain, and so this was the starting point for defining simulator facility system requirements. An environmentally controlled building envelope would have to enclose the terrain to enable year-round testing regardless of weather and to control temperature and humidity of the simulated-terrain materials so as to emulate the dry conditions

Copyright © 2008 by the authors. Published by the American Institute of Aeronautics and Astronautics, Inc., with permission.

FIGURE 1 ExoMars rover (courtesy of ESA).

of the Martian surface. The terrain would need to be of sufficient depth to achieve varied topography, with a deep pit to enable testing the ExoMars rover's drilling apparatus to a depth of 2 m. Simulated Martian terrain would need to be exchanged for simulated lunar, cometary, or NEO terrain for the facility to test rovers and landers for those bodies or other planets and Moons.

Of equal importance is the need to simulate Martian, lunar, and other sky conditions using a terrain lighting system that can be controlled and programmed to change the conditions without major refit. Planetary and lunar skies have widely different light source motion, light intensity, light spread, and sky color characteristics. Sky conditions vary from simulation of a bright, clear sun against a black sky in the case of the Moon to a hazy, more distant sun in a pink/orange sky in the case of Mars. Sun paths vary from a Martian midlatitude sol in which the sun crosses the sky daily in a similar manner to Earth to a lunar day/night combination that varies dramatically from the equator to the poles.

Another major requirement determined by ESA is for 100% access across the 30-m-diameter terrain arena from above to retrieve a broken-down rover or lander during testing without touching the surrounding terrain surface. It became clear during the study that the only way that this can be achieved economically is by means of an articulated boom crane. Stationed around the terrain perimeter, the crane would deploy a work platform across and above the terrain surface with two people who manually retrieve the rover and lift it onto the platform for transfer back to the perimeter. This requirement results in the need to provide crane access at several points around the terrain perimeter.

As well as requirements for the simulation testing environments, the study produced requirements covering support accommodation in the facility, some of which were initially determined by ESA. These included provisions for a perimeter service aisle around the terrain arena for staff and equipment access to all points; three sealable airlocks for the articulated boom cranes needed to place/retrieve rovers or landers; a laboratory for preparing and maintaining the rover and lander test models; a workshop for producing various simulated terrain and topographical features to insert into the terrain arena; adequate room for environmental control systems such as circulation fans, heaters, cooling towers, dehumidifiers and ducts; storage for terrain materials not in use; separate staff and visitor entrance lobbies; a visitor observation gallery; and various amenities.

A further set of requirements governs the facility's construction and appearance. Increasing concern over ecological and environmental issues has led to new initiatives and policies worldwide in sustainable building design and engineering, so that the ESA facility would require an environmentally responsible approach to choice of materials, energy consumption, waste treatment, and beneficial landscaping. Use of photovoltaic panels on the facility's exterior could help to generate electrical power to operate lighting and environmental control system, depending on orientation and exposure. The facility's appearance would require careful treatment because of the obvious size of a building requiring a clear-span enclosure of at least 30 m diameter. The facility would be located in a suburban, coastal setting. Its exterior architectural treatment would require balancing the need to insert the building carefully and sympathetically into an existing context, where it would dominate the skyline, with the need to have architectural identity as a space exploration landmark.

CONCEPT

The design concept evolved from development and refinement of ESA's notional concept for the facility at the beginning of the study. ESA proposed a hemispherical dome enclosing a circular terrain area 30 m in diameter from wall to wall, with a small laboratory blockhouse attached to the perimeter. This notional concept was inadequate for proper facility use and operation, and the requirements drove the preparation of a comprehensive architectural program for the building with the following major features:

- Circular perimeter aisle for 360-deg access to the terrain arena for terrain materials and small Earth-moving equipment during terrain installation and forming work, during rover and lander test operations, and during terrain removal and change-out for other applications
- The terrain arena formed as a shallow trough to hold the terrain material, with a 3-m-deep pit at the center to permit testing the ExoMars and other drilling equipment
- Computer-controlled, multiple-lamp, variable-intensity-color lighting system around the inside of the dome to simulate sun and sky lighting conditions varying from lunar to Martian
- Three airlocks located 120 deg around the perimeter with inner and outer sets of doors to accommodate storage and passage of cherry-picker cranes and other mechanical equipment
- Laboratory for preparing and maintaining rover and lander test prototypes
- Workshop for crafting and forming special terrain features such as rocks, outcrops, cliffs, crevices, exposed strata, and craters, etc.
- Storage for terrain materials and features not in use
- Staff offices, meeting rooms, and entrance lobby
- Visitor entrance lobby and observation gallery
- Operations control room
- Staff and visitor washrooms and other basic amenities and service areas
- Full environmental control of the terrain arena, airlocks, and other accommodation areas where applicable
- Dust-storm chamber for testing rovers and landers under simulated Martian dust-storm conditions

The architectural program provided the basis for development of a generic design concept for the facility (Figure 2). Three airlocks occur at 120-deg intervals around the perimeter of the central dome, projecting outwards. The airlocks create three segments of volume between them around the perimeter. These segments contain support facilities such as laboratories, workshops, and staff amenities. The segments could either be fully built out during initial construction to provide the complete inventory of accommodations, or built in phases if the initial budget precluded complete construction of the facility at the outset. Three construction phases would therefore be possible, with phase 1 comprising the terrain arena, airlocks and minimal support facilities, and phases 2 and 3 adding support facilities as affordable and needed. Figure 3 shows the integration of interior lighting units and other mechanical and electrical outfitting into the dome structure design.

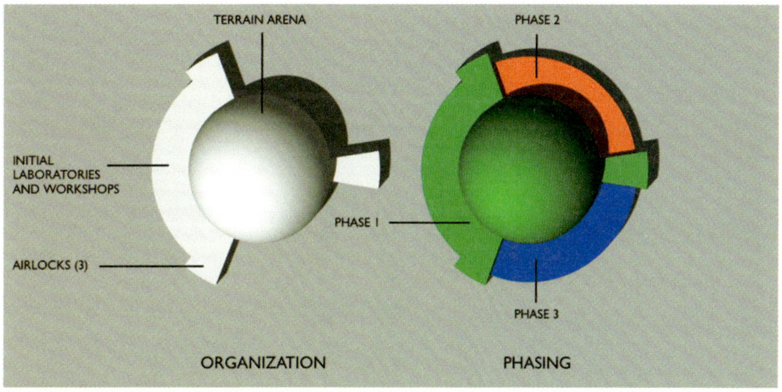

FIGURE 2 Design concept facilitates phased construction.

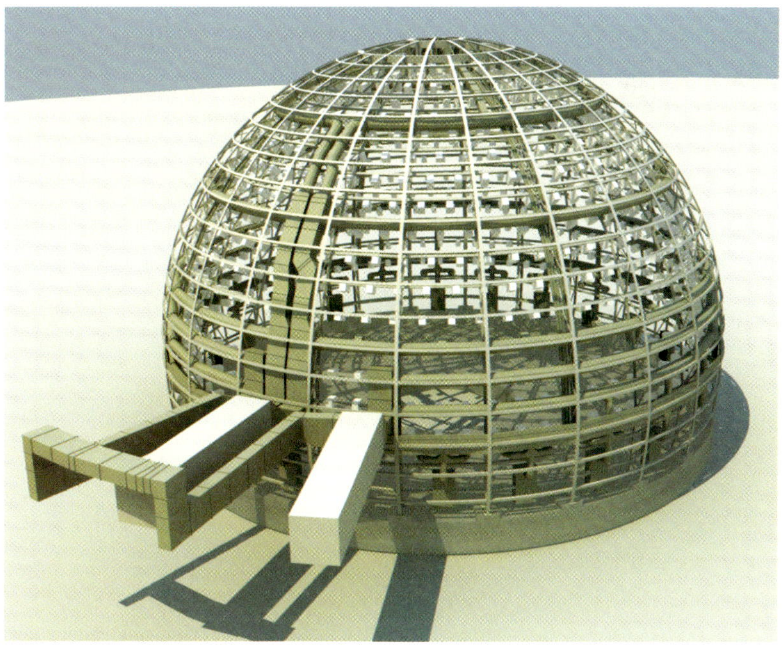

FIGURE 3 Integrated CAD model of dome structural, mechanical, and electrical systems.

PHASE 1 DESIGN

Discussions with ESA highlighted the need to limit initial construction cost of the facility and focus on an initial operational capability (IOC) covering ExoMars rover testing activities, with expanded support facilities deferred. Accordingly, the design concept evolved into schematic plans and sections incorporating major structural and environmental control elements in sufficient detail to enable preliminary cost estimating. Figures 4 and 5 respectively show the IOC facility plan and section. The three airlocks lead from the building exterior into the terrain arena. An articulated boom crane can pass through the airlock double doors to three circular pads at the edge of the terrain arena, from where the crane can reach any point on the terrain surface. Environmental control systems equipment is located above each airlock. A staff entrance leads via a corridor to a laboratory, workshop, store, boiler room, and basic amenities.

SITE

ESA is considering four possible sites for the building at its European Space Technology Center (ESTEC) in The Netherlands. Two of these are within the present ESTEC campus, and the other two are in a new space

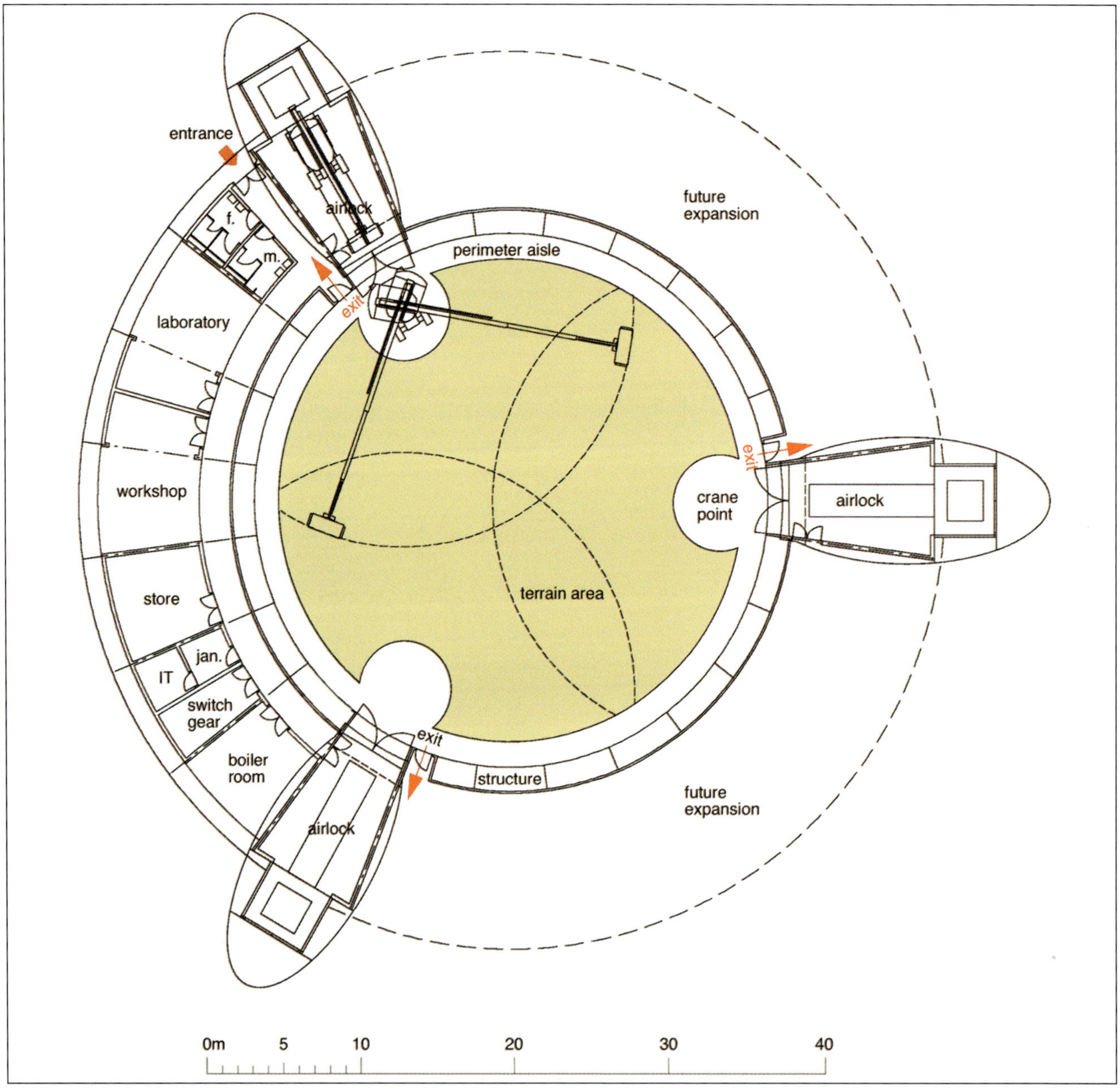

FIGURE 4 Planet surface simulator phase 1 plan.

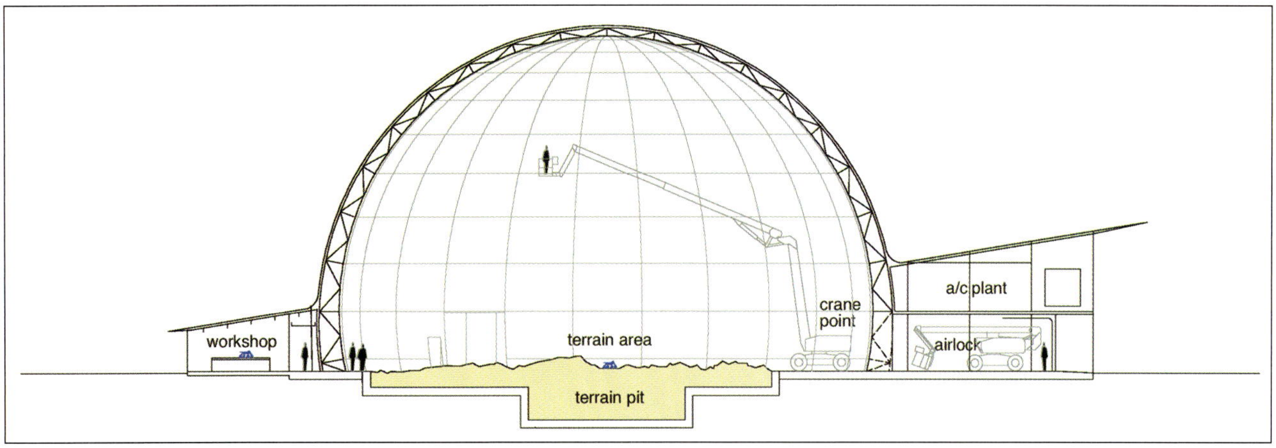

FIGURE 5 Planet surface simulator phase 1 section.

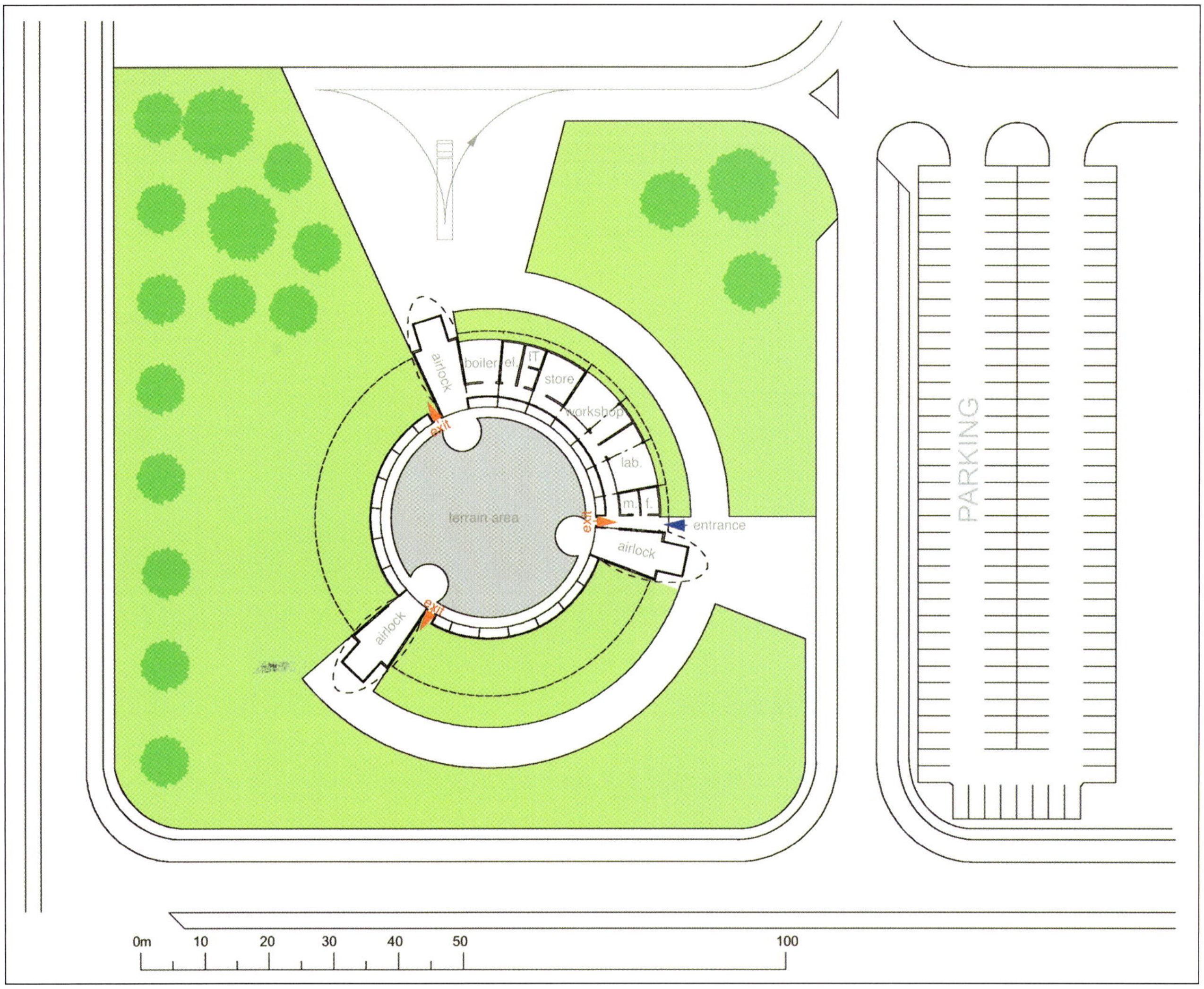

FIGURE 6 Simulator site layout.

industrial park being built adjacent to the campus. Figure 6 shows a generic site layout. The building occupies the center of the site, surrounded by landscaping. A service vehicle access and turning area is opposite one airlock to support all service and operational deliveries. A second airlock is opposite a pedestrian forecourt leading to the staff entrance to the building at the airlock's side. All three airlocks are connected by a peripheral access lane that enables the cherry-picker crane to enter any airlock. The rest of the site is landscaped. Parking is located off site, not included in the 1-hectare site area.

FIGURE 7 Exterior appearance evokes planetary exploration.

APPEARANCE

The building is highly visible because of its shape and size. Domed buildings of these proportions are often associated with nuclear energy plants, and so it is important that any negative association be precluded by the dome's exterior treatment. The graphic treatment of the curved zinc dome surface, based on cloud belts of the gas-giant planets, clearly signifies this dome's association with something quite different—space exploration (Figure 7). The building's apparent horizontal plane is deliberately tilted to reinforce its association with planetary exploration by echoing the way most planets' rotation axes are tilted away from the ecliptic normal.

Bibliography

Clancy, P., Nixon, D., and Larter, N. (2006), "Conceptual Design Study of DOME—A Facility for Design and Optimization of Moon/Mars Explorers," 4CON Space Ltd., ESTEC Contract 19642/05/NL/CP, Contractor Report DOME-ES-01, Ennis, County Clare, Ireland, Oct.

29

spaceport design

CONSTANCE M. ADAMS
AND
GEORGI PETROV

INTRODUCTION

WHILE RECENT YEARS HAVE WITNESSED MARVELOUS breakthroughs in private space initiatives and in commercial support for spaceflight, specific issues inherent in the design of spaceports require special examination. The airport paradigm evolved over the course of the 20th century from its basic functionality of airfield and hangar to a complex transportation nexus of rail, air, road, and sea transit, supported by various hospitality, training, and other logistical functions. To date, spaceflight has been restricted to government programs that can accommodate internal planning and the ability to leverage other facilities, both military and civilian, to meet its needs. The advent of commercial spaceflight will effect a radical change in operational requirements for launch facilities, and along with these operational shifts come associated technical complexity that cannot always be met by leveraging publicly owned resources. In addition, there is every reason to expect that facilities supporting commercial space access will follow a pattern of formal transformation similar to the history of air access, as technologies emerge and mature and as the market grows from an initial handful of wealthy suborbital tourists to a mass market we can now only imagine.

Based on this supposition, we infer that a financially and technically viable commercial spaceport will need to be capable of supporting not only the same strong and unambiguous connections and ground functions a regional airport can accommodate but also the logistical and strategic resources necessary to enable flexible support of a changing array of space-access technologies (e.g., launchpads, fuel facilities, runways for horizontal takeoff and landing, efficient means of transporting resources across the site, etc). Along with these physical aspects of the specific architecture necessary to add the spaceflight element to a transportation facility are the invisible architectures dictated by launch range safety and ground operations (telemetry; communication; guidance, navigation and control; recovery range; etc.).

This chapter reviews the history of air-access facility types in the context of the dual evolution of commercial market and technology; identifies aspects of the mature airport that offer useful paradigms for spaceport planning; and suggests a probable pattern of typological evolution for the spaceport that might permit more accurate phasing of development for new facilities.

PRECEDENTS: THE AIRPORT

On 6 August 2006, the *New York Times* Sunday Style section headline proclaimed, "My Other Vehicle is a Gulfstream" (Trebay 2006). The accompanying photo showed a regional airport runway covered chock-a-block with private jets of varying sizes, parked so closely together that very little tarmac was visible between the splayed wings and fuselages. As the story asserts, private air travel has now become so affordable that "even the merely rich can afford private flights." To most of us living today, this image is not terribly startling; yet 80 years ago it would have been considered fantasy. Even as Fritz Lang, Frank Lloyd Wright, and other early visionaries of the 20th century were depicting large urban or suburban tracts whose airspace was peppered with personal aircraft, the reality of air travel

Copyright © 2008 by the authors. Published by the American Institute of Aeronautics and Astronautics, Inc., with permission.

prior to Charles Lindbergh's transatlantic solo in 1927 was very different. "Air mail" postal service was established in 1918, and prototype "aerodromes" featuring a hangar and open landing field began to spread in earnest in the mid-1920s. Until the frenzied explosion in airfield construction that immediately followed the Lindbergh flight, only a few handfuls of humans had either the resources or inclination to engage Orville and Wilbur Wright's wondrous technology and experience flight for themselves. Those who did were predominantly aviation enthusiasts, relatively well versed in the engineering of their aircraft and in the theory behind their function (Gordon 2004). In this sense, the period was not unlike today's tiny but avid community of space voyagers, and the evolution of commercial spaceflight can reasonably be expected to continue to follow similar patterns.

The design of air-traffic facilities represents one of the greatest challenges in the design field. In the complexity of embedded systems and the critical importance of integration with regard to multiple operational requirements, the airport is a technology unto itself. During its experimental phases, it underwent aggressive testing and adapted new solutions in response to failures in every area. It principally involves harmonizing the connection among four principal, disparate, and complex sets of operational requirements (i.e., architectural programs) discussed next; and the strategies for achieving this reconciliation evolved through several phases of technical, economic, social, and political transformation. Although a view of the development of air travel through each of these lenses can be extremely illuminating, for purposes of this analysis we trace the development of these programs as the principal operational arenas that comprise an airport, to remain focused on the goal of establishing useful planning and design principles for the first generation of commercial spaceports.

Two programs present themselves even from the earliest days of commercial flight, although it is only since the 1970s that it has become common in aviation-planning parlance to refer to "air-side" and "land-side" facilities and operations. Looking carefully at both programs serves to demonstrate the complexity of the overall airport plan. In simplest terms, land-side functions support the passenger, and air-side functions support the aircraft. But at greater detail, the appearance of clarity slips away.

Land-side operations include the processing of passengers and their baggage,

> without being physically close to the aircraft. In addition, sufficient ancillary facilities, such as concessions, are located in landside facilities to provide amenities facilitating a pleasurable experience for the passenger. Airside facilities . . . focus on the efficient servicing of aircraft, including fueling, loading and unloading. (Wells and Young 2004)

Land-side facilities are served by road and rail, carrying automotive and livery traffic as well as subway or other local forms of rail. Not only must passengers and ticketing personnel be efficiently delivered to the land-side facility, but food, drink, goods, and cleaning services as well. Passenger baggage is taken and commuted to the air-side program independently of the passenger. Outside the airport terminal, but still part of land-side operations, are extended elements such as car rental, shuttle buses, hotels, and parking lots. Commercial spaceports will also need to be capable of receiving passengers from numerous locations and accommodating them on site for training, medical checks, and tourism.

Air-side facilities support equipment and personnel for fueling, inspecting, and maintaining aircraft; flight personnel; and baggage handling and transfer. Gates must deliver passengers to the aircraft, fuel storage facilities must be supportable and serviceable, and aircraft must be able to park and taxi to and from the runways as quickly and safely as possible. How spacecraft will taxi from their assembly and processing facilities to the launch site is one of the primary engineering and planning decisions any spaceport designer must undertake. Finally, the invisible architecture of flight operations—airport management and flight control centers—must be accommodated in every airport. Flight operations for a commercial spaceport will require all of the equivalent infrastructure as well as space communications capability and facilities for training and certifying flight directors.

The demonstrated tendencies of transportation technologies to change, and of demographic groups to shift patterns, have been nowhere more visible than in the century-long history of the airport. To meet future changes, flexibility in sizing and layout of both air-side and land-side operations is desirable. Best practice in airport planning today tends to separate facilities designed to serve the land-side program from those supporting air-side activities to the greatest extent possible. Taxiing aircraft have very different requirements than pedestrians and ground transportation; the difference in scale of operations alone strongly suggests different design responses to each. However, attention to air traffic and ground traffic does not completely encompass the airport's operational requirements. Two other form-driving programs emerge: cargo and security.

Processing and transferring nonbaggage commercial and noncommercial cargo independently of passenger operations has played a large role in commercial flight from its earliest days, when commercial aviators relied upon delivering the mail to stay afloat. Today, the growing reliance on air transport of carriers such as FedEx, DHL, and UPS has pushed major air cargo companies to take over less-used regional airports or to build private air facilities at major hubs to efficiently serve their specific requirements. Nonetheless, virtually every airport must accommodate air cargo transfers both by and for their commercial airlines and by air cargo carriers. Therefore, cargo operations share runways and taxiways, and even aircraft, with passenger air-side operations, even though their needs are different from passenger air-side requirements. Also, cargo does not tend to overlap with passenger-based land-side facilities but does require access to the same roads and rail lines that serve passenger terminals.

Cargo will continue to play a large role in commercial spaceport operations, both in the need to support the ground facilities and also to support delivery and processing of spacecraft components and hardware for launch. Because spacecraft cannot deliver themselves to their launch site, runways must be capable of supporting a fully loaded Airbus Beluga, Boeing SuperGuppy, or other spacecraft-carrying planes, the equivalent of landing an Airbus 380, a 747, or a C5.

Although the cargo function has in a sense coexisted with passenger land-side and air-side facilities, it was not until the most recent phase of airport typological evolution that a programmatic element appeared at the boundary otherwise separating land side from air side. This function, security, arose in response to the early wave of hijackings in the 1970s and has reached a new level of significance since 2001. One way of looking at the security function is its focus on "sanitizing" air-side operations to prevent contaminants (in the form of threatening persons or objects) from entering aircraft. Thus, the security cordon absolutely separates primary land-side from primary air-side operations. Once passengers have cleared security, they have essentially left all links to the city behind and are prepared for flight. On a secondary level, however, the presence and decreasing porosity of the security function force a transfer of operations such as concessions, vending, and other passenger amenities from land-side to air-side facilities. Security has always played an important role in federal space launch facilities, with the public kept at a considerable distance from ground and space operations. This is even more exaggerated in the Russian program, whose very launch facility was purposely mismarked on maps until the fall of the Soviet Union. As commercial spaceflight becomes real, the security cordon will necessarily move somewhat inward to permit greater involvement on the part of potential investors, suppliers, customers, and families, but will remain an important part of the plan for any new spaceport.

In addition to the four operational functions represented by air-side and land-side requirements, a physical airport must also accommodate the technical functions that serve the facility's "metabolic" program. This metabolic program includes all equipment and utilities that support the principal land-side and air-side functions, such as power, concessions, baggage handling from ticketing to aircraft to carousel, aircraft and facility maintenance and servicing, and so forth. Proper treatment of this program in a discrete manner—properly integrated with the air and land elements—has resulted in dynamic, highly efficient facilities capable of high volume and long-term growth. Placement of metabolic elements like the security function can dramatically improve the efficiency and long-term flexibility of a site. Whether the airport's planners anticipate the security cordon truncating the land side closer to the land-side metabolic stream or farther from it is an additional discriminator in the overall design and efficiency of the facility.

Several design principles have emerged over the 80-year evolution of airport design by trial, error, and success. First, all successful design solutions have chosen one of four fundamental typologies; second, each of these types tends to favor one of the programs over the others; and third, even those that appeared well-balanced evinced some signs of dysfunctionality. How these functional or programmatic elements—the land-side, air-side, and metabolic functions including cargo, security, and utilities—have been treated in the development of formal typologies responsive to the airport paradigm is a rich discussion beyond the present scope. Here we summarize the history of the airport paradigm and proffer a reasonable projection of what the next few decades hold for spaceport development.

BRIEF HISTORY OF THE AIRPORT ARCHETYPE

The airport evolved over the past century in parallel with two concurrent (but not always harmonious) phenomena: changes in social patterns related to air travel and changes in aviation technology. Both of these evolutionary streams have imposed new requirements upon all of the fundamental components that comprise an air-traffic facility, though not in equal degrees. For instance, the growing popularity of air travel influenced the design of land-side facilities to accommodate passengers. Eventually the greater rate of growth in land-side requirements in turn levied demands on air-side assets and on the overall metabolic functionality of the facility. On the other hand, although the advent of the jet engine forced new requirements on such air-side facilities as runways and airspace, eventually this new technology also revolutionized land-side activities by enabling a dramatic drop in the cost of air travel relative to the standard of living.

There is no reason not to anticipate a similar duality of impacts caused by evolution in both the market and technology engaged in commercial spaceflight. Studying the evolution of the airport archetype offers potential cost savings to spaceport developers by planning these very costly and technically complex facilities for long-term flexibility and robustness in response to changing external requirements. Reviewing airports over the past century, we can characterize three fundamental phases of their evolution.

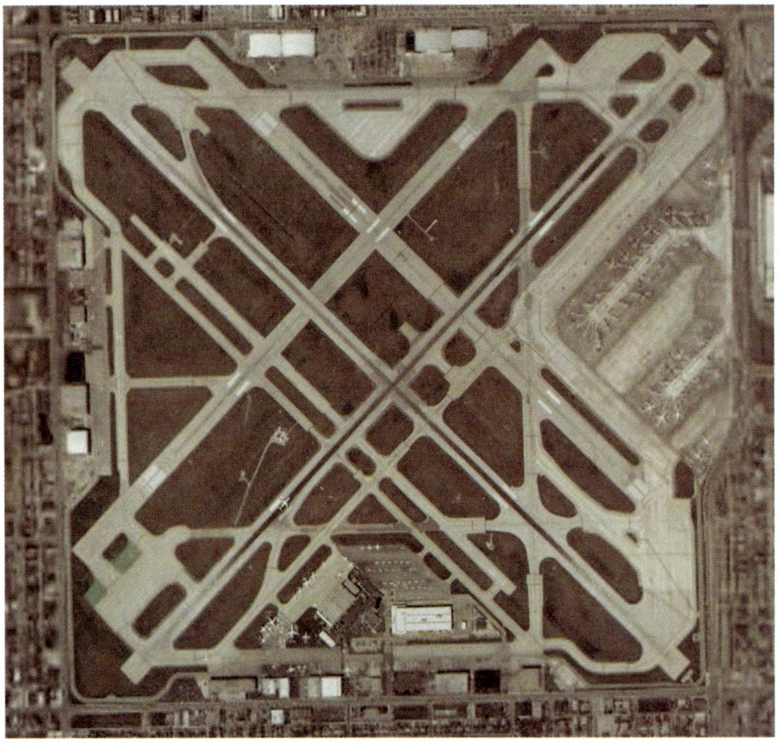

FIGURE 1 Chicago Midway, aerial view.

FIGURE 2 Bob Hope Airport, Burbank, California.

Airport Period 1: Evolution of Flight (1910–1930)

At the very beginning of passenger air travel, individual aviators were still making historic journeys, and those lucky few who were able to afford passenger travel did so with a sense of adventure accompanied by expectations of luxury similar to those onboard ocean liners despite the obvious hardships of flying. Elegance surrounded the passenger's ticketing and waiting area, and the limited baggage permitted was handled by the service providers from curbside to curbside. "Airfields" were located close to the center of town (Figure 1). Ideas about the eventual ease and universality of air access included plans to build airstrips across neighborhood rooftops to enable residents to commute to work by air. Reality, however, meant that most airfields were cow pastures with one or more hangars and a small passenger building at one edge (Figure 2). After Lindbergh's flight in the late 1920s, terminals were built up, and some runways were paved to support larger planes using prevailing winds.

The earliest type of airport, "point access," assumed a relatively direct point relationship between the land-side access (passengers) and the aircraft. In most cases the single, centralized passenger terminal similar to the railroad terminal served as a processing and waiting area; when a flight was ready for boarding, passengers would depart the terminal at one point and walk out onto the tarmac and climb stairs to their waiting aircraft.

Chicago's Midway airport is an excellent example of this type, and many major cities still have small commuter airports dating from this era, when easy access to the central business district was desirable and the (often unpaved) airfield enabled omnidirectional takeoff and landing to meet the requirements of small propeller-driven aircraft.

Advantages of point access type were very simple and direct access. Disadvantages of point access type were unsuitability for large volume of travelers or jet aircraft due to bottlenecks in both terminal and apron.

Airport Period 2: Evolution of Air Carriers (1930–1970)

By 1930, the effects of the Air Commerce Act of 1926 were being felt in the United States. As the market for commercial flight picked up, many companies already in the transportation business (including several railroad magnates and Henry Ford himself) decided to invest in air travel. Private air carriers sprang up in large numbers across the United States, Europe, and Latin America, and the market went through a rapid life cycle between 1930 and WWII. The formation in 1938 of the Civil Aeronautics Board (CAB) introduced regulations for air travel and reduced the number of new carriers entering the market. Many regional airports were laid out during this period, for the first time with sufficient acreage to permit growth. Thanks to improvements made in aviation technology during WWII, and the development in 1949 of the DeHavilland Comet—the first commercial jet aircraft—air travel became more available, and the carriers providing air service developed stronger brand identities. Every decade's projections for the growth of the industry were consistently outpaced by reality. As the volume of customers and the availability of aircraft grew, so did the need for good air-side logistical support and maintenance. Air carriers began to build their own facilities at municipal airports to accommodate air-side operations as well as to accommodate passengers and instill a sense of brand loyalty.

As air travel increased and airports grew, it was clear that point-access typology could not accommodate the volume of traffic or the apron room needed to support jet technology. In addition, the economics of commercial air travel generated an increasingly competitive atmosphere among carriers. Multiple terminals began to spring up between the access road and the runways so that the land-side access became decentralized despite air-side activities still being centrally controlled. Hybrid airports of the mid-20th century arose in the era of heavy corporate identity and loyalty such that each airline would build and maintain its own terminal, with the airport authority playing a planning role. However, in the 1960s the Federal Aviation Administration (FAA) and regional airport authorities began exerting greater influence over airports, insisting on uniform terminals and traffic conforming to centralized master plans. This development added considerably to safety and usability, as well as to economic stability as airline revenues rose and fell.

New York's John F. Kennedy airport (Figures 3 and 4) is the prime example of the agglomerated hybrid type, whereas at Chicago's O'Hare (Figure 5) and Los Angeles' LAX, greater uniformity of style was imposed on the carriers' terminals.

FIGURE 3 JFK layout in New York: each airline developed its own terminal.

FIGURE 4 TWA terminal, JFK: the pinnacle of air-carrier architectural branding.

FIGURE 5 Uniform multiterminal layout at O'Hare International in Chicago.

Advantages of the Hybrid type are that air carriers enjoy greater control over the passenger's total experience and pay to keep their assets in good working condition. Diversity of formal solutions results in uneven experiences but enables greater range of experimentation. Disadvantages of the hybrid type are optimized for point-to-point travel. Difficulty ensuring continuity of service on both air side and land side meant that transferring between air carriers left passengers and their baggage in logistical limbo. Additionally, security and other services must be separately addressed in each terminal.

Airport Period 3: Evolution of Market, Technology, and Regulation (1970–present)

On 22, January 1970, the first Boeing 747 "jumbo jet" landed at Heathrow Airport with a full complement of 324 passengers onboard, momentously ushering in the modern era of mass air travel (BBC News 1970). Jet engines had a dramatic effect on airport design in two ways. First, the greater speed and noise of the aircraft required longer runways than the older airfields and sufficient buffer to mitigate noise pollution on residential neighborhoods. Second, the increase in range and capacity enabled virtually anyone to fly and opened up the skies to practically any destination the public desired. The need to invest in significant new infrastructure to support this phenomenon enabled local, state, and federal governments to apply new air travel regulations and introduced the monolithic airport whose identity is more informed by local culture than by corporate branding. After the Air Deregulation Act of 1978, air carriers engaging in price wars began to establish "hub" airports and point-to-point travel succumbed to hub-based transfers that increased revenues. The impact of transfer flights on land-side operations and design requirements was dramatic. Formerly efficient point-to-point airports became nightmares for passengers desperate to make connections because they had to move perpendicularly to the planned land-side accommodation. This period gave rise to two principal types: "bar" and "hoop."

PRINCIPAL TYPE "BAR" All land-side functions are laid out in a bar shape parallel to air-side functions. Land-side operations are isolated at one end of the facility, at a central ticketing and processing terminal; passengers are then shuttled to remote gate concourses situated in the middle of an articulated apron that connects directly to taxiways with minimal taxi time between gate and runway. This type is characterized by long, continuous bar-shaped concourses with land-side traffic in the center and air-side traffic all around on both sides. This is a later evolution based on moving ticketed passengers smoothly and efficiently from processing terminal to gate on parallel concourses without crossing air-side traffic flow.

The first bar-type airport was Dulles International (Figure 6). The distance it posed from curbside to jetway was alienating prior to the security challenges of the 1970s; but once hub travel began, Atlanta Hartsfield's new bar plan revived the type and demonstrated its efficiency transferring passengers (or at least their baggage) between flights (Figure 7). The only major U.S. airport built in the past decade, Denver International Airport, replicates the Atlanta bar typology.

Advantages of the bar type are that functionality can expand by adding bars; also, flow of certain air-side operations is very good, and taxi time is minimized. Disadvantages of the bar type are that flow of land-side operations between concourses is badly congested; also, the metabolic utility system must be doubled, or more, to accomplish its task. As security is tightened, access from the processing terminal to concourses is a major bottleneck, with only a single access point for all airport passengers.

PRINCIPAL TYPE "HOOP" Land-side/air-side interface describes an arc, with land-side access traffic forming a single arterial core in the center. The smaller inner circumference better accommodates foot passengers and slower-speed traffic such as cars. With air-side functions on the outside of the arc, the broader space along each radian easily accommodates larger, faster vehicles and permits parking aircraft with

FIGURE 6 Washington Dulles International Airport: a) aerial view and b) view of terminal building interior.

broad wing spans. It has been argued (Wells and Young 2004) that this type is a projection of multiple point-access solutions around what is often referred to as a "chain of pearls" (Callahan and Birk 1969). In general, the geometry of this type balances air-side functions against land-side functions with interesting results. As each terminal fills out into a total circle, the more luxurious the air-side access is on the one hand but the less efficient the land-side and metabolic functions are; in full-circle terminals, air-side functions require more infrastructure than is efficient or necessary to do the job, and security becomes an

FIGURE 7 Hartsfield Jackson Atlanta International Airport: extensible bar type.

extremely costly operation. Also, although this type offers a very efficient experience for the destination traveler, those passengers who must change planes—particularly between terminals—find themselves confronted with a Herculean dash and the likelihood of missed flights. These conditions taper off in extremity as the arc of the hoops softens; Charles de Gaulle Airport (CDG), for instance, is much easier to traverse on the land side than Dallas/Fort Worth Airport (DFW) even if the ratio of open tarmac to aircraft is somewhat reduced.

There are different versions of this type. The Kansas City Airport features a cloverleaf with circular terminals that encompass all landside parking in their center and allow aircraft total range around the exterior of the circle (Figure 8). DFW has a modified version of this layout, with semicircular terminals split along a bar of car traffic, and aircraft navigating the outer field on both sides (Figure 9). In the slenderest adaptation of this form, CDG Terminal 2's split-lozenge shapes offer fewer gates at the exterior but greater concentration of activity on the land side (Figure 10).

Advantages of the hoop type are that the shallower hoop type evolved in recent years has spun off a separate curvature for metabolic functions, thus achieving the greatest integration and optimization across systems so far. Passenger transfers are relatively straightforward in a shallower hoop airport. Disadvantages of the hoop type are that deeper hoops are more costly to support per gate and risk stacking delays as more transfer flights fill the profile. Transfer time on land side rises considerably as the arc deepens.

FIGURE 8 Kansas City Airport: ideal point-to-point facility.

APPLICABILITY TO FACILITIES FOR SPACE ACCESS

As a complex technology supporting other complex technologies and multiple operational profiles, the airport is generally understood to have entered its operational phase. Therefore, facilities built to support commercial spaceflight as the next generation of transportation should be planned from the outset with a keen awareness of the airport as prior art, including its history and full range of operational and technical requirements. It is likely that facilities built to accommodate carriers pioneering space access using X-prize technologies will most closely resemble the point-access type of airport in overall layout. However, the cost and effort of constructing physical and operations facilities for spaceflight are so significant that a phased approach to spaceport master planning, taking into account potential evolution into a more sophisticated and efficient paradigm over time, can significantly enhance investor value, and reduce technical and financial risk, as strategies mature. Table 1 summarizes operational requirements (program) for both airport and spaceport facilities.

The spaceport functions represented here are taken from ASTWG (2003), a study done by the NASA Kennedy Space Center's Advanced Spaceport Technology Working Group. What is most interesting about this comparison is the extent to which the fundamental operations and requirements of a space-access facility mirror those of an airport. Even where extra functions are necessary to support space operations (for example, activities shown in Figure 11), these functions are additive to basic day-to-day airport operations, that is, they are of a specialized but not exotic type.

Such strong correlation between the operational requirements (or program) of the two applications supports consideration of the airport as prior art for planning commercial spaceports. Consistent with design engineering best practices, spaceport planning has access to a large database of lessons learned and alternate paradigms (or planning strategies) valuable for risk reduction and management.

FIGURE 9 Dallas/Fort Worth: a challenge for passengers transferring between terminals.

FIGURE 10 Charles de Gaulle Terminal 2: shallower hoops make land-side transfers easier.

TABLE 1 Comparison of airport and spaceport operational requirements.

OPERATIONAL REQUIREMENTS	
AIRPORT	**SPACEPORT**
Ground access—road, rail	Ground access—road, rail, water
Runways for aircraft (private jets to cargo planes)	Runways for aircraft (private jets to jumbo cargo aircraft)
Runway maintenance	Runway maintenance
	Launch pads—vertical, horizontal
	Launch range surveillance and support
Taxiways	Transportation to launch facilities
	Ground systems restoration/turnaround
Fuel farm—jet fuel, gasoline, diesel	Fuel farm—jet fuel, gasoline, diesel, solid and liquid propellants
Operations support	Operations support
Flight crew	Flight crew, trainers
Ground vehicle management	Ground vehicle management
Control tower	Control tower
Air traffic control	Mission operations center
	Ground operations, vehicle telemetry, flight ops
Airport management (strategic, operational, and fiscal)	Spaceport management (strategic, operational and economic)
Airport engineering	Spaceport engineering
Aircraft ground operations	Spacecraft preflight operations
	Spacecraft postflight operations
Aircraft maintenance facility	Aircraft maintenance facility
	Spacecraft element assembly and integration
	Spacecraft checkout
Cargo and baggage handling	Payload handling and assembly
Cargo and baggage loading and unloading	Payload integration
	Passenger cargo integration
Airport logistics	Spaceport logistics
Passenger drop-off, care	Passenger processing
	Passenger training
	Medical screening and support
Emergency medical services	Emergency search-and-rescue, launch range safety and medical services
Security—federal, port authority	Security—corporate, port authority
Conference facilities	Conference facilities
Air carrier offices	Launch carrier offices
Hotels	Passenger hostel facilities
	Tourist facilities
Restaurants	Commissary/restaurants
Supply management	Supply management
IT and logistics information systems	IT and logistics information systems
Passenger services	Customer services for commercial astronauts, families, tourists, commercial space entrepreneurs

(a)

(b)

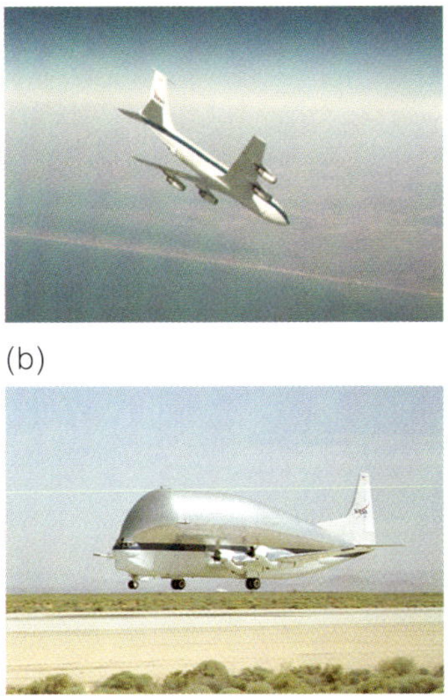

FIGURE 11 Spaceport-unique functions include:
a) reduced gravity training flights and
b) logistics delivery of rocket stages.

EYES TOWARD THE HEAVENS

Applying the air-travel evolution pattern just described, we can expect similar stages in technology and social patterns as space travel evolves and prepare to meet them by phasing the facilities that support air and spaceflight. The projected pattern might look a bit like the following.

Air/Spaceport Period 4: Evolution of Spaceflight (1960–2005)

DRIVERS AND CONSTRAINTS Just as the first period of air travel involved evolution from a risky endeavor undertaken by only a few experts, so has human spaceflight in its first phase been restricted to government-run programs whose goal was to achieve and then sustain human presence in space (Figure 12). Unlike early attempts at flight, spaceflight access evolved in an atmosphere of intense concern for national security that fostered the research and design (R&D) investment necessary to develop the technologies yet rendered the parallel U.S. and Soviet programs all but inaccessible to the public. Therefore spaceports were designed neither to accommodate commercial activity nor parallel technological efforts; in addition, they were developed one program at a time, leveraging existing resources where possible and without the ability to anticipate requirements future generations of spaceflight technology might levy. Safety has played a heavy role (Figure 13) in restricting further development of launch facilities because of the need for a huge launch safety range within anticipated low-elevation rocket

(a)

(b)

FIGURE 12 Government launches: a) *Apollo 11* and b) *Soyuz*.

(a)

(b)

FIGURE 13 Large potential for public safety concerns: a) crash of M2F2 test vehicle; b) Texas National Guardsmen with a piece of OV-101 *Columbia*, February 2003.

and spacecraft trajectories. Likewise, the relationship between orbital inclination and latitude governs desirability of specific locations for space launch.

RESOURCE Because of these constraints and the large cost of building new vertical-launch facilities, both U.S. and Soviet spaceflight programs found ways of reusing their infrastructure from generation to generation. Existing runway resources such as military airfields were leveraged to address three principal operations areas: development and testing of horizontal takeoff and landing programs such as the X-15, lifting-body X-plane, and operational flight systems (Figure 14); delivery of cargo and equipment including spacecraft modules; and training of crew members and flight-support specialists. In addition, existing vertical-launch resources such as missile test ranges were leveraged in Kazakhstan and Florida and converted into long-term launch facilities (Figures 15 and 16). The original Soviet approach, relying on rail lines for transporting vehicles and major cargo across the launch facility, proved to be extremely robust (Figure 17). Major areas of improvement indicated for future facilities principally involve testing, processing, and maintaining spacecraft and spacecraft components. Ground processing and logistics issues can be pinpointed as the source of most, if not all, space fatalities to date,

(a)

(b)

FIGURE 14 Runways to support spaceflight: a) X-15 and b) Space Shuttle Orbiter horizontal landings.

FIGURE 15 Kennedy Space Center launchpad 39A, with shuttle encased in launch-support infrastructure.

FIGURE 16 Baikonur LS250 launchpad with *Energia*.

(a)

(b)

FIGURE 17 Constrasting approaches for vehicle transportation to the pad: a) *Soyuz* rail transport; b) shuttle "stacked and packed" on crawler.

and remain largely unaddressed as an open source of risk for future programs.

ACCESS Even after easing political hostilities (Figure 18), both spacefaring superpowers retained control over their equipment and programs. In the last few years of this phase, the Russian program began to make noncritical portions of their program available for commercial exploitation, including training and flying private passengers to the International Space Station for brief durations known as "taxi flights" (Figure 19). This phase may be thought to have ended with disbursement of the first X-Prize for commercial achievement of suborbital flight. However, in terms of infrastructure, even commercial efforts stimulated by the X-Prize challenge have tended to leverage existing publicly owned infrastructure (including Baikonur Kosmodrom, Dnepr launch range, and the Dryden Air Force Base facilities at Mojave Airport) rather than to develop their own facilities

FIGURE 18 Aleksei Leonov and Deke Slayton during *Apollo–Soyuz* Test Project—detente in space.

for development, support, and launch of vehicles, and for training and support of crew members.

MARKET During this early stage, there is very little market for commercial human spaceflight because of the cost and complexity of putting people in orbit. Many "blue sky" ideas are proposed, and the desire to stimulate a viable space access market begins to grow.

Air/Spaceport Period 5: Evolution of Commercial Access (2005–2035?)

DRIVERS AND CONSTRAINTS Beginning with disbursement of the first X-Prize for commercial suborbital flight to Burt Rutan in 2005, the era of commercial human access to space officially arrived (Figure 20).

In the early phase of commercial human spaceflight, the most critical constraints are technical and economic. Even as some city-states and regional governments begin to invest public money to purchase and plan tracts for spaceport development, the wide range of suborbital and orbital technologies in development and test stages means a risky period as

FIGURE 19 Dennis Tito, first space passenger tourist.

FIGURE 20 X-Prize winner *Space Ship One* mounted to *White Knight* carrier airplane.

FIGURE 21 Sea Launch infrastructure solution is both flexible and constraining in unique ways.

FIGURE 22 Mojave spaceport.

infrastructure providers bet on the viability of particular systems to plan their support facilities (Figure 21). Technologies that are still experimental or developmental cannot attract a mass market; so in this phase the field winnows as a few systems mature toward commercial human spaceflight, and the options are downselected. The space carrier industry starts out allied with technology developers (e.g., with Blue Origin, Virgin with Scaled Composites, Kistler with Space Adventures, PlanetSpace with Canadian Arrow), and each lives or dies according to its system's success rate. Economics and regulatory guidelines also drive the space carrier industry in this stage. Once governments establish guidelines to support and regulate space commerce, we can anticipate an upward tick in the market of would-be passengers because of improved perceptions of safety.

RESOURCES As in the early days of flight, when landing fields developed to support WWI airpower were converted into regional "aerodromes," the first commercial phase of spaceflight leverages existing, publicly owned, and underutilized facilities for private events. Already the former Dryden test flight field at Mojave, California, has hosted large audiences for the spectacle of the first private suborbital flight (Figure 22), and continuing X-Prize competitors gather annually at the future site of Spaceport America (SA) near White Sands, New Mexico, to give public demonstrations of recent technical achievements. Like Mojave Airport, SA leverages airspace already cleared by and for a nearby government facility. Similarly, Bigelow Aerospace and other private enterprises have leveraged launch services assets formerly belonging to the Soviet military. Thus, commercial space enterprises overcome the economic hurdle of initial infrastructure investment by leveraging former government resources for private enterprise. Until the market boom that defines entry into the third, mature phase of commercial human spaceflight, spaceport resources combine such leveraged activity with joint enterprises like public-private partnerships. Regional authorities, such as Oklahoma, New Mexico, and Florida, and similarly sized city-states like Dubai or Singapore, have all declared intentions to build commercial spaceports. In the United States then, early "astropreneurs" share a portion of their risk (the spaceport) with the tax-paying public. Just as early airport construction often met with local resistance and scorn on presumption of wasted expenditure of public funds, spaceport developers can expect to have to woo local communities. Regional spaceport authorities that have not planned carefully for long-term technical maturation and facility flexibility face economic challenges during this growth period.

ACCESS Because of lingering technical risk and economic pressure, access to space remains limited by economic factors. However, during this phase, access broadens tremendously. As it does, commercial spaceports develop long-stay hostels and facilities for medical observation and training of the new cadre of space travelers, as well as tourism and grandstand facilities for public observation of launches and landings.

MARKET The early commercial market includes "extreme space tourists" like Dennis Tito and Mark Shuttleworth, along with a small cadre of notable daredevils and wealthy enthusiasts. Just as Charles Lindbergh personally led a charge into the air and WWI flying ace Eddie Rickenbacker formed Eastern Airlines, so will the first leaders of this market be individuals of high profile who themselves want to fly to space. This category already includes individuals like PayPal inventor Elon Musk, Virgin records founder Sir Richard Branson, and Amazon.com founder Jeff Bezos. However, unlike the early days of air travel, today's culture of litigation and liability likely slows this process until technical maturity challenges are largely overcome (Figure 23).

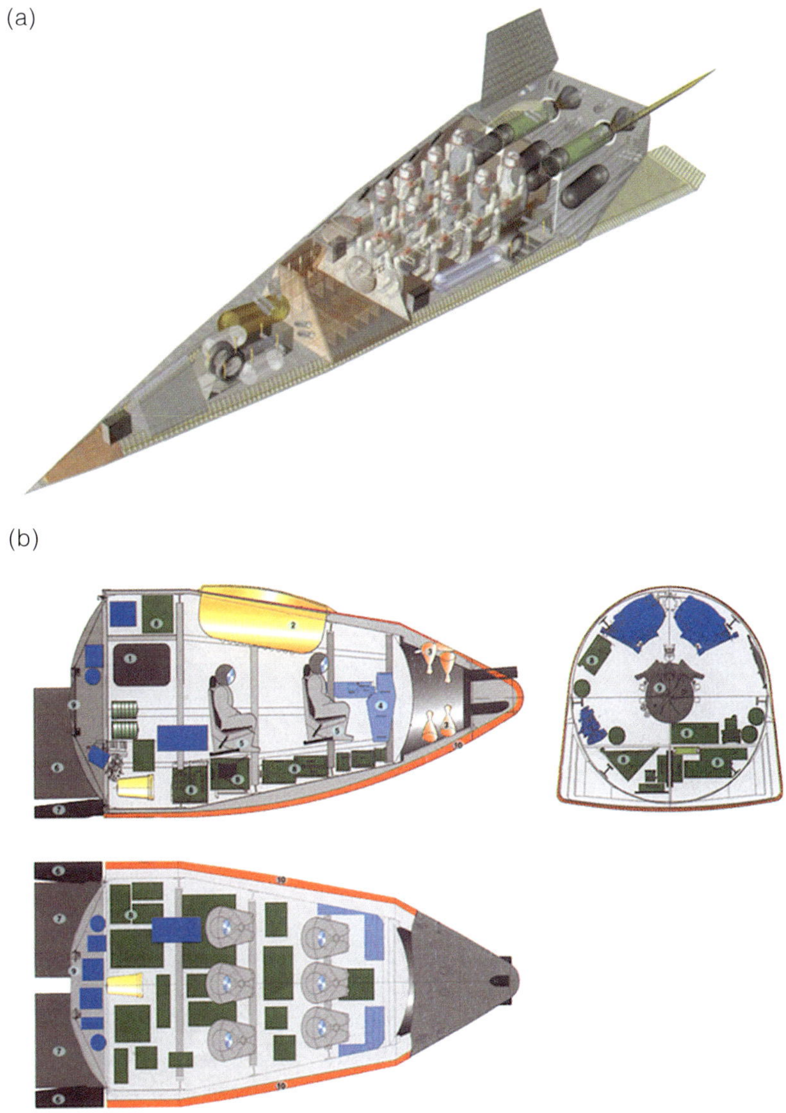

FIGURE 23 Market sorts out would-be approaches based on flight success rate and economic viability: a) Silver Dart and b) Kliper.

Spaceport Period 6: Evolution of Space Travel Market, Commerce, and Regulation (2035?––?)

DRIVERS AND CONSTRAINTS This phase does not begin until at least one system for space access is fully operational—beyond "experimental" as defined by the Columbia Accident Investigation Board (CAIB 2003), and the risk to carriers of economic repercussions from failures is brought under control. Once this phase begins, regulations to protect launch range safety continue to drive spaceport site selection and viability. So does the ability of the spaceport to accommodate both large numbers of passengers and participants, and several simultaneous, discrete operations such as training, vehicle integration, and passenger preparation for more than one space carrier. Later in this phase, space carriers and consortia construct orbital or lunar-based tourist facilities. Building and supporting these facilities requires significant new ground infrastructure, ideally closely linked to operations centers for processing, cargo, and launch.

RESOURCES Proper planning by spaceport developers in the prior phase makes resources available during this phase to expand their facilities to accommodate mass-market access to space.

ACCESS Space access grows from a small market of wealthy enthusiasts to a broad, deep market.

MARKET The drivers, constraints, and paradigms just listed suggest that the market for space access remains small for an extended period of time, but once opened expands at a rate beyond early expectations (Figure 24).

FIGURE 24 Toroidal space colony—an orbiting destination and spaceport.

CONCLUSION

Long-term mass utilization of human space passenger travel is difficult to envision, but the precedent of air travel strongly suggests that it should be anticipated. The airport is a complex technology that has passed its experimental phase and is a valid example of "prior art" useful for planning commercial spaceports. When contemplating principles that will govern robust and efficient planning of spaceports, the lessons of nearly a century of air-traffic planning and design can be applied with confidence so long as fundamental ideas are understood and translated with enough flexibility to anticipate growth and change over time. Based on projecting operational requirements, preliminary planning concepts for commercial spaceports can take into account requirements of various launch platforms, local or regional benefits, and different rates of growth. Although a commercial spaceport designed for today's market would be necessarily rather modest, the cost of infrastructure that must be invested in undertaking it means that it is financially prudent to begin development based on a sustainable plan for clear and phased growth over the next 50 years or more.

References

Advanced Spaceport Technology Working Group Technology Team (2003), "V1 Generic Spaceport Enabling Operations" and "Generic Spaceport Operations Model," edited by Keith Britton, Rev. B. 17, April.

BBCnews (1970), "On This Day 22 January 1970: Heathrow welcomes first 'jumbo jet;" http://news.bbc.co.uk/onthisday/hi/dates/stories/january/22/newsid_3725000/3725963.stm.

Callahan, R. H., and Birk, R. F. (1969), "Airports for Future Aircraft—A Planning Guide," AIAA Paper 69–808, July, p. 2.

Columbia Accident Investigation Board (2003), "*CAIB Report*," Vols. I–VI, NASA, Washington, D.C., Aug.–Oct., http://caib.nasa.gov/default.html.

Gordon, A. (2004), *Naked Airport: A Cultural History of the World's Most Revolutionary Structure*, Metropolitan Books, New York, Chap. 1.

Trebay, G. (2006), "My Other Vehicle is a Gulfstream," *New York Times*, Sunday, Aug. 6.

Wells, A., and Young, S. B. (2004), *Airport Planning and Management*, 5th ed., McGraw–Hill, New York, p. 207.

Bibliography

Adams, C., Rivera, A., and Wenda, A. (2002), "Built for Flight" Poster, World Space Congress Space Architecture Symposium. International Astronautical Federation, Houston, TX, Oct.

Future Interagency Range and Spaceport Technologies [FIRST], (2004), Advanced Spaceport Technology Working Group Technology Team, "Spaceport Concept of Operations: A Vision for Spaceports and Future Space Transportation Systems," edited by Keith Britton, NASA Kennedy Space Center, Cape Canaveral, FL. Nov., http://astwg.ksc.nasa.gov/Media/Final_Draft_Spaceport_CONOPS_V2.pdf.

Garcia, J., Berlanga, A., Molina J. M., Besada, J. A., and Casar, J. R. (2002), "Planning Techniques for Airport Ground Operations," Inst. of Electrical and Electronics Engineers, Paper 2002 0–7803–7367–7/02.

Siddiqi, A. (n.d.), "Deregulation of Flight and Its Consequences," U.S. Centennial of Flight Commission, http://www.centennialofflight.gov/essay/Commercial_Aviation/Dereg/Tran8.htm.

Spaceport Systems International (1996), "The California Spaceport Integrated Processing Facility Safety Plan," Document No.: IPF 96-SA01, Rev. A, Lompoc, CA, Nov. 21, http://www.calspace.com/IPFSAFE.PDF.

Zukowsky, J. (ed.) (1996), *Building for Air Travel: Architecture and Design for Commercial Aviation*, Catalog, Art Inst. of Chicago, Prestel, New York.

Zukowsky, J. (ed.) (2001), *Building for Space Travel*, Catalog, Art Inst. of Chicago, Harry N. Abrams, Inc. New York.

INTRODUCTION

SINCE THE BEGINNING OF THE SPACE AGE in the late 1950s, an understanding has grown of our planet as an extraordinary cradle of life in the vast, dark universe. But we also are learning to see the limits of the Earth system, the fragility of our atmosphere, and the influence human population has on this closed-loop system that receives its driving energy from the sun.

Yet we exploit our planet with unappeasable hunger: the wealthiest 20% of people account for 86% of the world's total private consumption. If every person alive today consumed at the rate of an average person in the United States, three more planets would be required to fulfill the demand. This by itself is alarming. Even more appalling is that despite the accelerating accumulation of wealth in industrial societies, and despite real-time news distribution and high-speed transportation by airplanes and trains, it nonetheless seems unachievable to prevent people dying from hunger or lack of clean water. The State of the World report concludes balancing population growth vs poverty is our number-one challenge (Starke 2003). Almost a quarter of the planet's population, 1.2 billion people, are classified by the World Bank as living in "absolute poverty," defined as living on less than 1 U.S. dollar a day (World Bank 2008).

On the other hand, we are flying to space, and in so doing realizing how valuable life's most basic resources are. To launch 1 kg of water into low Earth orbit costs about 20k euro. In contemplating human missions to Mars, transporting all supplies for over two years appears impractical. Systems would have to be developed to recycle air and water, and perhaps food, and to conserve energy. The end state of such development is eventually a closed-loop system powered by the sun: a microcosm of our "mother ship" Earth.

30

space architecture for the mother ship: bringing it home

ANDREAS VOGLER
AND
ARTURO VITTORI

COWBOYS, SPACEMEN, AND PROSUMERS

Kenneth Boulding (1966) stresses the need to change from a "cowboy economy" to a "spaceman economy." He describes the cowboy economy as an open system with plenty of world. Its measure of success is high throughput. The spaceman economy is a more closed system in a narrow world. Its measure of success is quality and complexity of stock (i.e., human bodies and minds).

One of the key problems in developing countries is lack of infrastructure and high costs to introduce it. Industrial societies rely on costly infrastructure to bring fresh water, electricity, heating, and telecommunications directly into apartments and houses and to remove waste seemingly easily. However, there is no such thing as a municipal water supplier in a spaceship. On a spaceship, the concept of waste is redundant; instead, there are resources in different states of processing (just like in a natural system). Low-energy and lightweight systems are needed to keep the systems going, make potable water from liquid waste, and make food using solid waste. Such systems do not yet fully exist, but much research is heading in this direction. Eventually this technology will enable spaceships to become independent of Earth-based resupply.

This technology will be costly initially, but the spin-off will address one of the largest markets of the world: private housing. The Fraunhofer Institute (2004) forecasts a strong, growing market for autonomous systems in private housing, with people investing in independence from municipal infrastructure and ever-growing prices for water, waste and energy. (Note: throughout this chapter, the

Copyright © 2008 by the authors. Published by the American Institute of Aeronautic and Astronautics, Inc., with permission.

word "autonomous" specifically denotes such independence from societal infrastructure.) This competitive market would make the systems cheap and eventually enable developing countries to leapfrog the whole process of investing in and maintaining expensive centralized infrastructure. A remarkable example of such a technology jump is the mobile phone, which is currently spreading successfully in developing countries. Obviating the construction of landlines, cell phones are proliferating currently in rural areas of China, Ghana, and India.

The underlying thesis of this analysis is that once we can think in terms of decentralized, autonomous, solar-powered, recycling space systems, we can imagine such technology jumps for terrestrial applications. Once we imagine them, they become more likely to happen. The oft-heard argument that technology is too expensive and that there is no money in developing-world markets, has been dramatically disproved by examples like the motorbike and mobile phone. And the need is real: we live in a time when despite unprecedented amounts of money being available, we have not figured out how to help the world's poor lead a decent life.

Promising crossover fields from space technology to development aid include 1) mobility and communication; 2) closed-loop autonomous systems; and 3) *in situ* resource utilization (ISRU). These are elaborated and illustrated by examples given in what follows. Recent projects show enormous potential for such development. The company Architecture and Vision has proposed several projects suggesting implementation of space technology in developing countries. Unfortunately many architects left this field after the 1960s when frustration developed that design and architecture could not "change the world." Whereas the earlier belief that it could, might have been naive, our task as architects and engineers is to try staying current with advanced technology and implementing it into dignified housing for human beings under given environmental, economic, and social conditions. Whether in space, Europe, or Africa, this challenge is largely invariant.

In Alvin Toffler's "Third Wave" society (Toffler 1980), highly infrastructure-dependent consumers eventually become self-producing and independent "prosumers" who help to produce and maintain a clean and healthy environment as they consume. Through easy access to communication they can identify ad hoc business opportunities, using information to offset material resource needs as in the mobile phone example.

MOBILITY AND COMMUNICATION

Communication can be described as mobility of information. Mobility of hardware and software is a key to information and knowledge. Most developing countries are severely deprived of such mobility. Long distances must be covered by foot every day, to work, to get water, and to visit relatives and friends. Transportation and access to information, Internet, or even telephone are quite limited. Compared to the Western world's access to knowledge and information, these cultures are literally disconnected. However, the abundance of technology and some consequences of globalization open possibilities for poor nations and even individuals to leapfrog development.

Digital Divide

Increasingly the world divides not only into rich and poor, but into those who have access to information technologies and those who do not. The term "digital divide" came into regular usage in the mid-1990s (Wikipedia 2008), when the Internet began wide distribution into Western households. To cross this divide, Massachusetts Institute of Technology (MIT) Media Lab launched a research initiative to develop a "$100 laptop" now named the XO-1, a technology that could revolutionize how we educate the world's children. To achieve this goal, a new, nonprofit association, One Laptop per Child, was created. "If you take any world problem, any issue on the planet, the solution to that problem certainly includes education," says Nicholas Negroponte, the initiator of the project (Bullis 2005). He learned from previous work with schools in Senegal, Costa Rica, India, and other countries that simply providing access to a computer is key to leveraging a child's innate creativity and curiosity. "Even in the developing parts of the world, kids take to computers like fish to water," Negroponte notes.

The XO-1 (Figure 1) is designed to have much lower cost (market price is about $200 U.S.) and be much longer lived than typical laptops. It uses flash memory instead of a hard drive, runs the Linux operating system, and uses the Sugar user interface. Mobile ad hoc networking based on the 802.11s wireless mesh network protocol allows students to collaborate on activities and share Internet access from one connection. The wireless networking has much greater range than typical consumer laptops. For use at home and where power is not available, the XO-1 can be solar or foot powered. It comes with at least two of three options: crank, pedal, or pull-cord. Children could have a second battery for group charging at school while using their laptop in class (OLPC 2008).

FIGURE 1 XO laptop (courtesy of fuseproject).

FIGURE 2 Already today India has more cellular phone connections than landlines (courtesy of Hunger Project).

Children could take a computer with them wherever they go, learning languages, math, science, geography, and economics as well as playing games and chatting online with friends. They could draw and compose music. Negroponte's original plan was aggressive: to produce 100 to 150 million laptops by 2007. By mid-2008, 667,000 orders were confirmed. Uruguay was the first country to purchase a full order: 100,000 laptops in October 2007. With another 200,000 laptops Uruguay can cover all public school children between 6 and 12 years old. Also participating in the project are Afghanistan, Cambodia, Ethiopia, Colombia, Haiti, Mexico, Mongolia, Papua New Guinea, Peru, Rwanda, and the United States of America.

Negroponte's emphasis on education is visionary. Although its impact on developing nations might occur only over the long term, we should not underestimate people's inventiveness once technology is easily accessible. Yet *The Economist* (2005a) sees the digital divide not as a problem in itself, but a symptom of deeper, more important divides: income, development, and literacy. Fewer people in poor countries own computers and have access to the Internet simply because they are too poor, are illiterate, or have other more pressing concerns such as food, healthcare, and security. So even if it were possible to cause a computer to appear in every household on Earth, this would not necessarily achieve very much: a computer is not useful if you have no food or cannot read.

Leapfrogging the Digital Divide

One interesting effect of digital technology (and improved batteries to make devices long-running) is that it not only quickly opens the gap between the people who have access and those who do not, but it also provides the means to leapfrog that gap. India has the second-largest mobile phone market in the world (Hindu Business Line 2008) (Figure 2). This means the turnover is already fast underway.

But technology can also have immediate effects on the economic prosperity of even poor people. Evidence suggests mobile-phone technology has the greatest impact on development. Mobile phones raise long-term growth rates, their impact is twice as big in developing nations as in developed ones, and an extra 10 phones per 100 people in a typical developing country increases GDP growth by 0.6 percentage points (*Economist* 2005b).

When it comes to mobile phones, there is no need for intervention or funding from the United Nations (U.N.); even the world's poorest people are already rushing to embrace mobile phones because their economic benefits are so apparent. Mobile phones do not rely on a permanent electricity supply and can be used by people who cannot read or write. Phones are widely shared and rented out by the call, for example, by "telephone ladies" in Bangladeshi villages. One person in a village buys a mobile phone, perhaps using a microcredit loan. Others then rent it out by the minute; the small profit margin enables its owner to pay back the loan and make a living. When the phone rings, its owner carries it to the home of the person being called, who then takes the call. Other entrepreneurs are "text message interpreters," sending and receiving text messages (which are generally cheaper than voice calls) on behalf of their customers, who might be illiterate. So although the number of phones per 100 people is low by rich-world standards, they still make a big difference. As in industrial societies, children are very open to, and quickly adapt, new technologies (Figure 3).

FIGURE 3 African boy with mobile phone dummy built of clay (courtesy of *The Economist*).

Farmers and fishermen use mobile phones to call several markets to work out the best price for their produce. Small businesses use them to shop around for supplies. Mobile phones are used to make cashless payments in Zambia and several other African countries. In Argentina, poor people who live by searching household trash for valuables use mobile phones to have an advantage over competitors. Again, mobile phones have a dramatic impact despite their low per-capita numbers: by reducing transaction costs, broadening trade networks, and reducing the need to travel (of particular value for people looking for work). So it is no surprise that people in poor countries spend a larger proportion of their income on telecommunications than those in rich countries.

The digital divide that really matters, then, is between those with access to a mobile network and those without. But this gap is closing fast. The UN set a goal of 50% access by 2015, but a new report from the World Bank notes that 77% of the world's population already lives within range of a mobile network (*Economist* 2005b).

Mobile HIV/AIDS Health Clinic

Education, communications, and health have been described as the key factors to help developing nations. They have to go together: experience shows distribution of medicines or condoms without providing long-term education does not lead to improvements. The tragedy AIDS has introduced into Africa shows there is urgent need for combining education, communication, and health.

The facts speak for themselves (UNAIDS 2007):

- *Global*: Sixty-five million people worldwide are infected with HIV, of whom 25 million have died. In 2000, approximately 5.3M people were newly infected with HIV, and 600,000 of them were children. A child is orphaned because of AIDS every 14 seconds—one-third of these is younger than five. Ninety-five percent of those infected with HIV live in the Global South.
- *Africa*: Sub-Saharan Africa is home to over 25 million cases, about 70% of the world's total. Fifty-five percent of the HIV-positive people in sub-Saharan Africa are women. Six of seven HIV-positive children in Africa are girls. In eight African countries, over 15% of adults are infected. Around a third of today's 15-year-old Africans will die of AIDS.
- *South Africa*: With over five million infected people, it has the largest number of people living with HIV/AIDS in the world. One of four South African women between ages 20 and 29 is infected with the virus.

These shocking statistics led the nonprofit organization "Architecture for Humanity" to launch an international competition for a mobile HIV/AIDS clinic for Africa. Architecture and Vision proposed a clinic that would be built up on a standard 2.55-m-wide truck (Figures 4 and 5). The deployable roof is equipped with solar cells, which allow constant daytime charging of batteries and fuel cells. Preferably the truck

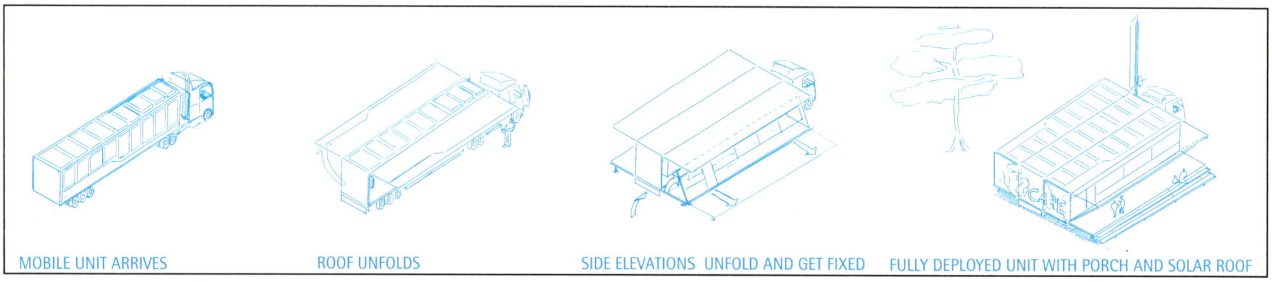

FIGURE 4 Mobile health clinic travels compactly on a truck and can be deployed at desired location (courtesy of Architecture and Vision).

FIGURE 5 Mobile health clinic comes fitted onto a standard truck (courtesy of Architecture and Vision).

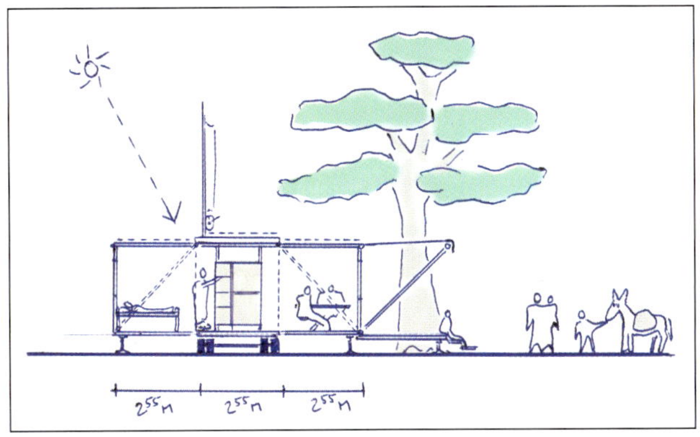

FIGURE 7 Mobile health clinic has two fold-out sides: one provides patient beds, and the other reception and consultancy. Veranda is a shading device and stage for presentations, lectures, or educational plays (courtesy of Architecture and Vision).

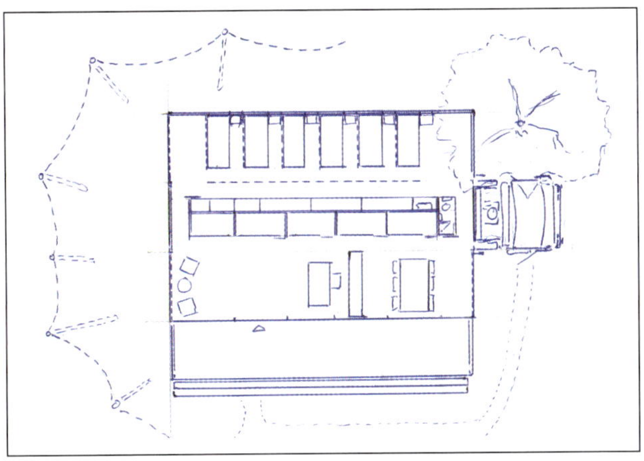

FIGURE 6 Plan of deployed clinic. The space created by unfolding the container can be extended by additional canvas structures (courtesy of Architecture and Vision).

FIGURE 8 Six-meter-diameter vacuum chamber in Houston where NASA has tested autonomous life-support systems for up to 90 days (courtesy of NASA).

would be operated by a hybrid engine to promote alternative energy use in developing nations. Figure 4 shows how the Health Clinic is set up. The container's side walls fold up to form the roof. In the same way floor panels are folded down. An extra outdoor veranda is provided as well. A flagpole provides visibility from afar. A central core contains the storage, facilities, water supply, and toilet needed for a doctor, a nurse, and up to six bed-ridden patients (Figure 6). Six beds can be set up in the part separated by the central storage unit. The side adjacent to the veranda serves as reception and consultation room. The veranda has a double function: first, it protects activities from sunlight and rain and offers a shaded entrance; second, it serves as a stage to perform educational plays informing people about AIDS prevention (Figure 7).

CLOSED-LOOP AUTONOMOUS SYSTEMS

Today spaceflight is still dependent on resupply from Earth. The International Space Station (ISS) is resupplied by Progress and automated transfer vehicle (ATV) capsules, which bring consumables like water, food, and oxygen and are filled with waste for burnup during reentry into Earth's atmosphere. To go far beyond Earth orbit, increasing loop closure will be necessary.

Several research programs are moving in this direction including space-simulation chamber experiments at the NASA Johnson Space Center (JSC) (Figure 8). The primary goal of the Lunar-Mars Life Support Test Project (LMLSTP) conducted from 1995 through 1997

FIGURE 9 Plants provide oxygen, clean the water, and yield food. They also support crew psychological health and will become an important element of deep-space missions (courtesy of NASA).

at JSC was to test an integrated, closed-loop system that employed biological and physicochemical techniques for water recycling, waste processing, and air revitalization for human habitation. Conditions of isolation and confinement enable studies of human factors, medical sciences (both physiology and psychology), and crew training (see Chapter 25). Study results provide a wealth of data important for long-duration space missions and extreme environments here on Earth. The longest simulation in the JSC study was by a crew of four for 90 days, using wheat to revitalize the air and a bioreactor for water recycling, which used microbes to clean the water. An incinerator was used in the solid-waste processing system to turn crew fecal matter into ash and gaseous carbon-dioxide products for uptake by the wheat (Lane et al. 2004).

Bioregenerative life-support systems will have to become more lightweight, power-efficient, and reliable in the future to be incorporated as critical systems for long-duration space missions. The deep-space astronaut of the future will likely be a "bionaut" as well, living in balance with controlled plant, animal, and microbial systems in very confined spaces (Figure 9).

The traditional farm on Earth has been fairly autonomous (as defined in this chapter) based on ISRU: firewood from the forest and food and materials from crops and animals. With the growth of cities and the buildup of modern infrastructure, this model has declined. Most buildings are fully dependent on utilities supplying water, electricity, and heating energy. With the first oil crisis in the 1970s, better insulation and solar energy started to reduce the energy demand of houses. Continued sharply rising costs for energy, water, and waste management, as well as homeowner desire to become more independent, will open a large market for autonomous systems in western societies. In 2004, the Fraunhofer Institute predicted a substantially growing market for passive houses over the succeeding 15 years (Fraunhofer Institute for Solar Energy 2004).

However, a more dramatic beneficiary of such systems would be the developing world, which lacks the sophisticated infrastructure of the first world. According to U.N. reports, 1.1 billion people do not have direct access to drinking water, and 2.4 billion do not have access to basic sanitation (U.N./WWAP 2003). Developing countries often do not have the means for extensive infrastructure. Mobile and low-energy water recovery systems can help improve this situation and possibly allow a technology jump as happened with the mobile phone. Analogous to the mobile-phone market, the large housing market in wealthy countries could help motivate a cheap mass product for the future based on currently expensive technology.

The UN also estimates 50% of the world's drinking water is transported on women's heads. Sixty percent of these carriers are more likely children (Figures 10 and 11). In many countries it is the women's and

FIGURE 10 A girl helping her sick grandmother to carry water in Changara district, Tete province, in March 2004 (© UNICEF/MOZA0239/G.Pirozzi).

FIGURE 11 Boy and a young girl on their way to carry water in Gondola, Manica, March 2004 (© UNICEF/MOZA0521/G.Pirozzi).

children's tasks to collect water from sources often up to two hours walking distance away. According to the World Health Organization, in the Sudan the energy used to tote water from rivers and other water sources accounts for one-third of a woman's daily calorie intake (Starke 2003).

The facts again speak for themselves (WaterAid 2008):

- Roughly one-sixth of the world's population, 1.1 billion people, are without access to safe water. About two-fifths of the world's population, 2.4 billion people, are without access to adequate sanitation.
- There are 2.2 million people in developing countries, most of them children, dying each year from diseases associated with lack of access to safe drinking water, inadequate sanitation, and poor hygiene. Six-thousand children are dying every day from these causes, equivalent to 20 jumbo jets crashing every day.
- Half of the world's hospital beds are occupied by patients suffering from water-borne diseases. Two-hundred million people in 74 countries are infected with schistosomiasis; 20 million suffer severe consequences. A 77% reduction of incidence is achievable through well-designed water and sanitation interventions alone.
- The average walking distance for women in Africa and Asia to walk to collect water is 6 km. The weight they carry on their heads is equivalent to your airport luggage allowance (20 kg).
- The average person in the developing world uses 10 liters of water a day. The average person in the United Kingdom uses 135 liters.

Eco-Unit

The Architecture and Vision Eco-Unit responds to the water situation by providing a communal unit for water collection, water recovery, hygiene, and controlled disposal of human wastes, as would a spaceship (Figure 12). Water is collected over the roof or from a nearby water source. Solar cells and biogas created by human waste provide power to recover both potable water and grey water for the toilet.

The roof shape is inclined to collect water when it rains and provide a good angle for the solar cells. The box has a 2.5-m transportation width and is deployed with inflatable walls after installation, providing the most economic use of space for transportation and local use. Its construction uses a mix of hard and soft materials. The design conception derives from current aerospace technologies, although materials will be more conventional and cheap. Hard materials include wood and protected cardboard. From the hard core a soft inflatable section can be deployed, which minimizes volume for transportation to the site.

FIGURE 12 Mobile Eco-Unit powered by solar energy could provide safe sanitation for a local village community (courtesy of Architecture and Vision).

Eco-Unit technology would be adaptable, meaning it could be high tech and expensive or low tech and inexpensive, and so implementation in poor countries would be possible. The main problem is not the technology, but its implementation and the education of people to use it correctly. For most villages in poor countries, large-scale infrastructure for fresh water and waste water recovery is much too expensive. The Eco-Unit concept offers a decentralized approach that would avoid infrastructure and downstream health costs and immediately provide better living conditions for people. It would especially relieve women and children in developing countries from the unergonomic burden of carrying water.

FIGURE 13 Introduction of *in situ* biogas not only relieves people from buying expensive firewood, but yields a clean, soot-free kitchen (courtesy of Ashden Awards for Sustainable Energy/Martin Wright; www.ashdenawards.org).

Technology Jump: From the Middle Ages into Modern Times

Often new technologies are around for a considerable time before their exponential adoption begins. The challenge is how to identify the promising ones early and control the parameters necessary for their commercial introduction. Successful new technologies are mostly connected to considerable improvement of the quality of life. For example, the Frankfurt Kitchen introduced in 1929 was heavily criticized in the beginning, but became a game-changer that saved women a lot of time and allowed them to become more independent. Often the introduction of technology, even if quite low tech, can solve many problems simultaneously. Once the introduction is completed, many people ask why it was not done long before.

A good example for such a technology jump can be found in Nepal, where the majority of people still cook over open fires on the kitchen floor. Firewood is getting rare and more expensive as Nepal suffers from severe deforestation. The introduction of biogas into households starts solving many problems with simple, low-tech, decentralized technology. Many people in rural areas have one or two cows. Feces of people and animals are collected in a concrete biogas tank, which provides enough gas for clean, soot-free cooking (Figure 13). The tank residue is excellent fertilizer (Figure 14). Taking into account the costs of wood, the biogas system pays for itself within two years. Part of its successful introduction is the use of micro-finance-style loans (Ashden 2005). This example raises the question, "Why hasn't this been done

FIGURE 14 Biogas production technology is low tech. Residue from the concrete tank yields fertilizer (courtesy of Ashden Awards for Sustainable Energy/Martin Wright; www.ashdenawards.org).

earlier?" Sufficient wood supply? No financing for farmers? The technology, and the possibility to apply it cheaply, had been around for over 30 years.

IN SITU RESOURCE UTILIZATION

In situ resource utilization (ISRU) is a very old concept that has shaped cultures and buildings over centuries (Figure 15). The industrial revolution introduced an increasing independence from local energy and resources. Transportation became cheap. Globalization economics creates the absurd situation in which an apple from the neighborhood can (apparently) be more expensive than one imported from Argentina or elsewhere. The high degree of infrastructure and logistics has even made us forget the value of local resources. This value has begun to be reestablished from an unlikely direction: technology development for human planetary exploration. ISRU techniques are being developed in laboratories to extract oxygen propellant from lunar regolith and methane propellant using carbon dioxide in the Martian atmosphere. Such technologies are economically enabling for advanced missions. But they also sharpen our sense of how to use even small amounts of local resources to best benefit, which is a real "spaceman" mentality as defined earlier. Spaceman technological thinking brings control over the environment back into the hands of people, yielding a powerful lever.

One of the most complex challenges for long-duration human spaceflight is radiation protection, as discussed throughout this volume. Current shielding concepts require bulk mass, which is why use of local regolith has been proposed (Figure 16).

FIGURE 15 Swiss mountain hut built from stones found on the site (courtesy of Book SolarPower).

FIGURE 16 Nader Kahlil's sand-based domes at Cal Earth Institute suggest a building method for planet surfaces using their local resources (courtesy of Cal Earth Institute).

Christopher Alexander, the father of architecture pattern language, defined the leapfrog opportunity by using local resources, perhaps knowingly. In a 1996 lecture (Gupta 2004), he commented to a group of software programmers in San Jose that

> In traditional society where lay people either built or laid out their own houses, their own streets, and so on, the adaptation was natural. It occurred successfully because it was in the hands of the people that were directly using the buildings and streets. So, with the help of the shared pattern languages, which existed in traditional society, people were able to generate a complete living structure.
>
> In our own time, the production of environment has gone out of the hands of people who use the environment. So, one of the efforts of the pattern language was not merely to try and identify structural features which would make the environment positive or nurturing, but also to do it in a fashion which could be in everybody's hands, so that the whole thing would effectively then generate itself.

Advanced technology can provide new solutions for the simple problems of developing countries. One example is the LifeStraw by Torben Vestergaard Frandsen, a 25-cm-long plastic pipe filter that turns dirty water into clean, potable water (Figures 17 and 18). Sucked-up water meets two textile filters that remove particles including even clusters of bacteria. Then the water enters a chamber of iodine-impregnated beads, where bacteria, viruses, and parasites are killed. The second chamber is a void space, where the iodine

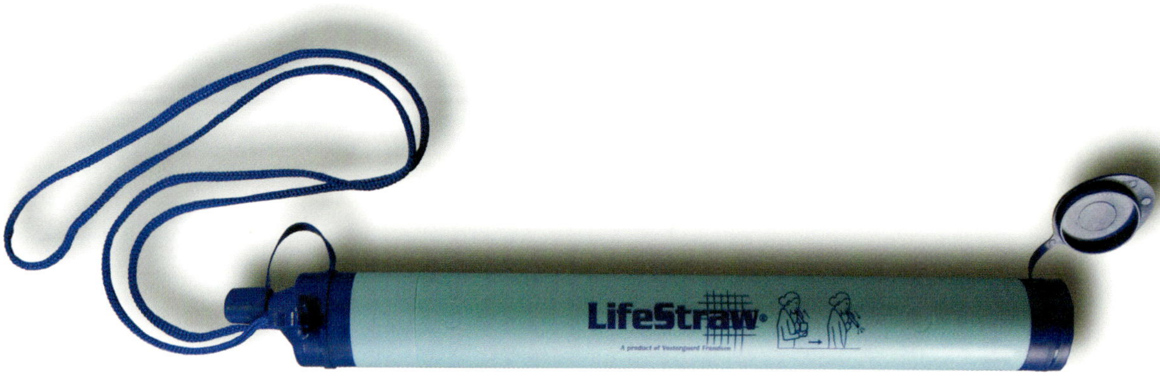

FIGURE 17 LifeStraw uses textile filters to make water safe for drinking on the go (courtesy of Vestergaard Frandsen).

FIGURE 18 LifeStraw is likely to be one of the inventions with a very high impact on the situation in developing countries (courtesy of Vestergaard Frandsen).

can maintain its killing effect. The last chamber contains granulated active carbon that removes most of the bad smell of iodine and parasites that have not been taken by the prefilter or killed by the iodine. LifeStraw lasts for one person's annual needs of clean water and costs $5. The costs of a water treatment plant and infrastructure would be far higher and probably still not reach all people. Nobody needs to die from diseases originating from unsafe water resources.

FIGURE 19 Desert Seal is a space technology transfer project that uses local energies to cool a tent in desert regions (courtesy of Céline Laurière).

Another example is Desert Seal, a tent for hot, arid climates that uses local energies for cooling itself (Figure 19). The tent resulted from a study by Architecture and Vision for the European Space Agency (ESA), investigating use of space technologies for inflatable tents in hot regions (European Space Agency 2005).

Desert Seal is designed specifically for hot, arid environments where the air becomes considerably cooler higher above the Earth's surface. During the day, temperatures can easily go above 60°C at ground level, but remain 40°C lower just 2–3 m above the ground (Figure 20). Traditional Persian buildings already use this effect (Figure 21), and camels are evolved to benefit from this effect with their long legs and neck (Figure 22).

During the day, an electric fan in the top of the tent, 2.26 m above the ground, constantly blows cooler air inside, thus reducing the temperature

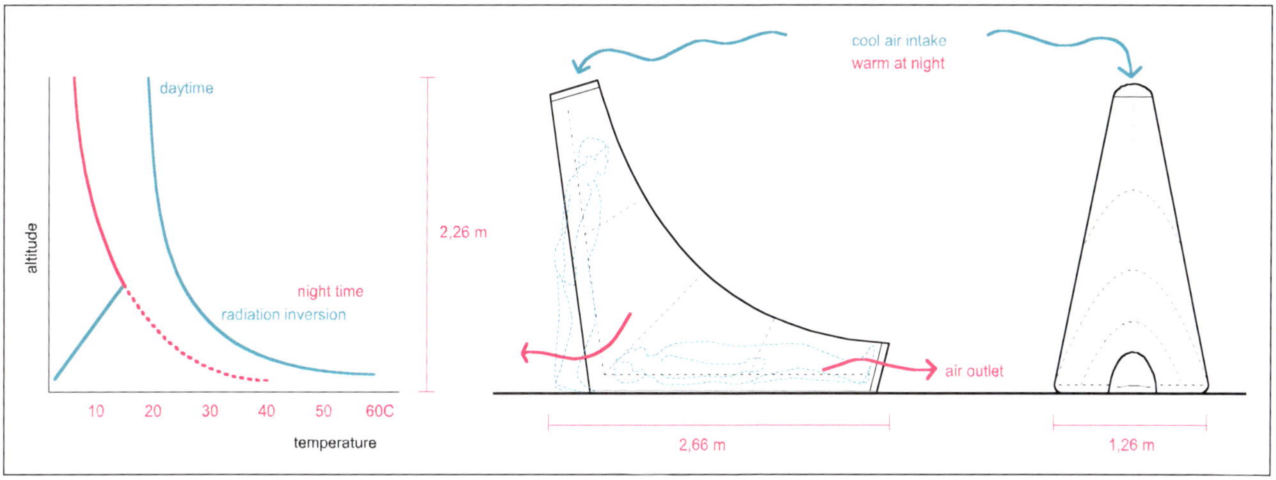

FIGURE 20 Desert Seal makes use of desert temperature profile. Cooler air is blown into the tent from the top by a solar-powered electric fan (courtesy of Architecture and Vision).

FIGURE 21 Wind scoop tower in the historical district of Dubai (courtesy of Architecture and Vision).

FIGURE 22 Desert Seal tent collects cooler air at higher levels (courtesy of Architecture and Vision).

FIGURE 23 Desert Seal heat-reflecting silver-coated awning and flexible solar array to power the fan (courtesy of Céline Laurière).

inside. The fan is powered by batteries charged by flexible solar panels mounted outside the tent (Figure 23). During the night, the desert ground radiates heat to the dark sky and quickly reaches temperatures below 0°C. Because air acts as a good insulator, at higher levels it stays considerably warmer. The fan on top, now running on batteries, blows warmer air into the tent. The tent consists of an air-beam structure made of polyethylene-coated material. It has an awning of silver-coated, high-strength textile to reflect heat and provide protection from direct sunshine. The L-shaped tent allows upright entry and minimizes aerodynamic loading.

CONCLUSION AND CALL TO ACTION

The space age created the concept of Earth as our mother ship. The concept leads to understanding our planet as an interdependent system of life and matter, technology as a tool for interacting with the planet and its crew, and the value each crew member adds to the whole mission.

The space age also yields technologies we can use to maintain and protect the mother ship and to provide higher living standards for the third world. The case studies presented in this chapter illustrate the tremendous potential for creative use of modern technologies to help developing countries. As technology becomes ubiquitous and cheaper, old preconceptions—for example, that technology cannot help poor people because it is too expensive—must and can be overcome.

The mobility of knowledge (computer), communications (mobile phone), and health services (mobile clinic) can positively affect individual lives. Closed-loop systems and intelligent use and recovery of local resources can make people more independent and yield free time for self-organization, education, and working to improve quality of life. Modern lightweight and low-energy technologies can unlock the potential of using resources locally and individually, as with LifeStraw. Large, centralized infrastructure is not only expensive but also hinders self-organization. Monsoon and earthquake disasters demonstrate that centralized infrastructure, once destroyed, yields more dire postdisaster disruption. The more decentralized the provision of basic needs, the less vulnerable the population becomes.

Space architects are uniquely positioned to participate in the development of approaches for solving tough problems of human spaceflight, to integrate those solutions into comfortable, attractive, safe, and productive human environments for space missions, and then to see how those solutions can be adapted to benefit indigent populations and compromised environments on Earth. In many cases, we already have technologies and capacity to solve very practical yet devastating problems. The architect's most fundamental motivation—to use good design to improve the world—has never been needed more. |

References

Ashden (2005), "NEPAL - Biogas Sector Partnership," The Ashden Awards for Sustainable Energy, http://www.ashdenawards.org/finalist05_1.html [Retrieved 17 Aug. 2005].

Boulding, K. E. (1966), *The Economics of the Coming Spaceship Earth*, Sixth Resources for the Future Forum on Environmental Quality in a Growing Economy, Johns Hopkins Univ. Press, Baltimore, MD, pp. 3–14.

Bullis K. (2005),"A Hundred-Dollar Laptop for Hungry Minds," *Technology Review*, Vol. 28, Sept. http://www.techreview.com/articles/05/09/wo/wo_092805bullis.asp. [retrieved 16 Jan. 2005]

Economist (2005a), "Calling Across the Divide," *The Economist*, 374 Issue 8417, March 12, p. 74.

Economist (2005b), "The Real Digital Divide," *The Economist*, 374 Issue 8417, March 12, p. 9.

European Space Agency (2005), "Space Concepts Improve Life in the Desert, http://www.esa.int/esaCP/SEM0TB6Y3EE_index_0.html [retrieved 20 Oct. 2005].

Fraunhofer Inst. for Solar Energy (2004), "Enormous Potential for Passive and Lowest-Energy Houses," Press Release 06/04, Fraunhofer, Freiburg.

Gupta, R. (2004), "Switching Sides—Leapfrog Nations—Emerging Technology in the New Developing World," http://www.worldchanging.com/archives/001408.html [retrieved 14 Dec. 2005].

Hindu Business Line (2008), "India Becomes Second Largest Mobile Market in the World," *The Hindu Business Line*, 25 April, http://thehindubusinessline.com [retrieved 20 July 2008].

Lane, H. W., Sauer, R. L., and Feeback, D. L. (Eds.) (2004), "Isolation: NASA Experiments in Closed-Environment Living," *Advanced Human Life Support Enclosed System*, Vol. 104, Science and Technology Series, AAS, San Diego, CA.

NASA (2004), "ALS: Lunar Mars Life Support Test Project, http://advlifesupport.jsc.nasa.gov/lmlstp.html [retrieved 14 Dec. 2005].

OLPC (2008), "Laptop: a Learning Tool Created Expressly for the Children in Developing Nations," One Laptop Per Child, Cambridge, MA, July 20.

Starke, L. (ed.) (2003), "State of the World 2003: Worldwatch Institute Report on Progress Toward a Sustainable Society," W.W. Norton Com., New York, Jan., p. 49.

Toffler, A. (1980), *The Third Wave*, William Morrow and Co., Inc., New York.

UNAIDS (2007), *AIDS Epidemic Update December 2007*, UNAIDS, Geneva, pp. 1–50.

UN/WWAP (2003), *Water for People, Water for Life: The United Nations World Water Development Report,* United Nations/World Water Assessment Program, UNESCO (United Nations Educational, Scientific and Cultural Organization) and Berghahn Books, Paris, pp. 5–23.

WaterAid (2008), "Statistics, Facts About Water, Sanitation, Hygiene and WaterAid's Work," http://www.wateraid.org/uk/what_we_do/statistics/default.asp [retrieved 18 July 2008].

Wikipedia (2008), "Digital Divide," *Wikipedia, The Free Encyclopedia*, July 14, http://en.wikipedia.org/w/index.php?title=Digital_divide&oldid=225687168 [retrieved 20 July 2008].

World Bank (2008), *Global Economic Prospects 2008: Technology Diffusion in the Developing World*, The World Bank, Washington, D.C., p. 17.

Index

A

B

C

D

E

F

G

H

K

L

M

N

O

P

Q

R

S

T

U

V

W

X

Z

contributors

Constance M. Adams, *Synthesis International*
Juergen F. Bahr, *ERC Inc.*
Haym Benaroya, *Rutgers University*
Melissa Ann Borrego, *MEI Technologies Inc.*
Hugh Broughton, *Hugh Broughton Architects*
James Lee Broyan, *NASA Johnson Space Center*
Stephen D. Capps, *Boeing Company*
Dave Cadogan, *ILC Dover*
Paul Clancy, *4CON Space S.A.R.L.*
Marc M. Cohen, *Northrop Grumman Aerospace Systems*
Susan M. Fairburn, *Fibre Design Inc.*
Paola Favata, *architect*
Ian Gibson, *National University of Singapore*
Loris Granziera, *independent consultant*
Brand Norman Griffin, *Gray Research Inc.*
Theodore W. Hall, *University of Michigan*
Matthew Herman, *Buro Happold Architects and Engineers*
A. Scott Howe, *NASA Jet Propulsion Laboratory*
Barbara Imhof, *LIQUIFER Systems Group*
Rod Jones, *NASA Johnson Space Center*
Kriss J. Kennedy, *NASA Johnson Space Center*
Bas Lansdorp, *Ampyx Power*
James D. Lowe, *independent researcher*
Susmita Mohanty, *Earth 2 Orbit*
David Nixon, *Altus Associates and Astrocourier (Ireland) Ltd.*
Jun Okushi, *Space Projects Group, Okushi Architects*
Truls Ovrum, *Altus Associates*
Georgi Petrov, *Synthesis International*
Stephen Ransom, *independent aerospace consultant*
Craig Scheir, *ILC Dover*
Brent Sherwood, *NASA Jet Propulsion Laboratory*
Ross A. Tisdale, *architect*
Larry Toups, *NASA Johnson Space Center*
Arturo Vittori, *Architecture + Vision*
Andreas Vogler, *Architecture + Vision*
Kristian von Bengtson, *Copenhagen Suborbitals*

supporting materials

Many of the topics introduced in this book are discussed in other AIAA publications. For a complete listing of titles in the Library of Flight, as well as other AIAA publications, please visit http://www.aiaa.org.

AIAA is committed to devoting resources to the education of both practicing and future aerospace professionals. In 1996, the AIAA Foundation was founded. Its programs enhance scientific literacy and advance the arts and sciences of aerospace. For more information, please visit www.aiaafoundation.org.